The Lineman's and Cableman's Handbook

Linemen transferring conductors on vertical corner pole to switch on new pole. Work is being completed with hot-line tools from pole and bucket truck. Cover-up material is being used to provide protection from energized conductors. *(Courtesy A. B. Chance Co.)*

The Lineman's and Cableman's Handbook

Edwin B. Kurtz, E.E., P.E., Ph.D.

(Deceased)

Emeritus Professor of Electrical Engineering and
former Head of Department, The University of Iowa
Formerly Member of the Educational Department, The
Wisconsin Electric Power Company; Fellow, IEEE
Fellow, AAAS; Member, ASEE

Thomas M. Shoemaker, P.E.

B.S.E.E. The University of Iowa; Manager, Distribution
Department, Iowa Illinois Gas and Electric Company
Formerly Captain, Signal Corps, U.S. Army; Member
Transmission and Distribution Committee, Edison
Electric Institute; Senior Member, IEEE

SIXTH EDITION

McGraw-Hill Book Company

New York   St. Louis   San Francisco   Auckland
Bogotá   Hamburg   Johannesburg   London   Madrid   Mexico
Montreal   New Delhi   Panama   Paris   São Paulo
Singapore   Sydney   Tokyo   Toronto

Library of Congress Cataloging in Publication Data

Kurtz, Edwin Bernard, date.
  The lineman's and cableman's handbook.

  Includes index.
  1. Electric lines—Handbooks, manuals, etc .
  2. Electric cables—Handbooks, manuals, etc.
  I. Shoemaker, Thomas M., date.   II. Title.
  TK3221.K83   1981   621.319′22   81-5985
                                            AACR2

1234567890   KPKP   8987654321

ISBN 0-07-035678-5

The editors for this book were Harold B. Crawford and Margaret Lamb,
the designer was Blaise Zito Studios, and the production
supervisor was Teresa F. Leaden. It was set in Caledonia
by University Graphics, Inc.

Printed and bound by The Kingsport Press.

Contents

Preface

This Handbook is written expressly for the apprentice, the lineman, the cableman, the foreman, the supervisor, and other employees of transmission and distribution departments. It is primarily intended to be used as a home-study book to supplement daily work experiences. Of the 50 sections in the Handbook, 12 sections are devoted to a general understanding of electricity, electrical terms, and electric power systems; 30 sections are devoted to the actual construction of overhead and underground distribution and transmission lines and to maintenance procedures; and although all the material has a relationship to safety, 7 sections are specifically devoted to that subject. The final section is intended for the reader's self-examination of the information presented.

The following 10 sections are new in this edition: Section 3, Substations; Section 4, Transmission Circuits; Section 6, Construction Specifications; Section 7, Wood-Pole Structures; Section 8, Aluminum, Concrete, Fiberglass, and Steel Structures; Section 13, Insulators; Section 14, Line Conductors; Section 26, Grounding; Section 27, Protective Grounds; and Section 40, Voltage Regulation.

A special effort was made to present all discussions clearly and in simple language; in fact, a reading knowledge of the English language is all that is required to understand the book. As in former editions, a large number of illustrations showing the various steps in the construction and maintenance process are provided to assist the reader in better understanding the text; the illustrations appearing in the Handbook should be considered as much a part of the book as the words themselves. Illustrations bring out many details that would require many additional words to express. Many of the photographs were taken specifically for use in this edition. They therefore portray the latest practices in use today by some of the foremost electric utility companies in this country.

Methods of transmission-, distribution-, and rural-line construction have become quite standardized since the first edition of this Handbook was published in 1928. The construction procedures described and illustrated are therefore in most instances representative of general practice. While each operating company has its own standards of construction which its linemen and cablemen must adhere to, the procedures described explain why things are done in a given way. Such basic knowledge should be helpful to the lineman or cableman who is interested in learning the whys and wherefores of doing things one way or another.

Safety is again emphasized throughout the book. Of course, understanding the principles involved in any operation and knowing the reasons for doing things a given way are the best aids to safety. Nevertheless, the opinion has become quite firmly established that a man is not a good lineman unless he does his work safely. It therefore behooves those engaged in electrical work to become familiar with the safety rules and the precautions applicable to their trade and to make their observance an inseparable part of their working habits.

Emphasis is also placed on the National Electrical Safety Code, the National Power Survey, OSHA Regulations, ANSI Standards, and ASTM Standards. The important requirements of the code are reprinted and incorporated in the text where corresponding topics are discussed.

In this way the lineman becomes acquainted with the minimum construction requirements that will ensure safety to the public and to the lineman. If more code information is desired, a copy of the *National Electrical Safety Code* ANSI C2, can be secured for a charge from the Institute of Electrical and Electronics Engineers, Inc., 345 East 47th Street, New York, NY 10017.

The *National Electrical Code*® details the rules and regulations for electrical installations except those under the control of an electric utility. It excludes any indoor facility used and controlled exclusively by a utility for all phases from generation through distribution of electricity and for communication and metering, as well as outdoor facilities on a utility's own or a leased site or on public or private (by established rights) property.

Reference material includes *Standard Handbook for Electrical Engineers*, edited by Donald G. Fink and H. Wayne Beaty, and published by McGraw-Hill Book Company; Edison Electric Institute Publications; *Underground Power Transmission* by Arthur D. Little, Inc., for the Electric Power Research Council; *Electric Power Transmission and Environment*, published by the

Federal Power Commission; *IEEE Standard Dictionary of Electrical and Electronic Terms;* and *User's Manual for the Installation of Underground Plastic Duct,* published by the National Electrical Manufacturers Association.

The author and editors are well aware that one cannot become a competent lineman or cableman from a study of the pages of this book alone. However, diligent study along with daily practical experience and observation should give the apprentice an understanding of construction and maintenance procedures—and a regard for safety—that should make his progress and promotion rapid.

Thomas M. Shoemaker

Lineman and cableman, long-established and still current terms in the industry are beginning to be replaced by nonsexist titles in official documents and government publications. Both men and women are employed in these capacities in the military and in the industry. To avoid awkwardness, the Handbook uses the masculine pronoun, but it in no way implies that the jobs involved are held only by men.

Acknowledgments

The editors wish to express their sincere appreciation to the many companies and their representatives who kindly cooperated in supplying illustrations, data, and valuable suggestions. We are especially grateful to L. C. Hansen, W. R. Hooper, and R. L. Ambre of the A. B. Chance Company; to Robert Kinsinger of the Commonwealth Edison Company; to J. A. Pulsford and H. M. Spooner of the Public Service Electric and Gas Company of New Jersey; to D. S. Page of the Asplundh Tree Expert Company; to D. L. Jansa of the General Electric Company; to H. M. Butzloff of the Westinghouse Electric Corporation; to W. E. Mueller of the I.T.T. Blackburn Company; to L. J. Dylewski of the Anaconda Wire and Cable Company; to A. M. Petersen of the Petersen Company, Incorporated; to R. Gagnon of the Preformed Line Products Company; to William Martin of the Homac Manufacturing Company; to Neal Bodin of the Ultig Engineers, Inc.; to A. J. Samuelson of the American Electric Power Corp.; to Wilford Caulkins III of Sherman & Reilly, Inc.; to Daniel Bye of The Okonite Company; to Ray G. Brown of The Ohio Brass Company; to Mort Hough of the Nebraska Public Power District; to Fred E. Wherry of the Timberland Equipment, Limited; to James D. Brame of the Pettibone Corporation; to Robert L. Mann of the McGraw-Edison Company; to Steven L. Thumander of the Joslyn Mfg. & Supply Co.; to Francis X. Hennessy of the AMP Special Industries; to Ruth E. Shoemaker; and to others.

The editors are also indebted to the Institute of Electrical and Electronics Engineers, Inc., publishers of the National Electrical Safety Code, *Electric World, Electric Light and Power,* the Rural Electrification Administration, and the Edison Electric Institute for permission to reprint various items from their literature.

EDWIN B. KURTZ
Author and Editor
(Deceased)

THOMAS M. SHOEMAKER
Editor

Introduction

Linemen and cablemen construct and maintain the electric transmission and distribution facilities that deliver the electrical energy to our homes, factories, and commercial establishments. The linemen and cablemen provide important and skilled services to the electrical industry—important because the health and welfare of the public is dependent on reliable electric service.

When emergencies develop as a result of lightning, wind, or ice storms, the linemen and cablemen respond to restore electric service at any time of the day or night.

An understanding of electrical principles and their application in electrical construction and maintenance work is essential to completing the work safely, efficiently, and reliably. The new equipment and the public's increased dependence on continuous electric service requires that all linemen and cablemen be highly skilled.

The time and effort spent studying these pages will increase the reader's knowledge of electric distribution facilities and improve his skills. Every lineman, cableman, and groundman should develop the highest possible level of skills so that he will be able to meet the challenges of the work and be qualified for promotion when the opportunity becomes available.

The Lineman's and Cableman's Handbook

Elementary Electrical Principles

Electron Theory The basis of our understanding of electricity is the electron theory. This theory states that all matter, that is, everything that occupies space and has weight, is composed of tiny invisible units called atoms. Atoms in turn are subdivided into still smaller particles called protons, neutrons, and electrons. The protons and neutrons make up the central core, or nucleus, of the atom, while the electrons spin around this central core in orbits as illustrated in Fig. 1-1.

The protons and electrons are charged with small amounts of electricity. The proton always has a positive charge of electricity on it, while the electron has a small negative charge of electricity on it. The magnitude of the total positive charge is equal in amount to the sum of all the negative charges on all the electrons. The neutron has no charge on it, either positive or negative, and is therefore neutral and hence called neutron.

Atoms differ from one another in the number of electrons encircling the nucleus. Some atoms have as many as 100 electrons spinning around the nucleus in different orbits. The atom of hydrogen gas has only 1 electron. The atom of lead has 82 electrons.

Positive and negative charges of electricity attract each other; that is, protons attract electrons. But the atom does not collapse because of this attraction. The spinning of the electron around the nucleus causes a centrifugal force that just balances the force of attraction and thus keeps them apart.

Electric Current The electrons in the outermost orbit of an atom are usually not securely bound to the nucleus and therefore may fly off that atom (see Fig. 1-2) and move into an outer orbit of another atom. These relatively free electrons normally move at random in all directions. However, when an electrical pressure (voltage) is applied across a length of wire, the free electrons

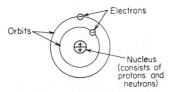

Fig. 1-1 Typical atom consisting of nucleus and revolving electrons. The nucleus is composed of protons and neutrons. The protons carry a positive electrical charge, the electrons carry a negative electrical charge, and the neutrons are neutral; that is, they carry neither a positive nor a negative charge.

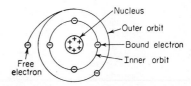

Fig. 1-2 Atom showing electron in outer orbit leaving atom. The atom then has more positive charge than negative charge. The nucleus will therefore attract some other free electron that moves into its vicinity.

in the wire give up their random motion and move or flow in one general direction. This flow of free electrons in one general direction, shown in Fig. 1-3, is called an electric current or simply current.°

Conductors and Insulators   Materials having many free electrons (Fig. 1-4), therefore, make good conductors of electricity, while materials having few free electrons (Fig. 1-5) make poor conductors. In fact, materials that have hardly any free electrons can be used to insulate electricity and are called insulators. Samples of good conductors are copper and aluminum. Samples of good insulators are glass, mica, porcelain, rubber, paper, polyethylene, and fiberglass.

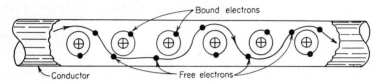

Fig. 1-3   Flow of free electrons in a conductor. Only electrons in outer orbit are free to move from one unbalanced atom to another unbalanced atom. This flow or drift of free electrons is called an electric current.

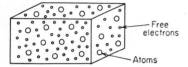

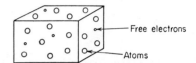

Fig. 1-4   Material having many free electrons makes a good conductor. Copper and aluminum have many free electrons and therefore are widely used as conductors of electricity.

Fig. 1-5   Material having few free electrons makes a good electrical insulator. Porcelain, glass, mica, rubber, polyethylene, fiberglass, cotton, silk, etc., contain few free electrons and, therefore, are widely used as insulators of electricity.

Uses of Electricity   Electricity today is used for many different purposes. In fact, the uses to which electricity can be put are so numerous that one can hardly count them all. Several general applications are itemized in Table 1-1.

Electric Light   Electricity, for example, is used to furnish light to nearly every home in our cities, to every shop and factory, and to the busy streets of our cities. Everybody is familiar with light fixtures using the fluorescent lamp shown in Fig. 1-6. In an electric lamp, the electricity is changed to light.

Electric Heat   Electricity is used to furnish heat. The electric iron to press clothes is a device in which electricity is changed to heat. Figure 1-7 shows a bread toaster, another device in which electricity furnishes heat. Figure 1-8 is a picture of a heat pump, an efficient way of converting electrical energy into heat. The average ratio of efficiency is about two to one, depending on

Table 1-1   Electric Use Classifications

Used as light in	Used as heat in	Used as power in	Used as a communication medium in
House lamp	Range	Rapid transit	Radio
Street light	Iron	Factory motor	Television
Sign light	Portable heater	Elevator	Computer
Floodlight	Toaster	Refrigerator	Telephone
Headlight	Water heater	Washing machine	Teletype
Movie projector	Furnace	Vacuum cleaner	Etc.
Flashlight	Heat pump	Dishwasher	
Etc.	Clothes dryer	Fan	
	Etc.	Pump	
		Etc.	

° By convention, however, the flow of current in a circuit is taken to be opposite in direction to that of the electrons.

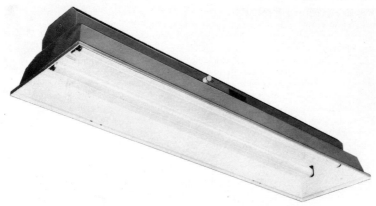

Fig. 1-6   Typical fixture with fluorescent lamps used to produce light when energized by an electric current. *(Courtesy Westinghouse Electric Corp.)*

Fig. 1-7   Typical bread toaster, in which electricity is changed to heat. *(Courtesy McGraw-Edison Co.)*

Fig. 1-8   Heat pump used to electrically heat in the winter and cool in the summer. *(Courtesy Westinghouse Electric Corp.)*

winter climate, compared to electric-resistance heating. The heat pump extracts some of the heat from the outdoor air, with temperatures as low as $-20°F.$, and moves it indoors. In warm weather, the heat pump reverses itself, cooling and dehumidifying the inside air.

Electric Power   The largest use to which we have been able to put electricity is to furnish power. A good illustration of a case in which electricity is changed to power is that of electrically driven rapid-transit systems. Everybody knows that it takes force to move a heavy transit car on a track, and the power required for this is obtained from the electric motor. Such a motor is shown in Figs. 1-9 and 1-10. When riding in a rapid-transit car one generally does not see this motor, but it is there. It is mounted under the floor of the car and is geared to the axle of the car.

Machine shops use electric motors to drive the lathes, millers, planers, drills, punch presses, etc. One large factory may have several hundred motors installed like the one shown in Fig. 1-11. The parts of this motor are shown in Fig. 1-12.

Fig. 1-9   Rapid-transit car motor, with commutator covers removed, in which electrical energy is changed into mechanical power. *(Courtesy General Electric Co.)*

Fig. 1-10   Heavy-duty direct-current traction motor for use on diesel-electric locomotives. *(Courtesy General Electric Co.)*

Fig. 1-11 Typical electric motor of the alternating-current type. *(Courtesy Westinghouse Electric Corp.)*

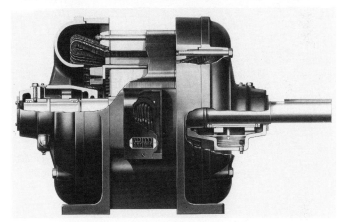

Fig. 1-12 Alternating-current motor cut away showing parts. *(Courtesy Allis-Chalmers Mfg. Co.)*

Electric Communication Electricity is the source of power for communication mediums. All of us are dependent on radios, telephones, televisions, computers, and teletypes to keep us informed. The communication devices use large amounts of electricity.

Elec ric Circuit Compared with Water Circuit An electric circuit is the path in which the electric current flows. While the flow of electricity in a wire is actually the simultaneous motion of countless free electrons in one direction, it is often compared with the flow of a liquid like water. Electricity can then be said to flow in a wire as water flows in a pipe. Take a simple water circuit like the one shown in Fig. 1-13. By noting the resemblance between this pipe circuit and a typical electric circuit shown in Fig. 1-14, one can get a real understanding of the flow of electric currents.

In Fig. 1-13 water flows around the pipe circuit in the direction shown by the arrows. It is evident that this current of water flows because of a pressure which is exerted on it. This pressure is produced by the rotary pump, often called "centrifugal pump," which is driven by a gasoline engine. On the end of the pipeline, a water motor is connected, and, therefore, all the water that flows around the circuit must pass through the motor. It is plain that in so doing it will cause the motor to revolve and, therefore, deliver power to the shaft and the rotating equipment connected to the shaft. Similarly, when an electric current flows in a wire, it flows because an electric pressure causes it to flow. Thus, the current in Fig. 1-14 is made to flow because of the electric pressure produced by the dynamo, or electric generator, which is driven by a gasoline engine. As the electric current flows along the wire, it will be forced to flow through the electric motor. This motor will begin to revolve as the electricity begins to flow through it and thus also will deliver power to the shaft and the rotating equipment connected to the shaft.

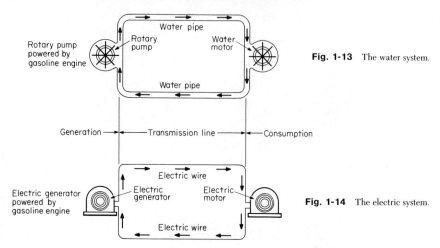

Fig. 1-13 The water system.

Fig. 1-14 The electric system.

Series Circuit   An electrical circuit can be arranged in several ways, just so the path for the electric current is closed. The simplest arrangement is the so-called series circuit. In the series circuit all the elements of the circuit are connected onto each other as illustrated in Fig. 1-15. The same current from the battery flows through all the lamps. If for any reason one of the lamps burns out, all the lamps will go out because the circuit is no longer closed.

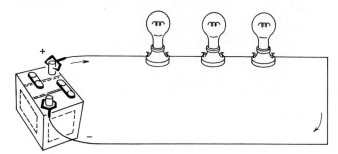

Fig. 1-15 A series circuit. The same current flows through all the lamps.

Parallel Circuit   Another arrangement is the so-called parallel or multiple connection. Instead of all the lamps being connected onto each other and then onto the battery, each lamp is individually connected across the battery as shown in Fig. 1-16. The lamps are now said to be in parallel with each other. If any given lamp burns out in such an arrangement, the remaining lamps will continue to burn, as the path for the current through each of them is still closed.

Series-Parallel Circuit   A third arrangement is a combination of the series and the parallel circuit. In such a circuit part of the circuit is in series and part in parallel as shown in Fig. 1-17. In the series part all the current flows through each element, and in the parallel part the current divides and only a portion of the current flows through each of the parallel paths.

Electric System   Each of the circuits in Figs. 1-13 and 1-14 can be seen to consist essentially of three main divisions. The section where the pressure is produced, that is, where the engine drives the generator, is called the "generator section." That part of the circuit which furnishes the path for the current from the place where it is generated to where it is used is the transmission section, and the section where the electricity is used or consumed is the conversion division. Conversion means "change," and it is here that electricity is changed to the light, heat, power, or communication medium. In an actual electric circuit like the one shown in Fig. 1-18, the three parts of the electric circuit are generating station, transmission line, and factory. The factory is the place where the power is consumed. Such an electric circuit is often called an "electric system."

The wires of the system serve to carry the electricity just as highways carry automobiles and railroad tracks carry trains. The reason one does not see the electricity moving along the wires is

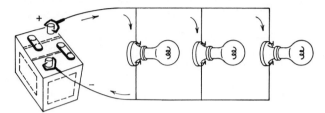

Fig. 1-16   A parallel circuit. Each lamp is independent of the other lamps and draws its own current.

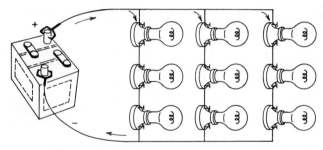

Fig. 1-17   A series-parallel circuit. The lamps in each branch are in series, and the three branches are connected in parallel.

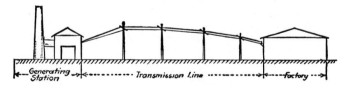

Fig. 1-18   An actual electric system.

because it is invisible. And while the wires and transformers appear lifeless, they are very much alive and ready to do almost any work for us.

One should look upon the generation, transmission, and distribution of electrical energy as one does upon the manufacture, shipment, and delivery of goods. In comparing electricity to a manufactured product, like shoes, one should note one important difference. The manufacturer of shoes can estimate the demand for shoes and then manufacture them in advance and put them in a warehouse. Electricity, however, cannot be generated in advance and stored in large quantity and delivered when needed. Electricity has to be manufactured or generated at the very instant that it is wanted. When the customer snaps a light switch or turns on the electric range, an order is flashed back through the distribution system (the retail outlet), the substation (the warehouse), the transmission line (bulk transportation), to the generator (the factory), and before the customer can release the switch, delivery is made.

It is well to observe that the current path or the transmission line must have a return wire just as the water must have a return pipe. As is clear from Fig. 1-13, the water passes out along the pipe in one side of the circuit, through the water motor, where it does its work, and then returns to the rotary pump in the other pipe. In the same way, in Fig. 1-14, the electricity passes out along one wire to the motor, does its work, and then returns to the generator in the other wire.

Electric Current   In the foregoing, it was pointed out that the flow of electricity in a wire is similar to the flow of water in a pipe. When water flows in a pipe, one speaks of a current of water or a water current. Similarly, when electricity flows in a wire, it is called an "electric current."

Ampere   Generally we want to know how much water is flowing in the pipe, and we answer by saying "10 gal per sec." In the same way we can express how much electricity is flowing in a

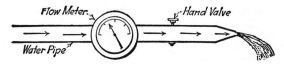

Fig. 1-19   Flowmeter in water pipe.

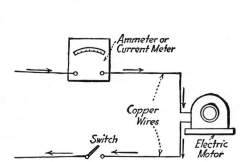

Fig. 1-20   Ammeter in electric circuit.

Fig. 1-21   Typical direct-current switchboard ammeter. *(Courtesy General Electric Co.)*

wire by saying "25 amp." The ampere is the unit of electric current. One can learn how much an ampere is by watching what it can do. Thus an ordinary 60-watt incandescent lamp will require about ½ amp. That means that ½ amp is flowing through it all the time that it is glowing.

A lamp of the type used for street lighting requires about 10 times as much current as the little house lamp. A medium-weight iron requires about 5 amp. A cooling-fan motor takes about 1 amp. A rapid-transit car motor requires about 200 amp. A good sized factory requires several thousand amperes. The total flow of current from a large central generating station or generating plant may be as large as 10,000 amp. These figures will give one a fairly good idea of the size of an ampere and what it will do.

Ammeter   If we wish to measure the current of water flowing through a pipe, we place a meter right in the pipeline. A meter for measuring the flow of water in a pipe is called a "flowmeter." When such a meter is placed in the line, water flows through it, as shown in Fig. 1-19, and the meter indicates the number of gallons per second which pass through it. It is clear that the meter must be inserted in the pipe so that the water flows through it. In exactly the same way, the number of amperes of electric current flowing in a circuit can be measured by connecting a current meter or an ampere meter in the circuit, as shown in Fig. 1-20. Since such a meter is to read amperes, it is called an "ampere meter" or an "ammeter." It should be noted that the ammeter is inserted in the line in order that all the current taken by the motor may pass through the meter. Figure 1-21 shows a typical direct-current switchboard ammeter.

Electric Pressure   Whenever a current of water is flowing in a pipe we know that a pressure is behind it which makes it flow. It may be well to note that the pump in the water circuit does not make water; it only produces a push or pressure. In the electric circuit, the generator does not create electricity; the generator merely produces an electrical pressure that causes electricity to flow. If there is no electrical pressure, no current will flow.

If a hand valve is put in the water circuit and the flow of water is stopped, the water pressure will still be there, but there will be no flow of water through the water motor and the pipe. This everybody knows to be the case, for everybody has turned off the faucet in the kitchen sink. The water flow ceases, but the pressure is still there. In the electric circuit; if a switch is placed in the circuit shown in Fig. 1-20 and opened, the electric pressure will be there as long as the generator is being driven by the engine. Thus it is clear that there can be pressure and no current.

Volt   To learn something about electrical pressure, one must be able to talk of its strength. This requires a unit with which to measure it; that unit is the volt. We can learn how much a volt is by observing what it can do. We can note, for example, how much pressure or how many volts are required, in general, to force a current through a doorbell, an electric light, an iron, a washing machine motor, a small factory motor, and a large factory motor. The most common values are as follows:

An electric doorbell requires 2 to 5 volts.
An electric light requires 120 volts.
An electric iron requires 120 volts.
A washing-machine motor requires 120 volts.
An electric range requires 240 volts.
An electric rapid-transit car motor requires 600 volts.
A small factory motor requires 480 volts.
A large factory motor requires 2400 volts.

These values simply mean that so many volts are required to push or force the working current through the devices or machines.

One can also gain an idea of the strength of electric pressure by observing how much of a shock is received when one puts his hands across the two wires of a circuit. A person cannot detect or

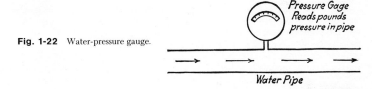

Fig. 1-22   Water-pressure gauge.

feel as little as 5 volts of electric pressure but can, however, feel 50 volts. And 120 volts will give nearly everyone a very unpleasant shock even when a very light and brief contact is made with the wires. If a firmer contact with the wires is made, it may prove fatal. For example, if the hands should be moist or wet and if a firm grasp is made, death is likely to result. All voltages, therefore, starting with 120 volts should be considered as dangerous and handled with great care. Even with light contact 240 volts is more dangerous, and 480 and 2400 volts as well as all higher voltages should be handled only using proper hot-line work procedures. Such voltages are always well guarded, and no one except an authorized person has any business getting near them.

Voltmeter   If we wish to measure the pressure in a water circuit, all we do is to tap a pressure gauge onto the pipe line as shown in Fig. 1-22. Everyone is familiar with such a gauge. The few points to be noted are that the gauge is simply tapped on the pipeline at the point at which the pressure is wanted, so that the pressure at that place can get up into the gauge and make it indicate. It is also evident that the flow of water in the pipe is not disturbed by the insertion of the gauge.

In the same manner, we can measure the electric pressure. We simply connect the two leads from a voltmeter across the line, as shown in Fig. 1-23. The current through the voltmeter will then vary directly with the voltage, and the meter can be made to read volts. It is to be noted that the current which flows to the motor does not flow through the voltmeter. This is because the voltmeter is not a part of the circuit as the ammeter is. Figure 1-23 shows the two meters, the ammeter and voltmeter, properly connected. The ammeter reads the flow of current, and the voltmeter reads the pressure which causes the current to flow. Figure 1-24 shows a typical direct-current switchboard voltmeter.

Waterpower   We have likened an electric current to a current of water. When a current of water flows in a pipe in a simple circuit, as shown in Fig. 1-13, power is delivered to the water motor. We know this because the water motor revolves and can do work. The power delivered depends upon the amount of water flowing and the pressure under which it flows. This is self-evident, for more power will be developed if 50 gal per sec flows through the water motor than

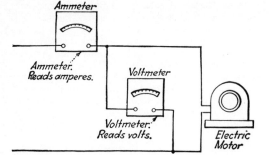

Fig. 1-23   Ammeter and voltmeter correctly connected.

if only 25 gal per sec flows through it. Furthermore, more power will be developed with 100 lb of pressure than with only 50 lb of pressure. The power delivered in the pipeline to the motor thus depends on the amount of water flowing and on the pressure.

Electric Power   In exactly the same way as in waterpower, the amount of power delivered by an electric circuit to an electric motor depends upon the number of amperes flowing and the number of volts of pressure. The greater the current, the larger the number of amperes, the greater will be the amount of power developed by it; and the greater the pressure, the more effect the current will have. The actual value of power in a direct-current circuit (not true for an alternating-current circuit) is equal to the product of volts times amperes, thus

$$\text{Power} = \text{volts} \times \text{amperes}$$

Watt   The unit of power in an electric circuit is the watt. An ordinary electric lamp when connected to an electric circuit, as in Fig. 1-25, will draw about 150 watts from the circuit. An ordinary iron when connected to a circuit will draw about 550 watts of power from it. The motor shown schematically in Fig. 1-26 will draw about 5600 watts of power. It is plain that when we come to large machines the number of watts runs up quickly. So we have chosen to call 1000 watts a kilowatt, generally abbreviated and written kW. A kilowatt is equal to about 1⅓ hp. The motor in Fig. 1-26 thus draws 5.6 kW of power, or 7½ hp.

Fig. 1-24   Typical direct-current switchboard voltmeter. *(Courtesy General Electric Co.)*

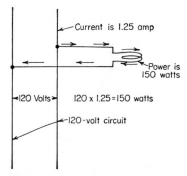

Fig. 1-25   Electric lamp taking 1.25 amp and 150 watts from 120-volt circuit.

Fig. 1-26   Motor drawing 23.4 amp of current and 5.6 kW of power.

Wattmeter To know how much power any electric device or apparatus is drawing from the line, we have to measure it. This is done by the use of a wattmeter. Such a meter registers watts or kilowatts, and by reading it one can at once tell how much any piece of apparatus is consuming. Since the amount of electrical power delivered by a circuit depends upon the amperes flowing and the volts of pressure, the meter must be so connected that the entire load current flows through it and the voltage pressure is across it. The connections are as shown in Fig. 1-27. A wattmeter is essentially a combination of two instruments, an ammeter and a voltmeter. It has an ammeter coil of low resistance, which is connected into the circuit, and a voltmeter coil of high resistance, which is connected across the circuit. A wattmeter will thus have four terminals or binding posts, two for the current-coil leads and two for the voltage-coil leads.

Electric Energy In order that electricity may do useful work, it must act for a period of time. The power expressed in watts tells how much electricity is working, and the hours express the time during which it acts. The product of these two factors gives the amount of work done. Thus,

$$\text{Power} \times \text{time} = \text{energy}$$

or

$$\text{Watts} \times \text{hours} = \text{watthours}$$

Since the unit of power, the watt, is a rather small unit, the larger unit equal to 1000 watts is used. The unit for 1000 is called kilo, therefore, 1000 watts is called 1 "kilowatt." This unit is abbreviated kW. Likewise,

$$1000 \text{ watthours} = 1 \text{ kilowatthour}$$

which is abbreviated 1 kWh. A kilowatthour can thus be thought of as

$$1000 \text{ watts acting for } 1 \text{ h} = 1 \text{ kWh}$$

Any other combination of volts, amperes, and hours whose product is 1000 would give 1 kWh of energy.

Watthour Meter The total amount of electrical energy consumed over a period of time, such as a day or month or year, is indicated by a watthour meter. We are all familiar with the common electric house meter. This is nothing other than an integrating watthour meter which shows how much energy the lights, the iron, the toaster, the washing machine, etc., consume in the course of a month. Such a meter is shown in Fig. 1-28.

Conductors In the water circuit the pipes merely furnish the path for the flowing water. If we were to be more exact, we would say that the hole in the pipe was the path of the water, for we know it does not flow through the iron shell of the pipe. It actually passes through the opening in the pipe inside the iron shell. The hole or opening in the pipe may then be directly compared

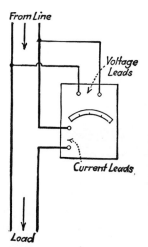

Fig. 1-27 Wattmeter correctly connected into circuit.

Fig. 1-28 Typical single-phase alternating-current watthour meter. *(Courtesy Westinghouse Electric Corp.)*

to the electric wire. The electricity, however, flows right through the hard wire. It does not need an opening. It passes right through the metal. But we find that it can travel with less difficulty through some metals than through others. It travels easily through copper; therefore we often use copper wire. Copper is therefore called a good conductor. Any substance that offers little resistance to the flow of electricity is called a "good conductor." Electricity passes still more easily through silver, but silver is too expensive to be made into wires. Aluminum has been used extensively on long transmission lines, because of its rather good conduction and its light weight, making long spans between poles or towers possible. It is now in general use in distribution lines as well. Iron is not a good conductor but is used as a supporting wire in parallel with or coated by copper or aluminum.

Insulators   In the above paragraph it was pointed out that the hole in the pipe serves as the path for the water. This does not mean to imply that the shell of the pipe is useless and unnecessary. It is very necessary. The shell serves to hold the water in its path. A pipe without a shell would never conduct any water. There would not be any pipe, and the water would flow everywhere. It is well to remember, then, that the shell keeps the current of water in its path.

In an electric circuit, there must be something to keep the current from leaving the wire. The metal of the wire is its path, but there must be something to keep it from leaving the metal. Generally, a shell is put around the wire just as in the water pipe. This shell, however, is not iron but is usually rubber, polyethylene, polyvinylchloride, silk, cotton, or paper. The layer of rubber, polyethylene, polyvinylchloride, silk, cotton tape, or cambric is called "insulation," and a wire so covered is called an "insulated" wire. Figures 1-29 and 1-30 show conductors covered with different kinds of insulations. Silk and cotton are used to insulate wire when the voltage of the circuit is low. For example, cotton is used on doorbell circuits when the voltage is about 5 volts.

Fig. 1-29 Three-conductor (triplex) self-supporting cable commonly used by electric utilities for low-voltage power distribution secondaries and services.

Fig. 1-30 Three-conductor lead-sheathed paper-insulated power cable showing various layers of shielding and insulation. *(Courtesy the Okonite Co.)*

Nonemetallic-sheathed cable—which is an assembly of two or more insulated conductors having an outer sheath of moisture-resistant, flame-retardant, nonmetallic material—is usually used for house wiring where the voltage is 120 or 240 volts. Higher voltages are insulated with rubber, polyethylene, or several layers of paper or cambric, or cotton tape; some even have a steel layer around the outside for protection. This is generally the case with ocean telegraph cables where protection from animals and decay is required. Cables shown in Fig. 1-31 are constructed for direct-buried installation.

Wires are usually mounted on poles where they rest on glass or porcelain insulators, shown in Figs. 1-32 and 1-33. Glass and porcelain are good insulators of electricity. When wires are brought into a station, they are carefully mounted on glass or petticoat porcelain insulators (named because of their semblance to a skirt) in order that the wire may not touch anywhere, thus preventing the electricity from leaving the wire.

On high-voltage transmission lines, a number of porcelain insulators are connected in series, making a string from which the line conductor is suspended. Such a string is shown in Fig. 1-34. The greater the number of insulators in a string, the higher the voltage it can withstand.

Resistance   One can draw up an ice-cream soda faster and with less exertion through a short and large straw than through a long and narrow straw. Consider again the water circuit of Fig. 1-13. It is evident less water will flow if the pipe is long and has a small opening than if it is short and has a large opening provided the pressure is the same in both cases. This is to be expected, for in the long pipe the friction is greater because of the greater length, and if the pipe is small in diameter the friction will be still greater because of the smaller space through which the water must be forced. It may be said that a long narrow pipe offers more resistance to the flow of water than a short and wide pipe.

Fig. 1-31 Cables on reels mounted on back of truck for underground distribution installation. Cable in foreground has an aluminum conductor with high-voltage solid-dielectric (polyethylene) insulation and a copper concentric neutral. Cable in background is three conductor (triplexed) aluminum conductors, with low-voltage polyethylene insulation.

Fig. 1-32 Glass-pin insulator used on low-voltage circuits to insulate the conductors from each other and from the ground. *(Courtesy Kimble Glass Co.)*

Fig. 1-33 Two-layer porcelain-pin insulator for use on medium-voltage circuits. The layers are cemented together. *(Courtesy Ohio Brass Co.)*

In an electric conductor, the resistance a conductor offers to the flow of current will depend on its length and its thickness or diameter. If the wire is very long, more friction must be overcome than if it is short; if it is of small cross section, it will take still more effort to crowd the current through the wire.

Ohm   The unit of resistance for wire is the ohm (symbol Ω). We can picture how much an ohm is by noting how many feet of wire of a given size it takes to make an ohm of resistance. Wires used for electric purposes are supplied on the market in regular sizes of specified diameter. The No. 10 copper wire has a diameter of about ⅒ in and has 1 ohm of resistance for each 1000 ft of length. Thus 5000 ft of this wire would have 5 ohms resistance. Another wire whose cross section is one-half as large would have a resistance of 2 ohms for every 1000 ft of length.

Ohm's Law   From the foregoing it is clear that resistance will reduce the amount of current that will flow. The number of amperes that will flow in a circuit will not be determined wholly

Fig. 1-34   Three-unit string of suspension insulators for use on high-voltage circuits. *(Courtesy Locke Dept., General Electric Co.)*

by the voltage or pressure which causes the current to flow but by the amount of friction or resistance in the wires. Thus, with a given voltage, the greater the resistance of a circuit, the smaller the current will be that flows, and the smaller the resistance, the greater the current will be. This general relation between voltage, current, and resistance is commonly called "Ohm's law." It is a law because it has been found to hold in every case. It is written thus:

$$\text{Current} = \frac{\text{voltage}}{\text{resistance}} \quad \text{or} \quad \text{amperes} = \frac{\text{volts}}{\text{ohms}}$$

$$\text{Resistance} = \frac{\text{voltage}}{\text{current}} \quad \text{or} \quad \text{ohms} = \frac{\text{volts}}{\text{amperes}}$$

$$\text{Voltage} = \text{current} \times \text{resistance} \quad \text{or} \quad \text{volts} = \text{amperes} \times \text{ohms}$$

The law as stated above applies only to direct-current circuits, circuits in which the current continuously flows in one direction in the wire. The law is quite obvious, because it agrees with the common principle with which all are familiar, namely, that the result produced varies directly in amount with the magnitude of the effort or force and inversely with the resistance or opposition encountered.

Rheostats   Often it is necessary to reduce or increase the current in a circuit. This can be done if it is possible to increase or decrease the resistance in the circuit. This can be arranged by connecting into the circuit a resistor having a variable resistance, that is, one that can be changed at will. Resistance boxes, or resistors, are shown in Figs. 1-35 and 1-36. By turning the handle of the hand-operated resistor, more or less resistance is put into the circuit and the current accordingly decreases or increases. Such resistors are commonly called "rheostats" and are used wherever it is necessary to control or regulate the current flowing in a circuit.

Current Direction   Up to this point in this discussion of basic principles, a water circuit has been used in which the flow of water was always in the same direction around the circuit; for it is evident, from Fig. 1-13, that if the engine continues to rotate in the same direction the rotary

Fig. 1-35 Hand-operated field rheostat used for controlling the field current in the field circuits of motors or generators. *(Courtesy General Electric Co.)*

Fig. 1-36 Motor-operated field rheostat. The motor mounted at the top through the worm drive moves the contact arm up or down as desired. *(Courtesy General Electric Co.)*

pump will rotate in the same direction, and the current of water will continue to flow in the same direction. Such a current is called a "direct current." The principal use of direct current is for elevators, electric furnaces, electroplating, etc.

    Alternating Current   If the rotary pump is replaced with a reciprocating pump, like the one shown in Fig. 1-37, the current will not flow in one direction continuously; instead the current of water will first flow in one direction, and then as the piston moves in the opposite direction, the water will also flow in the opposite direction. From Fig. 1-37, it can be seen that when the piston is moving upward the flow of water will be as shown by the arrows in the pipe circuit, and when the piston moves downward the direction of the flow of water will have to reverse. It is to be

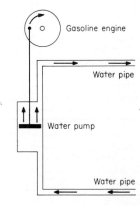

Fig. 1-37   Alternating-current water pump.

noted that in such an arrangement the engine still keeps on revolving in one direction as before. The pump, however, being of a different type, causes the current to flow in one direction for a very short time and then reverses it and makes it flow in the other direction for a very short time. This changing of direction continues all the time. A current that flows first in one direction in a pipe and then in the other direction is said to alternate in direction and is called an "alternating current."

Cycle   Such a water pump causes the current to change its direction two times for each stroke of the pump. When the piston moves up it has a clockwise direction around the circuit, and when the piston moves down it has a direction opposite to the hands of a clock. When the piston comes up again, it has the clockwise direction again. So the current reverses its direction two times for each complete stroke of the pump. Such a complete stroke of the pump with its two reversals of flow is called a "cycle." A cycle is, therefore, something which repeats itself.

Frequency   Frequency is the number of complete cycles made per second. If the pump made 25 complete strokes per sec, the pump could be said to have a frequency of 25 cycles per sec; and

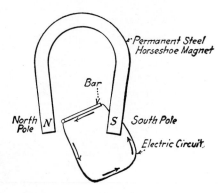

Fig. 1-38 Elements of an alternating-current generator.

if it had a frequency of 60 strokes per sec, it could be said to have a frequency of 60 cycles per sec. Frequency is expressed in hertz (Hz), which is a unit of frequency equal to 1 cycle per second. It is indeed interesting to know that almost all the electricity generated in the world is of the alternating-current type. That is to say, most of the electricity generated and consumed flows first in one direction in the wires and then in the other direction, just as the water does in Fig. 1-37, and most of this alternating current is 60-Hz (or 60-cycle) alternating current. The currents in the house lamp, the toaster, the kitchen range, and the fan motor all flow for a very short time in one direction around the circuit and then quickly change and flow in the other direction. This continues all the time and at a very high rate. Indeed, these reversals generally take place at the rate of 120 per sec or 7200 per min, making the frequency 60 per sec or 3600 per min. This is so fast that one cannot notice any flicker in the light of the house lamp at all. Such currents are, therefore, known as 60-Hz (or 60-cycle) alternating currents.

Outside the United States, the prevailing frequency is 50 Hz. Furthermore, some of the first U.S. hydroelectric plants—like those at Niagara Falls and Keokuk, Iowa—generated 25 Hz. There is no particular advantage associated with the lower 50-Hz frequency.

Alternating-Current Generator   Under the discussion of electrical pressure or voltage, it was stated that a generator or dynamo produces electric pressure which causes current to flow. And now, since most of the electricity used flows in the form of alternating currents, a description of the type of generator that produces an alternating pressure will be taken up. It must be clear that, if a current alternates, the pressure which causes it to flow must also alternate.

Elementary Type Alternating-Current Generator   The simplest arrangement for generating such an alternating pressure is shown in Fig. 1-38. It consists of nothing more than a steel horseshoe magnet and a bar of copper to the ends of which is attached a wire, forming the circuit shown. When this bar is moved down between the poles of the magnet so that it passes through the magnetism at the ends of the poles, an electric pressure or voltage is generated in the wire. If the bar is now raised or moved upward so that it passes through the magnetism in the other direction, a voltage will again be generated in the wire, but this time the voltage will be in the opposite direction. To obtain an alternating voltage, all that is needed is a machine in which a wire moves through magnetism first in one direction and then in the other and arrangements for connecting this moving wire to the outside circuit. Such a device is shown in Fig. 1-39 and is called an "alternating-current generator." In this elementary alternating-current generator the wire has been bent into the form of a loop or coil so that it can be rotated continuously in one direction, instead of having to move it up and down. The two sides of the coil will then alternately pass through the magnetism at the north pole and then through the magnetism at the south pole. If this loop is revolved in the direction of the arrow, the side of the loop marked *AB* will generate a voltage in a direction from *B* to *A* and the other side of the loop, marked *CD*, will generate a

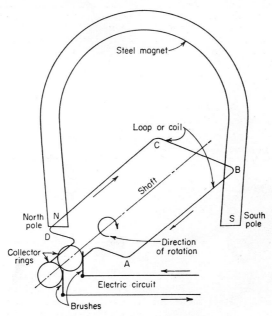

Fig. 1-39   Elementary alternating-current generator.

voltage in a direction from *D* to *C* at the instant shown. If the loop is connected to a closed circuit through rings at the ends of the loop by means of sliding-contact brushes, a current will flow.

One-fourth of a revolution later, when the loop is in the vertical position, the sides of the loop are not passing through any magnetism, and during that instant there will be no voltage generated and also no current flow. When the loop has advanced one-half of a revolution from the first position, the wires are again passing through magnetism, causing a voltage to be generated and current to flow if the circuit is closed. This time, however, the current will flow in the opposite direction, because the wires are passing different poles. An alternating current can thus be taken from the rings. If the values of generated voltage are plotted against the corresponding positions of the coil, the curve shown in Fig. 1-40 results. This is the familiar wave of an alternating voltage. Brushes sliding on the collector rings connect the loop to the electric circuit. Such a machine is then a generator of alternating currents.

The voltage in the coil will reverse its direction twice for every revolution of the coil, just as the water changed its direction twice for every stroke of the pump. The frequency in cycles per second will then be the same as the revolutions of the coil per second. If the loop revolves 60 times per sec, the frequency of the voltage will be 60 Hz. The time required for each revolution and for each cycle will be ⅟₆₀ sec.

Single-Phase Alternating-Current Generator   Such a generator as has just been described is shown in Fig. 1-41. This generator has only two slip rings which connect to a simple circuit of two wires. A generator with only two slip rings is known as a "single-phase" generator, single phase meaning that the circuit has only two wires.

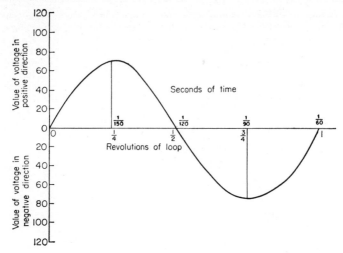

Fig. 1-40 Alternating-voltage wave. This chart gives the positive and negative values of voltage for each position of the loop. Zero corresponds to vertical position of loop.

Fig. 1-41 Telephone magneto alternating-current generator. *(Courtesy Kellogg Switchboard and Supply Co.)*

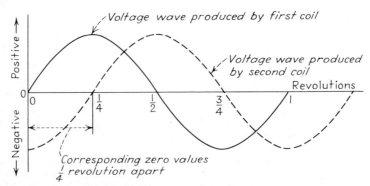

Fig. 1-42 Voltage waves of two-phase generator.

Two-Phase Alternating-Current Generator   If another coil or loop is placed on the rotating shaft so that the wires of this loop lie halfway between those of the loop already described, and if the ends of the second loop are also brought out to another pair of slip rings, such a machine will generate two voltages, one in each loop. There would be this difference: one loop would be passing through magnetism when the other would be midway between the poles, and so the voltage in one would be a maximum while the other was zero. The voltage in one loop would therefore be ahead of the other by the time necessary for the shaft to turn through the space separating the coils, or in this case one-fourth of a revolution (see Fig. 1-42). Such a generator, generating two voltages, is called a "two-phase" generator. There would be four wires leading from the machine making up the two circuits.

Three-Phase Alternating-Current Generator   It is not difficult to picture three coils or loops equally spaced on the shaft as were the two coils of the two-phase generator. If such a machine were built with three coils and three pairs of collector rings, one would obtain three alternating voltages in the three circuits, but each of these voltages would be a little ahead of the others (see Fig. 1-43). Such a machine would then be called a "three-phase" generator. And if the machine were of the simple type with only two poles, the cycles per second would again be equal to the revolutions per second. Such a machine would have six slip rings and six wires leading from it, making up the three circuits.

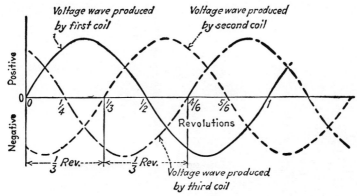

Fig. 1-43   Voltage waves of three-phase generator. The corresponding zero values are one-third revolution apart.

It has been found, however, that the sum of the currents at any instant in three of the six wires is zero. That is to say, some of the currents are flowing away from the generator and others are flowing toward the generator at the same instant, and the net sum of the currents in three wires is zero. This fact makes it possible to do away with three of the six wires, thus making a three-phase line a line with only three wires. The generator will only have three collector rings, and only three wires are connected to the machine.

Revolving-Field Alternating-Current Generator   In the alternators so far described, the field poles were stationary and the conductors were arranged to move past the poles. In large-sized alternators, it is better construction to have the armature conductors stationary and the poles revolve. Such alternators are called "revolving-field alternators" instead of revolving-armature alternators. Figure 1-44 illustrates this type of construction in a hydraulic turbine-driven generator, and Fig. 1-45 in a steam turbine-driven machine. The principal advantages of the revolving-field construction are (1) the armature conductors can be more securely fastened and better insulated, and (2) the amount of material in the rotating part is considerably reduced.

Three-Phase Connections Alternating-Current Generator   The three coils of the three-phase alternator can be connected in two ways, one to form a Y connection and the other to form a delta (Δ) connection. In the Y connection one set of corresponding ends of each of the three coils of the alternator are connected together as shown in Fig. 1-46. The name "Y" is taken from the appearance of the connection when shown as a diagram. The three free ends of the coils connect to the three wires of the three-phase line.

In the delta connection, the three coils are connected in series as shown in Fig. 1-47. The name again is taken from the appearance of the connection when shown as a diagram. The line wires are connected to the junction points or the corners of the delta.

Transmission Lines   Since most of the electricity used in the world is of the alternating-current type, and since the most economical transmission is three phase, it would, therefore, be expected that there would be three wires on the top of long-distance transmission towers or poles.

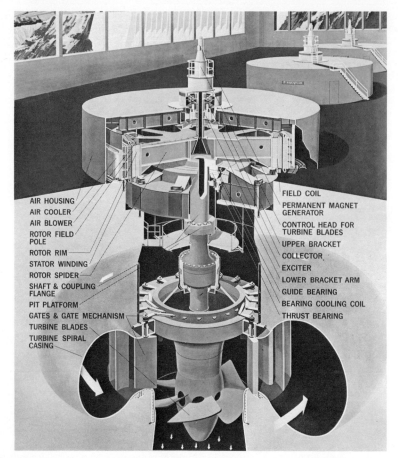

AIR HOUSING
AIR COOLER
AIR BLOWER
ROTOR FIELD POLE
ROTOR RIM
STATOR WINDING
ROTOR SPIDER
SHAFT & COUPLING FLANGE
PIT PLATFORM
GATES & GATE MECHANISM
TURBINE BLADES
TURBINE SPIRAL CASING

FIELD COIL
PERMANENT MAGNET GENERATOR
CONTROL HEAD FOR TURBINE BLADES
UPPER BRACKET
COLLECTOR
EXCITER
LOWER BRACKET ARM
GUIDE BEARING
BEARING COOLING COIL
THRUST BEARING

Fig. 1-44 Three-phase revolving-field type hydroelectric generator. Shaft is vertical and connects water turbine inside casing with revolving field on floor level. *(Courtesy Westinghouse Electric Corp.)*

Fig. 1-45 Cutaway view of steam turbine direct connected to three-phase revolving-field alternating-current generator. Tandem-Compound, Four-Flow, 3600 rpm turbine-generator unit rated 800 MW. *(Courtesy Westinghouse Electric Corp.)*

A typical three-phase transmission line is shown in Fig. 1-48. In many cases, electric light and power companies provide lines with two sets of three wires each as shown in Fig. 1-49.

Such double circuits increase the amount of power transmitted over the line, make more effective use of the right of way, and also improve the probability of ensuring continuous service.

Transmission lines have occasional breakdowns due to strong windstorms which tear down the poles, sleet storms which cause heavy loads on the wires and break them, and line insulators which

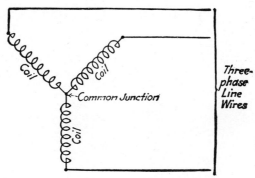

Fig. 1-46   Y connection of three-phase alternating-current generator.

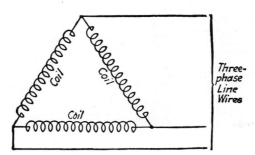

Fig. 1-47   Delta connection of three-phase alternating-current generator.

fail and put the line out of use. Modern and well-designed transmission lines, however, seldom fail because of the elements.

Power Factor   When an alternating voltage and the current which it causes to flow rise and fall in value together and reverse in direction at the same instant, the two are said to be "in phase" and the power factor is unity. This condition is illustrated in Fig. 1-50. The same formula for power as was discussed for direct current holds for alternating current when the power factor is unity. Thus

$$\text{Power in watts} = \text{volts} \times \text{amperes}$$

In most cases, the current and voltage waves are not in phase; that is, they do not rise and fall in value together, nor do they change in direction at the same instant, but instead the current usually lags behind the voltage. This condition is illustrated in Fig. 1-51 and represents the usual condition in transmission and distribution circuits. The current and voltage are now said to be "out of phase." The current drawn by idle running induction motors, transformers, or underexcited synchronous motors lags even more than the current shown in the figure.

It is only occasionally that the current leads the voltage. An unloaded transmission line or an overexcited synchronous motor or a static condenser takes leading currents from the line. In all

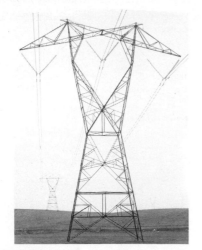

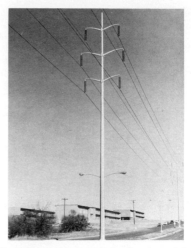

Fig. 1-48 Single-circuit 500,000-volt three-phase high-voltage transmission line. Note the long strings of suspension insulators needed to insulate the conductors. The insulators are installed in a V shape to maintain adequate air clearance between the conductors and the grounded steel tower by preventing the line conductors from swinging too close to the tower. The two ground wires at the top of the steel tower are used to shield the line from lightning.

Fig. 1-49 Double-circuit high-voltage transmission line. The three conductors on each side of the steel pole constitute a single three-phase circuit. The ground wire on top of the steel pole is for protection against lightning. The steel pole construction eliminates congestion in urban areas. Street lights may be mounted directly on the transmission line pole as shown. *(Courtesy A. B. Chance Co.)*

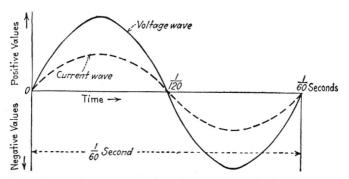

Fig. 1-50 Voltage and current waves in phase. The power factor under these conditions is unity.

cases when the current leads or lags the voltage, the power in the circuit is no longer equal to volts times amperes but is now

$$\text{Watts} = \text{volts} \times \text{amperes} \times \text{power factor}$$

From this the

$$\text{Power factor} = \frac{\text{watts}}{\text{volts} \times \text{amperes}}$$

and the power factor can thus be defined as the ratio of the actual power to the product of volts times amperes. The latter product is generally called "volt-amperes" or apparent power. The value of the power factor depends on the amount the current leads or lags behind its voltage. When the lead or lag is large, the power factor is small, and when the lead or lag is zero, as when the current and voltage are in phase, the power factor is unity. This is the largest value that the power factor can have. In practice the power factor is usually between 0.70 and 1.00 lagging. An average value often taken in making calculations is 0.80 lagging. Table 1-2 gives the power factors of various types of electrical equipment.

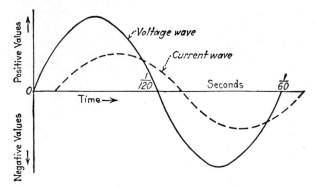

Fig. 1-51 Current wave lagging behind the voltage wave. This is the usual condition in transmission and distribution circuits.

Table 1-2   Power Factors of Electric Equipment

Name of equipment	Power factor, percent	Lagging or leading
Lightly loaded induction motor	20	Lagging
Loaded induction motor	80	Lagging
Incandescent lamps	100	In phase
All types of heating devices	100	In phase
Neon-lighting equipment	30–70	Lagging
Static condenser	0	Leading
Overexcited synchronous motor	Varies	Leading
Underexcited synchronous motor	Varies	Lagging

Low Power Factor   The cause of low power factor is an excessive amount of inductive effect in the electric consuming device, be it motor, transformer, lifting magnet, etc. Induction motors when lightly loaded exhibit a pronounced inductive effect. Idle transformers likewise have a strong tendency to lower the power factor.

Capacitance   Capacitance is the direct opposite of inductance, just as heat is the opposite of cold, sweet the opposite of sour, and day the opposite of night. Capacitance is the property of a condenser, and a condenser is a combination of metal plates or foil separated from each other by an insulator such as air, paper, or rubber. The capacitance, or the capacity of the condenser to hold an electric charge, is proportional to the size of the plates and increases as the distance between the plates decreases.

Power Factor Correction   Since capacitance is the opposite of inductance, and since too much inductance is the cause of low power factor, the manner of raising the power factor is therefore to add capacitors to the circuit. Such capacitors are installed underground on underground distribution circuits, as illustrated in Fig. 1-52, or mounted on poles, as shown in Figs. 1-53 and 1-54. The capacitors can be directly connected to the circuit or switched on and off as needed.

The capacity of the underground bank shown in Fig. 1-52 is 600 kilovolt-amperes (kVA) capacitance. The pole-mounted three-phase bank of capacitors illustrated in Figs. 1-53 and 1-54 is rated 1200 kVA capacitance, 13,200 Y/7620 volts.

Capacitor Switching   A switched capacitor bank is a large capacitor bank which is energized or connected to the line on a given schedule. Because of its size, it is sometimes switched in a number of steps. The schedule is determined by time, circuit load, circuit voltage, or other consideration. It is therefore switched on and off to meet these requirements. The switching can

Fig. 1-52 Submersible capacitor equipment installed in a 36-in diameter vault for use in underground distribution systems. The equipment includes oil switches for energizing three- to six-capacitor units rated 2400 or 7960 volts, 150 kVA capacitance each, for a three-phase bank rating of 450 to 950 kVA capacitance. *(Courtesy Westinghouse Electric Corp.)*

Fig. 1-53 Cluster-mounted bank of four three-phase 300 kVA capacitance, 13,200Y/7,620-volt grounded wye capacitors complete with switches, fuses, and lightning arresters. *(Courtesy Westinghouse Electric Corp.)*

Fig. 1-54 Pole side view of cluster-mounted three-phase 1200 kVA capacitance capacitor bank. *(Courtesy Westinghouse Electric Corp.)*

be done either manually or automatically. This is in contrast with a nonswitched capacitor bank that is connected to the line at all times and is disconnected only for maintenance purposes.

Capacitor Connections   Figure 1-55 shows the detail connections for connecting a capacitor bank to a three-phase four-wire feeder. On installations greater than 180 kVA capacitance, an oil or vacuum switch is necessary. Each phase of the capacitor bank is fused. The cases of the capacitors must all be grounded. Figure 1-56 shows a typical installation of a pole-type capacitor on a three-phase four-wire feeder. Figure 1-57 shows a close-up view of a three-phase capacitor installation connected to the three-wire primary line.

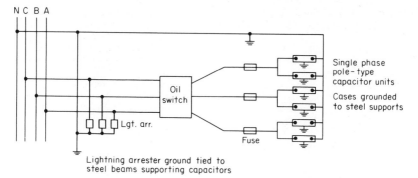

Fig. 1-55 Typical connection diagram of three-phase capacitor bank. *(Courtesy Consolidated Gas Electric Light and Power Co.)*

Fig. 1-56 Typical installation of cluster-mounted pole-type three-phase capacitor for power-factor improvement. Note fuse cutouts and lightning arresters. *(Courtesy Joslyn Mfg. and Supply Co.)*

Fig. 1-57 Twelve units totaling 600 kVA capacitance of switched capacitors installed in a cluster-type rack on a single pole.

Capacitor Precautions   Capacitors and transformers are entirely different in their operation. When a transformer is disconnected from the line, it is electrically dead. Unlike the transformer and other devices, the capacitor is not "dead" immediately after it is disconnected from the line. It has the peculiar property of holding its charge because it is essentially a device for storing electrical energy. It can hold this charge for a considerable length of time. There is therefore a voltage difference across its terminals after the switch is opened.

Capacitors for use on electrical lines, however, are equipped with an internal-discharge resistor, as shown in Fig. 1-58. This resistor being connected across the capacitor terminals will gradually discharge the capacitor and reduce the voltage across its terminals. After 5 min the capacitor can be considered as fully discharged.

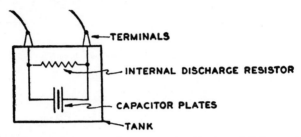

Fig. 1-58   Internal connections of a static capacitor. Note internal-discharge resistor.

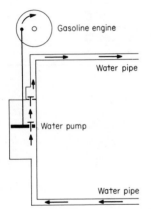

Fig. 1-59   Direct-current water pump.

To be perfectly safe, however, proceed as follows: Before working on a disconnected capacitor, wait 5 min. Then test the capacitor with a high-voltage tester rated for the circuit voltage. If the voltage is zero, short-circuit the terminals externally using hot-line tools and ground the terminals. Now you can proceed with the work.

Direct-Current Generator   We have carefully studied how an alternating current of water could be produced by the pump shown in Fig. 1-37. We observed that the current of water circulated around the pipe circuit first in one direction and then in the opposite direction. And we also observed that the water current reversed its direction twice for each complete stroke of the pump. All this we have carefully compared with the action of a real electric generator and found that the processes are quite alike. This much is fundamental in the generation of electricity, whether it is alternating-current or direct-current electricity. To obtain alternating current we simply fastened slip rings to the ends of the loop and connected those rings to the circuit by means of sliding brushes.

Direct-Current Water Pump   To obtain a direct current of water, which is a current that flows continually in one direction, we simply have to add a set of valves to our water pump, as shown in Fig. 1-59. The water will now not be able to flow backward after it is once forced through the valve and, therefore, the direction of the water flow will always be the same. As the piston moves up, it forces the water through the top valve, and as the piston drops, the top valve closes and the lower valve rises and allows the water to fill the cylinder. When the piston moves upward again, the cylinder is full of water and more water will be forced into the circuit through the top valve as before.

Elementary Type Direct-Current Generator To obtain a direct current of electricity from the machine shown in Fig. 1-39, we simply replace the slip rings with a *commutator*. The commutator is the device which changes an alternating voltage to a direct voltage. It corresponds to the valves in the hydraulic direct-current generator. The process is called "commutation." In this simple case, the commutator consists of two bars of copper connected to the two ends of the loop or coil. These two copper bars revolve with the shaft as did the slip rings. Upon these bars rest brushes which connect the machine to the electric circuit. The general arrangement is shown in Fig. 1-60.

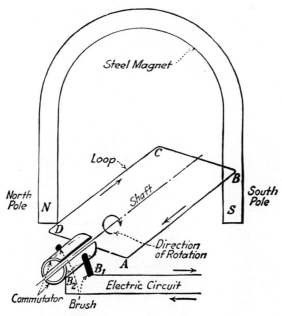

Fig. 1-60 Elementary direct-current generator.

Commutation, Direct-Current Generator From Fig. 1-60, it is clear that the commutator will first connect one side of the loop to one side of the circuit, and then as the shaft revolves it will connect the other end of the loop to the same side of the circuit. If the brushes are correctly set, this change will take place when both sides of the loop are not generating any voltage, that is, when the coil is midway between the poles. This is the point at which the voltage in the loop changes direction. When A side of the loop is passing the south pole, the voltage is in a given direction and feeds into brush B_1, and D side of the loop feeds into brush B_2. But when the loop has revolved to its midway position, the direction of the generated voltage in the loop is about to change, and just at this instant the ends of the loop are also changed to the opposite sides of the circuit, so that when the loop revolves still further, the voltage into the circuit will remain the same. In this way, the voltage that alternates in the loop is made to produce a pressure in the same direction in the outside circuit. Such a machine with its commutator is called a "direct-current generator."

Commercial Direct-Current Generator The principal parts of a commercial direct-current generator are shown in Fig. 1-61. These parts are the frame, the poles, the armature, the commutator, and the brushes. The purpose of the frame is to support the north and south poles. The armature supports the conductors which constitute the armature winding. The commutator, which in an actual machine is made up of many copper segments, is also supported by the armature. A special brush rigging holds the brush holders and brushes. Figure 1-62 shows these parts assembled and placed in proper relation to each other.

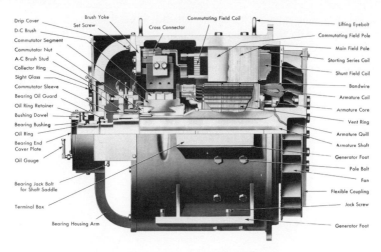

Drip Cover
Brush Yoke
Set Screw
Cross Connector
Commutating Field Coil
Lifting Eyebolt
D-C Brush
Commutating Field Pole
Commutator Segment
Main Field Pole
Commutator Nut
A-C Brush Stud
Starting Series Coil
Collector Ring
Shunt Field Coil
Sight Glass
Commutator Sleeve
Bandwire
Bearing Oil Guard
Armature Coil
Oil Ring Retainer
Armature Core
Bushing Dowel
Vent Ring
Bearing Bushing
Oil Ring
Armature Quill
Bearing End
Cover Plate
Armature Shaft
Oil Gauge
Generator Foot
Pole Bolt
Bearing Jack Bolt
for Shaft Saddle
Fan
Flexible Coupling
Terminal Box
Jack Screw
Bearing Housing Arm
Generator Foot

Fig. 1-61 Sectional view of a direct-current generator showing details of construction. Generator is rated 100 kW. *(Courtesy Allis-Chalmers Mfg. Co.)*

Fig. 1-62 Direct-current generator assembled. This machine is rated 250 kW, 250 volts, and runs at 225 rpm. *(Courtesy Allis-Chalmers Mfg. Co.)*

Direct-Current Circuits   Direct-current circuits, as we have seen, require only two wires, one for the outgoing current and the other for the return current. Overhead circuits carrying direct current will, therefore, consist of only two wires.

In general, the use of direct current is confined to special applications, such as electroplating, battery charging, and elevator operation, or where fine speed control of industrial motors is required. High-voltage direct-current circuits are used to transmit large amounts of power over long distances. Direct-current transmission lines transmit power from remote generating stations, often located adjacent to a large coal mine or at a water reservoir, to large load areas.

Transformers   The voltages necessary to obtain economical transmission are higher than can be directly generated by an alternator. Also, the voltage at which electric power is used in motors and lamps is less than that required for distribution. It is, therefore, necessary to raise the voltage at the generating station to the value required for transmission and to lower it at the point of consumption to the values required by the motors and lamps on the system. In the first case, the voltage must be *stepped up,* and in the latter it must be *stepped down.* The transformer is the apparatus used to make these changes in voltage.

Hydraulic Transformer   To make clear how this is done, the water circuit shown in the left of Fig. 1-63 will be referred to. Since this is an alternating-current circuit the water flows to and fro or back and forth in the circuit. As it does this, it causes the piston in *B* to move up and down. But piston *B* is connected to piston *C* by a rod pivoted at *P,* so that when *B* moves up and down *C* moves down and up. The pressure in *B* is very high and the current of water is small, just like the voltage and current in a high-voltage transmission line. In *C* it is clear that the pressure will be low and the current of water large, because the size of the piston is so much larger. Such a mechanism could thus be used to transform or step down a water circuit having a high pressure to one having a low pressure. The current on the low-pressure side would be much larger than the current on the high-pressure side.

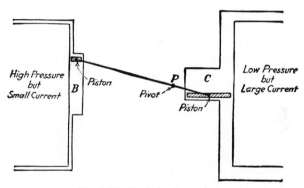

Fig. 1-63   The hydraulic transformer.

It is well to note that the frequency of the current has not been changed. The only changes are the decrease in pressure and the increase in current. It should also be noted that the total power delivered has not been changed, for under the discussion of power it was pointed out that power is dependent upon the product of voltage and current. The same power can be obtained from a circuit of high voltage and small current as from a circuit that has a low voltage and a large current.

Electric Transformer   An electric transformer operates in the same manner as the hydraulic transformer. On the high-voltage side the electrical pressure is high and the current is small. On the low-voltage side the pressure is low and the current is large. One can, in general, easily tell the high-voltage from the low-voltage side of a transformer by observing the size of the insulators or bushings on the top of the transformer case. The high voltage must be better insulated than the low voltage. One generally finds large porcelain bushings projecting from the case on the high-voltage side and smaller bushings on the low-voltage side. Figure 1-64 shows the core and windings of a single-phase transformer removed from its tank. The high- and low-voltage leads are brought out on top of the tank.

In Fig. 1-65 are shown the elementary parts of an electric voltage transformer. Sheet-iron plates are so arranged as to form a closed magnetic circuit. Upon these plates are placed two coils of insulated wire, one of which has many turns of small wire, and the other has few turns of heavy, coarse wire. The coil with the many turns is the high-voltage coil and is called the "primary winding" in a step-down transformer. The other coil with the few turns is the low-voltage coil and is called the "secondary winding." Thus, the winding into which current is brought is the primary, and the winding from which current is taken is the secondary. The coils correspond to the two pistons in the water transformer, and the magnetic core acts as the coupler between the two pistons.

Actually, what goes on in the transformer is somewhat as follows: The voltage applied to the primary causes a current to pass through the primary coil. This current creates a magnetic flux in the core. The flux in the core cuts both the primary and the secondary coils. This cutting of the primary coil creates a countervoltage in the primary coil which very nearly equals the primary

Fig. 1-64 Single-phase transformer with case removed showing core and windings. High-voltage winding is rated 66,000 volts, and low-voltage winding is rated 13,200 volts. Hand-operated tap changers are shown in center foreground. *(Courtesy General Electric Co.)*

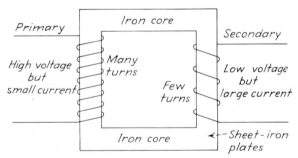

Fig. 1-65 Electric voltage transformer.

voltage applied. Thus, the current in the primary at no load is only great enough to magnetize the core at no load. The secondary coil being cut by the flux will have voltage at no load, but there is no current. Now, let us apply load on the secondary. The load current in the secondary will create a counterflux in the core which reduces the magnetic flux in the core. The reduction in magnetic flux in the core reduces the primary countervoltage. The reduction in primary countervoltage increases the difference between the applied voltage and the countervoltage. More current will therefore flow into the primary, thereby increasing the magnetic flux to its former no-load value. This again raises the secondary induced voltage to its original value. All these adjustments within the transformer take place automatically and instantaneously.

It should be noted that in a modern transformer the primary and secondary windings are not placed on separate legs, as shown in the elementary diagram of Fig. 1-65, but, instead, each winding is generally divided into two parts, and one-half of each winding is placed on each leg. This gives a more constant voltage with changes in load. Figure 1-66 shows a typical transformer from which the case has been removed. It shows clearly the magnetic core, the coils, and leads.

Transformer-Core Construction   The core of a transformer can be built in various shapes. The following types are in general use.

Core Type   The core is in the shape of a rectangle. Coils are placed on two legs of the core.

Shell Type   The core is rectangular in shape with a central leg in addition to the core first described. Coils are placed on the central leg.

Fig. 1-66 Three-phase distribution transformer with case removed, showing core, windings, and high- and low-voltage leads. Transformer is rated 37.5 kVA, 4160Y/2400 to 416Y/240 volts, 60 Hz. *(Courtesy General Electric Co.)*

Fig. 1-67 Wound-core type of distribution transformer. *(Courtesy Westinghouse Electric Corp.)*

Cruciform Type   The cruciform core is a modification of the shell type. It is the same as if two shell-type cores were set at right angles, using a common central core for the windings.

Wound-Core Type   In this type, the core is made of a ribbon of sheet steel wound in a spiral. After the core is wound and annealed, the coils are wound on the core. In the smaller sizes, this type may supersede all other core types eventually. (See Fig. 1-67.)

Transformer Cooling   When the voltages of the primary and secondary windings are both low and the transformer is of very small size, the transformer, that is, the iron core with its two windings, is placed in a metal case merely to keep out dirt and moisture. But if the voltage exceeds 2000 volts, the core with its two coils is usually placed in a tank filled with oil. The oil is a good insulator and also helps to cool the windings. It will perhaps be remembered that whenever current flows in a wire there is a friction loss in heat. An illustration of this is found in the toaster or the iron. The same is true in a transformer; whenever currents are flowing through the primary and secondary windings, heat is given off which must be carried away so that the transformer may not become too hot and char or burn the insulation on the wire of the coils. If it becomes too hot, the transformer will be ruined just as a person will die from running a high temperature for a long time.

Because a transformer must not get too hot, more power can be safely drawn from it in cold weather than in hot weather. Moreover, more power can be taken off for short periods of time than can be taken off continuously. Any transformer too hot to touch is overheated because of an overload and should be replaced with a larger transformer before it becomes damaged.

Methods of Cooling Transformers   The oil in the tank is the principal cooling agent. It keeps the windings cool by carrying the heat from the coils to the surface of the tank. Many times the transformer case itself is built so that it has much more surface for cooling than if it were smooth. The metal sides of the tank are bent into ruffles (Fig. 1-68), or pipes are connected to the top and bottom of the tank (Fig. 1-69), so that whatever oil is in the pipes is exposed on all sides. The object in either case is to increase the surface from which the heat can radiate. This is the same principle as is used in the automobile radiator.

Some very large transformers have coils of pipes, placed in the top of the tank, through which cold water is circulated by a pump. The cold water in the pipes cools the oil in the tank. The water pipes are placed in the top of the tank because the hottest oil is always at the top. An illustration of this method of cooling is shown in Fig. 1-70.

Fig. 1-68 Single-phase transformer showing the tank corrugated or ruffled to increase the cooling surface. Also note oil-level gauge on right. Transformer is rated 250 kVA, 60 Hz, 4160Y/2400 on high-voltage side and 240/120 volts on low-voltage side. *(Courtesy General Electric Co.)*

Fig. 1-69 Transformer tank showing pipes connected to top and bottom of tank to aid in the cooling of the oil. Oil temperature gauge is located centrally near the top of the tank. This transformer is entirely air cooled. This three-phase oil-immersed power transformer is rated 1000 kVA, 60 Hz, 13,800 volts on high-voltage side, and 480 volts on low-voltage side. *(Courtesy General Electric Co.)*

Transformer Temperature Limits In general, the maximum safe temperature for such insulating materials as cotton, silk, cambric, tape, fabric, etc., is 105°C, which corresponds to 221°F. It is impossible, however, to measure the actual temperature in the transformer because the exact location of the hottest spot is not known; nor is it accessible. The temperature of the oil plus a 10 or 15° correction is usually taken as an indication of the temperature of the windings. In large power transformers the temperature of this oil is usually obtained by some form of thermometer, such as that shown in Fig. 1-69.

In distribution transformers devices are now being used which indicate the maximum load on the transformer in percent of the transformer capacity. One such device, shown in Fig. 1-71, is known as the "thermotel," meaning that it tells the temperature. The rapid growth of loads on distribution systems makes it necessary, in many cases, to make a check of the loads on transformers in order to determine whether the transformer is still large enough to carry the load. Another overload indicator, illustrated in Fig. 1-72, makes use of a bimetallic thermostat to give

Fig. 1-70 Interior of case of water-cooled transformer showing cooling coils in which cold water is circulated. The cooling coils are located in the top of the case because the hottest transformer oil is at the top. This transformer is rated 23,333 kVA, 60 Hz, 115,000Y/66,420 volts on high-voltage side and 13,200 volts on low-voltage side. *(Courtesy General Electric Co.)*

Fig. 1-71 Thermotel being installed on distribution transformer to indicate the largest load that the transformer has carried since the last reading. It indicates the load in percent of transformer rating. *(Courtesy General Electric Co.)*

Fig. 1-72 Self-protected pole-type distribution transformer with low-voltage circuit breaker and overload indicating light. Red light indicates transformer has been overloaded. *(Courtesy Westinghouse Electric Corp.)*

an indication of the winding temperature. It turns on a red light which is visible from the ground if the transformer is overloaded. A reset device makes it possible to turn the light off.

Three-Phase Transformer A three-phase transformer is really three single-phase transformers in one case, using a single combined core. The core has much the same shape as the shell-type single-phase transformer. However, in the three-phase, instead of having only a winding on the central leg, the two outside legs have windings also (see Fig. 1-66). Each winding acts for a separate phase, but the common core acts to supply flux for all three phases.

A three-phase transformer weighs about two-thirds as much as the same capacity in single-phase transformers. Their biggest disadvantage lies in the fact that if one phase fails the entire transformer must be taken out of service.

Fig. 1-73   Cutaway view of a shell-form substation transformer showing core, coils, internal bridge structure, bushings, and cooler with fans and oil pump mounted below. *(Courtesy Westinghouse Electric Corp.)*

Large substation transformers employ forced-air and forced-oil cooling in order to increase the output rating. By forced circulation of the oil through the external radiators and by blowing air against the radiators, the heat is more rapidly removed from the hot oil. A typical transformer equipped with forced-air fans is shown in Fig. 1-73.

Electric System

Electric System—General The electric systems in the United States and Canada are inter-connected forming a network of electric facilities all across the continent. It is difficult to separate the electric system of one company from neighboring utilities. Figure 2-1 is a single-line diagram showing schematically a portion of an elementary electric system. Most electric utility companies will have many generating stations supplying a very large complex and intricate network of trans-mission and distribution circuits.

The electric system in Fig. 2-1 has nuclear, fossil-fired, and hydroelectric generating stations. The bulk of the electric energy in the United States is generated in fossil-fired steam turbine-driven generating stations using coal, natural gas, and oil fuel. Hydroelectric generating stations using water pressure to turn a turbine have been constructed at most of the sites where there is sufficient waterfall to economically generate electricity. Nuclear generating stations use the heat from an atomic chain reaction to generate steam to turn a turbine generator.

Most electric energy is generated at a voltage in a range of 13,200 to 24,000 volts. The voltage is raised to transmission levels in a transmission substation located at the generating station. Trans-mission substations N, F, and H, shown schematically (Fig. 2-1) serve this purpose. Transmission circuits or lines operating at 138,000 to 765,000 volts deliver the energy to the load area. Trans-mission substations A, B, and C (Fig. 2-1) adjacent to the load area reduce the voltage to subtrans-mission levels 34,500 to 138,000 volts. The subtransmission circuits or lines extend through met-ropolitan areas to distribution substations located in the area of the load to be served. Distribution substations 10, 11, and 12 (Fig. 2-1) reduce the voltage from subtransmission levels to distribution range usually 4,160Y/2400, 12,470Y/7200, 13,200Y/7620 24,940Y/14,400 or 34,500Y/19,920 volts.

Distribution circuits radiate from the distribution substations to supply the customer's load. Some large industrial customers may be served directly from transmission or subtransmission cir-cuits. These customers will normally be metered by a high-voltage or primary-meter installation. The large industrial customer will have a substation or substations to reduce the voltage to the desired level for distribution throughout the plant area. Utility owned distribution circuits will normally exit from a distribution substation underground and then rise and connect to overhead distribution circuits. Underground exit cables are being extended several thousand feet in some cases to minimize the overhead circuits in the vicinity of a substation. Some companies are build-ing new distribution circuits completely underground. The dashed lines extending from substation 11 (Fig. 2-1) indicate underground circuits. The solid lines indicate overhead circuits.

Branch circuits tap to the main distribution feeder circuits through high-voltage fuses com-monly called "distribution cutouts" on overhead lines. The branch circuits may be either overhead or underground, as shown schematically. These circuits are often referred to as "primaries."

Distribution transformers connect to the branch circuits or primaries. Pole-type transformers are used on overhead circuits and padmounted, submersible, or direct-buried transformers are used on underground circuits. The distribution transformers reduce the voltage to utilization level

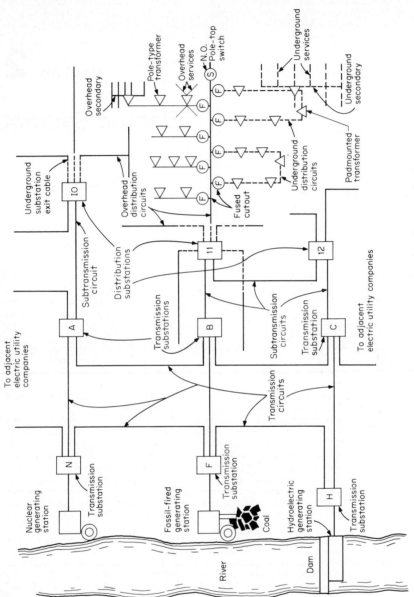

Fig. 2-1 Single line schematic diagram of a portion of an elementary electric system.

usually 120/240 volts single-phase for residential service and 208Y/120 or 480Y/277 volts three-phase for commercial or industrial service. Some small commercial installations may be served by 120/240 volts single-phase. The low-voltage service to the customers may be either overhead or underground connecting directly to a distribution transformer or a secondary extending from the distribution transformer. Both types of installations are shown schematically in Fig. 2-1.

The main distribution feeder circuits often have normally open switches connecting to other main distribution feeder circuits to provide an alternative source for emergency operation. A switch serving this purpose is shown schematically in the circuit fully illustrated extending from distribution substation 11 (Fig. 2-1). Other switches may be installed in main feeder circuits to permit sectionalizing the circuit if a component of the circuit fails electrically.

Figure 2-2 is a schematic drawing illustrating a portion of an electric system pictorially. The

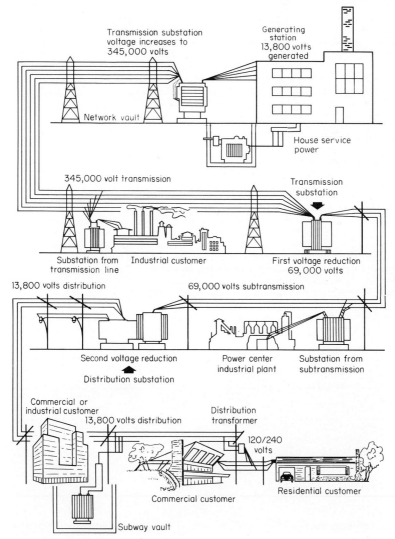

Fig. 2-2 Portion of a typical electric system showing generating station, step-up transmission substation, transmission line, step-down transmission substation, subtransmission line, distribution substation, distribution system, and industrial commercial and residential customers.

generators in a power plant will normally be tied to the transmission lines in a transmission substation located at the generating station. Power transformers in the transmission substation raise the voltage from the generated level to the transmission level. Circuit breakers and other equipment in the transmission substation isolate defective circuits or equipment from the electric system. Substations are used to connect the transmission lines to the distribution system. The purpose of the distribution substation is to provide equipment to protect the complete electric system from trouble on any distribution feeder. The substation is almost always equipped to isolate the feeder in trouble from the other feeders radiating from the substation. Where load and voltage conditions require it, the substation has the additional function of regulating the voltage on the feeders of the distribution system.

GENERATING STATIONS

A generating station is a plant in which the energy of coal, oil, natural gas, waterfall, or atom is changed into electrical energy. Central generating stations have come into existence because it is inefficient and uneconomical for every user of power to generate electricity. Generating stations produce alternating current. The frequency of the alternating-current power generated in the United States and Canada is generally 60 Hz (hertz). In most other countries nearly all the power is generated at a frequency of 50 Hz. Where direct current is required, it is usually obtained by converting alternating current to direct current.

Generating stations may be any of the following types:
1. Hydroelectric
2. Steam Turbine
   a. Fossil fired
   b. Nuclear
3. Combustion turbine
4. Internal-combustion engine
5. Geothermal
6. Solar

Hydroelectric Generating Stations   If a generating station is a hydroelectric station, it must be located at a waterfall. This is sometimes far removed from the territory to be served. Figures 2-3 and 2-4 show the exterior and interior of typical hydroelectric plants. Such plants cost little to run because they do not burn fuel, but the interest and taxes per year are higher than for steam plants because of the larger investment in a dam, land flooded by backwater, etc.

Fossil-Fired Steam-Turbine Generating Stations   Figure 2-5 is a typical fossil-fired steam turbine-driven generator power plant. The generating station contains three generating units. The three units are rated 147 MW (megawatts), 330 MW, and 520 MW. The generating station is located near a river.

The boilers burn low-sulfur coal. The coal is mined in Wyoming and delivered to the station by 100-car unit trains hauling 10,000 tons. Thawing sheds, heated electrically, are used to thaw

Fig. 2-3   Hydroelectric plant located on Tennessee River at Wheeler Dam. The plant contains four generating units, each rated 36,000 kVA, 13,800 volts, 60 Hz, and run at 85.7 rpm. The rotor of each generator has 84 poles. *(Courtesy General Electric Co.)*

Fig. 2-4 Powerhouse of Bonneville hydroelectric plant located on Columbia River in Oregon. The plant contains 10 vertical water-wheel generating units, each rated 60,000 kVA, 13,800 volts, 60 Hz. and run at 75 rpm. Each generator rotor has 96 poles. *(Courtesy General Electric Co.)*

Fig. 2-5 Fossil-fired steam-turbine generating station with three generators. The boilers use low-sulfur coal for fuel. The plant buildings in the foreground house equipment with a capacity of 520 MW. *(Courtesy Iowa Public Service Co.)*

the coal frozen in transit during the period of cold winter weather. The unit train hopper cars pass over a pit where the bottom of the cars are opened automatically and unloaded in 2 min. A conveyor belt transports the coal from a hopper pit to the coal storage area or the silos in the plant at the rate of 3000 tons of coal per hour. The coal storage area has a capacity of 800,000 tons.

The boiler, for the 520-MW generator unit, develops steam with a pressure of 2500 lb/in^2 and a temperature of 1005°F. The boiler is installed in the highest building, 240 ft, in the background of Fig. 2-5. The lower structure between the boiler enclosure and the 400-ft chimney is the electrostatic precipitator. The precipitator removes 99.2 percent of the particulates from the boiler gases before they are emitted to the atmosphere through the chimney. The electrostatic precipitator occupies an area about two-thirds the size of a football field.

The 520-MW turbine generator is located in the lower building in front of the boiler enclosure. The turbine generator (Fig. 2-6) is 18 ft high and 132 ft long. The turbine spindle has rows of blades that stem out from a shaft at the center. Steam from the boiler strikes the blades, like wind on a windmill, to spin the turbine shaft at 3600 rpm. The turbine shaft turns the generator rotor, a strong electromagnet, to create the electrical energy at 24,000 volts. Water from the river next to the generating station is used to condense the steam after it leaves the turbine. The water condensed from the steam is returned to the boiler feed pumps. The generating station is located on a 122-acre site.

Single steam-turbine generator units (Fig. 2-7) are built with a capacity of as much as 1200 MW. The majority of the steam-turbine generators are smaller units with a capacity of 25, 50, 90, 125, 250, 375, 450, 525, 600, 800, or 1000 MW. A typical fossil-fired steam turbine-driven generating station is shown schematically in Fig. 2-8.

Nuclear Generating Stations   Generating stations using nuclear energy are constructed similar to fossil-fired steam turbine-driven generating stations except the boiler is replaced with a nuclear reactor. Nuclear plants do not discharge sulfur dioxide and particulates into the atmosphere like plants burning fossil fuel, thus, minimizing air pollution.

Fig. 2-6   Steam-turbine generator unit rated 520 MW. The turbine shaft and the generator shaft are coupled directly together. The turbine is in the foreground and the generator in the background. *(Courtesy Iowa Public Service Co.)*

Fig. 2-7   Perspective rendering of a large steam-turbine generator unit. *(Courtesy General Electric Co.)*

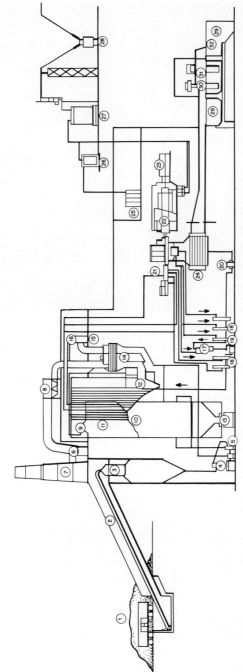

Fig. 2-8 Fossil-fired steam-turbine generating station. (1) Reserve coal supply—300,000-ton capacity; (2) coal conveyor system—360-ton/h capacity; (3) coal bunker—1750-ton capacity; (4) coal pulverizer; (5) exhauster; (6) induced-draft fan—510,000 ft³/min; (7) Stack—reaching 346 ft above the ground; (8) electrostatic precipitator; (9) boiler drum; (10) steam generator—including furnace 133 ft high—860,000 lb of steam per hour capacity; (11) superheated steam ready to go to turbine; (12) economizer containing boiler feedwater; (13) ash sluice system; (14) air preheater; (15) forced-draft fan—340,000 ft³/min; (16) air intake; (17) deaerating open feedwater heater; (18) closed feedwater heaters; (19) boiler feed pump; (20) condensate pump; (21) reheat steam turbine; (22) electric generator—125,000 kW rating—138,000 kW net capability; (23) exciter; (24) steam condenser; (25) house service switchgear; (26) house service transformer; (27) 69Y/39.8-14.4 kV—main transformer—145,000 kVA; (28) circuit breaker; (29) river; (30) circulating pumps—65,000 gal per min; (31) traveling screen; (32) debris rack.

Fig. 2-9 Model of Carolinas Virginia Nuclear Power Associates Plant at Parr, S.C., showing cutaway view of reactor area. The nuclear reactor in which the atomic fission takes place is housed under the metal dome for protection. Heat produced by fission changes water to steam which drives a conventional steam-turbine generator. *(Courtesy Stone and Webster Engineering Corp.)*

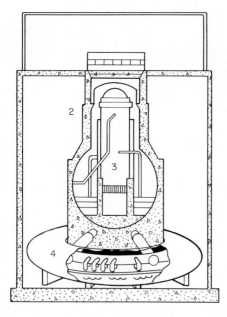

Fig. 2-10 Boiling water reactor (BWR). (1) Concrete reactor building; (2) steel-lined concrete drywell; (3) steel reactor vessel; (4) suppression pool.

Radiation, measured in millirems, is present everywhere. Natural background radiation in Iowa and Illinois along the Mississippi River is 100 to 130 millirems every year. If you lived next door to a nuclear power station, you would receive less than 5 millirems of additional radiation annually, about the same amount you would receive on a round-trip coast-to-coast jet flight.

A nuclear generating station is shown in Fig. 2-9 and a boiling water reactor in Fig. 2-10. Nuclear fission process is used to produce heat in the reactor. Fissioning occurs when a subatomic particle, known as a neutron, strikes the nucleus of an atom with sufficient force to split the nucleus. When an atom splits, energy in the form of heat is released. If enough neutrons are released, they will continue to strike and split other atoms causing a chain reaction. Each separate fission releases only a very small quantity of heat. In a nuclear reactor, many trillions of controlled fissions take place every second. The fissioning process is maintained and controlled in the reactor. Heat from fissioning produces steam, which is piped to large turbines (see Fig. 2-11).

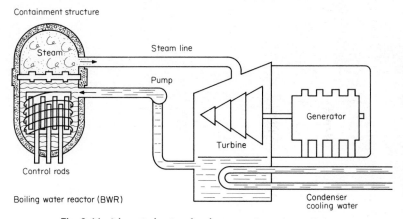

Fig. 2-11   Schematic drawing of nuclear generating station equipment.

Everything in the world is made of atoms—the ground, your body, the food you eat. Certain materials, however, lend themselves to being split, or fissioned, more readily than others. Uranium is such a material, and at the present time it is the most commonly used nuclear fuel.

Combustion-Turbine Generating Stations   Generating stations using combustion turbines as prime movers for the generators can be located in an area with other industrial type installations. Locating the generation near major electrical loads improves reliability and may reduce investment in transmission and distribution equipment.

A combustion-turbine generator facility (Figs. 2-12 and 2-13) will include a starting motor, a rotary-type compressor, combustors, ignitors, a multiple-stage turbine, an alternating-current generator, control devices, and necessary accessories. Fuel used in the turbines may be natural gas or oil in crude form or refined state.

The starting motor drives the compressor to produce pressurized air which is mixed with the fuel in the combustor and ignited. The hot gases flow through the turbine causing it to accelerate. As the turbine power and speed increases, the starting motor is disengaged and the electric generator is synchronized to the system and connected to the load. The hot gases produced from the burning natural gas or oil expanding through the turbine produce the force to rotate the generator and produce electrical energy.

Combustion-turbine generating stations usually provide peaking power or emergency power. The capacity of the generating units are usually in the range of 15 to 95 MW. The fuel cost for combustion turbines is high, but they have the advantage of low capital cost, automatic operation, fast starting, and minimum electric transmission and distribution investment.

The hot exhaust gases from a combustion turbine can be used to produce steam in a waste heat boiler. The steam can be used to operate a turbine generator unit improving the efficiency of the operation and reducing the operating costs. This type of installation is called a combined cycle plant.

Internal-Combustion-Engine Generating Stations   Generating units driven by internal-combustion engines of the diesel type are usually located in remote areas, small towns, and sites requiring standby or emergency power. The engines use oil or natural gas for fuel. The diesel engine-driven generating units have a low installed cost, can be designed for automatic operation, and can start without an alternating-current source. The fuel cost for the units is high compared

Fig. 2-12   Combustion-turbine generator installation. Units can be used for peak-load, emergency or base-load generation. High cost of fuel limits base-load operation. *(Courtesy Westinghouse Electric Corp.)*

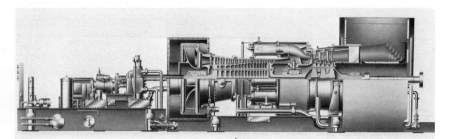

Fig. 2-13   Sectional view of combustion turbine for 50 MW generator. *(Courtesy General Electric Co.)*

to coal- or nuclear-fired stations. Diesel engine generating stations have a capacity in the range of 500 to 5000 KW. Construction of a diesel engine generator unit is shown in Fig. 2-14.

Geothermal Generating Stations   Geysers located north of San Francisco, Calif., produce steam that is piped to turbines to generate electricity. Geothermal steam produces electricity without consuming fuel. It is estimated the geysers will be capable of generating 1000 to 2000 MW. Other geothermal energy sources may be found in southern California. The costs of exploring for and purifying the steam from geothermal fields may make it uneconomical for electrical energy generation.

Solar Generating Stations   The sun can be used to generate electricity by using solar cells that produce small amounts of electric energy. The capital costs for this method of generation exceed the practical limit except for special uses, such as electrical energy for space ships. Steam can be generated by using mirrors to direct the solar energy to a boiler mounted on a tower. The steam created can be used to operate a turbine to generate electricity. The capital cost of this procedure is excessive. Solar energy can be used to heat water economically. Some homes and commercial buildings use solar panels to produce hot water for heating.

Electric Generation   The sources of oil and natural gas are diminishing. This makes it essential to expand the generation of electricity using coal and nuclear fuels. Solar-generated electrical

Fig. 2-14 Diesel gas-engine-driven central station. Alternator is direct connected to the engine. The exciter is the small machine mounted on the floor in the foreground. *(Courtesy Nordberg Mfg. Co.)*

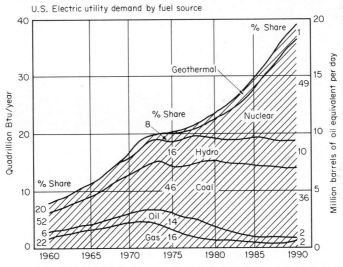

Fig. 2-15 Historical and predicted electrical generation by fuels. *(From Edison Electric Institute, "Electrical Perspectives.")*

energy will probably not be significant before the year 2000. Geothermal energy sources are limited. The Edison Electric Institute predicts electric generation will consist of the following in 1990:

Energy source	%
Solar	less than 1
Geothermal	1
Natural gas	2
Oil	2
Coal	36
Hydro	10
Nuclear	49

Figure 2-15 shows historical and predicted electrical generation by fuels.

TRANSMISSION CIRCUITS

A set of conductors energized at a high voltage and transmitting large blocks of electrical energy over relatively long distances is called a "transmission circuit" or "transmission line." Transmission circuits are constructed between transmission substations located at electric generating stations or switching points in the electric system. Transmission circuits may be overhead, with structures supporting conductors attached to insulators, or underground, with conductors surrounded by insulation, shielding, and a sheath or a jacket to form a cable. The distinguishing characteristics of transmission circuits are that they are operated at relatively high voltages, transmit large blocks of electrical power, and extend over considerable distances.

Overhead Transmission Circuits   Transmission lines or circuits are usually constructed overhead for economic and capability reasons. The lines extend for long distances in rural areas. Alternating-current (ac) transmission circuits generally operate in the voltage range of 138,000 to

Fig. 2-16 Linemen changing insulators on a 345,-000-volt three-phase bundle conductor transmission line on a wood pole structure. *(Courtesy A. B. Chance Co.)*

Fig. 2-17 Transmission line ± 250 kV dc, 456 miles long, 1000 amps, 500 MW capacity located between Minnkota Power Cooperative, Center, North Dakota, and Minnesota Power and Light Co., Duluth, Minnesota. *(Courtesy Ulteig Engineers, Inc.)*

765,000 volts. Transmission circuits energized at 345 kV, 500 kV or 765 kV are referred to as extra-high-voltage (EHV) lines. Experimental alternating-current ultra-high-voltage (UHV) transmission circuits in the voltage range of 1100 kV to 1500 kV are being developed. A 345,000-volt ac overhead transmission circuit is shown in Fig. 2-16.

Porcelain-disk insulators in strings insulate the two bundled conductors for each phase of the three-phase line from the wood-pole structure. The small wires at the top of the structure are so-called static wires or ground wires. They are not insulated and directly connect to the metal hardware on the structure and the ground. Their purpose is to shield the line conductors from lightning.

Overhead direct-current (dc) transmission circuits are used where it is desired to transmit large amounts of power over long distances without tapping the circuit at intermediate points. The circuits are usually several hundred miles in length. A circuit in operation in the western United States extends from northern Oregon to southern California, approximately 850 miles, and operates at 400 kV. Direct-current high-voltage (dc HV) line conductors are either positive or negative polarity. A 400-kV line would operate with one conductor 400 kV positive and one conductor 400 kV negative or ±400 kV dc. Some of the voltages used for HV dc circuits are ±250 kV ±400 kV and ±500 kV. Figure 2-17 shows a HV dc transmission-line tower.

Underground Transmission Circuits   In some areas of large cities it is not practical or permissible to build high-voltage overhead lines. All high-voltage electric transmission and distribution circuits must, therefore, be installed underground. Underground transmission circuits may be constructed for ac or dc operation. Alternating-current underground transmission circuits are costly to install and limited in length by charging current. Circuits over 25 miles long must be compensated by using shunt reactors to absorb the capacitive charging current, or the cable capacity will be used without transmitting any useful power. Alternating-current underground transmission circuits are operating at voltages through 500 kV. Direct-current underground transmission circuits are economically attractive for transmitting large amounts of electrical energy where underground construction is required for distances in excess of 25 miles. Underground cables (Fig. 2-18) transmit ac power between substations in a large metropolitan area.

Overhead Subtransmission Circuits   Large blocks of power are transmitted from generating stations to one or more transmission substations through high-, extra-high-, or ultra-high-voltage lines. Intermediate-voltage lines or circuits transmit alternating current from transmission substations to subtransmission or distribution substations. These subtransmission lines carry medium-sized blocks of power and usually form a grid between substations in a metropolitan area. Intermediate voltages often used for these so-called ac subtransmission lines are 34,500, 69,000, and 115,000 volts.

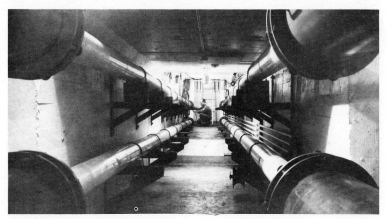

Fig. 2-18   High-voltage underground pipe-type cables installed in a tunnel. Each pipe contains three cables with paper insulation and metallic shielding surrounded and impregnated with insulating oil. The cables are capable of transmitting large quantities of three-phase ac power.

Figure 2-19 illustrates a 69,000-volt three-phase subtransmission circuit constructed along an urban street to provide a source for a distribution substation. Armless-type construction is utilized to gain public acceptance. The lower circuit is a three-phase distribution feeder operating at 13,200Y/7620 volts.

Substations   A high-voltage electric system facility used to switch generators, equipment, and circuits or lines in and out of the system, change alternating-current voltages from one level to another, and/or change alternating current to direct current or direct current to alternating current is called a "substation." Equipment in substations is used to connect electric generators to the system and to terminate and interconnect transmission, subtransmission, and distribution circuits. Power transformers in the substations change the voltage levels to permit interconnecting the generators and circuits forming the electric system.

Substation Types   Transmission substations are constructed at generating stations and at switching points in the transmission system (Fig. 2-1). The transmission substations near generating stations (Fig. 2-20) provide a means to connect the electric generators to the system and change the voltage to the appropriate magnitude to transmit the power the required distance. The power is usually generated at 25,000 volts ac or less and stepped up by power transformers to the transmission voltage level of 138,000, 161,000, 230,000, 345,000, 500,000, or 765,000 volts or higher. The power transformer connected to each generator and the transmission lines are switched in and out of service with circuit breakers.

The transmission substations located at switching points (Fig. 2-21) connect various circuits to the network and can serve as a source to subtransmission circuits. Power transformers installed at the substations can change the transmission system voltage from one level to another and the

Fig. 2-19 Modern 69,000-volt subtransmission line with 13,200Y/7620-volt distribution line underbuild. The picture illustrates armless-type construction utilized to enhance the appearance of the line.

Fig. 2-20 A 345,000-volt transmission substation located adjacent to the Quad Cities 1600-MW nuclear generating station. Four 345,000-volt transmission circuits originate at the substation. *(Courtesy Iowa Illinois Gas & Electric Co.)*

Fig. 2-21   A transmission substation containing 345,000, 161,000, and 69,000-volt equipment. Microwave reflector antenna used for transmitting high-frequency electric signals to control the substation and for communications is mounted on the control house in the foreground.

Fig. 2-22   Large distribution substation rated 69,000-13,200Y/7620 volts, three phase. Step-down power transformer has a capacity of 20/26/33 MVA.

transmission voltage to subtransmission level. The subtransmission circuits originating at a transmission switching substation will serve as a source to distribution substations.

Distribution substations are located near the utilization point in the electric system. The distribution substations change the subtransmission voltage to lower levels, providing sources for the distribution circuits supplying power for short distances to the customer's premises. Figure 2-22 illustrates a large distribution substation operating with a subtransmission voltage source of 69,000 volts and a distribution circuit voltage of 13,200Y/7620 volts. Figure 2-23 illustrates a small distribution substation operating with a source voltage of 13,200 volts and a distribution circuit voltage of 4160Y/2400 volts.

Fig. 2-23   Small distribution substation rated 13,200-4160Y/2400 volts, three phase. Step-down power transformer has a power capacity of 5/6.25 MVA. All circuits entering and leaving the substation are underground.

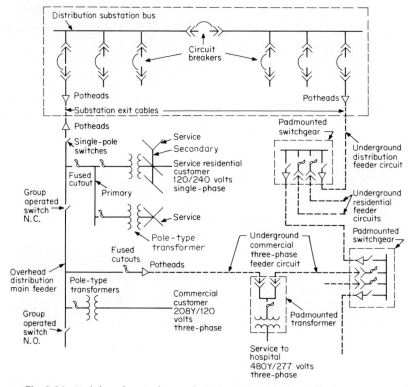

Fig. 2-24   Single-line schematic diagram showing a portion of an electric distribution system.

DISTRIBUTION SYSTEM

The electric facilities connecting the transmission system to the customer's equipment is called the distribution system. An electric distribution system (Fig. 2-1) consists of several parts: (1) Distribution feeder circuits, (2) switches, (3) protective equipment, (4) primary circuits, (5) distribution transformers, (6) secondaries, and (7) services.

Distribution Feeder Circuits   The conductors for a distribution feeder circuit originate at the terminals of a circuit breaker in a distribution substation. The circuit usually leaves the substation underground and may be installed entirely underground. The underground cables, commonly called "substation exit cables," normally connect to an overhead distribution feeder near the substation. Installing the distribution feeder circuits underground near the substation eliminates multiple circuits on the pole lines adjacent to the substations, thus improving the appearance of the substation and surrounding facilities.

The distribution feeder circuits serve as a source to branch circuits and are often referred to as "distribution main feeders," "express feeders," or "primary main feeder circuits." Several distribution feeder circuits will orginate at a distribution substation and extend in different directions to serve the customers (see Fig. 2-24). Distribution feeder circuits will usually have a capacity of 10 megavolt-amperes (MVA) or more. The substation exit cables, or the underground main feeder cables, for a 13,200Y/7620-volt three-phase feeder circuit could have 750 mcm aluminum conductor with solid dielectric insulation for each phase and a 400-mcm bare copper neutral conductor. The overhead phase conductors might be 795-mcm bare aluminum or aluminum conductor steel reinforced (ACSR) with a 336.4-mcm aluminum or ACSR bare neutral conductor.

Fig. 2-25   Overhead distribution primary main feeder in background supplied from substation by underground cable circuit.

Figure 2-25 shows a distribution substation with an overhead main feeder distribution circuit in the background connected to the substation by underground exit cables. A three-phase overhead distribution main feeder circuit or express feeder is shown in Fig. 2-26. The top three conductors mounted on porcelain line post insulators are energized at 13,200Y/7620 volts. The neutral conductor is mounted on a small white spool-type insulator below the phase conductors.

Switches   Distribution feeder circuits normally have switches installed in the main portion of the circuit to permit sectionalizing the circuit in case the feeder is damaged. Substation exit cables are usually terminated in potheads on the terminal pole below the overhead conductors. The cables are protected by lightning arresters. The terminals on the potheads are often connected to single-pole disconnect switches which connect to the overhead line conductors. Figure 2-27 is a typical installation of equipment on a riser pole for substation exit cables near a substation.

The single-pole disconnect switches provide a means to isolate the underground cables from the overhead circuit if the cables fail or maintenance work is required. The lineman shown in Fig. 2-28 is operating single-pole disconnect switches using a hot-line tool and a "Loadbuster." The Loadbuster provides a means to interrupt load current. Single-pole switches are normally operated with the circuit deenergized or when the current to be interrupted is limited to very small values.

Group-operated switches with load-break devices are usually installed at appropriate locations in distribution main feeder circuits. These switches provide a means for the lineman to sectionalize the feeders or to connect the feeders to other circuits by manually moving the grounded operating handle of the switches. Figure 2-29 illustrates a pole-top group-operated switch with the contacts closed.

Figure 2-30 shows a padmounted switching cubicle used in an underground distribution main feeder circuit. The large cables terminated in the switchgear cubicle are shown in Fig. 2-31. The equipment is protected by lightning arresters mounted behind the potheads. The solid round conductors above the potheads provide a means to test and ground the cables if they are deenergized. The insulators at the top of the illustration rotate to operate the switch contacts when the switch operating mechanism is actuated.

Fig. 2-26 Overhead distribution main feeder circuit showing primary or branch circuit tapped to main circuit through a fused cutout.

Fig. 2-27 Riser pole located near a substation with power cables terminated in potheads. Lightning arresters protect the cables and single-pole disconnect switches provide connection to overhead distribution feeder circuit.

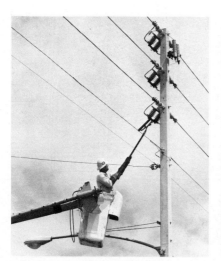

Fig. 2-28 Lineman operating single-pole disconnect switches in series with overhead distribution line. Top conductor of circuit is protected by lightning arresters.

Fig. 2-29 Pole-top group-operated three-phase disconnect switch installed in a 34,500Y/19,920-volt overhead distribution feeder circuit. Switch is designed to be compatible with armless-type construction to enhance the appearance of the line. *(Courtesy A. B. Chance Co.)*

Fig. 2-30   Padmounted switchgear cubicle. Equipment provides switches for sectionalizing main feeder circuit and fuses for connecting branch underground primary circuits to the main feeder circuit. *(Courtesy S & C Electric Co.)*

Fig. 2-31   Padmounted switchgear switching compartment with door open to show cables, potheads, lightning arresters, testing and grounding terminals, and three-phase group-operated switch.

Protective Equipment Circuit breakers in the substations serving as a source to distribution circuits are actuated by protective relays that recognize short circuits and grounds on a circuit or line. If a circuit breaker opens, the complete distribution circuit is deenergized and a large number of customers will have their electric service interrupted.

Lightning arresters and fuses are installed on distribution circuits to minimize the customer outages. Figure 2-32 shows a valve-type distribution lightning arrester similiar to those mounted on a pole and connected to the top conductor of the circuit, shown in Fig. 2-28. The lightning arresters are designed to bypass electrical surges on the circuits protecting the insulators and equipment from over voltages. Fuses are installed in the circuits to isolate branch primary circuits and equipment from the main distribution feeder circuit if a fault or short circuit develops on one

Fig. 2-32 Valve-type distribution lightning arrester equipped with mounting bracket designed for use on overhead distribution lines and equipment. *(Courtesy Westinghouse Electric Corp.)*

Fig. 2-33 Fused cutout equipped with bracket for mounting on a crossarm.

Fig. 2-34 Fused cutout is being opened by lineman using hot-line tool and "Loadbuster." Lightning arresters installed on brackets attached to pole provide electrical surge protection. *(Courtesy S & C Electric Co.)*

of the taps to the circuit. It is better to have a fuse blow and interrupt service to a small number of customers than to have the substation circuit breaker open and deenergize all the customers served from the circuit.

Figure 2-33 shows a fused cutout designed to be used on an overhead distribution circuit. The lineman in Fig. 2-34 is operating a fused cutout connected between the overhead main feeder circuit conductor and a pole-type distribution transformer. Figure 2-35 shows fuses mounted in padmounted switchgear. The fuses are connected between the bus and the underground primary feeder cables to protect the main feeder circuit for faults on the branch circuits or equipment. Faults or short circuits usually result from cable insulation failures frequently caused by excavating equipment hitting a cable. The fuses are operated or replaced by linemen or cablemen using hot-line tools or hot sticks.

Fig. 2-35   Padmounted switchgear with doors to fuse compartments open. Barriers have been removed from one compartment exposing the high-voltage equipment and providing access to the fuses. *(Courtesy S & C Electric Co.)*

Primary Circuits   All high-voltage distribution circuits are often referred to as "primaries" by the lineman. The branch circuits that connect to the main, or primary, distribution feeder circuit are properly called primary circuits or primaries. The overhead primary circuits normally connect to the distribution feeder circuit through fused cutouts. The primary circuit may be three-phase (Fig. 2-36) or single-phase (Fig. 2-37).

Figure 2-35 shows a three-phase primary underground circuit orginating at fuses in a pad-mounted switchgear cubicle. The fuses connect to a bus in the switchgear cubicle in series with the underground distribution main feeder circuit. The primary branch circuits serve the customers in a small area. The primary conductors are normally small with a capacity of 200 amps or less. The size of the fuses protecting the main distribution feeder circuit from a fault on the branch primary circuit will limit the capacity of the primary.

Figure 2-38 illustrates a three-phase underground primary circuit originating on a riser pole. The underground cables are supplied through fused cutouts connected to the overhead distribution feeder circuit. The primary branch circuits provide service to small industrial, commercial, and residential customers. The small industrial and commercial customers will normally use three-phase power. The residential customers will normally be provided single-phase service.

Distribution Transformers   The function of the distribution transformer is to reduce the voltage of the primary circuit to the utilization voltage required to supply the customer. The primary voltages vary from 34,500Y/19,920 to 4,160Y/2400 volts. A three-phase circuit with a grounded neutral source operating at 34,500Y/19,920 volts would have three high-voltage conductors and one grounded neutral conductor, a total of four conductors or wires. The voltage between the insulated, or phase conductors, would be 34,500 volts, and the voltage between one-phase conductor and neutral, or ground, would be 19,920 volts. Similarly, a 13,200Y/7620-volt

Fig. 2-36 Three-phase primary or branch circuit is connected to distribution feeder circuit through fused cutouts. Main distribution circuit is armless-type construction.

Fig. 2-37 Single-phase primary or branch circuit is connected to three-phase distribution feeder circuit through fused cutout. Lightning arrester is installed ahead of cutout to protect primary from electrical surges.

Fig. 2-38 Underground primary cables are terminated in combination fused cutout terminator assemblies called isolating terminators. Lightning arresters are installed on the mounting frame. The arresters provide surge protection for the underground primary cables and the overhead main distribution feeder circuit. *(Courtesy A. B. Chance Co.)*

Fig. 2-39 Typical distribution transformer installation on pole. The primary wire to which the primary winding is connected is carried on the insulator on top of the pole. The transformer is a self-protected type complete with a primary lightning arrester, an internal fuse, and a low-voltage or secondary circuit breaker. The transformer is rated 7620–120/240 volts, 15 kVA. *(Courtesy Westinghouse Electric Corp.)*

distribution feeder would have 13,200 volts between phase conductors, and 7620 volts between a phase conductor and neutral, or ground.

Residential customers are normally provided 120/240 volts single-phase service. The single-phase service would have three wires. Two of the wires would be insulated, and one would be grounded. The insulated wires would have 240 volts between the conductors and 120 volts between the insulated wires, or hot wires as they are often referred to by the lineman, and the ground or neutral wire. Some small older homes may have 120-volt service which would be provided by one insulated, or hot, wire and the ground or neutral wire. A typical single-phase transformer installed on a pole along the rear lot line of a residential area is shown in Fig. 2-39. The

Fig. 2-40 Single-phase padmounted transformer installed on rear lot line of a residential subdivision. Transformer is rated 7620–120/240 volts, 50 kVA. Cableman is installing terminations on primary cable. *(Courtesy A. B. Chance Co.)*

Fig. 2-41 Dual three-phase padmounted transformers serving a commercial customer. The transformers are rated 13,200Y/7620–480Y/277 volts, 500 kVA.

center low-voltage wire connected to the bushing on the side of the transformer tank is the neutral wire. Padmounted transformers are frequently used in underground residential distribution systems. Figure 2-40 shows a single-phase padmounted distribution transformer installed along the rear lot line of a residential subdivision. The primary cables connect to the transformer with load-break terminators or elbows. The single-phase primaries normally feed through the transformer to form a loop. This permits a defective primary cable to be isolated between transformers.

Commercial and light industrial customers normally require three-phase service. The voltage supplied is usually 480Y/277 volts or 208Y/120 volts. Three-phase padmounted transformers (Fig. 2-41) are used with underground primary circuits. Dual transformers have been used in the installation shown to provide adequate capacity and partial service if one of the padmounted transformers fails. The three-phase underground primaries are installed to form a loop to permit isolating defective underground cables without a long outage to the customer. Three single-phase pole-type transformers can be cluster-mounted on a pole to provide three-phase service (Fig. 2-42). The transformers illustrated are single–primary-bushing transformers connected line to ground.

Network service is provided to many business areas with highly concentrated electric loads. Network transformers are designed for installation in a vault usually located under the sidewalk. The network transformers are designed for submersible operation (Fig. 2-43). The primary underground cables terminate in a switching compartment. The primary switch provides a means to deenergize the network transformer without deenergizing the primary circuit that serves as a source to other network transformers. The secondary voltages are usually 208Y/120 volts or 480Y/277 volts. The secondary terminals of the network transformer connect to a network protector.

Fig. 2-42 Three-phase transformers bank rated 13,-200Y/7620–208Y/120 volts, 150 kVA. Primary wires are connected to transformers through fused cutouts. Lightning arresters connect to primaries to provide surge protection. Low-voltage connections to transformers terminate on secondary racks mounted on pole below transformers. Quadriplex service to customer's building terminates on secondary racks.

Fig. 2-43 Network transformer complete with primary switch and accessory devices. Transformer rated 13,200Y/7620–208Y/120 volts, three-phase, 1000 kVA. *(Courtesy Westinghouse Electric Corp.)*

The network protector is a low-voltage circuit breaker and fuse assembly complete with protective relays and control devices designed to isolate the low-voltage network from a transformer or primary cable fault. The network is supplied from several underground primary feeder cables.

Secondaries   The conductors originating at the low-voltage secondary winding of a distribution transformer that extend along the rear lot lines, alleys, or streets past the customer's premises are called secondaries. The secondaries in residential areas are three-wire single-phase circuits

(Fig. 2-44). The secondary conductors for overhead circuits are strung below the primaries and are cabled (Figs. 2-45 and 2-46) or lie in a vertical plane (Fig. 2-47). The secondaries for underground circuits are normally direct-buried in the same trench with primary cables. The cables are laid in the trench without separation (Fig. 2-48 and 2-49).

Services The wires extending from the secondary, or distribution transformer, adjacent to the customer's property to the customer's premises is called a service. An overhead triplex cable service extending to a customer's premises is shown in Fig. 2-50. The transformer reduces the

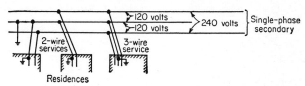

Fig. 2-44 A 120/240-volt three-wire single-phase secondary supplying two-wire 120-volt and three-wire 120/240-volt services.

Fig. 2-45 Distribution transformer with primary connected phase to ground. The single-phase primary is supported on the insulator on top of the pole. The wires connecting to the secondary terminals of the transformer terminate on a secondary rack and connect to a triplex secondary and three triplex services. The bare conductor serves as the grounded neutral for both the primary and secondary circuits.

Fig. 2-46 Triplex secondary extends from transformer pole to pole in foreground of picture. Secondary connects to residential services and conductors for street light.

voltage from 7620 volts to 120/240 volts. The triplex service cable has two insulated aluminum conductors and a bare aluminum steel reinforced neutral wire that provides physical support for the insulated conductors. Termination of the service cables at the customer's premise is shown in Fig. 2-51. A quadriplex cable providing overhead 208Y/120-volt three-phase service to a commercial establishment is illustrated in Fig. 2-52. The underground services extend from a secondary enclosure (Fig. 2-53) or padmounted transformer (Fig. 2-54) to the customer's premises. The cables are installed in a trench (Fig. 2-55) or plowed in (Fig. 2-56).

Fig. 2-47 Distribution transformer mounted on pole with primary winding connected to primary, carried on horizontal crossarm, through a fused cutout. The secondary mains supplied from the transformer are carried on a rack mounted in a vertical plane on the pole. *(Courtesy General Electric Co.)*

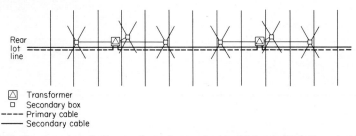

Rear
lot
line

△ Transformer
□ Secondary box
---- Primary cable
—— Secondary cable

Fig. 2-48 Plan view schematic diagram of an underground residential electric distribution system for installation along the rear lot line. Schematic locates padmounted transformers, primary cable, secondary enclosures, secondary cables, and service cables.

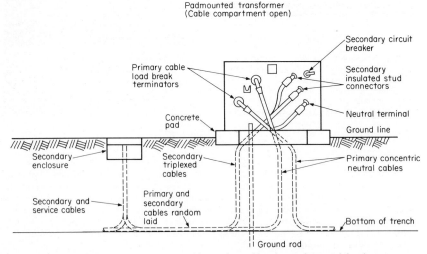

Padmounted transformer
(Cable compartment open)

Secondary circuit breaker

Primary cable load break terminators

Secondary insulated stud connectors

Neutral terminal

Concrete pad

Ground line

Secondary enclosure

Secondary triplexed cables

Primary concentric neutral cables

Secondary and service cables

Primary and secondary cables random laid

Bottom of trench

Ground rod

Fig. 2-49 Sectional view locating primary and secondary cables of an underground distribution system.

Fig. 2-50 Service wires to customer's premise terminate on pole below distribution transformer. Transformer is connected line to ground from one phase of three-phase primary circuit. Service extends perpendicular from primary circuit.

Fig. 2-51 Cable service drop connected to residential service entrance. *(Courtesy Kansas City Power and Light Co.)*

Fig. 2-52 Quadriplex service cable terminates on wooden mast adjacent to customer's service entrance conduit. Cables enter conduit through a weatherhead.

Fig. 2-53 Connections for underground services to the secondary in the secondary enclosure. *(Courtesy ITT Blackburn Co.)*

Fig. 2-54 Linemen completing connections for underground service to a residential customer in compartment of padmounted transformer. Note the trencher-backhoe equipment in background of picture. *(Courtesy Iowa Illinois Gas and Electric Co.)*

Fig. 2-55 Cableman and helper digging trench and laying cable in trench to provide service to new home.

Fig. 2-56 Lineman using Ditch Witch vibrating plow to install underground service to a street light in a new subdivision. *(Courtesy The Charles Machine Works, Inc.)*

EMERGENCY GENERATION

Electric customers offering specialized services may require very reliable electric service. Customers of this type can be provided dual electric service from two distribution feeder circuits. The dual electric service is normally supplied from one circuit with facilities to switch to the alternative circuit in case of a failure of the principal source of electrical energy. Customers are required to pay the extra cost associated with the extra electrical distribution equipment. Hospitals and similar types of customers will be equipped with emergency electric generators in addition to dual service from the electric system. The emergency generators are designed to provide sufficient capacity to supply critical loads, such as operating rooms, emergency rooms, and minimum lighting in halls and stairways. Figure 2-57 shows an emergency electric generator. The emergency electric generators are designed to start automatically when the normal source of electrical energy is lost. The equipment is installed to ensure that the generator does not feedback into the electric distribution system. It is essential for the lineman and the cableman to protect themselves for a possible malfunction of the equipment which would permit a feedback through the customer's service to the distribution facilities.

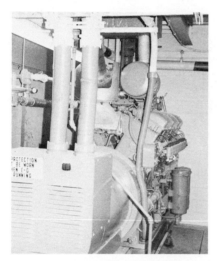

Fig. 2-57 Emergency electric generator rated 500 kVA, three-phase 480Y/277 volts. Generator is driven by diesel-type engine using natural gas for fuel.

Fig. 2-58 Control center building designed to withstand tornado winds.

ELECTRIC SYSTEM CONTROL

The modern electric system is operated from a control center. Equipment in the control center is used to analyze the electric system operation continuously. The output of the electric generators is varied automatically to match the energy supplied with the customer's demand and the purchases and sales to interconnected electric systems. The control center is located at a strategic point in the electric system service area. The control center building (Fig. 2-58) is designed to protect the operating personnel and control equipment from severe weather conditions. The operating personnel responsible for the electric system use schematic diagrams drawn on a map (Fig. 2-59) mounted on a large display board to give them a picture of the entire electric system.

Instruments mounted on a panel (Fig. 2-60) provide essential information concerning the electric system, such as system net load, generation, net tie line loads, area control error, and frequency. Cathode ray tubes (CRTs) display schematic diagrams of the electric system including generating station equipment, transmission circuits, substation equipment, and distribution circuits. The CRTs connect to digital computers that link up with supervisory control equipment at the generating stations and substations. Electrical energy generated and distributed throughout the system is continuously displayed on the schematic diagram appearing on the CRT. The electrical equipment, such as circuit breakers, is operated by action of a light pen at the appropriate location on the schematic diagram appearing on the CRT. As the circuits and generators are switched in and out of service, the change in electrical energy flows are made automatically by the computer to correctly display the information on the CRT.

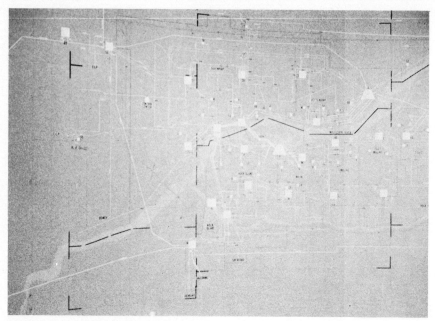

Fig. 2-59 Map showing location of electric facilities mounted on a specially constructed support. The scale of the map is large enough to permit identification of the electric system from operating positions.

Fig. 2-60 Control panel with instruments supplying essential information on the electric system operation.

Fig. 2-61 Operating station with control panel and electric system map in background. CRT with light pen control, alarm logger, and alphanumerical CRT are readily available to operating personnel.

Fig. 2-62 Dual CRTs with light pen control in front of electric system operator. Alarm logger is located adjacent to the CRTs behind the system operator. Normally one CRT will display the transmission system and the second CRT will show the distribution system.

Alarm logger and alphanumerical CRT equipment (Figs. 2-61 and 2-62) provide a record of electric system operations. The displays on the alphanumerical CRT can be printed by proper inputs to the computer equipment to operate a special high-speed printer. The computers provide signals that continuously change the output of the electric generation to meet the requirements of the electric system. The electric system operators use two-way radio equipment to communicate with linemen and cablemen working on the electric transmission and distribution equipment.

Section **3**

Substations

SUBSTATION TYPES

An electrical substation is an installation with electrical equipment capable of interrupting or establishing electrical circuits and changing the voltage, frequency, or other characteristics of the electrical energy flowing in the circuits.

Transmission Substations   Figure 3-1 shows a transmission substation located at an appropriate point in a network of transmission lines. The substation contains equipment used to sectionalize the electric transmission system when a fault or short circuit develops on one of the circuits. The defective circuit is switched out of service automatically and deenergized, protecting the remaining portion of the transmission network from trouble. The electric transmission system is designed to permit the deenergizing of any one circuit without interrupting electric service to customers.

Fig. 3-1   Transmission substation with 4 to 445 MVA single-phase 500/230/34.5-kV autotransformers. One transformer is a spare. Circuit breakers in the substation structure are used to switch circuits and the power transformers in and out of service. The shunt reactors in the foreground limit the fault current in the 34.5-kV circuits. *(Courtesy Westinghouse Electric Corp.)*

Fig. 3-2 Power transformers in transmission substation located adjacent to generating station. Power transformers raise the generator voltage from 25 to 765 kV. Each single-phase transformer has a capacity of 317 MVA. The three-phase bank capacity is 950 MVA. Fourth transformer serves as spare. (*Courtesy Westinghouse Electric Corp.*)

Fig. 3-3 Subtransmission substation equipped with oil circuit breakers, lightning arresters, disconnect switches, and potential transformers. Substation is located in the fringe area of a large city and serves as a switching center for the subtransmission circuits.

Transmission substations located at electric generating stations (Fig. 3-2) have power transformers designed to raise the voltage from generation value, usually in the range of 13,200 to 25,000 volts three-phase, to transmission level, from 161,000 to 765,000 volts or higher. Circuit breakers in the transmission substation are used to switch generating and transmission circuits in and out of service.

Subtransmission Substations   Electric substations (Fig. 3-3) with equipment used to switch circuits operating at voltages in the range of 34.5 to 138 kV are referred to as subtransmission substations. Subtransmission circuits and substations are located near to high-load concentrations, usually in urban areas. Subtransmission circuits supply distribution substations.

Distribution Substations   Substations located in the middle of a load area are called distribution substations. These substations may be as close together as 2 miles in densely populated metropolitan areas. The substations contain power transformers that reduce the voltage from subtransmission levels to distribution level usually in the range of 4.16Y/2.4 kV to 34.5Y/19.92 kV.

Fig. 3-4   Single-phase voltage regulators in a substation installed in series with distribution circuits. *(Courtesy Allis Chalmers Manufacturing Co.)*

The transformers are normally equipped to regulate the substation bus voltage. Circuit breakers in the distribution substations are installed between the low-voltage bus and the distribution circuits. The distribution circuits may vary in capacity from approximately 5 to 15 MVA. Large distribution substations may have power transformer capacity as large as 100 MVA and serve as a source to as many as 10 distribution circuits. If the power transformers do not have automatic tap changing, it is usually necessary to install voltage regulators (Fig. 3-4) in series with each distribution circuit. Distribution substations (Fig. 3-5) may be constructed near a large manufacturing plant to serve a specific industrial customer. Distribution substations are normally unattended and operate automatically or are controlled remotely by supervisory control equipment from a central control center.

SUBSTATION FUNCTION

An electric substation is an installation in which the following functions can be accomplished for the power system:
  1. Voltage changed from one level to another level
  2. Voltage regulated to compensate for system voltage changes
  3. Electric transmission and distribution circuits switched into and out of the system
  4. Electric power quantities flowing in the transmission and distribution circuits measured
  5. Communications signals connected to the transmission circuits

6. Lightning and switching surges eliminated from the electric transmission system
7. Electric generators connected to the transmission and distribution system
8. Interconnections between the electric systems of various companies completed
9. Reactive kilovolt-amperes supplied to the transmission and distribution circuits and the flow of reactive kilovolt-amperes on the transmission and distribution circuits controlled
10. Alternating current converted to direct currrnt or direct current converted to alternating current
11. Frequency changed from one value to another

Fig. 3-5   Distribution substation located at an Aluminum Company of America large rolling mill plant. The source to the substation is a double 69,000-volt circuit contructed on steel poles. The power transformers are each rated 20/26 MVA foa, three-phase, 69 to 13.2 kV. The small structure has circuit breakers for the plant distribution circuits.

SUBSTATION EQUIPMENT

Electric substations contain many types of equipment. The major components of equipment consist of the following:

Concrete foundations	Potheads
Duct runs	Metal-clad switchgear
Manholes	Shunt reactors
Steel superstructures	Capacitors
Bus support insulators	Synchronous condensers
Suspension insulators	Grounding transformers
Lightning arresters	Grounding resistors
Oil circuit breakers	Control house
Air circuit breakers	Control panels
Disconnect switches	Meters
Power transformers	Relays
Coupling capacitors	Supervisory control
Potential transformers	Power-line carrier
Current transformers	Batteries
High-voltage fuses	Battery chargers
High-voltage cables	Conduits
Rectifiers	Control wires
Frequency changers	

Power Transformers   Transformers are used to change the voltage from one level to another, to regulate the voltage level, and to control the flow of reactive kilovolt-amperes in the power system. Power transformers installed in transmission substations will normally be constructed for and operated at the higher voltages in the range of 138,000 to 765,000 or higher volts. Figure 3-6 shows a large power transformer located in a transmission substation.

Most substations will have three-phase transformers. Some substations will have three single-phase transformers installed in a bank (Fig. 3-7). Often a fourth single-phase transformer will be

Fig. 3-6   A power transformer installed in a transmission substation. The transformer is equipped with radiators for cooling the insulating oil, oil-level gauge with alarm contacts, dial hot-spot-winding temperature equipment, no-load tap changer, nitrogen cylinder cabinet, sudden pressure relay, bushings, and lightning arresters. *(Courtesy Westinghouse Electric Corp.)*

located at the substation to serve as a spare. If one of the transformers connected in the bank fails, the spare is energized to expedite restoration of electric service.

The capacity of the transformers will normally be in the range of 1,000,000 to 50,000 kVA. Power transformers installed in distribution substations will normally be constructed for and operate at the lower voltages and smaller capacities. The voltages will generally be in the range of 161,000 to 4160 volts and the capacities in the range of 50,000 to 5000 kVA. Figure 3-8 shows a power transformer that is part of a distribution substation.

Steel Structures   The substation's steel structures (Fig. 3-9) provide support for insulators, which are used to terminate lines and support busses that interconnect the equipment. Disconnect switches and other equipment are mounted on the steel structure (Fig. 3-10).

Lightning Arresters   The lightning arresters protect the electric system and substation equipment from lightning strokes and switching surges. The lightning arresters are installed in a substation near the termination of aerial circuits and close to the more valuable pieces of equipment such as power transformers. Lightning arresters can be seen in Fig. 3-6 mounted on the power transformer adjacent to the transformer bushings. Figure 3-11 shows an installation of three lightning arresters in a substation on the terminals of a transmission line.

Lightning arresters generally contain semiconductor blocks that act as a valve, allowing large surge currents to pass harmlessly to ground and limiting the power follow current to low values. Gaps are connected in series with the valve blocks internal to the arrester to prevent the arrester

Fig. 3-7 Transmission substation with three single-phase power transformers connected into a three-phase bank. *(Courtesy Westinghouse Electric Corp.)*

Fig. 3-8 Distribution power transformer with tap changing underload equipment in compartment in foreground. The distribution voltage connection to the transformer is completed through the bus duct on the right side of the picture to metal-enclosed switchgear. *(Courtesy Westinghouse Electric Corp.)*

from conducting current under normal voltage conditions and to interrupt the power follow current after the surge current has flowed to ground. The gap structure utilizes a magnetic force to move the arc when the arrester operates. This is done to prevent damage to the gaps and help extinguish the follow current. The internal components of a lightning arrester are shown in Fig. 3-12.

Circuit Breakers   Oil, air, gas, and vacuum circuit breakers are used to switch electric circuits and equipment in and out of the system. The contacts of the circuit breakers are opened and closed by mechanical linkages manufactured from insulating materials and utilizing energy from compressed air, electric magnets, or charged springs. Some of the high-voltage circuit breakers utilize compressed air to operate the contacts and interrupt the current flow when the contacts open. Operation of the circuit breakers is initiated, utilizing direct-current (dc) circuits, by manually operating a switch, by remote operation of supervisory control equipment, or by relays that automatically recognize predetermined conditions or electrical failures in the system. Various types of circuit breakers can be seen in Figs. 3-13 to 3-16.

Fig. 3-9 Linemen installing tubular bus conductor on top of insulators supported by steel substation structure. *(Courtesy L. E. Meyers Co.)*

Fig. 3-10 Distribution substation with a lattice-type steel structure. A 69,000-volt subtransmission circuit in the background provides a source to the station. Underground cables connect the power transformer to the metal-clad switchgear and the circuit breakers in the switchgear for the distribution circuits.

Fig. 3-11 Three lightning arresters installed on a substation steel structure and connected to the terminals of a transmission line originating at the substation.

Fig. 3-12 Cutaway view of a lightning arrester. The gap structure is located in the center and the valve blocks are above and below the gaps. *(Courtesy Westinghouse Electric Corp.)*

Fig. 3-13   Three-phase high-voltage oil circuit breaker with center-break group-operated disconnect switches installed as isolators. *(Courtesy Allis Chalmers Mfg. Co.)*

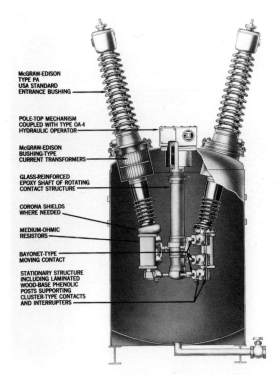

McGRAW-EDISON
TYPE PA
USA STANDARD
ENTRANCE BUSHING

POLE-TOP MECHANISM
COUPLED WITH TYPE OA-4
HYDRAULIC OPERATOR

McGRAW-EDISON
BUSHING-TYPE
CURRENT TRANSFORMERS

GLASS-REINFORCED
EPOXY SHAFT OF ROTATING
CONTACT STRUCTURE

CORONA SHIELDS
WHERE NEEDED

MEDIUM-OHMIC
RESISTORS

BAYONET-TYPE
MOVING CONTACT

STATIONARY STRUCTURE
INCLUDING LAMINATED
WOOD-BASE PHENOLIC
POSTS SUPPORTING
CLUSTER-TYPE CONTACTS
AND INTERRUPTERS

Fig. 3-14   Cutaway view of one pole of a three-phase high-voltage oil circuit breaker showing contact structures and operating linkage. *(Courtesy McGraw Edison Co.)*

Fig. 3-15 Air-blast three-phase circuit breaker for use with extra high voltages. *(Courtesy General Electric Co.)*

Fig. 3-16 Three-phase single-tank 69,000-volt oil circuit breaker. *(Courtesy McGraw Edison Co.)*

Disconnect Switches   High-voltage equipment is usually installed in substations with disconnect switches in series with it to facilitate isolating the equipment for maintenance. The disconnect switches are normally operated by linkages connecting to a lever or geared-type handle. The switches are designed to interrupt charging current but—unless they are specially constructed—not load current. The disconnect switches shown in Fig. 3-13 on each side of the oil circuit breaker can be operated after the circuit breaker contacts are opened to visibly isolate the circuit breaker. The disconnect switches are operated by turning the crank handle, which moves the interphase rods through a gear arrangement and rotates the insulators, causing the switch blades to move and opening the contacts. The disconnect switch in Fig. 3-17 is in the open position.

Fig. 3-17   Center side-break group-operated disconnect switch in partially opened position. *(Courtesy Allis Chalmers Mfg. Co.)*

Coupling Capacitors   Communication signals in the form of high-frequency voltages are transmitted to the transmission lines through coupling capacitors. Some of the coupling capacitors are equipped with potential devices which make it possible to measure the voltage on transmission-line circuits. The coupling capacitor potential devices are accurate enough to be used for supplying voltages to protective relays but—unless they are specifically compensated—not accurate enough to supply voltages for meters designed for billing purposes. Figure 3-18 shows a coupling capacitor.

Potential Transformers   High voltages are measured by reducing the voltage proportionately with a potential transformer (Fig. 3-19), which has its high-voltage winding connected to the transmission or distribution circuit and its low-voltage winding connected to a meter or relay or both. Potential transformers are required to provide accurate voltages for use with meters used for billing industrial customers or connecting utility companies. If single-phase transformers are used, three transformers are generally required to measure power on a three-phase circuit.

Current Transformers   The primary winding of a current transformer is connected in series with the high-voltage conductor. The magnitude of amperes flowing in the high-voltage circuit is reduced proportionately to the ratio of the transformer windings. The secondary winding of the current transformer is insulated from the high voltage to permit it to be connected to low-voltage metering circuits. A bushing-type current transformer is shown in the cutaway view of the oil circuit breaker (Fig. 3-14). Current and potential transformers supply the intelligence for measuring power flows and the electrical inputs for the operation of protective relays associated with the transmission and distribution circuits or equipment such as power transformers.

Fig. 3-18 Coupling capacitor potential device designed for use on 161,000 volts. *(Courtesy General Electric Co.)*

Fig. 3-19 A high-voltage single-phase potential transformer for connecting between one phase of a circuit and neutral or ground. The transformer is filled with insulating oil. *(Courtesy Westinghouse Electric Corp.)*

High-Voltage Fuses   High-voltage fuses are used to protect the electrical system from faults in such equipment as potential transformers or power transformers. Figure 3-20 shows a distribution unit substation with a line terminal structure designed for appearance as well as function. A group-operated high-voltage disconnect switch and three power fuses are mounted on the terminal structure. The fuses are connected in series with the power transformer that serves as a source for the switchgear.

Metal-Clad Switchgear   Outdoor metal-clad switchgear (Figs. 3-21 and 3-22) provides a weatherproof housing for air circuit breakers (generally rated 15 kV or below), protective relays, meters, current transformers, potential transformers, bus conductors, and other items necessary to provide electric system requirements. Indoor switchgear must be installed in a building for protection from the elements. Figures 3-23 to 3-27 illustrate metal-clad switchgear installations and some of the equipment—including air circuit breakers—normally installed with the switchgear.

Shunt Reactors   When the capacitive reactance from extra high-voltage transmission line circuits exceeds the ability of the system to absorb the reactive kVA (kilovars), shunt reactors are installed. The shunt reactor shown in Fig. 3-28 is a three-phase single-winding transformer which serves as an inductive reactance on the power system, thus neutralizing the capacitive reactance associated with the long extra high-voltage transmission line. Shunt reactors are usually located in substations and connected to a transmission line terminal through a disconnect switch.

Meters   Meters mounted on control panels are used to measure the energy flowing in a circuit or through a piece of equipment. Some of the meters are utilized to bill industrial customers and adjacent utility companies. Interstate flows of electrical energy are measured in the substations.

Relays   Protective relays installed on control panels are used to identify electrical failures on transmission and distribution circuits or in pieces of equipment, such as power transformers. Some

Fig. 3-20 Unit-type substation with high-voltage fuses to protect the electric system from power-transformer faults. *(Courtesy Allis Chalmers Mfg. Co.)*

Fig. 3-21 Outdoor metal-clad switchgear designed for appearance. *(Courtesy General Electric Co.)*

are also used to perform predetermined operations as required by system conditions. Relays automatically operate their contacts to properly identify the source of trouble and remove it from the electrical system.

Supervisory Control The supervisory control equipment permits the remote control of substations from a system control center or other selected point of control. This equipment is used to open and close circuit breakers, operate tap changers on power transformers, supervise the position and condition of equipment, and telemeter the quantity of energy flowing in a circuit or through a piece of equipment.

Capacitors Reactive kilovolt-amperes can be supplied to the electrical system by connecting banks of capacitors to the distribution circuits in substations or out on the distribution circuits to neutralize the effect of customer inductive loads. Capacitors used in this manner help to control voltages supplied to the customer by eliminating the voltage drop in the system caused by inductive reactive loads. Capacitors in substations are usually mounted in metal cubicles. The capacitors mounted on the racks in the cubicles are usually single-phase single-bushing units rated 100-, 150-, or 200-kVA capacitance, 60 Hz, 7620 volts. They are connected between each of the three

Fig. 3-22 Outdoor metal-clad switchgear with an enclosed aisle. The housing protects equipment and substation electricians performing maintenance work from the elements. *(Courtesy Westinghouse Electric Corp.)*

Fig. 3-23 Indoor switchgear with 15-kV vacuum horizontal-draw-out circuit breaker removed from cubicle. *(Courtesy General Electric Co.)*

Fig. 3-24 Indoor metal-clad switchgear. Top cubicles provide enclosures for potential transformer and panels for meters, protective relays, and control switches. Bottom cubicles house draw-out circuit breakers, bus current transformers, and distribution circuit underground cable terminations. *(Courtesy Westinghouse Electric Corp.)*

Fig. 3-25 Horizontal-draw-out 15-kV air magnetic circuit breaker. *(Courtesy Allis Chalmers Mfg. Co.)*

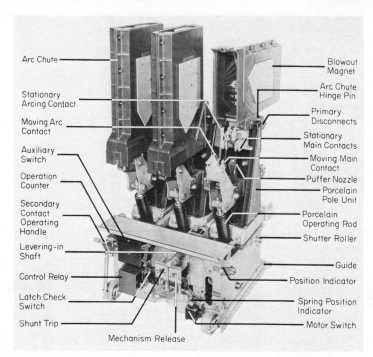

Arc Chute

Stationary
Arcing Contact

Moving Arc
Contact

Auxiliary
Switch

Operation
Counter

Secondary
Contact
Operating
Handle

Levering-in
Shaft

Control Relay

Latch Check
Switch

Shunt Trip

Mechanism Release

Blowout
Magnet

Arc Chute
Hinge Pin

Primary
Disconnects

Stationary
Main Contacts

Moving Main
Contact

Puffer Nozzle

Porcelain
Pole Unit

Porcelain
Operating Rod

Shutter Roller

Guide

Position Indicator

Spring Position
Indicator

Motor Switch

Fig. 3-26 Horizontal-draw-out 15-kV air-magnetic circuit breaker with barriers removed and one arc chute tilted to expose contacts. (*Courtesy Westinghouse Electric Corp.*)

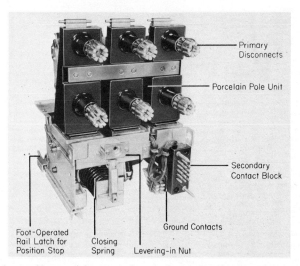

Primary
Disconnects

Porcelain Pole Unit

Secondary
Contact Block

Foot-Operated
Rail Latch for
Position Stop

Closing
Spring

Levering-in Nut

Ground Contacts

Fig. 3-27 Back view of horizontal-draw-out 15-kV air-magnetic circuit breaker with barriers and arc chutes removed. (*Courtesy Westinghouse Electric Corp.*)

Fig. 3-28 Shunt reactor installed in an extra-high-voltage substation rated 345 kV, 50 MVA.

Fig. 3-29 Metal-enclosed capacitor equipment for substation installation. *(Courtesy Westinghouse Electric Corp.)*

phases and ground. The equipment in Fig. 3-29 includes instrument transformers in the first cubicle, a three-phase vacuum switch in the second cubicle, and capacitor units in cubicles three, four, and five.

Control House   The substation control house (Fig. 3-30) is used to protect the control equipment, including switchboard panels, batteries, battery chargers, supervisory control, power-line carrier, meters, and relays from the elements. Underground and overhead conduits and control wire are installed to connect the controls of all the equipment in the substation to the control panels normally installed in the control house. Each substation usually contains several thousand feet of conduit and miles of control wire.

Control Panels   The control panels (Fig. 3-31) installed in a control house provide mountings for meters, relays, switches, indicating lights, and other control devices.

Carrier Equipment   The power-line carrier equipment provides high-frequency voltages to be used for transmitting voice communications or telemetering signals on high-voltage transmission line circuits. In the case of voice communications, the sound frequency modulates a high-frequency signal connected to the transmission-line circuit by means of coupling capacitors. This equipment permits using the line conductors for communication, relaying, supervisory control, and metering in addition to the transmission of electric power.

Batteries   Control batteries (Fig. 3-32) supply energy to operate circuit breakers and other equipment. It is necessary to use direct-current control systems with a storage battery as a source to make it possible to operate equipment during periods of system disturbances and system outage.

Fig. 3-30 Control house for distribution substation constructed as part of a decorative wall. *(Courtesy Iowa Illinois Gas & Electric Co.)*

Fig. 3-31 Control panels with meters, relays, and control devices mounted for operation of a substation. *(Courtesy General Electric Co.)*

Fig. 3-32 Control battery and solid-state-electronic battery charger installed in substation control house. Battery provides 125 volts for operation of control equipment.

Fig. 3-33   Mercury-arc rectifier installed in a distribution substation to convert alternating current to direct current for use in railway service. *(Courtesy General Electric Co.)*

Fig. 3-34   Typical forced-air-cooled silicon rectifier for electrochemical service rated 10,000 amp up to 1000 volts direct current. Left-hand section contains cooling fan. *(Courtesy Westinghouse Electric Corp.)*

Fig. 3-35 Frequency changer installed in a substation. Note direct-connected exciter in foreground. *(Courtesy Westinghouse Electric Corp.)*

Battery chargers are used to automatically keep the batteries charged completely to provide sufficient emergency power for all necessary operations.

Rectifiers   Equipment used to convert alternating current to direct current may be installed in distribution substations if the station is adjacent to the load utilizing the direct current. A mercury-arc rectifier installation is illustrated in Fig. 3-33; Fig. 3-34 shows a silicon rectifier assembly.

Frequency Changers   Substations are ideal locations for frequency changers. The frequency changer in Fig. 3-35 consits of an alternating-current synchronous motor of one frequency and a generator direct-connected to the motor designed for a different frequency.

Transmission Circuits

Alternating-current (ac) transmission is the movement of alternating-current electrical energy from one location to another, such as from a terminal at a generating station to a substation near a load center.

Direct-current (dc) transmission is the movement of direct-current electrical energy from one location to another. One terminal may be at a generating station. Direct current is normally used to transmit large blocks of power long distances when it is not necessary to tap the transmission line at intermediate points.

Transmission Circuits—History   The first electric generators produced direct current. The generators operated at a voltage matched to the load to be served. The invention of the ac transformer made it possible to step voltages up and down. Alternating-current generators used with transformers made it practical to locate the generators remote from the utilization area. Polyphase systems developed because of improved efficiency over single-phase and reduced costs for a given quantity of power. The early systems generated, transmitted, and utilized two-phase ac power. The two-phase system is now obsolete. Three-phase ac systems with a frequency of 60 Hz (hertz) are constructed throughout the United States and Canada. The development of water-power sites made it necessary to transmit electrical energy long distances from the generating stations to the factories located in the cities. High-voltage equipment was developed to permit the transmission of electrical energy from the hydroelectric generating stations to the load economically.

The development of transmission circuits across the continent has increased the reliability of the electric system, reduced the operating and standby reserve generating requirements, and made possible both the interchange sales of electric energy between utility companies, thus reducing costs, and the installation of large economical generating units.

The voltages used on transmission circuits have been steadily rising. A 2000-volt ac transmission line was constructed in Italy in 1886. The chart of Fig. 4-1 shows the historical development of maximum voltage of overhead ac transmission voltages. In 1902, the maximum ac voltage rose to 80,000 volts and remained there for 5 years. In 1913, it reached 150,000 volts and stayed there for 7 years. In 1922, it reached 220,000 volts, which remained the highest voltage until 1936, when the 286,000-volt line from Hoover Dam to Los Angeles was put into operation. In 1952, a 345,000-volt grid was put into operation in Ohio. Virginia Electric and Power Company placed in service a 500,000-volt line 350 miles long designed to transmit 1 million kVA in 1964. American Electric Power Service Corporation energized a 765,000-volt circuit (Fig. 4-2) in 1969 that now extends across the states of Indiana and Ohio.

Experimental transmission lines have been built and tested to operate at voltages in the range of 1100 to 1500 kV. It is predicted that ac transmission lines will be required to operate at 1000/1100 kV during the period from 1985 to 1995 to have sufficient capacity to carry the electrical energy necessary to supply the industrial, commercial, and residential customers. An experimental

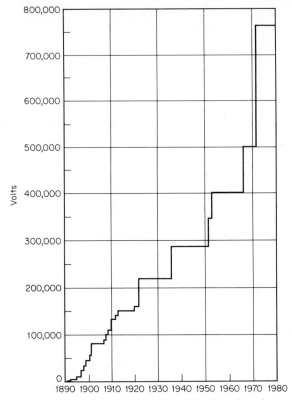

Fig. 4-1 Maximum alternating-current transmission-line voltages by years.

Fig. 4-2 Transmission-line tower supporting conductors energized at 765 kV. *(Courtesy American Electric Power Service Corp.)*

ac transmission line designed to operate at voltages as high as 2255 kV has been constructed (Fig. 4-3). Tests are being conducted to determine the radio-interference voltages and corona performance of the facilities.

Underground transmission circuits have been installed in large metropolitan areas and other places where overhead transmission lines are not practical. The first underground installations were developed for telegraph circuits. A cable insulated with strips of India rubber was used in 1812. Telegraph cables consisting of bare copper conductor drawn into glass tubes and joined together with sleeve joints and sealed with wax were used in 1816. Vulcanized-rubber-insulated cables proved to be successful.

Edison designed a buried system using copper rods insulated with a wrapping of jute. Two or three insulated rods were drawn into iron pipes and a heavy bituminous compound forced in around them. The insulated conductors were laid in 20-ft sections and joined together with specially designed tube joints from which taps could be taken if desired. The underground system

Fig. 4-3 Test station constructed to operate at voltages as high as 2255 kV. *(Reprinted from December 1, 1976, issue of "Electrical World." Copyright 1976, McGraw-Hill, Inc. All rights reserved.)*

was used to distribute energy for lighting in New York City. Overhead electric lines are prohibited in sections of Washington and Chicago. New York City enacted a law forcing the removal of all overhead wires from the streets in 1884.

The short life of the early underground systems led to the development of rubber-insulated lead-covered cables installed in ducts. Cables that were insulated with paper saturated with oil in a lead pipe were developed in England in 1890 and operated at 10,000 volts ac. There was approximately 5000 miles of underground cables constructed with copper conductor, insulated with paper saturated with oil, and covered with a lead sheath operating at voltages between 6.6 and 25 kV in the United States in 1921.

Underground transmission circuits are more expensive than overhead transmission lines. The high cost of underground installations limits their installation. Factors considered in determining the necessity to invest the larger amounts of money for an underground transmission circuit include:

1. City ordinances
2. The availability and costs of rights-of-way for overhead transmission lines compared to underground transmission circuits
3. The need to provide mechanical protection for the circuits
4. The congestion, and the appearance, of the area where the circuit is to be installed

New York City is supplied by an extensive underground transmission system with some cables operating at a voltage of 345 kV ac. Pipe-type cables (Fig. 4-4) are used extensively for underground transmission circuits. These cables operate at voltages in the range of 69 to 500 kV ac. Research is in progress on cables to operate at 765 kV ac. Charging current as a result of the capacitance between the phase conductors and their shielding conductors limits the power-carrying capacity of cable circuits. Uncompensated cables 26 miles long operating at 345 kV will have all the cable capacity absorbed by the charging current. Underground transmission circuits operating at 345 kV cost approximately 15 times more than overhead transmission lines.

Direct-current transmission was initiated in Germany in 1883. The system consisted of generators operating at 2400 volts dc transmitting approximately 5 kW over a distance of 8 km (kilometers). Transmission of electric energy by the Thury system was developed in 1902. Generators were connected in series to transmit 150 amps at a maximum voltage of 27,000 volts. The current

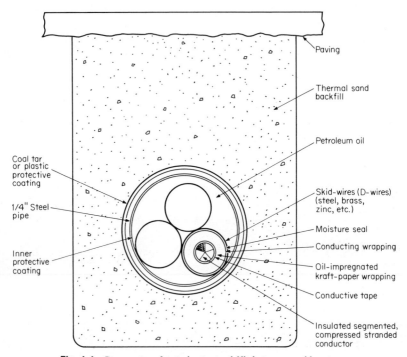

Fig. 4-4   Cross section of a single-circuit oil-filled pipe-type cable system.

in the transmission line was kept constant by varying the output voltage of the generators and switching generators in and out of service. The load was varied by bypassing and switching motors, connected in series, in and out of service.

Modern dc transmission lines have substations that convert high-voltage ac to dc, and invert dc to ac using rectifier circuits with mercury-arc valves or semiconductor elements. Direct-current transmission lines are being constructed to operate at high voltages. Some of the higher voltage transmission lines by operational date are

Year	Voltage, kV	Country
1950	200	Russia
1954	100	Sweden
1961	±100	England-France (Cross Channel)
1962	±400	Russia
1965	±250	Sweden
1970	±400	United States
1977	±450	Canada
1978	±400	United States
1981	±500	Canada

Direct-current transmission circuits are economical for installation whenever the overhead lines extend over 400 miles or underground cable circuits extend over 20 miles without a tap to the circuit.

Polyphase Transmission   Alternating-current voltages can be changed from one level to another easily by using power transformers. Direct-current transmission circuit voltages require converters to change the voltage from one level to another. The cost of converters limits dc transmission circuits to special applications. Single-phase ac transmission circuits would be similar to dc circuits except for the inductance and capacitance associated with ac transmission. Almost all ac transmission lines, from the smallest to the largest, are three-phase. Even some distribution lines built into rural communities to supply electricity to farmers are being built as three-phase lines. Two-phase lines are out of date, and single-phase lines are used only in distribution systems as branch lines and in some cases in connection with electrified railroads and rural electrification.

The two main reasons for building transmission lines three-phase are

1. A single-phase two-wire transmission line operating at a certain voltage requires a certain amount of conductivity in the wires to carry the load. A two-phase four-wire line would require

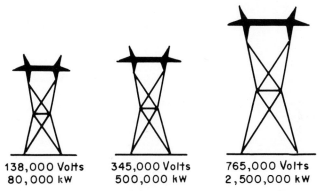

138,000 Volts   345,000 Volts   765,000 Volts
80,000 kW   500,000 kW   2,500,000 kW

Fig. 4-5   Comparison of power-carrying capability of three-phase alternating-current transmission lines at different voltages. *(Courtesy Edison Electric Institute.)*

the same total weight of conductor. There would be no gain, therefore, in building the two-phase line. An equivalent three-phase three-wire line, however, will require only three-fourths of the conductors. There will thus be a substantial saving in conductors if a three-phase system is selected.

2. Motors operating from a three-phase line are much cheaper and more satisfactory in operation than single-phase or two-phase motors. This reason is more important than the saving in conductors in the line.

Transmission Voltages   Transmission lines are operated at very high voltages to deliver more power with a particular size line conductor and the same line current. A three-phase ac line operating at 345,000 volts, for example, can carry 6.25 times as much power as a 138,000-volt line with the same current flowing in the lines. This is so because the power-carrying capacity of a line varies as the square of the line voltage. Even though the voltage of the 345,000-volt line is only 2.5 times as high as that of the 138,000-volt line, 2.5 squared—that is, 2.5 times 2.5—equals 6.25; therefore the higher-voltage line can carry 6.25 times as much power as the lower-voltage line. If we assume the 138,000-volt line to be carrying a load of 80,000 kW, the 345,000-volt line can carry 6.25 times 80,000 or 500,000 kW.

In like manner, the three-phase 765,000-volt ac line of the American Electric Power System, having a voltage 5.6 times as great as the 138,000-volt line, can carry 5.6 times or 31 times as much power; 31 times 80,000 equals 2,500,000 kW.

This tremendous increase in power-carrying capacity with the higher voltages makes it possible to send more power over the same right of way, thereby causing large savings in additional right of ways. It also reduces the number of parallel lines needed where heavy loads must be transmitted.

Figure 4-5 visually represents three high-voltage lines operating at 138,000, 345,000, and 765,000 volts respectively and their corresponding power-transmitting capacities of 80,000, 500,000, and 2,500,000 kW.

   Voltage levels of ac transmission circuits vary from 34,500 to 2,250,000 volts. Table 4-1 gives the preferred transmission-circuit voltages. Voltages originally used for transmission circuits below 34,500 have been converted to distribution circuits. Circuits with voltages in the range of 34,500 to 138,000 volts have been converted to subtransmission use. In many cases, these circuits will provide high-voltage service to large industrial customers. Extra-high voltages have been used extensively for transmission circuits in the United States. The ultra-high ac and highest dc voltages are still experimental.

   The determination of the most economical voltage to use requires careful analysis. Usually several voltages are considered. In each case the total annual operating costs, including interest,

Table 4-1   Preferred Transmission-Circuit Voltages

	Alternating current, three phase			Direct current
Subtransmission	Transmission	EHV* Transmission	UHV† Transmission	Transmission
34,500				
69,000				
				±100,000
115,000				
138,000	138,000			
	161,000			
	230,000			
				±250,000
		345,000		
				±400,000
				±450,000
		500,000		±500,000
				±600,000
		765,000		
			1,000,000	
			1,500,000	
			2,250,000	

*Extra-high-voltage
†Ultra-high-voltage

Table 4-2   Capacity of Typical Overhead ac Three-Phase Transmission Lines

Line voltage, kV	Circuits per tower	Conductors per phase	Conductor size ACSR, kcmil	Phase resistance, Ω/mi	Ampacity, A	Thermal line rating, MVA
138	1	1	795	0.140	880	210
138	2	1	795	0.140	880	420
230	1	1	1431	0.076	1220	490
230	2	1	1431	0.076	1220	980
345	1	2	795	0.070	1750	1050
345	2	2	1590	0.035	2640	3160
500	1	2	1780	0.031	2850	2470
765	1	4	954	0.028	3890	5150

depreciation, and taxes, plus power lost in the line, are estimated for each voltage studied. If no special factors have to be considered, the voltage which gives the lowest total annual operating cost is the one selected.

   The selection of the proper voltage is sometimes also governed by the voltage of existing power lines. If the lines are to be connected to other lines, as they probably will be, it is convenient to have the lines at the same voltage.

   Transmission-Line Capacity   Thermal conditions of the conductors is usually the load limiting factor of transmission circuits. Convection cooling by ambient air allows overhead conductors to carry higher currents than underground conductors. Alternating-current overhead transmission lines are limited in capacity by the thermal expansion and the resulting sag of the conductors. Short overhead transmission lines may be loaded continuously up to their thermal rating. Table 4-2 gives the thermal rating of typical overhead ac transmission lines installed in the United States.

The current flowing in a transmission-circuit conductor can be calculated if the voltage and the power in kilovolt-amperes (kVA) or megavolt-amperes (MVA) are known. If the power is expressed in kilowatts (kW) or megawatts (MW), the power factor must be known in addition to the voltage. Kilovolt-amperes and kilowatts can be calculated if the current, voltage, and power factor are known.

The current flowing in a single-phase two-wire circuit is given by the formula

$$\text{Current in amperes} = \frac{\text{apparent power in kilovolt-amperes} \times 1000}{\text{line voltage in volts}}$$

or

$$\text{Current in amperes} = \frac{\text{power in kilowatts} \times 1000}{\text{line voltage in volts} \times \text{power factor}}$$

To illustrate the use of these formulas, take the example of a single-phase line carrying a load of 100,000 kVA and operating at 80,000 volts. Substituting in the first formula gives

$$\text{Current} = \frac{100,000 \times 1000}{80,000} = 1250 \text{ amp}$$

The current just calculated is the current in each wire of the single-phase circuit.

If the kilowatts, power factor, and voltage are known instead of kilovolt-amperes and voltage, the second formula is used. For example, if an 80,000-volt circuit delivers 100,000 kW at 80 percent power factor, the current in the line will be given by

$$\text{Current} = \frac{100,000 \times 1000}{80,000 \times 0.80} = 1562.5 \text{ amp}$$

If this load of 100,000 kW were at unity power factor, instead of 0.80, the current would be less. The kilovolt-amperes and kilowatts would be the same and current would be 1250 amp.

The current flowing in a three-phase circuit is given by the formula

$$\text{Current in amperes} = \frac{\text{kilovolt amperes} \times 1000}{1.73 \times \text{line voltage in volts}}$$

or

$$\text{Current in amperes} = \frac{\text{kilowatts} \times 1000}{1.73 \times \text{line voltage} \times \text{power factor}}$$

To illustrate the use of the formula, take the example of a three-phase line carrying 100,000 kVA of load and operating at 138,000 volts. To find the current flowing in each line conductor, substitute in the formula. Thus

$$\text{Current} = \frac{100,000 \times 1000}{1.73 \times 138,000} = 418.9 \text{ amp}$$

If the amperes at unity power factor are known and it is desired to know what the current in amperes would be at any other power factor, the procedure is to divide the current at unity power factor by the power factor at which the current is desired. Thus

$$\text{Current at any desired power factor} = \frac{\text{current at unity power factor}}{\text{desired power factor}}$$

To illustrate: If the current at unity power factor is 100 amp, the current at 0.8 power factor is

$$\text{Current at 0.8 power factor} = \frac{100}{0.80} = 125 \text{ amp}$$

The formulas and equations presented are summarized and arranged for ready reference in Table 4-3. It might be pointed out as a matter of interest that power factor does not have to be considered in dc calculations as there is no lag or lead of current in such circuits. In ac circuit

Table 4-3 Summary of Circuit Calculations

To find	Direct current	Alternating current	
		Single-phase	Three-phase
Amperes when kilovolt-amperes and voltage are known	....	$\dfrac{\text{Kilovolt-amperes} \times 1{,}000}{\text{volts}}$	$\dfrac{\text{Kilovolt-amperes} \times 1{,}000}{1.73 \times \text{volts}}$
Kilovolt-amperes when amperes and voltage are known.	....	$\dfrac{\text{Amperes} \times \text{volts}}{1{,}000}$	$\dfrac{1.73 \times \text{volts} \times \text{amperes}}{1{,}000}$
Amperes when kilowatts, volts, and power factors are known	$\dfrac{\text{Kilowatts} \times 1{,}000}{\text{volts}}$	$\dfrac{\text{Kilowatts} \times 1{,}000}{\text{volts} \times \text{power factor}}$	$\dfrac{\text{Kilowatts} \times 1{,}000}{1.73 \times \text{volts} \times \text{power factor}}$
Kilowatts when amperes, volts, and power factor are known	$\dfrac{\text{Amperes} \times \text{volts}}{1{,}000}$	$\dfrac{\text{Amperes} \times \text{volts} \times \text{power factor}}{1{,}000}$	$\dfrac{1.73 \times \text{amperes} \times \text{volts} \times \text{power factor}}{1{,}000}$

calculations, the power factor is important. If its value is not known for a given circuit or line, the value 0.80 is often assumed as a basis for calculation. It will also be noted that the factor 1.73 is used in all three-phase calculations.

Transmission-Cable Capacity   Solid-dielectric cables are used extensively for subtransmission and transmission underground circuits. Research is being conducted in the extra-high-voltage range to develop solid-dielectric cables for operation at 345 and 500 kV. The insulating compounds used most extensively are crosslinked polyethylene (XLPE) and ethylene propylene rubber (EPR).

A typical extruded solid-dielectric power-transmission cable is shown in Fig. 4-6. The cable consists of a stranded copper or aluminum conductor covered by a thin extruded layer of semiconducting (carbon-black-filled) polyethylene. The semiconducting conductor shield provides a

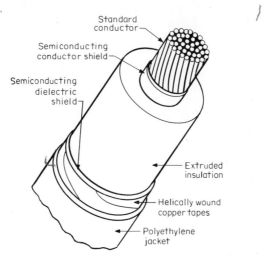

Fig. 4-6   Construction of solid-dielectric extruded-insulation, high-voltage transmission cable.

smooth conductor surface for the insulation and excludes air from the interface. The insulation is extended over the conductor shield. A thin semiconducting polyethylene shield is extruded over the insulation. The semiconducting shields over the conductor and the insulation prevent electric discharges, or corona, that would damage the insulation. A metallic layer on top of the dielectric shield provides a good conductor for the charging current, preventing damage to the semiconducting shield from the heat generated by the charging current. The metallic shield may be copper tapes, concentric wires, tubular corrugated copper, copper wires embedded in the semiconducting insulation shield, or a lead sheath. The metallic shield is usually protected by a polyethylene jacket. Extruded-rubber-insulated solid-dielectric cables operating at 69 kV ac are pictured in Fig. 4-7. Figure 4-8 is a cross-sectional view of the same cables.

Submarine cables constructed with solid-delectric insulations have been successful. Figure 4-9 illustrates a three-conductor solid-dielectric cable. Figures 4-10 and 4-11 show submarine cables being installed by cablemen and linemen.

Extruded solid-dielectric cables with a 2000 kcmil copper conductor installed direct-buried with the three cables in a flat configuration with one three-phase circuit per trench would have the following capacity:

Voltage, kV	Apparent power, MVA
138	289
230	470
345	700

High-pressure oil-filled (HPOF) paper-insulated cables in which the three-phase conductors are enclosed in a steel pipe containing oil are used extensively for underground cable-transmission circuits.

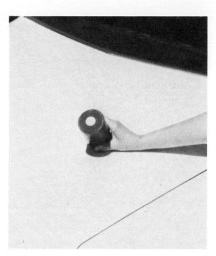

Fig. 4-7 Solid-dielectric three-phase subtransmission circuit cables operating at 69 kV ac terminated in a substation. The cable terminations consist of tape-formed stress cones with rubber rain shields above the stress cones.

Fig. 4-8 Cross-sectional view of 750 kcmil, 69 kV, solid-dielectric extruded-insulation, copper-tape shielded, polyethylene-jacketed cable.

Fig. 4-9 Three-conductor solid-dielectric insulated cable for submarine installation. The three insulated conductors are covered with jute and protected by polyethylene covered steel wires. *(Courtesy Okonite Co.)*

Fig. 4-10   Reels of submarine transmission cables mounted on a barge to permit installation under the water. *(Courtesy Okonite Co.)*

Fig. 4-11   Submarine cable in picture is being unreeled and placed in the water. *(Courtesy Okonite Co.)*

A cross-sectional view of a typical pipe-type cable installation is detailed in Fig. 4-4. The cables are terminated in potheads. Each phase conductor is installed in a nonmagnetic pipe (see Fig. 4-12) to a junction point called a trifurcator where the three-phase cables enter a common pipe. Shipping lengths of cable will normally be approximately 2500 ft for 138 and 230 kV, and 1750 ft for 345 and 500 kV. The cables are spliced in manholes. The steel pipe is usually direct-buried between the manholes. Oil pressure is maintained in the steel pipe with the use of pumps and

Fig. 4-12   Cablemen installing kraft-paper insulated copper-tape shielded cables in pipes at a terminal in a substation. The trifurcator, or junction device, is located below the terminal structure back of the reels. *(Courtesy Anaconda Wire and Cable Co.)*

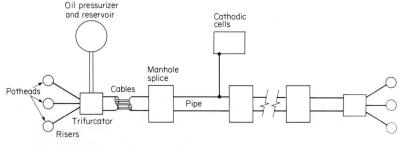

Fig. 4-13   Schematic drawing of a pipe-type cable transmission circuit.

reservoirs. The steel pipe is protected from corrosion by cathodic protection equipment. Alarms are installed on the system to detect low oil pressure. A schematic diagram of a pipe-type cable system appears in Fig. 4-13. The transmission circuit may consist of two cables installed in parallel. The two pipes can be placed in the bottom of the same trench separated as far as possible to minimize the installation cost. The dual facilities increases the reliability. The pressurized oil can be pumped through the pipe for one cable circuit and returned through the parallel cable circuit pipe. The capacity of oil-filled pipe-cable systems can be increased significantly by circulating the oil past the cable at several feet per second and cooling it externally by means of an oil-to-air heat exchanger (see Table 4-4).

Table 4-4   Capacities of Three-Phase (3 Ø) Oil-Filled Paper-Insulated Pipe-Type Cables

	2000-kcmil conductor				2500-kcmil conductor
Voltage, kV	138	230	345	500	765
Pipe size, nominal, in	8	10	10	12	14
Insulation thickness, in	0.505	0.835	1.025	1.25	1.56
Naturally cooled cables					
Capacity, MVA	224	344	453	571	645
Dielectric losses, watts/cond. ft	0.733	1.39	2.71	3.99	8.70
Resistive losses, watts/cond. ft	7.72	7.02	5.49	4.40	2.30
Pipe temperature, °C	62.0	59.2	58.4	57.1	64.7
Total 3 Ø reactive losses per cable mile, MVAR	4.7	8.9	17.3	31.8	69.4
Forced-cooled cables					
Capacity, MVA	426	688	941	1352	2137
Dielectric losses, watts/cond. ft	0.733	1.39	2.71	3.99	8.70
Resistive losses, watts/cond. ft	28.7	28.4	24.2	25.2	24.9
Pipe temperature, °C	50.2	37.8	38.8	32.4	29.1
Total 3 Ø reactive losses per cable mile, MVAR	4.7	8.9	17.3	31.8	69.4

Compressed-gas-insulated or gas-spacer cables using sulfur hexafluoride (SF_6) gas as the insulating medium are used for high-capacity insulated transmission circuits. The gas-spacer cable transmission system is an ac three-conductor isolated-phase design. The conductor can be copper or aluminum tube with a diameter of approximately 3 to 10 in, depending upon the current-carrying capacity required and the operating voltage. The conductor is surrounded by an outer sheath of extruded-aluminum tubing the diameter determined by the space required to provide insulation for the operating voltage. These diameters vary from 8.5 in at 138 kV up to 20 in at 500 kV and are approximately 30 in at 1000 kV.

The conductor is centered and maintained concentric with the outer enclosure by insulators installed in the enclosure at intervals along the cable (Fig. 4-14). The optimum insulating SF_6 gas pressure is approximately 50 psig. The cable assemblies are manufactured in approximately 40-ft lengths. The conductors are spliced together by plug-in type terminals or welding. The outer enclosures are welded together to form a continuous leak-proof pipe. The cables can be installed on top of the ground, overhead on supporting structures, or direct-buried. Direct-buried circuits

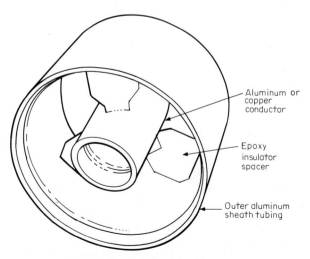

Aluminum or copper conductor

Epoxy insulator spacer

Outer aluminum sheath tubing

Fig. 4-14   Cross-sectional view of gas-spacer cable construction.

would require a trench 4 to 7 ft wide, depending on the operating voltage. Thermal sand is used for a back-fill material. Figure 4-15 shows a surface installation protected by concrete walls. The cables are terminated in potheads consisting of an extension of the outer sheath, the inner conductor, and a porcelain insulator. The capacities and charging currents of typical gas-spacer cable installations are itemized in Table 4-5.

Self-contained cables with a central duct (Fig. 4-16) using pressurized oil for forced cooling have been used in some locations for underground transmission circuits. The cables can be direct buried or installed in ducts. The capacity of the circuit is limited severely when these cables are installed in ducts as a result of the inability to dissipate the heat generated in the cable.

Fig. 4-15 Gas-insulated (SF$_6$) cables rated 3800 amps at 230 kV. Two transmission-cable circuits are located adjacent for economical installation. *(Courtesy Westinghouse Electric Corp.)*

Table 4-5   Single-Circuit Three-Phase AC Gas-Spacer Cable Characteristics

Operating voltage, kV	Capacity, MVA	Charging current, amps/mi	Uncompensated maximum length, mi
138	400	4.0	440
230	700	5.3	350
345	1200	7.4	295
500	2200	12.6	250
765	4200	20.0	
1000	6500	25.0	

Direct-current underground-transmission-cable circuits have advantages over alternating-current cables since they do not have induction losses or a continuous flow of charging current. Solid-dielectric extruded-polyethylene-insulated cables can be used on voltages up to 300 kV. Pipe-type high-pressure oil-filled cables with paper-tape insulation can be used for circuits with a voltage of ±600 kV. Self-contained oil-filled cables will operate satisfactorily for voltages of ±300 kV. Direct-current cable insulation must be thick enough to provide a safety factor of 3 to 5 with respect to dc insulation breakdown because of surges. The surges may be caused by malfunction of conversion equipment, operation of switches, accidental polarity reversal, or lightning strokes on overhead lines connecting to the underground cables. Table 4-6 itemizes the capacity of high-pressure oil-filled pipe-type 2000-kcmil conductor cable operating at various dc voltages. The high cost of converter stations limits the use of underground dc transmission circuits.

Transmission-Line Environmental Considerations Underground-transmission circuits would eliminate the environmental problems associated with electric power transmission. The large investment associated with underground cable installations and technical limitations to the length of ac underground-cable circuits prevents using underground installations to solve all the environmental problems.

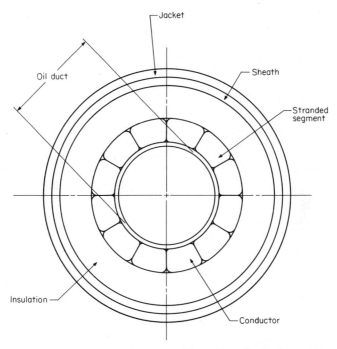

Fig. 4-16 Cross-sectional view of a self-contained cable used for underground-transmission circuits.

Table 4-6   Data for 2000-kcmil Conductor High-Pressure Oil-Filled Pipe-Type Underground Cable Installation Operating at Various DC Voltages

	Voltage dc, kV			
	± 180	± 300	± 450	± 655
Single-conductor loss, kW/mi	42.7	40.7	39.5	33.8
Maximum dc current, amps	1120	1095	1030	1000
Average electric stress, kV/in	356	360	440	525
Capacity, MVA	403	657	927	1307

Rights-of-Way Transmission-line environmental considerations include visual impact, natural land conditions, electrostatic effects, land use, corona generation and associated radio and television interference or audible noise, and the limiting of the voltages induced in objects located in the transmission line right-of-way. Rights-of-way for transmission lines should be selected with the purpose of minimizing the visual impact to the public. Existing rights-of-way should be given priority as the locations for additions to existing transmission facilities.

Where practical, rights-of-way should avoid the national historic places listed in the *National Register of Historic Places*, the natural landmarks listed in the *National Register of Natural*

Landmarks maintained by the Secretary of the Interior, and parks, scenic or wildlife areas, and recreational lands officially designated by duly constituted public authorities. If rights-of-way must be routed through such historic places, parks, scenic or wildlife areas, they should be located in areas or placed in a manner so as to be least visible from public view and as far as possible in a manner designed to preserve the character of the area.

Rights-of-way should avoid prime or scenic timbered areas, steep slopes, and proximity to main highways where practical. In some situations scenic values would emphasize locating rights-of-way remote from highways, while in others where scenic values are less important rights-of-way along highways in timbered areas would achieve desirable conservation of existing forest lands.

Where the transmission rights-of-way cross areas of land managed by federal or state agencies or private organizations, these agencies should be consulted early in the planning of the transmission project to coordinate the line location with land-use planning and with other existing or proposed rights-of-way.

In scenic and residential areas, clearing of natural vegetation should be limited to that material which poses a hazard to the transmission line. Determination of a hazard in critical areas, such as park and forest lands, should be a joint endeavor of the utility company and land manager in keeping with national, state, or other electric safety and reliability requirements.

Long tunnel views of transmission lines crossing highways in wooded areas, down canyons and valleys, or up ridges and hills should be avoided. This can be accomplished by having the lines change alignment in making the crossing, by concealment of terrain, or by judicious use of screen planting (Fig. 4-17).

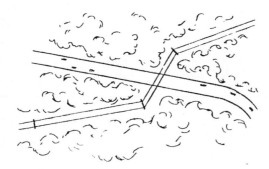

Fig. 4-17   Proper method for transmission line to cross highway to minimize visual impact of cleared right-of-way.

Rights-of-way clearings should be kept to the minimum width necessary to prevent interference of trees and other vegetation with the proposed transmission facilities. In scenic or urban areas trees which would interfere with the proposed transmission facilities and those which could cause damage if fallen should be selectively cut and removed (Fig. 4-18).

The time and method of clearing rights-of-way should take into account matters of soil stability, the protection of natural vegetation, and the protection of adjacent resources.

The use of helicopters for the construction and maintenance on rights-of-way should be considered in mountainous and scenic areas where it is consistent with reliability of service. This would permit rights-of-way to be located in more remote areas and would reduce disturbance of the ground and the number of access roads.

Poor example   Preferred method

Fig. 4-18   Right-of-way clearing.

Trees and other vegetation cleared from rights-of-way in areas of public view should be disposed of without undue delay. If trees and other vegetation are burned, local fire and air-pollution regulations should be observed. Unsightly tree stumps adjacent to roads and other areas of public view should be cut close to the ground or removed. Trees, shrubs, grass, and topsoil not cleared should be protected from damage during construction.

Rights-of-way should not be cleared to the mineral soil. Where this must occur in scattered areas of the rights-of-way, the topsoil should be replaced and stabilized without undue delay by the planting of appropriate species of grass, shrubs, and other vegetation which should be properly fertilized.

Soil that has been excavated during construction and has not been used should be evenly filled back onto the cleared area or removed from the site. The soil should be graded to comport with the terrain and the adjacent land, the topsoil should then be replaced, and appropriate vegetation should be planted and fertilized.

Scars on the surface of the ground should be repaired with topsoil and replanted with appropriate vegetation or otherwise conformed to local natural conditions. Grading generally should not be done on slopes where the scars cannot be repaired without creating an erosion problem.

Fig. 4-19   Right-of-way established to limit visual impact.

Terraces and other erosion control devices should be constructed where necessary to prevent soil erosion on slopes on which rights-of-way are located.

Where rights-of-way cross streams or other bodies of water, the banks should be stabilized to prevent erosion. Construction on rights-of-way should not damage shorelines, recreational areas, or fish and wildlife habitats. Care should be taken to avoid oil spills and other types of pollution while work is performed in streams.

In scenic areas visible to the public, rights-of-way strips through forest and timber areas should be deflected occasionally and should follow irregular patterns or be suitably screened to prevent the rights-of-way from appearing as tunnels cut through the timber (Fig. 4-19).

At road crossings or other special locations of high visibility rights-of-way strips through forest and timber areas should be cleared with varying alignment to comport with the topography of the terrain. In such locations also, where rights-of-way enter dense timber from a meadow or other clearing, trees should be feathered in at the entrance of the timber for a distance of 150 to 200 yards. Small trees and plants should be used for transition from natural ground cover to larger areas (Fig. 4-20).

Roads used during construction should be stabilized without undue delay by erosion control measures and by planting of appropriate grass and other vegetation. These roads should be designed for proper drainage, and water bars to control soil erosion should be installed. Access roads should not be constructed on unstable slopes. Where feasible, service and access roads should be used jointly.

If an overhead line must be routed across uniquely scenic, recreational, or historic areas or rivers, the feasibility of placing the lower voltage line underground should be considered. If the line must be placed overhead, it should be located on the right-of-way least visible from areas of public view.

Transmission facilities should be located with a background of topography and natural cover where possible. Vegetation and terrain should be used to screen these facilities from highways and other areas of public view (Fig. 4-21).

If transmission facilities must be placed on slopes that parallel highways or other areas of public view, they should be located approximately two-thirds the distance up the slopes where feasible. With the slopes as background, the presence of the facilities would be less noticeable (Fig. 4-22).

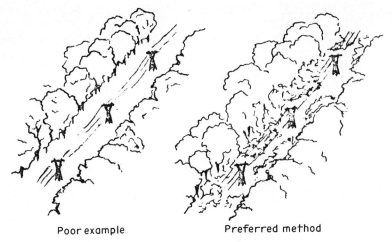

Poor example   Preferred method

Fig. 4-20   Right-of-way plantings.

Fig. 4-21   Use of vegetation to screen right-of-way.

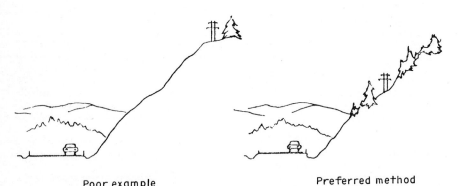

Poor example   Preferred method

Fig. 4-22   Location of transmission-line structures.

Transmission line rights-of-way should not cross hills and other high points at the crests, and when possible, placing a transmission tower at the crest of a ridge or hill should be avoided. Towers should be spaced below the crest to carry the line over the ridge or hill, and the profile of the facilities should present a minimum silhouette against the sky (Fig. 4-23).

Transmission lines should not cross highways at the crest of a road (Fig. 4-24). Long views of transmission lines parallel to highways should be avoided where possible. This may be accomplished by overhead lines being placed beyond ridges or timber areas (Fig. 4-25). Transmission lines should cross canyons up slope from roads that traverse the canyon basins if the terrain permits (Fig. 4-26).

When crossing canyons in a forest, high long-span towers should be used to keep the power lines above the trees and to eliminate the need to clear all vegetation from below the lines. Only as much vegetation as is necessary to string the line should be cut.

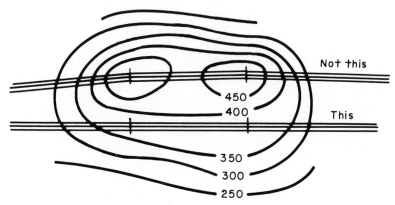

Fig. 4-23   Transmission-line structures should be located below crest of hill.

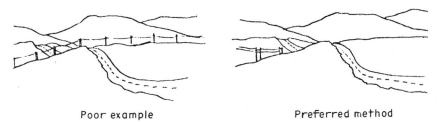

Poor example   Preferred method

Fig. 4-24   Transmission-line location to cross highway in rolling terrain.

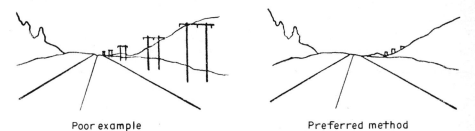

Poor example   Preferred method

Fig. 4-25   Transmission-line location when paralleling highways.

Where ridges or timber areas are adjacent to highways or other areas of public view, overhead lines should be placed beyond the ridges or timber areas. In forest or timber areas, high long-span towers should be used to cross highways in order to retain much of the natural growth beside them (Fig. 4-27).

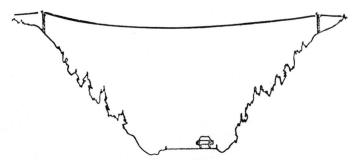

Fig. 4-26 Proper location of transmission-line towers when crossing highway in a canyon.

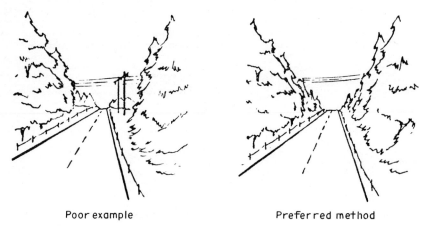

Poor example   Preferred method

Fig. 4-27 Transmission-line structures located adjacent to highway at crossing.

Native shrubs and trees should be left in place or planted at random, with the necessary allowance for safety, near the edges of rights-of-way adjacent to roads. Locating transmission lines near or crossing road interchanges should be avoided if possible.

The Federal Highway Administration and the State Highway Department should be consulted with respect to any applicable guidelines or regulations that they might have to govern transmission lines crossing highways.

One of the potential benefits of transmission-line routes is that clearings at safe distances adjacent to transmission facilities may be used for secondary purposes. If consistent with general safety factors, the following should be considered as possible secondary uses of rights-of-way to the extent permitted by the property interest involved:

Parks
Golf courses
Equestrian or bicycle paths
Picnic areas
Game refuges
Hiking trail routes

General agriculture
Winter sports
Orchards
Cultivation of Christmas trees, elderberry and huckleberry bushes, and other nursery stock

Electrostatic Effects   Electrostatic effects of transmission lines are not hazardous. Linemen regularly work on energized EHV and lower voltage transmission lines using hot-line tools and bare-hand methods with no ill effects. A study of 10 linemen who regularly worked on energized 345 kV transmission lines by personnel of Johns Hopkins University in Baltimore, Md., over a period of 10 years determined that the linemen were in excellent health.

Electric fields under EHV transmission lines are very low at ground level. Static charges on farm fences directly under the lines are usually drained to ground through the metal or wood posts or contacts with grass, weeds, or the ground. Vehicles, such as a tractor or combine, directly under an EHV line may become charged, and a farmer bridging the rubber tires of the vehicle might feel a tingle if the vehicle discharged through the driver's body to ground. The effect could be annoying, but it is not dangerous electrically, as linemen can verify. The annoying shock may startle a person, causing a fall or other dangerous act that could result in a personal injury. Vehicles operated under EHV transmission lines or facilities such as fences installed under EHV lines should be grounded to prevent annoying shocks and acts that might lead to injuries. All metal buildings or structures should be grounded whether they are built near a transmission line right-of-way or not. Grounding the building will eliminate any static charge.

Linemen use special conductive clothing, or electrostatic shields, while working on energized UHV transmission lines. It is necessary that special precautions be taken when UHV transmission lines are designed and constructed to protect the public from hazardous voltages. The major characteristics affecting a transmission line's electrostatic conditions are voltage, conductor type, phase configuration, phase spacing, and conductor height. Corona, visible at night as a luminous glow, is caused by ionization of the air close to the conductors. Corona, and resulting radio interference voltages and audible noise, can be controlled by proper transmission line design and construction precautions. The use of bundled conductors and appropriate hardware on EHV and UHV lines keeps corona to acceptable limits if the equipment is installed correctly. The conductors must be kept clean and free from damage that causes rough or pointed projections. Grease on a conductor can cause the generation of corona.

Induced voltages on objects in the right-of-way of ac UHV transmission lines can be controlled by the height of the towers or by shielding. The right-of-way for the transmission line can be purchased and fenced to prevent access to the land. It is important that the transmission lines be designed to minimize the cost and maximize the safety of the public.

Distribution Circuits

Distribution is that system conponent that delivers the energy from the generators, or the transmission system, to the customers. It includes the substations that reduce the high voltage of the transmission system to a level suitable for distribution and the circuits that radiate from the substation to the customers (Fig. 5-1).

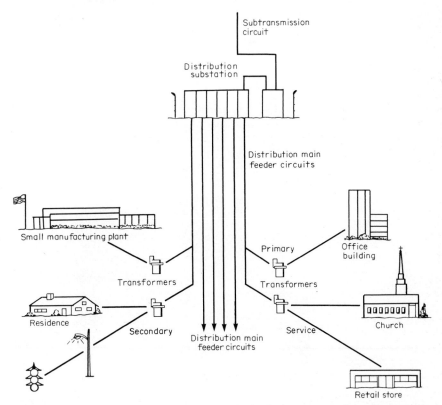

Fig. 5-1 Schematic diagram of a distribution system. The distribution circuits may be overhead or underground. The transformers may be pole-type, padmounted, submersible, or direct-buried.

The distribution circuits that carry power from the substations to the local load areas are known as "primary main circuits," "distribution main feeders," or "express feeders," and generally operate at voltages between 2400 and 34,500 volts. The loading of distribution circuits varies with customer load density, type of load supplied, and conditions peculiar to the area served. The nominal capacity of distribution circuits operating at different voltage levels is itemized in Table 5-1.

Table 5-1   Distribution Feeder Capacities of Distribution Circuits

Voltage class, kV	Capacity, MVA	
	Average	Maximum
4–5	2.5	3.5
12–15	7.5	15.0
20–25	12.0	20.0
30–35	18.0	30.0

The distribution circuits may be overhead or underground depending on the load density and the physical conditions of the particular area to be served. The vast majority are constructed overhead on wood poles that may support communication facilities and street lights. Street lighting is an important part of the distribution system.

Distribution transformers are installed in the vicinity of each customer to reduce the voltage of the primary circuit to 120/240 volts or other utilization voltage required by the customer. Secondary circuits and services carry the power at utilization voltage from the distribution transformers to the customers.

Distribution circuits originate in a distribution substation and include the conductors and equipment necessary to distribute electricity to the customer. The distribution substation may be located

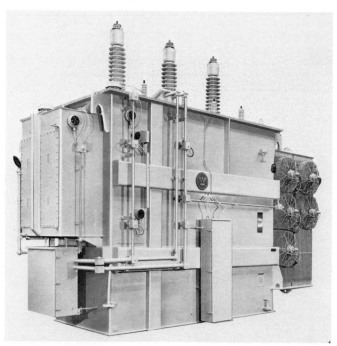

Fig. 5-2 Distribution substation power transformer equipped with tap-changing underload equipment. *(Courtesy Westinghouse Electric Corp.)*

at a generating station or at a point remote from the generation and connected to the generation by transmission and subtransmission circuits. Distribution substations are located near the customers utilizing the electricity to ensure good voltage and reliable service.

Distribution circuits can have a major impact on the customer's voltage and service continuity. The voltage must be maintained within a range that permits the customer's equipment to perform satisfactorily and within limits specified by regulatory agencies. If the customer's base voltage is 120 volts, then the minimum permissible service voltage would be 110 volts and the maximum permissible service voltage would be 127 volts. It is good practice to maintain the voltage as close to the base voltage as practical and to limit the range of the voltage variations from base. The maximum utilization voltage should not exceed 125 volts, and the minimum utilization voltage would normally be 110 volts with limited periods of operation as low as 106 volts.

It is normal to operate the distribution system at higher voltages during peak load periods to compensate for voltage drop in the distribution equipment and customer's wiring system. It is important that distribution circuit equipment be designed and installed in a manner to prevent objectionable light flicker as a result of momentary variations in voltage level.

The electric distribution system and distribution circuits are described in general in Section 2.

Three-Phase Three-Wire Distribution Circuits   Alternating-current distribution system feeders are commonly operated as three-phase three-wire or three-phase four-wire circuits. A three-phase circuit can be obtained from either a Y or a delta (Δ) connection of the three coils in the ac generator. Similarly, in distribution systems the Y or delta connection is obtained by the proper connection of the secondary coils of the transformers in the substation (Figs. 5-2 and 5-3).

Figure 5-4 shows a three-phase three-wire distribution feeder together with the Y and the delta connections of the secondary windings of the distribution substation step-down transformers. The transformer secondary connections are shown in the conventional and practical way of representation. It should be pointed out here that the Y connection is generally used on distribution feeders when the feeder is operated as a four-wire circuit. The voltage between any two of the three wires for either method of connection is the same; that is, the voltages between conductors A and B, B and C, and C and A are equal. In the delta connection this voltage between line wires is the same

Fig. 5-3   Distribution substation power transformer core and coil assembly. Coils are Y connected. *(Courtesy Westinghouse Electric Corp.)*

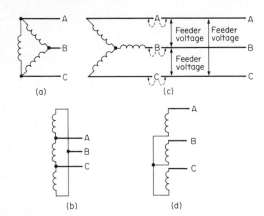

(a)　　　　(c)

(b)　　　　(d)

Fig. 5-4 Three-phase three-wire distribution circuit. (*a*) Conventional and (*b*) practical way of picturing delta connection of transformer secondaries. (*c*) Conventional and (*d*) practical way of picturing Y connection of transformer secondaries.

as the transformer-coil voltage, but in the Y connection the voltage between line conductors is $\sqrt{3}$ × the transformer-coil voltage.

Three-Phase Four-Wire Distribution Circuits The three-phase four-wire feeder circuit is in general use, especially in the larger cities. This circuit can be obtained only from a Y connection of the distribution substation transformer secondary. Figure 5-5 shows this type of circuit. The fourth wire is called the "neutral" wire because it is connected to the neutral point of the Y connection.

It will be noted that the voltage between any two of the line wires is no longer equal to E, the transformer secondary-coil voltage, but instead is $\sqrt{3}$ × E. The voltage E still exists between any line wire and neutral, but the voltage between any two line wires is $\sqrt{3}$ × E volts. If the voltage E between any wire and neutral is 7620 volts then the voltage between any two-phase wires is equal to $\sqrt{3}$ × E, or 1.732 × E which, in this example, would be 1.732 × 7620 or 13,200 volts. Single-phase voltages equal to 7620 volts would be available by using one-phase wire and the neutral conductor which is grounded.

The advantages of the four-wire feeder over the three-wire feeder are increased power-carrying capacity and better voltage regulation. Since the line voltage in the four-wire system is $\sqrt{3}$ = 1.73 times that of the three-wire system, its power-carrying capacity for the same current in the wires will also be 1.73 times as great; or for the same power transmitted, the current in the four-wire system will only be 58 percent as large. Moreover, the voltage regulation is also greatly improved. Many of the large cities have changed over to this system of feeders in order to get the benefit of this increased feeder capacity. Merely stringing the fourth wire and reconnecting the transformer secondaries from delta to Y make possible an increase in load-carrying capacity of 73 percent.

Primary distribution circuit voltage levels continue to move upward. In earlier years 2400 volts delta, 4160Y/2400, and 4800 volts delta were in wide use, amounting to approximately 60 percent of the systems in 1955. In 1969, usage of distribution voltages below 5000 volts was down to about 12 percent. The next higher levels, 12,470Y/7200, 13,200Y/7620, 13,800Y/7960, and, to a lesser extent, 12,000 to 15,000 volts delta are now the most prevalent.

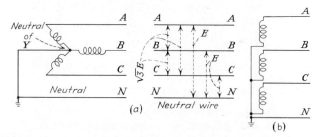

(a)　　　(b)

Fig. 5-5 Three-phase four-wire distribution circuit. (*a*) Conventional and (*b*) practical way of picturing Y connection of distribution substation transformer secondary with neutral brought out.

Higher primary distribution voltages are being used, including 24,500Y/14,400 volts and 34,-500Y/19,920 volts, with some exploration of even higher levels. There do not appear to be any insurmountable technical problems of producing suitable components of higher-voltage systems.

From an application standpoint, there are a number of factors to be considered. Factors favorable to the higher levels include the steady increase in consumer kWh, the permissible use of smaller conductors, better voltage regulation, and the availability of existing industrial feeders or subtransmission circuits at levels that are suitable, or could be made suitable, for the start of the

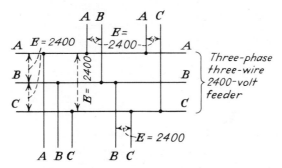

Fig. 5-6 ABC is a 2400-volt three-phase primary feeder, and AB, BC, and AC are single-phase 2400-volt primaries obtained by tapping the 2400-volt three-phase three-wire feeder.

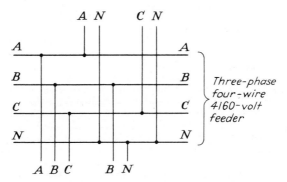

Fig. 5-7 ABC is a 4160-volt primary feeder obtained by tapping the three-phase wires, and AN, BN, and CN are single-phase 2400-volt primaries obtained by tapping a phase wire and neutral of the three-phase four-wire 4160-volt feeder.

higher-voltage distribution circuits. Some negative factors are the existence of a widespread use of the present levels and the possibility of an alternative solution of using more substations to supply present voltage levels as loads increase. At the increased voltage levels, more customers would be interrupted in case of system faults.

Single-Phase Primary Circuits A single-phase circuit is generally obtained from a three-phase circuit. It could be obtained from a single-phase ac generator, but such machines are very rare today. Figures 5-6 and 5-7 illustrate how single-phase primaries are obtained from three-phase three-wire and three-phase four-wire feeders. These single-phase primaries are tapped to the feeder circuit. By tapping two line wires in the three-wire feeder, the voltage of the primary main will naturally be the same as the voltage across the line wires, 2400 volts in this example. By tapping from line wire to neutral in the four-wire feeder, the voltage of the primary main will also be 2400 volts in this example.

An attempt is always made to keep the loads on the single-phase primaries connected to the three phases of the three-phase line balanced so that the currents in the wires of the three-phase feeder will also be balanced.

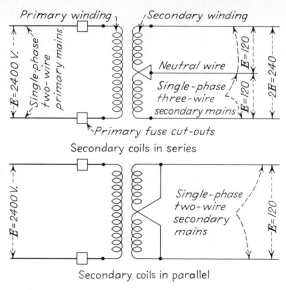

Secondary coils in series

Secondary coils in parallel

Fig. 5-8 Standard connections of common step-down single-phase distribution transformer.

Single-Phase Secondary Circuits Secondary mains, for residential customers, are generally single-phase three-wire circuits, although they may be only two-wire circuits. The secondary mains originate from the secondary winding of the distribution transformer. Figure 5-8 shows the single-phase two-wire primary main connecting to the primary winding of a single-phase transformer. The secondary winding of a single-phase distribution transformer is nearly always made up in two coils which can be connected either in series or in parallel, as shown. They are most often connected in series. Each coil has induced in it 120 volts so that when they are connected in series the voltage across the outside wires form a 240-volt single-phase two-wire circuit. When the neutral wire is used, either outside wire and the neutral form a 120-volt single-phase circuit. A consumer using lamps and small household appliances may have a two-wire service connected to the premises as the load would be small and would be at 120 volts. If the consumer uses an electric range for cooking, the service connection would consist of all three wires, because most electric ranges operate on 120 and 240 volts. Any consumer having a single-phase motor larger than ½ hp would likely operate it from the two outside wires of the secondary mains. Using higher voltage draws less current and keeps the line loss and the voltage drop at a minimum. Any office building or store using single-phase service would also take the three-wire service.

Three-Phase Secondary Circuits Commercial and light industrial customers usually require three-phase service to operate three-phase motors. Three-phase motors are more efficient and less expensive than single-phase motors. Large single-phase motors cause a sudden drop in the voltage when they start which appears as a voltage flicker that affects incandescent lights and distorts television pictures. Three-phase motors draw less current than single-phase motors when they start, thus minimizing voltage flicker. Installation of three-phase service helps to keep the current in each phase wire of the distribution circuit equal. The voltages normally used for three-phase secondary circuits are 208Y/120 volts four-wire, 240 volts three-wire, 480Y/277 volts four-wire, and 480 volts three-wire.

Figure 5-9 illustrates the proper connections for single-phase, pole-type transformers to obtain 208Y/120-volts service for a three-phase customer from a four-wire primary feeder circuit. The primary and the secondary transformer neutral connections should be connected to the distribution feeder circuit neutral and to ground. Transformer connections for a 480Y/277-volt three-phase four-wire service would be similar to those illustrated in Fig. 5-9.

Three-phase four-wire secondary voltages of 208Y/120 or 480Y/277 volts can be obtained by proper selection of pole-type transformers installed as shown in Fig. 5-10. Armless-type construction is used for the three-phase four-wire primary feeder circuit to enhance the appearance of the facilities. Using single primary bushing transformers with one side of the transformer primary winding connected to the transformer tank simplifies the installation and minimizes the cost. The tank of each transformer is connected to the primary neutral wire and ground.

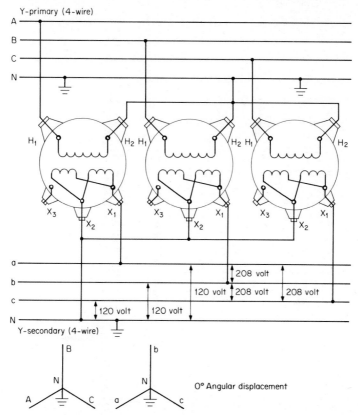

Fig. 5-9 Three-phase Y–Y pole-type transformer connections used to obtain 208Y/120 volts three-phase from a three-phase four-wire primary. If single-primary bushing transformers are used, the H_2 terminal of the transformer primary winding would be connected to the tank of the transformer.

Figure 5-11 illustrates the proper connections for single-phase pole-type transformers to obtain 240/120-volt service for a three-phase customer from a three-wire primary feeder circuit. The transformer providing the source for the single-phase load carries two thirds of the 240/120-volt single-phase load and one third of the 240-volt three-phase load. The other two transformers carry one third of the single-phase and three-phase loads. The transformers must have matched voltage ratios and impedances. Only one of the transformer's secondary bushings can be grounded. Grounding more than one transformer secondary bushing will cause a short circuit.

Figure 5-12 pictures a three-phase delta-delta transformer installation providing three-phase three-wire service. It is good practice to ground one of the transformer secondary bushings as illustrated in Fig. 5-11, even if single-phase service is not desired to limit the secondary voltage to ground.

Neutral Conductor The neutral conductor in the three-phase four-wire system and the neutral conductor in the single-phase three-wire system are always grounded, that is, connected to a low-resistance earth terminal. This is often accomplished by making a firm connection with water mains embedded in the ground. When this is not sufficient or available, metallic grids or plates are embedded in moist earth and an electrical connection is made with them. Frequently metallic rods driven deep enough to be in contact with moist earth are used.

In the case of the three-phase four-wire circuit the neutral of the Y-connected secondary coils of the transformer is grounded. When only one ground connection is made, the neutral conductor is called "unigrounded." General practice is to ground the neutral conductor at numerous points along its length; it is then called "multigrounded." When both primary and secondary circuits are

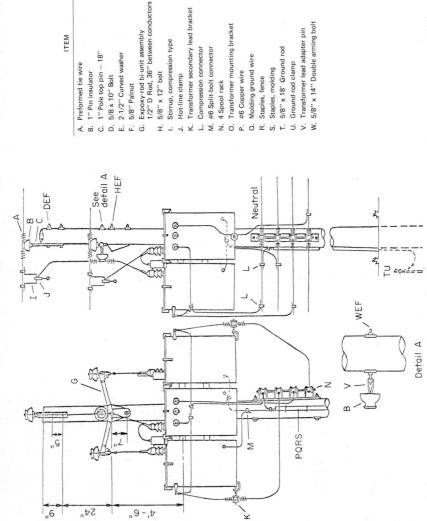

ITEM	No. required
A. Preformed tie wire	7
B. 1'' Pin insulator	4
C. 1'' Pole top pin – 18''	1
D. 5/8 x 10'' Bolt	2
E. 2 1/2'' Curved washer	8
F. 5/8'' Palnut	8
G. Expoxy-rod bi-unit assembly 1/2'' D Rod, 36'' between conductors	1
H. 5/8'' x 12'' bolt	4
I. Stirrup, compression type	3
J. Hot-line clamp	3
K. Transformer secondary lead bracket	2
L. Compression connector	5
M. #6 Split-bolt connector	3
N. 4 Spool rack	1
O. Transformer mounting bracket	1
P. #6 Copper wire	5#
Q. Molding ground wire	30'
R. Staples, fence	10
S. Staples, molding	12
T. 5/8'' x 18' Ground rod	1
U. Ground-rod clamp	1
V. Transformer lead adapter pin	1
W. 5/8'' x 14'' Double arming bolt	1

Fig. 5-10 Three-phase Y–Y pole-type single-primary bushing transformer installation details showing connections to provide three-phase four-wire low-voltage service from a three-phase four-wire primary.

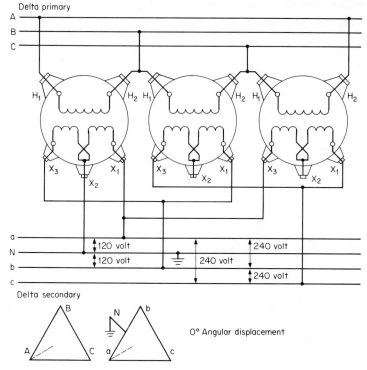

Fig. 5-11 Three-phase delta-delta pole-type transformer connections used to obtain 240/120 volts three-phase from a three-phase three-wire primary. The transformer with the secondary connection to bushing X 2 provides a source for single-phase 120/240-volt service.

Fig. 5-12 Three single-phase transformers connected to a 13,800-volt three-phase three-wire primary circuit to provide three-phase three-wire 480-volt service to a small manufacturing plant. Installation includes fused cutouts and lightning arresters to protect the installation. Current transformers and potential transformers installed above the distribution transformers provide currents and voltages for primary metering.

Y-connected four-wire circuits and both circuits run along the same route, a single neutral conductor can serve both circuits. Such an arrangement is called a "common neutral" system.

In the case of the single-phase secondary, either two- or three-wire, the neutral conductor is grounded at numerous poles. In addition, each service is grounded at the consumer's entrance. The reason for grounding these circuits is primarily for safety.

Distribution Environmental Considerations   The compatibility of distribution systems with the environments in which they are located has become an increasingly important consideration. The location and appearance of both substations and overhead lines have become primary criteria for the planning and design of distribution systems. A strong trend to use underground rather than overhead construction for distribution lines has developed.

Most of the existing distribution lines in the country are overhead, and the extent of these lines has been increasing annually. This increase will continue through 1990, but at a declining rate because of the trend to underground construction for new lines.

Techniques have been developed for improving the appearance of overhead lines by the selection of materials, structural shapes, and colors which are in harmony with the environment. The improved overhead lines should meet appearance objectives in many situations (Fig. 5-13).

Fig. 5-13 Three-phase four-wire primary feeder circuit armless-type construction to enhance the appearance of the distribution circuit. *(Courtesy A. B. Chance Co.)*

The use of underground construction for new distribution lines is increasing rapidly. It is estimated that in 1968 about 20 percent of new lines built in the country were underground. That figure is expected to increase to about 90 percent by 1990.

Underground distribution lines cost more than overhead. Since 1970, the cost difference has been much reduced by the development of new materials and installation techniques, particularly for lines to serve new residential subdivisions. It is expected that this cost differential will be further reduced, but it will not be eliminated in the foreseeable future.

Conversions of existing overhead distribution lines to underground is much more costly than is the underground construction of new lines. The investment required for a general conversion program to eliminate most of the existing overhead distribution lines by 1990 is estimated to be from 170 to 200 billion dollars.

The adverse visual impact of the existing overhead lines can be reduced to a significant extent by selective conversion programs of limited magnitude. The optimum environmental improvement per dollar expended for conversions can be obtained by selecting overhead lines that have a particularly adverse impact because of congestion of facilities, location, or exposure to public view.

Efforts to make distribution facilities more compatible with their environments by choice of location, consideration for appearance in the design of visible facilities, and the use of underground construction for new lines where economically feasible should be promoted and encouraged by all.

Section **6**

Construction Specifications

Many transmission and distribution lines are built by contracting companies employing linemen and cablemen. A contract is normally completed between the electric utility company and the contracting company to specify the responsibilities for completing the project. Construction specifications are included in the contract. These specifications are used by linemen and cablemen installing the facilities.

Construction Contracts

The contract describes the work to be completed, such as: construct a 161-kV wood-pole H frame transmission line to be owned by the utility company and to extend from Riverside Generating Station of the company in Pleasant Valley Township, Scott County, Iowa, to the Valley Substation, owned by the company, south of the town of Walcott in Blue Grass Township, Scott County, Iowa.

The scope of the project is described in the contract and would be similar to the following: The Contractor agrees to provide all necessary materials, tools, equipment, labor, and supervision, and to pay all the related costs listed in the contract necessary for the proper construction and completion of the transmission line described and specified in "Specifications for the construction of a 161-kV transmission line from Riverside Generating Station to Valley Substation" dated October 24, 1980, which specifications, including all exhibits thereto attached, are made a part of the contract. A hold harmless clause, a schedule for commencing and completing the work, and provisions for payment are usually included in the contract.

Specifications

General specifications would normally consist of the following:
Definitions
Items furnished by contractor
Materials furnished by others
Changed conditions
Changes in work
Inspection and testing of work
Correction of work
Guarantee
Contractor's right to terminate contract
Assignment
Separate contracts
Connection to work of others
Progress of work

Losses to be borne by contractor
Delays
Indemnity against loss of livestock and crop and fence damage
Indemnity against claims by subcontractors or miscellaneous third parties
Performance by owner of contractor's work
Owner's right to terminate contract
Subcontracts
Performance bond
Liens
Patents and royalties
Government labor regulations
Indemnity against loss due to damage or accidents
Labor
Progress payments
Final payment
Payments and payments withheld
The engineer
Drawings and specifications
Drawings furnished by the contractor
Schedule of work
Overtime
Use of premises
Accidents
Fire prevention
Contractor's liability insurance
Protection of and responsibility for work property
Safety and accident prevention
Right-of-way, easements, and permits
Public regulations, permits, and laws
Cleaning up
Acceptance
Contractor's responsibility

Special Conditions   Such conditions itemized in the specifications would probably consist of these items:

Definitions
Scope
Facilities at site
Standards
Inspection of the site
Surveys
Right-of-way
Cooperation with other contractors
Access to the work
Installing gates and maintaining fences
Underground structures
Damages
Loss or damage to livestock
Arrangement with owners of adjacent property
Roads
Use of explosives
Communication equipment
Warning to contractor's employees
Safety precautions
Equal opportunity
Insurance
Withholding of state income tax
Plan and profile, route map, and construction drawings
Special provisions
Cost accounting code
Commencement and completion of work

Technical Specifications   Such specifications for land clearing would cover the following:

(Scope of Work)   The work to be performed under the terms of this specification shall consist of clearing the owner's right-of-way for a new 161-kV transmission line. The land to be cleared will be from the Riverside Generating Station in Pleasant Valley Township, Scott County, Iowa, to the Valley Substation, south of the town of Walcott in Blue Grass Township, Scott County, Iowa. The width of the right-of-way is 150 feet.

(Conditions for Clearing)   The contractor shall clear the right-of-way of all obstructions that will interfere with the operation of the electric transmission line. The clearing to be performed, as a part of these specifications, shall be done in strict accordance with the contract documents. The actual work involved consists of clearing a strip of land 75 feet wide on each side of the centerline of the transmission line, by cutting and/or trimming of all trees and brush within the right-of-way limits. The work includes the removal or trimming of trees which the engineer may classify as "danger trees."

(Danger Trees)   The contractor must have written permission from the engineer before any danger trees are cut. The trees to be cut shall be marked with a distinctive blaze, by personnel furnished by the contractor, under the supervision of the engineer.

Clearing in developed areas, such as those planted for orchards or other shade, fruit, or ornamental trees, shall be conducted in accordance with directions given by the engineer. The contractor shall promptly notify the engineer whenever any objections to the trimming or felling of any trees or to the performance of any other work on the land are made by the landowner.

(Clearing)   All trees, brush, stumps, and other inflammable material, except grass and weeds, shall be removed unless the contractor is otherwise instructed by the engineer. All trees and brush shall be cut 3 inches or less from the ground line (measured on the uphill side). All stumps must be cut so that passage of trucks and tractors will not be hindered.

All trees that are located within highway rights-of-way and are to be removed shall be cut at the ground line or as near to the ground line as possible, in accordance with highway commission practices. The extent of clearing of brush, hedges, and the like within 150 feet of any public road shall be determined by the engineer. Wherever the right-of-way crosses orchards, parks, or indicated special areas, all clearing shall be done specifically as directed by the engineer.

(Disposal)   The contractor shall dispose of all trees and brush by either chipping and spreading or hauling away unless otherwise specified in the special provisions or as otherwise directed by the engineer. Disposal of any limbs or brush shall be conducted in conformance with state and local laws and applicable regulations and shall be accomplished in such a manner as not to create a hazard or a nuisance. The contractor shall obtain all necessary permits. No damage of any nature shall be inflicted by the contractor upon adjoining property by unwarranted entry or disposal upon said property. All logs and brush shall be the property of the contractor unless otherwise specified in the "special provisions."

(Stump Removal)   Any stumps that interfere with the installation of the structure shall be completely removed and disposed of by the contractor at no extra cost to the owner.

(Seeding Right-of-Way)   In wooded areas where the contractor may deem it necessary to bulldoze a road for construction purposes or in uncultivated areas where the ground is denuded of vegetation due to construction operations, the denuded areas shall be seeded with a mixture of two pounds of Alsike clover and six pounds of White Dutch clover per acre. The seed shall be furnished by the contractor and shall be well raked in to assure germination. Areas where germination fails shall be reseeded at the contractor's expense.

(Maintaining Fences during Clearing)   Where the removal of trees requires the removal of fencing along the owner's right-of-way, the contractor shall take steps to provide temporary fencing as directed by the engineer. Such temporary fencing shall remain the property of the contractor and shall be removed when permanent fencing is installed or when directed by the engineer. During disposal and/or removal of trees, the contractor shall keep the fences in good repair where the work has affected their condition and shall keep all gates and temporary openings closed except when in actual use. The contractor shall be responsible for any damage caused by failure to keep fences in good repair or gates closed. On completion of the work, the contractor shall restore to their previous condition all fences that have been damaged during clearing operations. This includes the removal of gates installed for the contractor's use if the landowner wishes those gates removed.

(Treatment)   Not less than 2 weeks before cutting, all trees and brush shall be given an adequate basal spray with a chemical that meets with the approval of the engineer. Treatment prior to cutting is required to allow absorption of the chemical by the tree and roots in order to prevent regrowth. This chemical shall be furnished by the contractor.

The basal spray shall saturate the bottom 2 feet of the tree or brush. It is important that the entire circumference be treated. Also, it is important that the ground line and all exposed roots be thoroughly saturated. Some species, such as box elder, willow, and osage orange, are very susceptible to chemical treatment while others, particularly ash and linden, are tolerant. Because of the difference of susceptibility, some species will require heavier applications than others to gain good kill. Among these are ash, linden, maple, hickory, and oaks.

In the event that the owner is not able to secure the right-of-way to do the work in time to allow the contractor to basal spray all trees and brush prior to cutting, all stumps shall be saturated thoroughly with chemical, especially at the ground line and on all exposed roots. Spraying after cutting may be done only when approved by the engineer. Care shall be used to prevent any chemical from contacting any foilage or farm land off the owner's right-of-way. Chemical spraying shall be prohibited on any state or federal conservation lands. Such lands are marked on the plan and profile sheets.

(Staking of Right-of-Way)   The owner will stake the right-of-way limits in advance of the contractor's clearing operations. If for any reason the stakes are removed or lost, the contractor shall inform the engineer who will arrange to have the stakes reset.

Engineering Specifications   Such specifications would include the following:

(Local Conditions)   The contractor must check all local conditions affecting the work and shall make a thorough examination of the route of the line, plans, specifications, and premises in order to be entirely familiar with the details and construction of the installation.

(Workmanship)   All work shall be executed in a neat and skillful manner as specified or detailed in these specifications and/or drawings as listed herein and in accordance with best construction practice.

(Drawings)   The character of the line, location of line, and details of various structures are shown on drawings listed herein. These plans and drawings are intended to be complete and final and shall be followed as closely as possible. Dimensions as shown on drawings are not guaranteed and contractor should check their accuracy before proceeding with the work.

(Pole Locations)   Stakes showing location of each pole will be set by the utility company. These stakes will be marked with structure number as shown on drawings. The size of poles required, class, height and other data for each structure number will be shown on supplementary tabulation sheets to be furnished to the contractor by the company.

(Depth of Pole Setting)   All poles to meet American National Standards Institute specifications: western red cedar, "Pentrex" butt treated with full-length life-span treatment or full-length creosote-pressure-treated fir, and shall be set in accordance with the following table except where directed otherwise in writing by the engineer:

Depth of setting

Height of pole, ft	In earth	In rock
50	7'0"	5'0"
55	7'6"	5'6"
60	8'0"	6'0"
65	8'6"	6'6"
70	9'0"	7'0"
75	9'6"	7'6"

Poles shall be set to stand perpendicular and in exact alignment—when the line is completed unless otherwise specifically called for. Pole-setting depths shall have a tolerance of not more than ±3 inches except as otherwise directed in writing by the engineer. Poles set partly in earth and partly in rock shall be set to the depth shown for earth.

At locations where the poles of any one structure are located at different ground elevations the bottoms of the poles shall be leveled by deepening the hole on the higher ground, the hole on the lower ground having the depth shown above. In no case shall any pole be set to any depth such that the top of the Pentrex treatment is less than 9 inches above ground level at the pole. In special cases it may be necessary for the contractor to excavate the shelf as indicated on the drawings in order to meet the above requirements. In special cases a longer pole may be required on the downhill side which shall be set at least as deep as shown in the table.

(Pole Selection)   The contractor shall select heavier poles for longer spans and angle points. The supplementary tabulation shows the class pole to be used for each structure. In addition, on this type of construction, the contractor shall match, as nearly as possible, the ground line and top dimensions of both poles in each structure. If possible, circumference at ground line of the two poles in each fixture shall not differ more than two (2) inches. A contractor who finds it impossible, because of variation in sizes of the poles received, to meet this requirement shall secure the written approval of the engineer before setting the poles.

(Pole Setting)   Holes shall be dug of sufficient size to permit free insertion of tamping bar on all sides of poles after poles are set. In backfilling, tampers shall continually tamp in earth until the hole is completely filled. No earth shall be added until that already in place is solid and tight. After hole is filled, excess earth shall be piled up and packed firmly around pole.

Line poles shall be set to stand perpendicular when line is completed. Poles at corners and angles shall be set with a rake of approximately one-fourth (¼) inch for each foot of height of pole above ground. No change from this specification shall be made except upon the written direction of the engineer. H-type fixtures are all double armed. Poles shall be set with gains at right angles to line, as no reaming or extra drilling of holes will be permitted to get arm bolts inserted. Poles shall be set not over two (2) inches more or less than the specified distance apart.

(Installation of Arms and Braces)   Arms on the two-pole fixtures are to be completely assembled. Poles are to be gained with bolt holes drilled so that the assembled double arm may be spread and slipped over top of pole and bolted in place. This may be done, at contractor's option, either before or after poles are set. Poles are to be drilled for shield-wire ridge iron, which the contractor will bolt in place. Arms must be level after installation. This leveling must be done by leveling the bottom of pole holes. It will not be permissible to level arms by drilling additional holes in poles. Poles are to be furnished with top X-brace bolt holes drilled. The X braces shall be installed so as to maintain the vertical alignment of the structures without causing undue permanent stresses in either poles or X braces. In some cases, the company may require the contractor to bore and gain certain poles, in which case contractor will be allowed a unit price shown in the quotation. All field gains and holes shall have brush or spray preservative treatment.

(Special Structures)   Poles for three-pole or special structures are to be furnished roofed only, with the contractor to do all boring and gaining. Structures are to be erected in accordance with drawings as furnished for each structure. All field gains and holes shall be brush or spray treated with preservative.

(Anchors and Guys)   Creosoted-oak railroad-tie anchors are to be used for all heavy guying. Expanding anchors are to be used only where specifically indicated on drawings or where approved by the engineer. Sizes of anchor rods, ties, expanding anchors, and guy wire are shown on drawings. Tie anchors shall be buried to the specified depth, anchor rods trenched out to lie in direction of strain, and holes backfilled and securely tamped. Thimble-eye guy rods will be furnished. At top end of guy, guy wire is to be bent around guy thimble of proper size. Guys shall be installed before wire is pulled, and pulled up to sufficient tension to pull pole over slightly. After wire is pulled, where two or more guys are used on one pole, they shall be inspected and equalized if necessary to see that each guy takes its proper portion of the strain. Tu-base-type two (2) bolt guy clamps shall be used for holding wire, and the loose end of the guy wire shall be fastened with a guy wire clip. Eight (8) foot guy guards will be installed on all anchor guys. The following numbers of guy clamps are to be used:

⅜″ EHS guy	3 clamps
½″ EHS guy	4 clamps

(Insulators)   Tangent line structures will be insulated with a string of ten 15,000# 10-inch ball-and-socket disk insulators, as shown by drawings; dead-end structures with a string of twelve 25,000# 10-inch clevis-type disk insulators. All cotter keys shall be bent sufficiently to prevent loss by vibration. All insulators shall be carefully handled to avoid chipping or breaking, and no chipped, cracked, or broken insulators shall be assembled on the line. It shall be the contractor's responsibility to inspect all insulators as taken from their crates and to replace all defective units.

(Splices)   Splices or joints may be made in the conductors and shield wires where necessary, except at crossings over railroads, important highways, and communication lines, and shall be at least 50 feet from the nearest pole or structure. All splices and dead ends shall be made up in strict accordance with the drawings and instructions furnished by the company, using steel-core aluminum sleeves and clamps and hydraulic press. Particular care shall be taken by the contractor in making up all splices or joints in order to see that the proper procedure is followed. The company reserves the right of inspecting any or all joints and rejecting those that do not meet with the approval of the engineer. All joints shall be made by one foreman and a crew especially trained for this work.

(Stringing Conductors and Shield Wires)   Conductors and shield wires shall be handled with care and must not be trampled upon or driven over. They shall be continuously inspected during installation, and any cuts, kinks, and other injured portions shall be cut out and the wire spliced,

said splices to be made strictly in accordance with the company's specifications and drawings. Conductors and shield wires may be strung out with reels in a stationary position with conductors supported on the structures in free running stringing blocks or paying off from reels towed along the line. Conductors or shield wires shall not be dragged out on the ground or dragged over the arms of the structures or over fences or foreign objects which may damage conductors. Each reel shall be equipped with a suitable braking device to keep the conductors or shield wires always under tension and clear of the ground.

(Sagging Conductors and Shield Wires)   The power-line conductors and the overhead shield wires shall be strung in accordance with sag and tension charts provided by the company. These sag charts are initial stringing sag charts, and no conductors or overhead shield wire shall be prestressed beyond that required by customary practice. The air temperature at time of sagging shall be determined with a certified etched glass thermometer. All conductor grips, come-alongs, stringing blocks, and pulley wheels shall be of a design which will prevent damaging the conductors or shield wires by kinking, scouring, or unduly bending them during stringing operations.

Sags shall be accurately measured in several spans approximately equal to the ruling span during stringing and checked against the sag charts for the prevailing air temperature. Sagging shall not be done during periods of high wind which, in the opinion of the engineer, might prevent accurate sag measurements. A tolerance of plus or minus one-half inch ($\frac{1}{2}''$) sag per 100 feet of span length, based on the sag and tension charts, will be permitted provided that all conductors in the same span assume the same sag and satisfactory clearances are obtained. The engineer will compute and check the sag at all points to be checked, and the contractor shall furnish the necessary crew for signaling and climbing purposes.

The methods for checking sag and the points at which the checks are to be made shall be agreed upon between the engineer and the contractor, the intent of these specifications being that the engineer shall be assured, by means of a sufficient and reasonable number of checks, that proper clearances are obtained at all points, that the proper tensions are being obtained, and that the general appearance of the line shall be satisfactory. The contractor shall make adequate provision to prevent pulling any structure out of line during the stringing and sagging operations.

(Dead Ends)   All dead ends of the power conductors will be made with Aluminum Company of America three-piece dead-end clamps. These three pieces are Catalog No. 1744.2 aluminum compression dead-end body, Catalog No. 772.1 compression steel clevis end, and Catalog No. A1763.2 aluminum compression jumper terminal. The shield wires will be dead ended by means of guy thimble or roller and three Tu-base clamps as shown on the drawings.

(Wire Attachment)   After sagging and dead ending conductors and shield wire, the contractor shall straighten any poles that may have been pulled out of plumb in the stringing and sagging operation. Conductors and shield wires shall then be securely clamped in their proper suspension clamps.

(Pole Grounding)   All H frame poles will have a No. 4 bare copper ground wire installed as shown on the drawing attached as an exhibit. Contractor shall securely staple 15 feet of No. 4 bare copper wire forming a flat spiral on the butts of all above poles, the free end to be stapled along the ground section and extending up 12 inches above the ground line. A solderless clamp connector shall be installed 6 inches below ground level as shown on the drawings, connecting the butt section of the ground wire with the down lead from the pole top. The down lead shall be connected, as shown on the drawings, to a $\frac{5}{8}$-inch $\times$ 12-foot Copperweld ground rod, and the two ground rods of each structure shall be connected by a No. 4 copper wire buried not less than 12 inches below ground. Eight feet of creosoted-wood molding shall be installed to protect the ground wire and shall extend 6 inches below grade. Ground molding shall be stapled to pole with four staples.

Exhibits   Exhibits attached to the specifications would include the following:

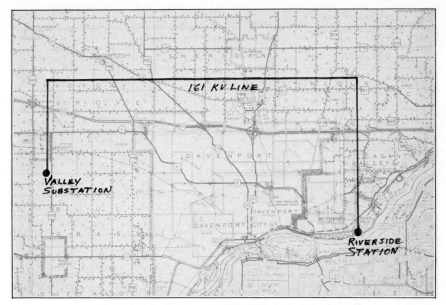

Fig. 6-1   Map showing route of 161 kV Riverside–Valley Line.

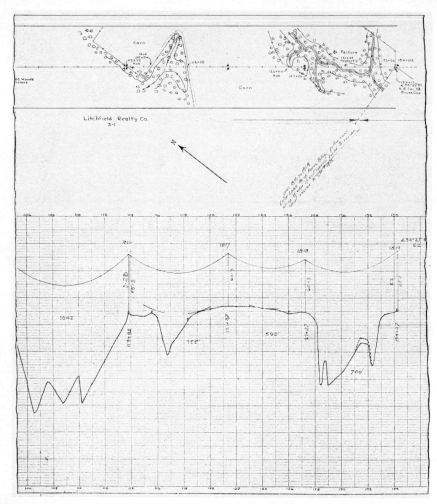

Fig. 6-2 Plan and profile sheets, 161 kV Riverside–Valley Line.

Item	No. required
A. Shield wire support (steel cross angle) with U-bolt, chain link and bonding bolt	1
B. Double crossarm assembly with 11" spacer fitting	1
C. Vee brace 9' 11 1/2" C-C hole 37 1/2° and 52 1/2° fittings	2
D. X-brace	2
E. Center clamp (for X-brace)	1
F. 7/8" x 12" Machine bolt	2
Curved washer, 15/16" hole	2
Standard nut	2
MF locknut	2
G. 7/8" x 12" Machine bolt	2
Curved washer, 15/16" hole	2
Standard nut	2
MF locknut	2
H. 7/8" x 14" Machine bolt	2
7/8" x 16" Machine bolt	4
Curved washer, 15/16" hole	4
Standard nut	4
MF locknut	4
I. 7/8" x 23" Threaded rod	2
Washer nut	4
Standard nut	4
J. Pole shim, 1" or 1 1/2" thick, as required	4

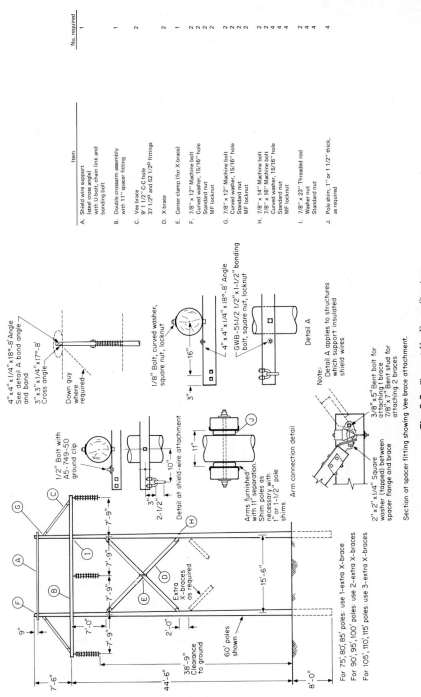

Fig. 6-3   Tangent H—Frame Structure.

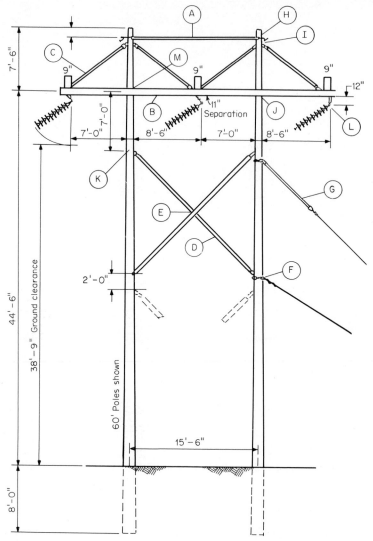

Fig. 6-4 Angle structure (½ to 6°).

Item	No. required
A. Shield wire support (steel cross angle) with U-bolt, chain link, and bonding bolt	1
B. Double crossarm assembly with 11'' spacer fitting	1
C. Vee brace 9' – 1 1/2'' C-C hole 37 1/2º and 52 1/2º fittings	4
D. X-brace	2
E. Center clamp (for X-brace)	1
F. Pole band	1 Set (4 Sections) 1 Set (4 Sections)
Stud bolt, 7/8'' x 8''	8
Links	2 Sets
Guy roller, 15/16'' hole	1
7/8'' x 3'' Machine bolt/MF locknut	1
G. Fiberglass strain insulator	1
H. 7/8'' x 12'' Machine bolt	2
Curved washer, 15/16'' hole	2
Standard nut	2
MF locknut	2
I. 7/8'' x 12'' Machine bolt	2
Standard nut	2
MF locknut	2
J. 7/8'' x 23'' Threaded rod	2
Washer nut	4
Standard nut	4
K. 7/8'' x 14'' Machine bolt	2
7/8'' x 16'' Machine bolt	2
Curved washer, 15/16'' hole	4
Standard nut	4
MF locknut	4
L. Swinging angle bracket	3
M. Pole shim, 1'' or 1 1/2'' thick, as required	4

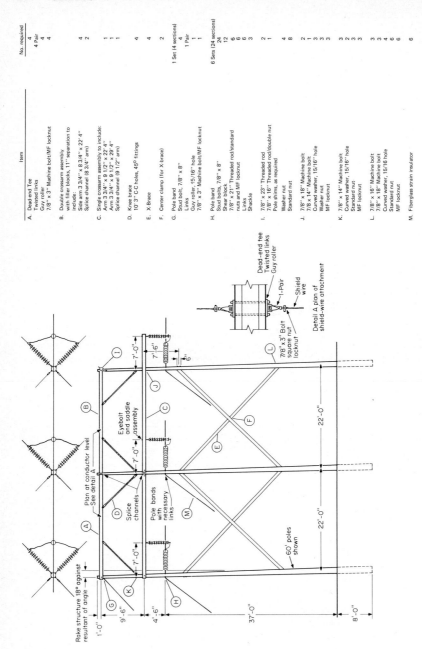

Item	No. required
A. Dead-end Tee	4
Twisted links	4 Pair
Guy roller	4
7/8" x 3" Machine bolt/MF locknut	
B. Double crossarm assembly	4
with filler blocks, 11" separation to include:	
Side arm 3 3/4" x 8 3/4" x 22' 4"	4
Splice channel (8 3/4" arm)	2
C. Single crossarm assembly to include:	1
Arm 3 3/4" x 9 1/2" x 22' 4"	1
Arm 3 3/4" x 9 1/2" x 29' 4"	1
Splice channel (9 1/2" arm)	1
D. Knee brace	4
10' 3" C-C holes, 45° fittings	
E. X-Brace	4
F. Center clamp (for X-brace)	2
G. Pole band	1 Set (4 sections)
Stud bolt, 7/8" x 8"	4
Links	1 Pair
Guy roller, 15/16" hole	1
7/8" x 3" Machine bolt/MF locknut	1
H. Pole band	6 Sets (24 sections)
Stud bolts, 7/8" x 8"	24
Shear block	12
7/8" x 21" Threaded rod/standard	6
nuts and MF locknut	6
Links	6
Shackle	3
I. 7/8" x 23" Threaded rod	2
7/8" x 16" Threaded rod/double nut	1
Pole shims, as required	
Washer nut	4
Standard nut	8
J. 7/8" x 18" Machine bolt	2
7/8 x 14" Machine bolt	1
Curved washer, 15/16" hole	3
Washer nut	3
MF locknut	3
K. 7/8" x 14" Machine bolt	3
Curved washer, 15/16" hole	3
Standard nut	2
MF locknut	3
L. 7/8" x 16" Machine bolt	3
7/8" x 18" Machine bolt	3
Curved washer, 15/16 hole	4
Standard nut	6
MF locknut	6
M. Fiberglass strain insulator	6

Fig. 6-5 Three pole, angle and dead-end structure. (60 to 90°).

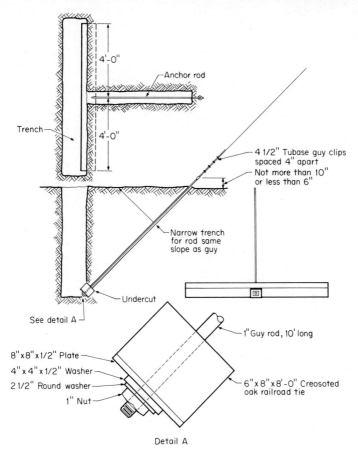

Fig. 6-6 Anchor installation details.

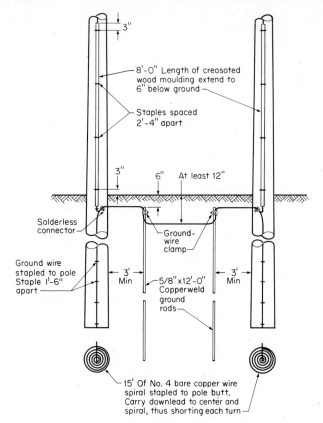

Fig. 6-7 Pole grounding installation details.

Wood-Pole Structures

Structures supporting electric lines must be designed to support the conductors, insulators, and shield wires in a manner that provides adequate electrical clearances and ample mechanical strength. Safe clearances must be maintained when the conductor temperature is elevated as a result of large currents flowing on a circuit and when the conductors are ice coated and strong winds are blowing. Common types of structures include wood poles, reinforced concrete poles, steel poles, steel towers, and aluminum towers. The total cost of the installed facilities determines the type of structure used. Lines built with wood-pole structures have generally proved to be the most economical where adequate physical and electrical requirements could be satisfied.

Wood-Pole Types The type of wood poles used in different parts of the United States is largely determined by the species of trees available in the locality. In the central United States, northern white cedar has been used because it is available in Minnesota, Wisconsin, and Michigan. This species is easily distinguished by the large butt and small top. On the Pacific coast, western red cedar is used because it grows in Washington, Oregon, and Idaho. A red cedar pole has very little taper, the butt being only a little larger than the top. In the eastern states, some chestnut is used because it grows in the Appalachian Mountains. The majority of poles used in the east are creosoted yellow pine poles. In the south, cypress and yellow pine are used. Figure 7-1 shows the location of the various species of timber used for poles in the United States.

Most of the poles set today undergo preservative treatment against butt rot before being set, thereby greatly increasing the life of the pole. It is claimed that proper treatment will almost double the life of a pole. Common preservatives are creosote oil and pentachlorophenol. Wood preservative treatment is specified in American Wood-Preserver's Association Standard CI-74, "All Timber Products—Preservative Treatment by Pressure Processes."

Pole Classification Wood poles are commonly classified by length, top circumference, and circumference measured 6 ft from the butt end. The lengths vary in 5-ft steps, and the circumference at the top varies in 2-in steps. Thus we have lengths, in feet, of 25, 30, 35, 40, etc., and minimum top circumferences, in inches, of 15, 17, 19, 21, etc.

The circumference measured 6 ft from the butt end determines to which class numbered from 1 to 10 a pole of a given length and top circumference belongs. The classification from 1 to 10 determines the strength to resist loads applied 2 ft from the top as given in Table 7-1.

American National Standard ANSI 05.1-1972, entitled "Specifications and Dimensions for Wood Poles," provides technical data for wood utility poles. The circumference of western red cedar, southern yellow pine, Douglas fir, and western larch poles at the top and at the ground line for various classes and lengths are given in Table 7-2. The weights of commonly used wood poles are given in Table 7-3.

Vertical pole loading capacities are shown in Table 7-4, which shows the maximum weight that may be installed on poles of different heights and classes. Weights were calculated for three-cluster or crossarm mounted transformers. Weights in the table would require adjustment for other material mounting configurations. Values in table are based on kiln-dried poles with fiber stresses

Table 7-1   Load Resistance 2 Ft from Top of Classified Poles

Class	Load pole must be able to withstand, lb force
1	4500
2	3700
3	3000
4	2400
5	1900
6	1500
7	1200
8 to 10	Not specified

Table 7-2   Wood Pole Classifications, Dimensions, and Breaking Strengths

Class			1H			1			2	
M.H.B.L.*			5400			4500			3700	
Minimum circumference at top, in			29			27			25	
		Minimum circumference (inches at ground line) †								
Length (ft)	Ground line distance from butt (ft)	WC	SP DF	WL	WC	SP DF	WL	WC	SP DF	WL
25	5	—	—	—	37.4	33.8	33.2	34.9	31.7	31.2
30	5.5	—	—	—	40.2	36.6	35.6	37.7	34.1	33.6
35	6	45.5	41.5	40.5	42.5	39.0	38.0	40.0	36.5	35.5
40	6	48.0	43.5	43.0	45.0	41.0	40.0	42.5	38.5	37.5
45	6.5	50.3	45.4	44.9	47.3	42.9	41.9	44.3	40.4	39.4
50	7	52.1	47.2	46.8	49.1	44.7	43.8	46.1	41.8	40.8
55	7.5	53.9	49.1	48.2	50.9	46.1	45.2	47.9	43.1	42.2
60	8	55.7	50.5	49.6	52.7	47.5	46.6	49.7	44.5	43.6
65	8.5	57.5	51.9	51.5	54.0	48.9	48.0	50.5	45.9	45.5
70	9	59.0	53.2	52.9	55.4	50.2	49.4	51.9	47.2	46.4
75	9.5	60.5	54.6	53.8	56.7	51.6	50.8	53.2	48.1	47.3
80	10	61.5	56.0	55.2	58.0	53.0	51.7	54.5	49.5	48.7
85	10.5	62.8	57.4	56.6	59.3	53.9	53.1	55.3	50.4	49.6
90	11	64.1	58.2	57.5	60.6	54.7	54.0	56.6	51.7	50.5
95	11	64.6	59.7	59.0	61.6	55.7	55.5	57.6	52.7	52.0
100	11	67.1	60.7	60.0	63.1	57.2	56.5	59.1	53.7	53.0
105	12	67.7	61.5	60.7	63.7	58.0	57.2	59.7	54.5	53.7
110	12	69.2	63.0	61.7	65.2	59.0	58.2	60.7	55.5	54.7

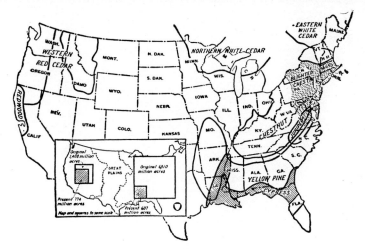

Fig. 7-1 Distribution of pole timber in the United States.

	3			4			5			6			7	
	3000			2400			1900			1500			1200	
	23			21			19			17			15	
WC	SP DF	WL	WC	SP DF	WL	WC	SP DF	WL	WC	SP DF	WL	WC	SP DF	WL
32.9	29.7	29.2	30.4	27.7	26.7	28.4	25.7	24.7	25.9	23.2	23.2	24.4	21.7	21.2
35.2	32.1	31.1	32.7	29.6	29.1	30.2	27.6	26.6	28.2	25.1	24.6	26.2	23.6	23.1
37.5	34.0	33.0	34.5	31.5	31.0	32.0	29.0	28.5	30.0	27.0	26.5	27.5	25.0	24.5
39.5	36.0	35.0	36.5	33.5	32.5	34.0	31.0	30.0	31.5	28.5	28.0			
41.3	37.4	36.9	38.3	34.9	33.9	35.8	32.4	31.4	32.8	29.9	28.9			
43.1	38.8	38.3	39.6	36.3	35.3	37.1	33.8	32.8						
44.4	40.1	39.7	41.4	37.6	36.7									
45.7	41.5	40.6	42.7	38.5	38.1									
47.0	42.9	42.0	44.0	39.9	39.0									
48.4	44.8	43.4	44.9	40.7	40.4									
49.7	45.1	44.3												
50.5	46.0	45.2												
51.3	46.9	46.1												
52.6	47.7	47.5												

SOURCE: Commonwealth Edison Co.
°Minimum horizontal breaking load applied 2 ft below top of the pole.
†This circumference at ground line for measurement in the field. Legend: WC = western red cedar; SP = southern yellow pine; DF = douglas fir; WL = western larch.

Table 7-3 Weights of Wood Poles That Are Used Extensively

Length of pole (ft)	Species*	Class				
		1H	1	2	4	5
		Average weight, lb per pole				
30	WC	1320	880	750	540	440
	DF	1760	1110	930	690	600
	SP	1884	1279	1082	784	660
	WL	1515	955	800	595	515
35	WC	1585	1055	880	660	570
	DF	2070	1435	1260	875	770
	SP	2223	1568	1343	1004	862
	WL	1780	1235	1085	750	660
40	WC	1760	1320	1145	790	705
	DF	2500	1760	1560	1120	920
	SP	2585	1884	1623	1219	1059
	WL	2150	1515	1340	965	790
45	WC	2025	1585	1365	1010	880
	DF	2860	2070	1845	1350	1125
	SP	2993	2223	1911	1444	1274
	WL	2460	1780	1585	1160	965
50	WC	2290	1760	1585	1230	1145
	DF	3360	2500	2150	1600	1300
	SP	3451	2585	2214	1687	1494
	WL	2890	2150	1850	1375	1120
55	WC	2815	2025	1760	1410	1410
	DF	3835	2860	2475	1815	1540
	SP	4015	2993	2567	1934	1719
	WL	3300	2460	2130	1560	1325
60	WC	3170	2290	1935	1670	
	DF	4340	3360	2820	2040	1740
	SP	4620	3451	2943	2186	1953
	WL		2890	2425	1755	
65	WC	3695	2815	2200	1935	
	DF	4800	3835	3250	2340	2015
	SP	5198	4015	3341	2457	2237
	WL		3300	2795	2010	
70	WC	4400	3170	2640	2290	
	DF	5240	4340	3640	2590	2240
	SP	6400	4620	3781	2732	2488
75	WC	4840	3695	3170	2640	
	DF	5780	4800	4050	2925	
	SP	7200	5198	4235	3021	
80	WC	5810	4400	3695	3080	
	DF	6300	5240	4400	3200	
	SP	8140	6400	5170	3615	
85	WC	6750	4840	3960		
	DF	6935	5780	4930		
	SP		7200	5745		
90	WC	7500	5810	4930		
	DF	7600	6300	5400		
	SP		8140	6405		
95	WC		6750	5950		
	DF		6935	5985		
100	WC		7500	6550		
	DF		7600	6700		

SOURCE: Commonwealth Edison Co.
*WC = western red cedar; DF = Douglas fir; WL = western larch; SP = southern pine.

Table 7-4 Maximum Weight That May Be Installed on Poles

Pole length (ft)	Mounting distance from pole top to uppermost attachment (ft)	Pole class				
		1H	1	2	4	5
		Maximum allowable weight (lb)				
30	1	6,900	4,650	3,500	1,700	730
	4	9,700	6,700	5,200	2,730	1,270
35	1	6,120	3,800	2,900	1,450	655
	4	8,200	5,240	4,000	2,070	970
	5	9,000	5,800	4,480	2,350	1,100
	8	12,000	7,900	6,200	3,360	1,630
40	1	5,100	3,200	2,300	1,100	590
	4	6,600	4,250	3,150	1,570	890
	5	7,120	4,670	3,500	1,770	1,020
	8	9,200	6,200	4,750	2,550	1,520
45	1	4,050	2,560	1,890	835	465
	4	5,200	3,400	2,550	1,170	670
	5	5,700	3,760	2,810	1,310	750
	(1) 8	7,400	4,950	3,780	1,830	1,070
	13	10,150	8,000	6,200	3,200	1,950
	16	14,800	10,600	8,350	4,500	2,820
	17	15,200	11,600	9,120	5,050	3,200
50	1	3,900	2,260	1,340	660	
	4	5,000	2,950	1,800	910	
	5	5,400	3,200	2,000	1,000	
	(1) 8	6,950	4,160	2,650	1,380	
	13	10,400	6,600	4,250	2,350	
	16	13,400	8,600	5,650	3,220	
	17	14,500	9,400	6,200	3,660	
	20	18,400	12,300	8,400	4,900	
55	13	9,000	5,750	3,600		
	16	11,400	7,300	4,700		
	(1) 17	12,300	7,900	5,100		
	20	15,500	10,200	6,700		
60	13	8,200	4,960	2,850		
	16	10,250	6,400	3,750		
	(1) 17	11,100	6,900	4,100		
	20	13,800	8,900	5,300		
65	13	6,700	3,750	2,550		
	16	8,400	4,850	3,300		
	(1) 17	9,080	5,260	3,600		
	20	11,300	6,800	4,700		

SOURCE: Commonwealth Edison Co.

(1) Mounting heights are provided for equipment located as under build on 34.5-kV lines. If higher poles are required for clearance (i.e., railroad x-ings), the preferred mounting heights for equipment is nearer the ground to reduce the stress on the pole.

for wood poles, published by the American National Standards Institute, degraded to 75 percent of values found by standard tests to allow for variation in fiber strengths of poles. The calculations, applying a factor of safety of 5 to 1, provide for two 250-lb linemen and their equipment on the pole and for stress caused by hoisting cables.

Operating Voltages Wood-pole structures are commonly used for most distribution line structures and are frequently used for subtransmission lines operating at voltages of 34.5 through 138 kV and for transmission lines operating at 138 through 230 kV. Some extra high voltage (EHV) transmission lines operating at 345 and 500 kV are constructed with wood-pole structures.

Distribution Structures Distribution circuit structures normally consist of single wood poles using line-post insulators for armless construction and pin-type insulators for crossarms. Armless construction is intended to create a pleasing appearance for overhead electric distribution systems (Fig. 7-2). Construction details and materials required for one type of three-phase 13.2Y/7.6-kV electric distribution line built on wood-pole structures using close space construction to enhance the appearance of the circuit is shown in Fig. 7-3.

Subtransmission Structures   Subtransmission lines operating at voltages of 115 kV and below can be built on single wood-pole structures. A single-pole structure for a three-phase ac circuit operating at 34.5 kV with a 4.16Y/2.4-kV three-phase distribution circuit underbuild is shown in Fig. 7-4.

"Wishbone" type construction of single-pole structures is popular for use on 69- through 138-kV subtransmission lines. This type of construction provides a symmetrical conductor arrangement which has been proved to be economical to build and reliable to operate. Figure 7-5 pictures a single-pole structure with "wishbone" type construction operating at 69 kV. Subtransmission structures have been simplified to make them as inconspicuous as possible to enhance the esthetic appearance and preserve the environment. This type of construction minimizes the width of right-of-way needed and reduces the number of trees that must be removed.

The structures may use light colored wood poles and gray colored post-type insulators (Fig. 7-6). Figure 7-7 is a drawing of a wood-pole tangent structure for a 69-kV three-phase ac circuit using horizontal post-type insulators with a triangular conductor configuration for installation in

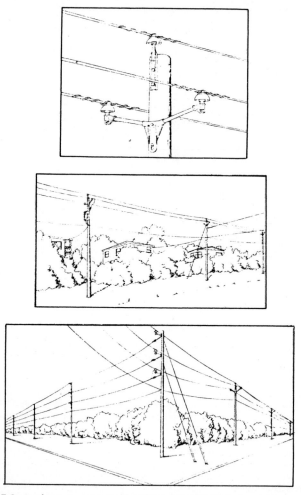

Fig. 7-2   Armless-type construction for 4-kV and 13.8-kV distribution circuit operation.

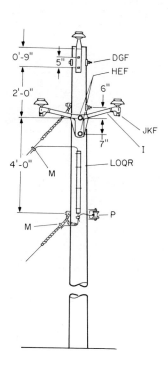
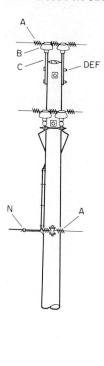

Item	No. required
A. Preformed tie wire size required	7
B. 1″ Pin insulator	6
C. 1″ Pole-top pin	2
D. 5/8″ x 10″ Bolt	3
E. 5/8″ Round washer	2
F. 5/8″ Palnut	9
G. 2 1/2″ Curved washer	1
H. 5/8″ x 12″ Bolt	2
I. Two-phase brackets 48″ spacing	2
J. 5/8″ x 14″ Double arming bolt	2
K. 2 1/4″ Square washer	8
L. #6 Copper wire	1#
M. Split-bolt connector, size required	3
N. Compression connector, size required	1
O. Molding, ground wire	5′
P. Single-clevis insulator	1
Q. Staples, molding	6
R. Staples, fence	6

Fig. 7-3   Three-phase 13.2Y/7.6-kV armless-type wood-pole structure, angle pole 6 to 20°.

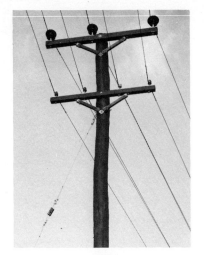

Fig. 7-4 Typical construction of wooden pole, wooden crossarms, and wooden braces. *(Courtesy Locke Dept., General Electric Co.)*

Fig. 7-5 Wood-pole line using wooden crossarms arranged to provide symmetrical arrangement of the line conductors. *(Courtesy Hughes Brothers.)*

Fig. 7-6 Post-type insulator in use in horizontal position. Note absence of crossarm and braces and narrow right-of-way required. *(Courtesy Lapp Insulator Co.)*

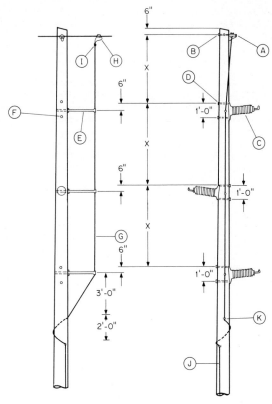

Item	No. required
A. Ground wire bracket	1
1/2" x 4" Lag screw	1
B. 5/8" x 14" mach. bolt	1.
3" Square curved washer/5/8" bolt hole	1
Standard nut	1
Locknut	1
C. 88 kV Horizontal post insulator, gray color	3
D. 3/4" x 12" Machine bolt	2
3/4 x 14" Machine bolt	3
3/4" x 16" Machine bolt	1
3" Square curved washer/3/4" bolt hole	6
Standard nut	6
MF Locknut	6
E. Fiber glass downlead bracket	3
F. 5/8" x 12" Machine bolt	3
Standard nut	3
2 1/4" Square Washer/5/8" bolt hole	6
G. 6A Copperweld groundwire	
H. Two-bolt connector	1
I. Split-bolt Connector	1
J. Ground-wire molding, 1/2" groove	
K. Bronze staple, for 1/2" plastic molding	

Note: Phase spacing (dimension X) varies with span lengths, conductor size, conductor sag, and midspan clearances. Spacing will be determined in accordance with appropriate design standard.

Fig. 7-7 A 69-kV single-circuit armless-type tangent structure.

a straight portion of the line. Figure 7-8 is a drawing of a 69-kV circuit single-pole structure designed for installation where it is necessary for the line to change direction. If a line is to be built in an area where appearance is not critical, such as an industrial location, crossarms are frequently used. The use of crossarms reduces the required pole height and permits the addition of a second circuit on the same wood-pole structures if the electrical system requires it at some future time. Figure 7-9 is a drawing of a typical 69-kV circuit tangent structure with crossarms and provision for a future second circuit. Figure 7-10 is a drawing of a 7 to 30° angle structure.

Transmission Structures   Two-pole wood structures commonly referred to as H-type structures are frequently used for three-phase ac transmission circuits operating at voltages of 138 through 230 kV. Most subtransmission circuits operating in the voltage range of 34.5 through 115 kV are constructed on single wood-pole structures, however, a few are installed on two-pole H-type wood structures. Extra-high-voltage circuits operating at 345 kV are frequently constructed on two-pole wood structures. A few 500-kV installations have been constructed with wood-pole structures. The wood-pole structures are esthetically acceptable in wooded areas and can be installed economically in many locations.

Figure 7-11 shows a typical tangent H-type two-pole wood structure commonly used for 161-kV three-phase ac circuits. The structure is usually built on 65-ft poles with some variation due to the terrain and has 15½ ft clearance between conductors. A complete structure with crossarms and insulators ready for erection will weigh from 5 to 6 tons. A three-pole, 60 to 90° angle, and dead-end structure can be seen in the background (Fig. 7-11). Figure 7-12 is a drawing of a two-pole H-type structure for a 161-kV line. Figure 7-13 is a drawing of a three-pole angle structure.

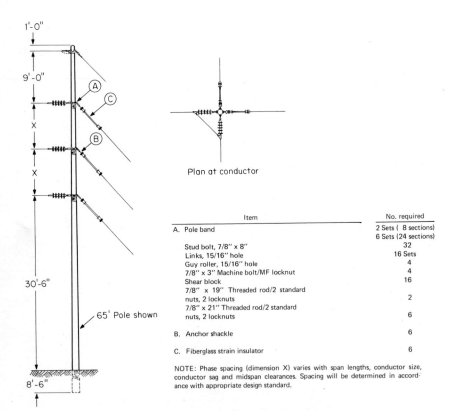

Plan at conductor

Item	No. required
A. Pole band	2 Sets (8 sections)
	6 Sets (24 sections)
Stud bolt, 7/8" x 8"	32
Links, 15/16" hole	16 Sets
Guy roller, 15/16" hole	4
7/8" x 3" Machine bolt/MF locknut	4
Shear block	16
7/8" x 19" Threaded rod/2 standard nuts, 2 locknuts	2
7/8" x 21" Threaded rod/2 standard nuts, 2 locknuts	6
B. Anchor shackle	6
C. Fiberglass strain insulator	6

NOTE: Phase spacing (dimension X) varies with span lengths, conductor size, conductor sag and midspan clearances. Spacing will be determined in accordance with appropriate design standard.

Fig. 7-8   A 69-kV single-circuit dead-end structure, 45 to 90° angle in line.

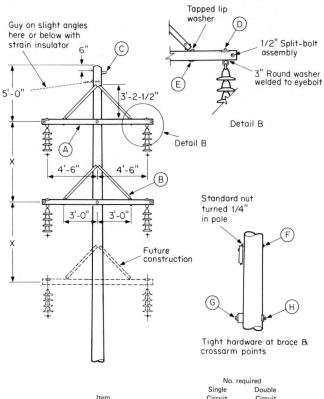

Item	No. required	
	Single Circuit	Double Circuit
A. Crossarm	2	3
B. Crossarm brace (1 left, 1 right)	2 Pair	3 Pair
C. Shield-wire support	1	1
Washer nut and standard nut	1	1
D. 5/8″ x 6 1/2″ Eyebolt	3	6
Saddle washer assembly		
Standard nut		
E. 5/8″ x 7″ Washer head bolt	3	6
Standard nut	3	6
MF locknut	3	6
F. 5/8″ x 12″ Washer head bolt	1	1
5/8″ x 14″ Washer head bolt	1	2
Standard nut	2	3
G. 3/4″ x 16″ Washer head bolt	1	1
3/4″ x 18″ Washer head bolt	1	2
H. 3/4″ Nut and locknut	2	3
Curved guide washer staked	2	3

NOTE: Phase spacing (dimension X) varies with span lengths, conductor size, conductor sag and midspan clearances. Spacing will be determined in accordance with appropriate design standard.

Fig. 7-9   A 69-kV double-circuit crossarm construction tangent structure, 0 to 30° angle.

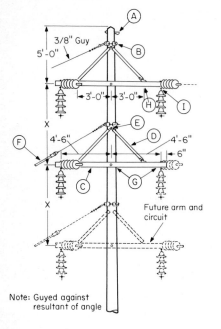

3/8" Guy

5'-0"

3'-0" · 3'-0"

X

4'-6"

4'-6"

6"

Future arm and circuit

Note: Guyed against resultant of angle

		No. required	
Item		Single Circuit	Double Circuit
A.	Shield-wire support	1	1
	Washer nut and standard nut	1	1
B.	Pole band	2 Sets (8 sections)	3 Sets (12 sections)
	Stud bolt, 3/4" x 6"	8	12
	Links	2 Sets	3 Sets
	Guy roller, 15/16" hole	2	3
	3/4" x 2 1/2" Machine bolt/MF locknut	2	3
C.	Double arm crossarm assembly	2	3
D.	Crossarm brace (1 left, 1 right)	4 Pair	6 Pair
E.	5/8" x 14" Double arming bolt	2	2
	5/8" x 16" Double arming bolt	0	1
	4 Standard nuts per bolt		
F.	Guy strain insulator, fiber glass	As required	
G.	3/4" x 28" Double arming bolts	2	3
	2 Washer nuts and 2 standard nuts per bolt	4	6
	3/4" x 28" Double arming bolts	8	12
	2 Washer nuts and 6 standard nuts per bolt	16	24
	Dead-end Tee	8	12
H.	5/8" x 6" Washer head bolt with lip washer	8	12
I.	5/8" x 6" Eyebolt with flange washer and standard nut	3	6

NOTE: Phase spacing (dimension X) varies with span lengths, conductor size, conductor sag and midspan clearances. Spacing will be determined in accordance with appropriate design standard.

Fig. 7-10   A 69-kV double-circuit crossarm construction, 7 to 30° angle structure.

Fig. 7-11   Two poles combined to make an H-frame structure. *(Courtesy Hughes Brothers.)*

	Item	No. required
A.	Shield wire support (steel cross angle) with U-bolt, chain link and bonding bolt	1
B.	Double crossarm assembly with 11" spacer fitting	1
C.	Vee brace 9' 1 1/2" C-C role 37 1/2° and 52 1/2° fittings	4
D.	X-brace	2
E.	Center clamp (for X-brace)	1
F.	Pole band	1 Set (4 sections)
	Stud bolt 7/8 x 8"	1 Set (4 sections)
	Links	8
	Guy roller, 15/16" hole	2 Sets
	7/8" x 3" Machine bolt/MF locknut	1
G.	Fiberglass strain insulator	1
H.	7/8" x 12" Machine bolt	2
	Curved washer, 15/16" hole	2
	Standard nut	2
	MF locknut	2
I.	7/8" x 12" Machine bolt	2
	Standard nut	2
	MF locknut	2
J.	7/8" x 23" Threaded rod	2
	Washer nut	4
	Standard nut	4
K.	7/8" x 14" Machine bolt	2
	7/8 x 16" Machine bolt	2
	Curved washer, 15/16 hole	4
	Standard nut	4
	MF locknut	4
L.	Swinging angle bracket	3
M.	Pole shim, 1" or 1 1/2" thick, is required	4

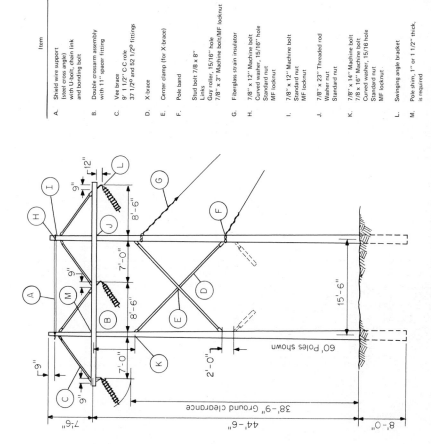

Fig. 7-12   A 161-kV construction angle (1½ to 6°) H-frame structure.

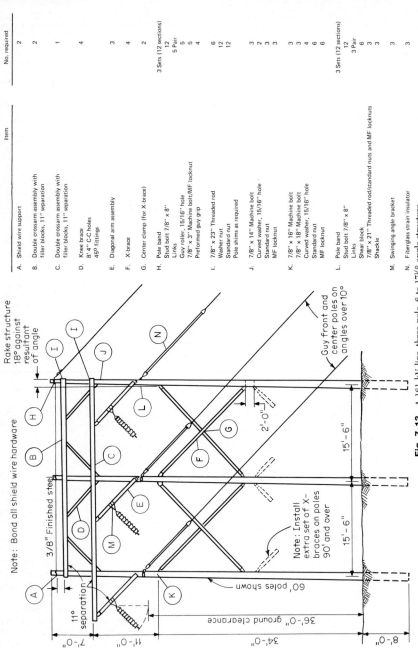

Item	No. required
A. Shield wire support	2
B. Double crossarm assembly with filler blocks, 11" separation	2
C. Double crossarm assembly with filler blocks, 11" separation	1
D. Knee brace 8' 4" C-C holes 45° fittings	4
E. Diagonal arm assembly	3
F. X-brace	4
G. Center clamp (for X-brace)	2
H. Pole band	3 Sets (12 sections)
Stud bolt 7/8" x 8"	12
Links	5 Pair
Guy roller, 15/16" hole	5
7/8" x 3" Machine bolt/MF locknut	5
Preformed guy grip	4
I. 7/8" x 23" Threaded rod	6
Washer nut	12
Standard nut	12
Pole shims as required	
J. 7/8" x 14" Machine bolt	3
Curved washer, 15/16" hole	2
Standard nut	3
MF locknut	3
K. 7/8" x 16" Machine bolt	3
7/8" x 18" Machine bolt	3
Curved washer, 15/16" hole	4
Standard nut	6
MF locknut	6
L. Pole band	3 Sets (12 sections)
Stud bolt 7/8" x 8"	12
Links	3 Pair
Shear block	6
7/8" x 21" Threaded rod/standard nuts and MF locknuts	3
Shackle	3
M. Swinging angle bracket	3
N. Fiberglass strain insulator	3

Rake structure 18° against resultant of angle

Note: Bond all shield wire hardware

3/8" Finished steel

11° separation

60' poles shown

36'-0" ground clearance

Note: Install extra set of X-braces on poles 90' and over

Guy front and center poles on angles over 10°

2'-0"

15'-6"

15'-6"

7'-0"

11'-0"

34'-0"

8'-0"

Fig. 7-13 A 161-kV-line, three-pole, 6 to 17½° angle structure.

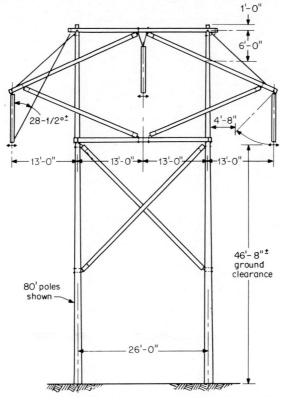

Fig. 7-14   A 345-kV EHV ac transmission-line tangent structure. *(Courtesy Hughes Brothers.)*

Figure 7-14 is a drawing of a double K-type two-pole tangent structure for an EHV ac 345-kV transmission line. A typical line for this use could have ⁷⁄₁₆-in extra high strength steel shield wires and two 795-kcmil 26/7 aluminum conductor steel reinforced (ACSR) bundled conductors per phase, with spans of 1000 ft between structures. Structure safety factors with ½ in of radial ice would be 4 for vertical and transverse loads. The structure vertical safety factor for 1½ in of radial ice would be 1.5, and the structure safety factor for 100 mph winds and bare conductors would be 2.3.

Aluminum, Concrete, Fiberglass, and Steel Structures

Environmental considerations have increased the use of aluminum, concrete, fiberglass, and steel to support electric transmission and distribution circuits. Wood-pole structures have generally proved to be the most economical for distribution and transmission lines operating at voltages through 345 kV. The design of structures to make them compatible with the environment has resulted in the use of materials and colors to comport with the natural surroundings. Aluminum, concrete, fiberglass, and steel structures, properly used, may reduce the width of right-of-way required for distribution and transmission lines.

Aluminum Structures   Ornamental street lighting is one of the most common applications of aluminum poles (Fig. 8-1). The hollow tubular street lighting poles give a pleasing appearance. The poles are light weight, making them easy to handle and install. The hollow pole facilitates the installation of the low-voltage cables supplying the lamps from underground electric distribution circuits.

Aluminum poles can be used for subtransmission circuits installed along urban streets (Fig. 8-2). Transmission lines operating at voltages through 500 kV have been constructed with H-frame (Fig. 8-3) and guyed aluminum towers (Fig. 8-4). Guyed aluminum towers have also been fabricated in Y and V shapes. The towers can be completely assembled on the ground at a convenient location and airlifted to the site for erection using a helicopter. The structures can be airlifted, complete with insulators stringing sheaves, tag lines for the conductor pulling, and guys. Airlifting the structures to the installation site can save time and money in inaccessible locations such as swamps, forests, or mountainous areas. Aluminun towers and poles do not need to be painted for protection, which minimizes the maintenance required.

Concrete Poles   Poles manufactured with concrete are used for street lighting and for distribution and transmission lines. Concrete poles have been used to improve the esthetics of overhead circuits and are preferred where the life of wooden poles is unduly shortened by decay and woodpeckers. Compared to wood poles, concrete poles have disadvantages of being more expensive, lower in insulation level, more difficult to climb, heavier to handle, and more difficult to drill when necessary to install equipment. Concrete has the advantages, compared to wood, of long life and availability on demand.

Hollow-type concrete poles are made by putting the concrete materials and steel reinforcing rods into a cylinder of the desired length and taper. Revolving the cylinder in a lathe-like machine for a period of 10 to 15 min then forces the concrete materials to the outside, thereby forming the hollow pole when the concrete cures. The hollow-type pole is lighter than the solid type and

Fig. 8-1 Ornamental street lighting poles made of aluminum.

Fig. 8-2 Octagonal aluminum transmission-line pole. *(Courtesy McGraw Edison Co.)*

provides a means for making connections through the pole to underground cables or services. The hollow-type poles are commonly used for ornamental street light installations served from underground electric circuits (Fig. 8-5).

The hollow-spun prestressed-concrete poles have a high-density concrete shell completely encasing a reinforcing cage containing prestressed high-tensile steel wires. Prestressing produces poles with a high strength-to-weight ratio that can be used for distribution and transmission lines. Hollow-spun prestressed-concrete poles equivalent to class 4 wood poles with lengths of 35, 40, 45, and 50 ft have been manufactured for electric distribution lines. Some 45-ft poles have a top diameter of 5.6 in, a bottom diameter 12.3 in, a wall thickness of 1.75 in, and a weight of 3300 lb.

Fig. 8-3 Linemen working from bucket truck on conductors supported by aluminum H-frame structures.

Fig. 8-4 Linemen assembling and preparing V-shaped guyed aluminum towers to be airlifted by helicopter to the installation site. *(Courtesy L. E. Myers Co.)*

Fig. 8-5 Hollow-type concrete street light pole.

Fig. 8-6 Square tapered ornamental street light pole installation in a residential area with an underground electric distribution system.

Square-tapered prestressed-concrete poles are constructed by placing the stressed reinforcing material in a form and pouring the form full of concrete. The poles may have a center void developed by use of a cardboard or plastic tube. A copper wire is usually cast in the pole for grounding purposes. A prestressed-concrete 50-ft distribution pole has been found to deflect 2 ft before cracks appeared and 6 ft before failure. The characteristics of square-tapered prestressed-concrete poles are itemized in Table 8-1.

Large numbers of the prestressed-concrete poles have been used in Florida for street lighting, electric distribution, subtransmission, and transmission circuits. Figure 8-6 pictures an ornamental square-tapered concrete street lighting pole. An overhead 34.5Y/19.9-kV three-phase distribution circuit utilizing square-tapered prestressed-concrete poles is illustrated in Fig. 8-7. The pole shown supports a street light and three single-pole switches. The installation is clean and neat, satisfying visual environmental requirements.

Florida Power and Light Company has constructed 138-kV transmission lines using line-post insulators on single square-tapered prestressed-concrete poles. Two-pole H-type structures have been used for double-circuit 240-kV transmission lines.

Composite concrete and steel telescoping poles manufactured with a series of nested steel tubes fabricated from high-strength steel and pumped full of concrete can be used to support transmission circuits. Pumping concrete into the pole raises it to its full height, ranging up to 150 ft. The assembled telescoping structures are manufactured to be transported to the installation site on a standard trailer. The steel-reinforced concrete foundation is constructed with a tube to accept the telescoping pole. The telescoped pole is installed on the base with crossarms, insulator strings, stringing sheaves, and tag lines in place. The pole is pumped full of concrete, using a commercially available pump and transit-mix concrete. A 100-ft pole can be pumped to its full height in less than 15 min.

Table 8-1   Tabulation of Square-Tapered Prestressed-Concrete Pole Characteristics

Length, ft	Nominal square pole top, in	Minimum breaking strength, #	Minimum load with no cracks, #	Repetitive testing load, #	Minimum crack closure after repetitive loading, #
25	4½	1200	510	850	255
35	5	1300	600	900	300
45	6	2600	1000	1300	560
45	6	2600	1000	2400	900
60	9	6000	2800	4200	1200

Fig. 8-7 Prestressed concrete pole supports three-phase 34.5Y/19.9-kV distribution line with single-pole sectionalizing switches and a street light supplied by a series street light circuit. *(Courtesy S & C Electric Co.)*

Fig. 8-8 Fiberglass pole. Note pole steps on entire length. Climbers cannot be used. Crossarm is also made of fiberglass and is attached with band. *(Courtesy Line Material Industries.)*

Fiberglass Poles   Poles made of fiberglass have been manufactured on an experimental basis. Fiberglass is immune to freezing, rotting, the pecking of birds, and damage from nails, all of which are injurious to wood poles. Fiberglass is an excellent insulator, even when wet from rain. Holes are drilled into the poles during manufacture for the insertion of pole steps. Figure 8-8 shows a typical fiberglass pole.

Linemen use bucket trucks to eliminate the need for pole steps when fiberglass poles are located in accessible places. Fiberglass poles are too expensive to be used extensively, but fiberglass rods have been used as insulators (Fig. 8-9), eliminating the need for crossarms. Armless construction, using fiberglass components, is used extensively to enhance the appearance of overhead lines.

Fig. 8-9 Armless-type distribution-line structure with fiberglass rods supporting porcelain insulators. *(Courtesy A. B. Chance Co.)*

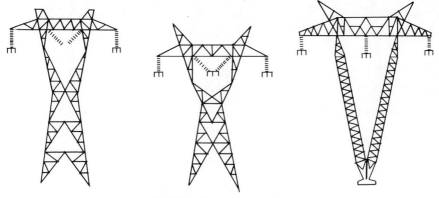

Fig. 8-10 Standard tower designs.

Fiberglass components are reasonably priced and economically acceptable where the appearance of the facilities is important.

Steel Structures   Steel towers have been used extensively to support subtransmission and transmission line conductors. Standard types of steel towers are shown in Fig. 8-10. Towers of this type are required for 765-kV three-phase ac transmission lines (Fig. 8-11). Specially designed steel structures have been created to blend with the natural beauty of the landscape (Fig. 8-12). The size of transmission-line towers should be kept to the minimum feasible to preserve the environment and minimize the conflict with other use of the land. Tubular-steel poles have been used to support subtransmission and transmission-line conductors energized at voltages of 345 kV and below (Figs. 8-13 and 8-14). Weathered galvanized-steel structures blend with the sky and reduce the silhouette appearance. Galvanized steel eliminates the need for paint, reducing maintenance costs, and steel is free from damage by lightning, woodpeckers, and bacteria.

Steel transmission-line towers are of either the rigid or the flexible type. The rigid tower is firm in all directions, whereas the flexible tower is free to deflect in the direction of the line. Rigid

Fig. 8-11 A 765-kV transmission-line angle tower in foreground and V-type guyed tangent tower in background. *(Courtesy American Electric Power Co.)*

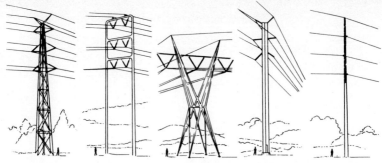

Fig. 8-12 Esthetically designed transmission-line structures.

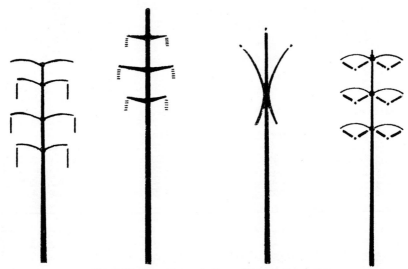

Fig. 8-13 Inconspicuous steel transmission-structure designs.

Fig. 8-14 Steel poles supporting double-circuit 161-kV transmission line.

Fig. 8-15 Four-legged rigid-type 230-kV transmission tower. *(Courtesy Pennsylvania Water and Power Co.)*

Fig. 8-16 Linemen performing hot-stick maintenance work on H-frame, two-pole steel flexible transmission-line tower operating at 161 kV ac. *(Courtesy A. B. Chance Co.)*

towers, in order to be rigid, must have three or more legs, but the flexible construction consists of only two legs. Figure 8-15 shows the four-legged rigid-type tower. Rigid towers can resist the same strain in all directions. Flexible towers (Fig. 8-16) can resist strain only transversely or across the line and have very little strength in the direction of the line. They will give or deflect if there is any unbalance in conductor pull; in fact, they depend largely upon the conductors to hold them in position.

Flexible H-frame steel structures are usually more expensive than wood pole structures. The steel H-frame structures have many advantages that have increased their use. A 345-kV three-phase ac transmission line constructed across agricultural land in a heavy loading district can have a ruling span of 1400 ft compared to 850 ft for wood-pole structures when the conductor is 954-kcmil 45/7 ACSR two-conductor bundle. The number of structures required is reduced considerably, minimizing the conflicts with cultivation of the land and reducing the cost of the right-of-way.

Fig. 8-17 Galvanized-steel structures supporting ± 250-kV dc transmission line constructed across cultivated land. *(Courtesy Ulteig Engineers, Inc.)*

Fig. 8-18 Galvanized-steel pole used with wood poles in a 69-kV three-phase subtransmission line to eliminate the need for guys and anchors.

The cost of structure assembly of steel H-frame structures is estimated to be approximately 50 percent of that for standard rigid latticework steel towers for this type of circuit; it is also less than that of wood-pole structures as a result of the reduced number of structures. The smaller number of structures minimizes the number of insulators and the line hardware required and reduces the labor costs to string, sag, and clip in the conductors. The total costs for the construction of this type of line is approximately the same for rigid and flexible steel towers. Galvanized-steel structures have been used for most high-voltage dc transmission lines (Fig. 8-17).

Steel poles can be used to establish a rigid structure in subtransmission and distribution lines to eliminate the need for guys and anchors (Fig. 8-18). Steel poles are frequently installed for angle structures in lines with wood poles used for tangent structures. Elimination of the guys and anchors improves the appearance of the facilities, enhances the environment, and solves right-of-way problems in urban areas. The use of line-post insulators and armless-type construction minimizes the silhouette established.

Section **9**

Locating and Staking Line

LAYING OUT THE LINE

Selecting the Route   The first step to be taken prior to the design or construction of any line is to conduct a survey and make a map of the country over which the line is to pass. The final location of the line is, of course, not known and, therefore, a wide strip of land must be included on the map. The main points between which the line is to be built will be known. The intervening territory should be laid out on a scale large enough to show clearly all division lines, towns, roads, streams, hills, ridges, railroads, and bridges and to permit a complete inclusion of all existing telephone, telegraph, and power lines.

With the map completed, the following principles should be used as guides in selecting the exact route:

1. *Select the Shortest Route Practicable.*   The shortest line naturally is the cheapest, other things being equal. This means that the line should be a straight line between the terminals of the line, for a straight line is the shortest distance between any two points.

2. *Parallel Highways as Much as Possible.*   This makes the line readily accessible both for construction and for inspection and maintenance. The hauling and delivery of materials will also be greatly facilitated if this can be done.

In the case of rural lines it has become quite common to locate lines a short distance from the highway in order to miss all trees. This practice has many advantages: the trees are preserved, tree-trimming expense is eliminated, there are no outages from trees falling into the line, no long poles are required to go over trees, side arms are not required, and stub guys are not required.

3. *Follow Section Lines.*   Doing this causes less damage to farmers' property and, therefore, makes it possible to purchase the right-of-way more cheaply. Paralleling railroads is desirable for the same reason, because the farms have already been cut and, therefore, the additional damage is negligible.

4. *Route in Direction of Possible Future Loads.*   If there is possibility of adding power loads, the route selected should be such as to come as as near as possible to such locations, provided that the additional cost is not excessive.

5. *Avoid Crossing Hills, Ridges, Swamps, and Bottom Lands.*   Hills and ridges subject the line to lightning and storms. Swamps and bottom lands subject the line to floods. Furthermore, delivery of material as well as the construction of the line becomes difficult. Extra guying and cribbing are often necessary in swamps.

6. *Avoid Paralleling Telephone Lines.*   Paralleling telephone lines causes disturbances or interferences in the telephone lines by induction and therefore requires transposition of the power conductors. This adds to the cost of the line. In some cases it is advisable to remove the telephone line.

7. *Preserve the Environment.*   Line routes should be selected which are the least visible from scenic locations or other areas where people pass or congregate. A simple test of visibility can be

carried out by touring roads and points close to a proposed right-of-way and noting the visual impact of the line prior to final selection.

Illustrations of line locations doing minimum damage to the landscape are shown in Figs. 9-1 to 9-3.

In Fig. 9-1, the line is run along valleys and the lower slope of hills.

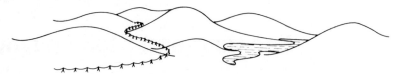

Fig. 9-1 Line is routed along the valley and the lower slopes of the hills or mountains and away from lake, thus causing minimum interference with natural landmarks. The route also avoids disruption of natural lines. *(Courtesy Hudson River Commission.)*

In Fig. 9-2, the line is straight across the plains, but it curves between mountains.

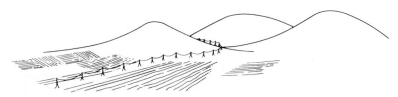

Fig. 9-2 Line is routed across plains but curves between hills or mountains, causing least disruption of natural lines. *(Courtesy Hudson River Commission.)*

In Fig. 9-3, the line crosses the highway at near right angle, thus reducing the sight distance along the right-of-way to a minimum.

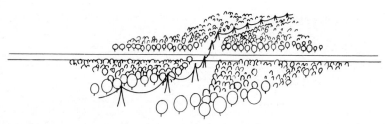

Fig. 9-3 Line crosses highway at close to a right angle, thereby reducing sight distance to a minimum. *(Courtesy Hudson River Commission.)*

Examples of poor selections of route are shown in Figs. 9-4 to 9-6.

In Fig. 9-4, the right-of-way produces a bad scar on the hill or mountain which is visible from great distances. It also permits soil erosion in the right-of-way.

Fig. 9-4 Line crosses hill or mountains, thereby causing a long visible scar. It also exposes the right-of-way to soil erosion. *(Courtesy Hudson River Commission.)*

In Fig. 9-5, the line should have been located next to the railroad right-of-way, by which the landscape was already cut up. It also leaves several small, irregular land areas which may become valueless.

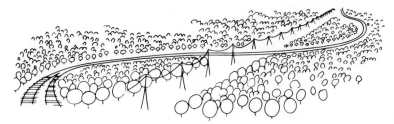

Fig. 9-5   Line should be located adjacent to railroad right-of-way if possible, thereby eliminating duplication of rights-of-way. *(Courtesy Hudson River Commission.)*

In Fig. 9-6, the two dotted routes are undesirable as they do not run along the edges of different land uses, particularly forests. The route selected causes minimum disruption of activities in each area.

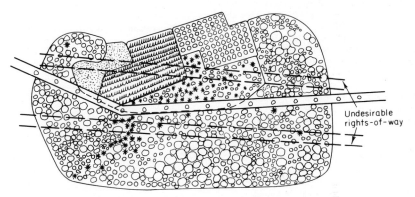

Fig. 9-6   Proposed route follows edges of different land use, thereby causing minimum disruption of activities in each area. *(Courtesy Hudson River Commission.)*

Another way of merging a right-of-way with natural scenery is illustrated in Fig. 9-7, where undergrowth is permitted in the right-of-way to the maximum height compatible with safe operating conditions. This is especially desirable when passing through a forest.

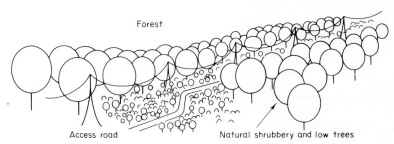

Fig. 9-7   Merging line with natural scenery by providing natural shrubbery and low trees. *(Courtesy Hudson River Commission.)*

Table 9-1   Approximate Range of Widths of Right-of-Way Required for Various Types of Lines

Pole lines	Feet
Single circuit	25 to 50
Double circuit	30 to 75
Two-pole lines	45 to 100
Tower lines	Feet
Single circuit	30 to 60
Double circuit	40 to 75
Two-tower lines	45 to 100

Width of Right-of-Way   When the route of the line has been selected, permission must be obtained to run the line, or land must be purchased. This is known as the "right-of-way." In this connection, it should be pointed out that it is well to select several tentative routes so that if difficulties arise in obtaining permits on any one route or the prices demanded become inflated, another route can be chosen. Most low- or medium-voltage pole lines do not have complete right-of-way, that is, the owners do not possess a continuous strip of land, but only the area on which the tower or pole is located. In the case of high-voltage lines, a continuous strip of land is often acquired. Table 9-1 gives a general idea of the width of right-of-way necessary for different lines.

The reason for acquiring a right-of-way much wider than the actual space necessary is to prevent tall trees from being blown into the line and to prevent damage from forest fires. Sometimes it is possible to purchase a narrower right of way and to obtain permission to clear back the remaining distance beyond the fences.

Clearing the Right-of-Way   Practically all lines will traverse through some brush or timberlands. Even lines built in the Middle West, an area generally thought of as level prairie, may pass through forests or hilly country covered with shrubs and underbrush. Lines built in the mountainous country of the west or the hills of the east naturally pass through country in which the right-of-way must be cleared before construction can be started.

In clearing the right-of-way, all stumps should be cut low (see Fig. 9-8). All logs and brush should be removed for a distance of at least 25 ft under each conductor so that there is ample room to assemble and erect poles or towers and later to string the line conductors. Any underbrush or piles of dead wood should be removed. Otherwise, they may catch fire and burn the line or anneal the conductors and cause them to sag abnormally. Figure 9-9 illustrates a properly cleared right-of-way.

Fig. 9-8   Right-of-way clearing operations getting under way. Note forestry crew felling a large hickory tree with a power saw. Tree stumps are cut off close to the ground line with a power saw later on. *(Courtesy Wisconsin Electric Power Co.)*

Fig. 9-9 A right-of-way for a new 132,000-volt H-frame power line properly cleared and ready for erection of H-frame structures. Note the stump in the left foreground has been cut off about 3 in above the ground line. The material for the H-frame structure is shown in the right background. It consists of two 80-ft class 1 fully creosoted poles for the uprights, one fully creosoted 35-ft pole for a crossarm, and four crossarm braces. There are also eighteen 10-in disk insulators for the three strings. *(Courtesy Wisconsin Electric Power Co.)*

If it is possible, permission should be obtained to cut any extremely tall trees immediately adjacent to the right-of-way. All dead limbs and branches in the adjoining strip should also be cut down as a high wind may blow them into the line.

Figure 9-10 is a plan view of a wooded section before and after clearing. The top of the figure shows an elevation of the line and bordering trees. The clearances indicated are the minimum requirements for a low-voltage rural line. In case of a high-voltage line it is more important to remove so-called danger trees along the side of the right-of-way as shown in Fig. 9-11. Trees which would reach within 5 ft of a point underneath the outside conductor in falling are known as "danger" trees.

Chemical Brush Control   Chemicals have recently become available by means of which the growth of brush may be controlled. The chemical is sprayed onto the basal area of the shrubs or small trees from the ground line up to a height of 12 to 15 in above ground. The rest of the shrub need not be sprayed. However, complete encirclement of the stem or trunk with the spray solution is necessary. When applied to stumps, the spray prevents resprouting.

Locating Pole Positions on Distribution Lines   The following principles should be followed in locating pole positions in distribution systems:

On public thoroughfares poles shall be placed on the side which is most free from foreign lines and trees, care being taken in cities and towns to keep the more important streets as free as possible from primary circuits.

The same side of the road shall be used throughout the length of the line wherever this is feasible. In cities and towns, lines along alleys or streets should occupy, so far as practicable, the same side of all streets or alleys in any given town. Poles in alleys should be set as close as possible to the side lines of the alleys.

The length of spans may be increased or reduced so as to make poles come in line with property lines or fences or to place them in positions satisfactory to adjoining property owners wherever this is practicable. In locating poles on lot lines, either in the street or in the alley, care must be taken not to block driveways or the entrances to garages. Furthermore, care should be taken not to obstruct doorways, windows, fire escapes, gates, coal holes, and runways.

Where lines cross private property, easements shall be obtained from the property owners before any work is started.

Poles should be set on the junctions of streets and alleys to facilitate the installation of branch lines or guys.

In locating lines along a curb, the poles shall be placed so that there is a distance of 2 ft between poles and the inside edge of the curb. Poles on suburban roads where there is no curb line should be set as specified above with respect to future curb lines wherever this is practicable.

On country roads, poles should be set 18 in outside fences, if straight; otherwise, poles should

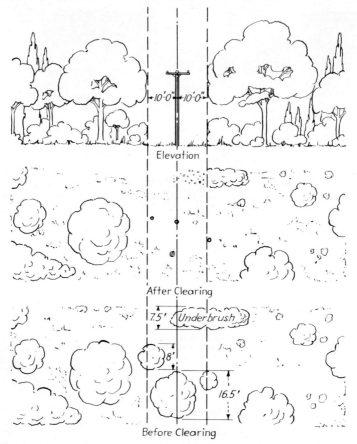

Fig. 9-10 Guide for right-of-way clearing. *(Courtesy Rural Electrification Administration.)*

be located to give the longest straight lines between angle points, using as much of the road as possible. However, written approval of proposed locations of poles with respect to property lines and highways shall be obtained from the district highway engineer or other responsible party.

Locating Pole or Tower Positions on Transmission Lines   In locating poles or towers on transmission lines, the following general principles should be kept in mind:

1. Select high places.
2. Keep spans uniform in length.
3. Locate to give horizontal grade.

By locating the poles or towers on knolls or high places, shorter poles or towers can be used and still maintain the proper ground clearance at the middle of the span. By avoiding ravines and low places, a better footing is usually obtained. In extremely hilly or mountainous country, towers are always located on ridges, thereby greatly increasing the span without greatly increasing the pull on the conductors. This is possible because the sag can be made very large and still maintain the required ground clearance.

In rolling country, the location should take into account the grading of the line. A well-graded line does not have any abrupt change, either up or down in the line. The conductors remain practically horizontal irrespective of the small changes in ground level. This is well illustrated in Fig. 9-12. Sometimes a change in pole position a few feet one way or another will make it possible to use the standard pole length where an odd size would otherwise be required.

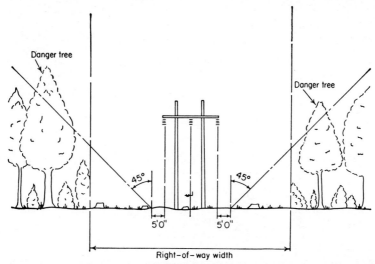

Fig. 9-11 Sketch indicating danger trees beside the right-of-way which should be removed or topped. Permission to do so must first be secured. Note that in approximately level terrain trees that would reach within 5 ft of a point underneath the outside conductor in falling are considered to be danger trees. Portions of the right-of-way must be cut as directed by the engineer so that stumps will not prevent the passage of tractors and trucks along the right-of-way. *(Courtesy Rural Electrification Administration.)*

In addition to the preceding general principles, some specific rules should be followed:

1. Locate poles, towers, and their guys and braces at least 3 ft from fire hydrants.

2. Locate poles, towers, and their guys and braces at least 2 ft (measured to the street side) from curbs.

3. Locate poles and towers at least 12 ft from the nearest track rail where lines cross railroad tracks. Allow a clearance of at least 7 ft at sidings.

Special attention should be given to the location of poles where the ground washes badly. Poles should not be placed along the edges of cuts or embankments or along the banks of creeks or streams. When it becomes necessary to set poles on the edge of a cut, the pole should be set deep enough to protect the line in case the bank washes or crumbles away.

STAKING AND GRADING THE LINE

After the exact positions have been fixed, a stake is driven to indicate the center of the pole or tower.

Staking Procedure on Rural Lines The general procedure in staking out a rural line is as follows:

1. Determine all fixed pole locations and the limits that poles can be shifted. These limits are usually very narrow. These fixed locations may be established by such crossings as power and communication lines, railroads and main highways, rivers and swamps, sharp high points in the profile, fixed angles in the line, and dead ends.

2. Locate all tentative pole locations and the limits that these poles can be shifted. These points

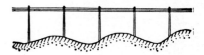

Fig. 9-12 A well-graded line. The changes in ground level do not show in the line conductors.

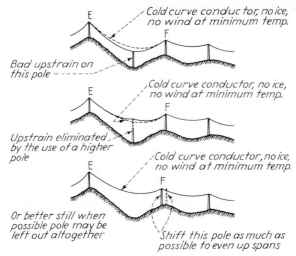

Fig. 9-13 Sketches illustrating methods of eliminating upstrain on poles or towers. *(Courtesy Cooper Wire Engineering Association.)*

are usually established by transformer and service locations, rounded hilltops, minor road crossings, trees and buildings, angles, fences, taps, etc.

3. Locate intermediate pole locations by dividing the distance between fixed locations, giving due consideration to tentative pole location limits, into such span lengths as will provide the most economical construction.

Grading the Line   The matter of locating the line supports so that the standard pole or tower height can be used has already been mentioned. In many cases, it will not be possible to keep the line wires on the same level by changing the location of the support. In such cases it will be necessary to select higher or lower supports than the standard for the lines (see Fig. 9-12). The selection of the proper support height to keep the conductors from making any abrupt change, either up or down, is called "grading" the line. The permissible difference in level between adjacent supports is usually limited to 5 or 10 ft for pole lines. A difference of 5 ft on spans of 150 ft or 10 ft on spans of 250 to 300 ft is allowed.

If it is necessary to locate a pole or tower in a low point or valley, the structure must be checked to see whether there is any upstrain or uplift, as shown in Fig. 9-13. If a reasonable increase in pole height does not eliminate the uplift, it is necessary to dead-end and down-guy the structure.

Section **10**

Unloading, Framing, and Hauling Wood Poles

Unloading Wood Poles   Poles are usually shipped on railroad flatcars (Fig. 10-1). Great care must be exercised while unloading poles from the flatcar to avoid all possibility of an accident. Before starting the unloading operation, the railroad car brakes should be set to prevent the car from moving. The wheels should be chocked as an added precaution. Flags and barricades should be set up to safeguard the area. Mechanized equipment reduces the time required and the number of workers needed to unload the poles safely (Fig. 10-2).

The lineman should inspect the stakes holding the poles to be sure the poles are secure before cutting the banding wires. The banding wires should be cut from the end of the poles cautiously by the lineman on the ground. After the banding wires are cut, equipment such as the Cary-Lift® Loader (Fig. 10-2) can be used to safely remove several poles at a time and place them on the pole storage pile (Fig. 10-3).

If mechanized equipment is not available, wood poles can be unloaded from a railroad flatcar with winch line on a line truck or with a mechanical hoist. After the railroad car brakes have been set, the wheels chocked, and the area barricaded, the linemen should restrain the poles on the railroad car by placing chains around the pile of poles near each end of the railroad car and secure each chain with a hoist (Fig. 10-4).

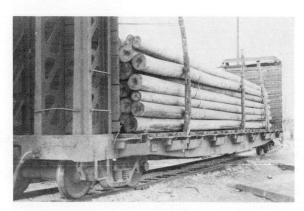

Fig. 10-1   Railraod carload of wood poles. Stakes and binding wires hold the poles in place for shipment.

Fig. 10-2 Cary-Lift® Loader being used by lineman to unload poles. The hydraulically controlled clamping device holds the poles firmly in place on the forks. *(Courtesy Pettibone Corp.)*

Fig. 10-3 Cary-Lift® Loader being used by lineman to place wood poles in storage crib. *(Courtesy Pettibone Corp.)*

Fig. 10-4 Poles on railroad car are secured with chains and chain hoists. Additional stakes have been placed in the pockets on the side of the railroad car opposite the unloading side.

When the poles have been secured, the linemen should install additional stakes in the empty pockets on the side of the railroad car opposite the area to receive the poles (Fig. 10-4). A long "V" sling should be attached to the railroad car approximately 8 ft from the ends of the poles on the unloading side (Fig. 10-5) and extended over the poles to the winch of the line truck or a mechanical hoist (Fig. 10-6). Unloading skids should be secured to the side of the railroad car and extended to the ground, or pole pile, to permit the poles to roll from the car when they are released (Figs. 10-7 and 10-8). Steel rails, or timbers, can be used for this purpose.

The linemen should check the chains and the "V" sling to be sure the poles are secure, and then, starting in the center and working to each end of the car, remove the stakes and cut the

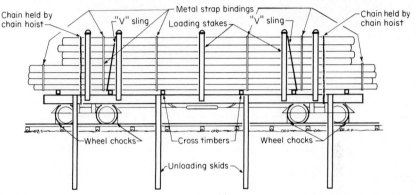

Fig. 10-5 Railroad car of poles viewed from unloading side after preparations for unloading have been completed.

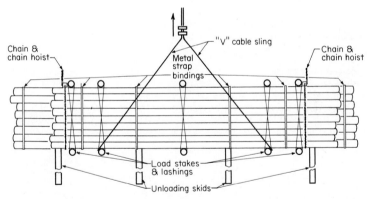

Fig. 10-6 Railroad car of poles (top view) after preparations for unloading have been completed.

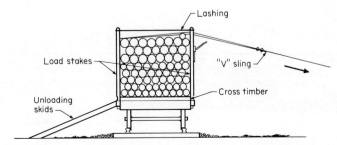

Fig. 10-7 Railroad car of poles (end view) after preparations for unloading have been completed.

Fig. 10-8   Steel rails secured to the railroad car and extended to the pole pile to permit the poles to roll from the car.

banding wires on the unloading side of the railroad car. When all personnel are in the clear, the chain hoists should be operated to provide slack in the chains. When the chains are slack, the "V" sling should be released slowly allowing the poles to roll off the railroad car onto the ground.

The remaining poles should be rolled off the car with a cant hook (Fig. 10-9). The lineman using the cant hook should stand at the end of the poles and remain alert in a position to move out of danger if the poles on the car should shift position unexpectedly. After all the poles have been removed from the railroad car, the skids and chocks should be removed and the area cleaned up disposing of the banding wires and stakes.

Framing Poles   Poles require several operations to be performed on them to prepare them for erection. These operations are often performed in the pole yard (Fig. 10-10) where the poles can be readily handled with mechanized equipment (Fig. 10-11).

The roofing operation consists in forming a roof on the top of the pole. The roof can be a simple slant cut as in Fig. 10-12(a) or a gable roof as in Fig. 10-12(b). Either type prevents water, snow, or ice from collecting on the top of the pole and causing decay. If poles are completely impregnated with wood preservatives, it is unnecessary to roof the pole. A simple cut at a 90° angle to the pole, as in Fig. 10-12(c), making a flat roof is sufficient.

To cut a gable roof, the pole should be raised into a framing buck or allowed to rest on two supports. The pole should then be turned so that the heaviest sag or curve will be nearest the ground. This is done by having one person turn the pole at the butt end with a cant hook while

Fig. 10-9   Lineman rolling poles from railroad car with a cant hook.

Fig. 10-10   Pole yard of a large electric utility company showing the manner of storing poles, reels, and crossarms. Poles are piled on 12 by 12 in creosoted timbers. Crossarms are piled neatly in background. The aluminum tag shown on butt end of some poles shows the length and the class of the pole. This tag is not removed when the pole is set. The darkened ends of the poles indicate creosote treatment. Crane for handling poles and cable reels is shown at right. The entire pole yard should be kept free from vegetation to prevent pole rot. *(Courtesy Wisconsin Electric Power Co.)*

Fig. 10-11   Cary-Lift® Loader being used by lineman to lift a pole from storage pile so that it may be framed. *(Courtesy Pettibone Corp.)*

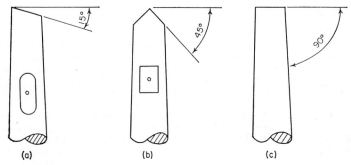

Fig. 10-12   Three-types of pole roofs. (*a*) Slant. (*b*) Gable. (*c*) Flat.

another sights along the pole and decides when the pole is in the right position. By putting the pole in this position, the direction of the roof will be up and down, and the gains will be on top. When this pole is erected in the line, the crook will be in the direction of the line. Looking along the line, the pole will appear straight. Figure 10-13 is a sketch of a pole provided with roof, and gains, and gives the correct relation of these to the crook or bend of the pole.

The gable roof is made by sawing the two sides of the pole top (Fig. 10-14) at an angle of 45° with the pole. One way of obtaining the 45° angle is to measure along the length of the pole a distance equal to one-half the pole-top diameter and to saw accordingly. This roof will shed the rain and prevent any accumulation of water, thereby decreasing possible decay.

A "gain" is the notch cut into the pole into which the crossarm is placed when mounted. Gains are cut to provide a flat surface to help maintain the arms in alignment. To cut a gain consists in

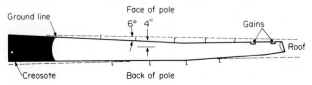

Fig. 10-13 Sketch of completely framed pole, showing relation of parts to curvature of pole.

Fig. 10-14 Pole top showing pole roof cut at 45° and two gains with crossarm through-bolt holes. *(Courtesy Wisconsin Electric Power Co.)*

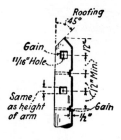

Fig. 10-15 Recommendations for the roofing and gaining of poles for low-voltage circuits.

first sawing along the upper and lower edges of the gain with a handsaw and then chiseling out the round portion and making a flat recess. It is good practice to hollow the gains out slightly in the center to ensure a snug fit of the crossarm, thereby preventing its rocking from side to side. There will be as many gains on a pole as the number of crossarms that the pole is to carry. In general, a pole is gained for all the crossarms whether all the crossarms are to be mounted at once or not.

The first crossarm gain is usually cut 12 in from the top of the pole. The succeeding gains are generally spaced 24, 30, or 36 in apart. The spacing selected depends largely on whether the pole is to carry buck arms or not. When buck arms must be provided, these are placed between the line arms at right angles. In order to permit the use of 28-in crossarm braces the spacing between line arm and buck arm should be at least 18 in. Gains should be square with the axis of the pole and about ½ in deep. The height depends on the height of the crossarm. The crossarm should fit snugly. Figure 10-15 gives suggested recommendations for roofing and gaining of distribution poles.

If several gains are to be cut, a gaining template (Fig. 10-16) is a great convenience. In using the template, the point of the roof of the template is placed over the roof of the pole and the template is shifted until its centerline lies in line with a string stretched from the point of the roof to the center of the butt indicated by a nail at the ground line. Then the outlines of the gains are marked with a sharp tool by marking along the side pieces of the template.

The vertical spacing of the crossarms, and consequently the gains, depends on the nature of the circuits and the voltages of the circuits carried by the arms. The specifications and standards for the type of construction to be used will show the details for gaining the poles.

One of the difficulties encountered in cutting more than one gain on a pole is to keep the flat surfaces of the gains parallel. This can be checked by resting or hanging two squares, one in each

Fig. 10-16   Template for use in marking gains.

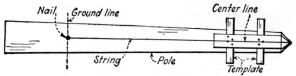

Fig. 10-17   Method of determining whether gains are cut parallel. Test is made by sighting over edges of squares. If gains are parallel, the edges of the two squares will be in line.

Fig. 10-18   Scheme for determining center for crossarm bolt hole. Two diagonal pencil lines are drawn as shown. The intersection of these lines is the bolt-hole center.

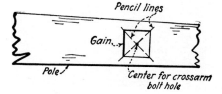

gain (Fig. 10-17). By sighting over them it is easy to see whether the gains are properly leveled. The square can also be used to advantage in getting buck-arm gains at right angles with the line gains.

The next operation consists in boring the holes for the crossarms. The size of the hole for the standard crossarm through bolt is ¹¹⁄₁₆ in. This is for the ⅝- or ¹³⁄₁₆-in through bolt. In the pole yard this operation is generally performed with a power drill. In the field, a brace and bit may be used.

A scheme often used for locating the center of the crossarm bolt hole is to draw two diagonal pencil lines across the gain (Fig. 10-18). The intersection of these lines determines the center of the bolt hole. In case the pole surface is not smooth, it is best to guess at the pole center but to measure for the vertical center.

Another scheme which takes less time is the use of a template. This consists of a short length of crossarm in which the hole has been bored. This is placed centrally over the pole in the gain, and the center is marked with a punch.

Wherever necessary, poles are painted. The desirability of painting is generally determined by appearance. Poles used in the streets of the larger cities are usually shaved and painted to improve

Fig. 10-19 Cary-Lift® Loader can rotate a pole and carry it parallel to the line of travel for easy maneuvering in the pole yard and efficiently loading the pole on a trailer for hauling to the work site. *(Courtesy Pettibone Corp.)*

Fig. 10-20 Pole on trailer connected to the back of line truck. Flag on the end of the pole is displayed to warn the public. *(Courtesy Iowa Illinois Gas & Electric Co.)*

Fig. 10-21 Pole is being lifted out of pile for loading on trailer with boom of line truck. Notice lineman controlling pole with cant hook. *(Courtesy A. B. Chance Co.)*

Fig. 10-22 Drawbar with eye attached to pole.

Fig. 10-23 Drawing eye coupled to truck.

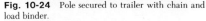

Fig. 10-24 Pole secured to trailer with chain and load binder.

Fig. 10-25 Rear view of trailer. Pole is secured with load-binder cable.

appearance. Some companies make a practice of painting the roof and the gains with one or more coats of approved paint.

Hauling Poles Loading of poles and hauling them from pole storage to the work site is hazardous to the linemen handling the poles and to the public. Moving poles short distances in the pole yard can be completed efficiently with mechanized equipment (Fig. 10-19). Poles are usually loaded on a trailer, and the trailer is pulled by a line truck to the work site (Fig. 10-20). The trailer is placed near the pole pile for convenience in loading the poles. The brakes on the trailer should be set and the wheels chocked while the poles are loaded on the trailer with mechanized equipment (Fig. 10-19) if available or with the boom of the line truck (Fig. 10-21).

If one of the poles is to be used as the tongue for the trailer, the pole selected should be of sufficient size and strength to carry the load. The drawbar is chained to the tongue pole (Fig. 10-22). The eye of the drawbar is engaged in the pintle hook on the back of the truck, and the safety chains on the truck are fastened to the tongue pole and drawbar (Fig. 10-23). The poles should be blocked into position as they are loaded on the trailer to keep them from shifting. The linemen must be careful not to overload the trailer or balance the load of poles so that the load is tongue heavy.

When all the poles have been loaded, they must be secured with cables or chains or both to keep the load from shifting in transit (Figs. 10-24 and 10-25). Figure 10-26 shows the overall view of a pole on a trailer complete with rigging. Flags should be installed on the end of the poles during daylight hours and warning lights after dark. The linemen must take care to protect the public while hauling poles on a highway. The driver should signal turns and stops early to warn other drivers. The poles may swing into adjacent traffic lanes when short turns are necessary. It is desirable for a lineman to use flags to warn the public if the area is congested. Figure 10-27 shows linemen and groundman having a tailboard conference at the work site to communicate the work plans to all members of the crew. The pole trailer in the picture has a metal tongue providing a mechanical coupling from the trailer to the line truck.

Fig. 10-26 Overall view of trailer. *(Courtesy Pacific Gas & Electric Co.)*

Fig. 10-27   Line truck is directly coupled to pole trailer with metal tongue. Crew leader is explaining work to linemen and groundman. *(Courtesy Iowa Illinois Gas & Electric Co.)*

Fig. 10-28   Transporting a fully framed pole by helicopter over rugged terrain. A 1-in manila rope sling 8 ft long was attached to each pole top with a spliced eye at the outer end for rigging to electrically operated sling hook suspended 4 ft below the helicopter. Each pole was numbered and keyed to a map. Pole holes were correspondingly identified by 2-ft square cardboard. *(Courtesy Pacific Gas & Electric Co.)*

Fig. 10-29   Helicopter transporting preassembled wooden pole H frame across the Roanoke River in Virginia. *(Courtesy American Electric Power System.)*

When the terrain is rugged and access roads are not available, poles can be transported by helicopter (Fig. 10-28). Each pole is completely framed in the pole yard or assembly point and numbered and keyed to a map. Pole holes are correspondingly identified so that each pole can be delivered to the correct location and lowered into the hole. Even a two-pole structure can be transported equally well as shown in Fig. 10-29, where a preassembled H frame is being carried by helicopter.

Section **11**

Erecting and Setting Poles

The specifications and drawings will give the location and size of each pole to be set. The pole locations are usually identified by stakes before the construction work is started. If one of the stakes has been removed or destroyed, the proper location can probably be determined by measuring and aligning from adjacent stakes. If several stakes have been lost or destroyed, it may be necessary to ask the foreman or engineer for assistance in resetting the stakes for the poles.

The holes for the poles are normally dug by a mechanical digger, sometimes referred to as an earth-boring machine. The mechanical digger consists of an auger, or helix, operated by a hydraulic system. The mechanical digger used to dig holes for large transmission-line poles is a complete unit normally mounted on a flatbed truck (Fig. 11-1). The casing around the auger operating shaft can be equipped with a pulley, for a winch line and pole grabber, to make it possible to set poles up to 65 ft long, which are commonly used for distribution lines, with the mechanical digger unit.

Fig. 11-1 Mechanical pole-hole digger mounted on a truck equipped with four-wheel drive. Unit is complete with motor, hydraulic-drive system, auger, and turntable to facilitate proper location and alignment of hole for pole.

Fig. 11-2 Boom- or derrick-mounted auger. The auger is supported by the line-truck boom and is operated by the truck's hydraulic system. *(Courtesy McCabe-Powers Body Co.)*

The standard way of digging pole holes for the smaller poles used for distribution lines is to use a derrick- or boom-mounted auger (Fig. 11-2). An auger attached to the boom of a corner-mounted derrick line truck is easily aligned to dig the pole hole in the proper location. A line truck, equipped with an auger, increases the versatility of the work that can be performed by the line crew.

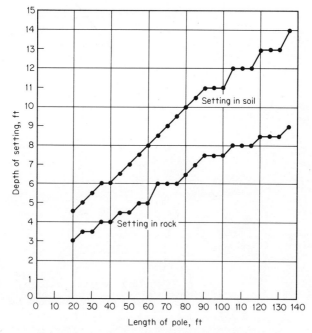

Fig. 11-3 Recommended pole-setting depths in soil and rock.

Pole-Hole Diameter   The diameter of the hole is determined by the size of the pole to be set. The hole should be large enough to allow plenty of space on each side of the pole for the free use of the tamping equipment, which requires at least 3 in all around the pole. The diameter of the hole at the top should not be greater than at the bottom, but rather it should be larger at the bottom than at the top. A slight increase at the bottom is often necessary to allow for the shifting of the pole in lining-in and to accommodate the larger diameter of the pole at the butt.

Pole-Hole Depth   The depth of setting is determined by the length of the pole and by the holding power of the soil or earth. The required average setting depths have been determined by experience. The recommended depths of setting in soil and rock are given in Table 11-1 and Fig. 11-3 for various pole lengths from 20 to 135 ft. It will be noted that a pole set in rock need not be set as deep as a pole set in soil.

A rule often followed for determining the setting depth in soil is to take 10 percent of the pole length and add 2 ft, with a minimum of 5 ft. The setting depth of a 40-ft pole would thus be:

$$10\% \text{ of } 40 = 4 \text{ ft} + 2 \text{ ft} = 6 \text{ ft}$$

This matches the table exactly. The pole-hole depths in Table 11-1 are for poles to be set in the straight portion of the line. Holes for poles that are to be set at angle or corner points in the line should be at least ½ ft deeper. If the soil is soft or marshy, dig the pole holes deeper. Holes for poles set on a slope should be ½ ft deeper than normal, measured from the lowest point on the ground.

Crushed rock is often used for backfill around transmission-line poles and distribution poles requiring extra stability (Fig. 11-4). The rock must be well tamped to obtain the desired strength.

Poles set in sandy or swampy ground should be set considerably deeper or be supported by guys, braces, or cribbing. One form of cribbing uses a barrel into which the pole is set. The barrel is then filled with concrete or small stones to make the pole secure. Another simple method of crib bracing is illustrated in Fig. 11-5.

When exceptional stability is required of a pole, an artificial foundation of concrete may be placed around the base (Fig. 11-6). This concrete filling should extend approximately 1 ft from the pole on all sides and should be carried ½ ft above the ground level. The top should be beveled to shed the rain. A good mix to use is 1 to 2½ to 5, that is, 1 part cement, 2½ parts of sand, and 5 parts broken stone or gravel. Enough water should be added to make the mixture flow freely and

Table 11-1   Recommended Pole-Setting Depths in Soil and Rock for Various Lengths of Wood Poles

Length of pole (ft)	Setting depth in soil (ft)	Setting depth in rock (ft)
20	4.5	3.0
25	5.0	3.5
30	5.5	3.5
35	6.0	4.0
40	6.0	4.0
45	6.5	4.5
50	7.0	4.5
55	7.5	5.0
60	8.0	5.0
65	8.5	6.0
70	9.0	6.0
75	9.5	6.0
80	10.0	6.5
85	10.5	7.0
90	11.0	7.5
95	11.0	7.5
100	11.0	7.5
105	12.0	8.0
110	12.0	8.0
115	12.0	8.0
120	13.0	8.5
125	13.0	8.5
130	13.0	8.5
135	14.0	9.0

take its place without tamping. The pole must be firmly braced in position, and the bracing must be left in position until the concrete is set. No line work should be done on the pole within a week after the concrete is poured.

Poles set on curves, corners, or at points of extra strain should be set at least 6 in deeper than the values given in Table 11-1. The side strain, however, should always be taken up with terminal, side, or line guys, leaving the pole to carry the vertical load only.

Poles subjected to slight side pulls can be kept erect by the use of pole keys (see Fig. 11-7). Very small angles in the line or service drops at right angles to the line produce slight side pulls. When

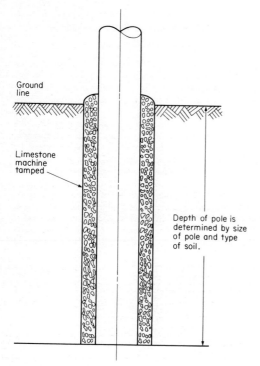

Ground line

Limestone machine tamped

Depth of pole is determined by size of pole and type of soil.

Fig. 11-4 Crushed rock used as backfill around pole to provide stability.

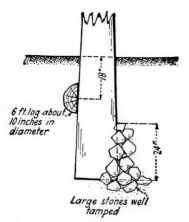

6 ft. log about 10 inches in diameter

Large stones well tamped

Fig. 11-5 A common and inexpensive form of crib bracing.

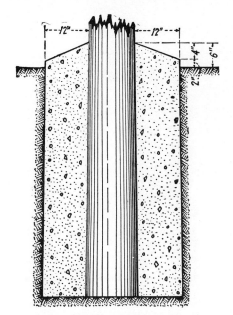

Fig. 11-6 Concrete foundation for wooden pole. The mixture should be made thin enough so that it will not require tamping. *(Courtesy Texas Power & Light Co.)*

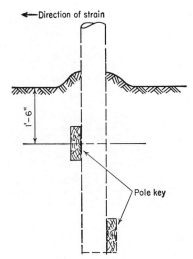

Fig. 11-7 Pole keys installed on pole which is subjected to slight side pull. When side pull is very slight, only the upper key is used. In loose soil both keys are used.

the poles are set in firm soil, one key placed 1 ft below ground surface on the strain side of pole is sufficient. In loose soil, however, a second key placed on the opposite side at the butt of the pole is required.

Pole-Hole Digging Mechanical diggers minimize the physical effort, and time, required to dig holes for poles. The lineman should locate the site where the pole is to be set and determine the best route to get the truck to the location. Special precautions must be taken if the hole is to be dug near an energized circuit. If adequate clearance cannot be maintained, protective equipment must be used to insulate the energized conductors to prevent accidental contact. If adequate protection cannot be obtained from existing conductors, the circuit should be deenergized, isolated, and grounded before the work is started. The truck should be located properly to permit accurate location of the auger.

It may be necessary to start angle-structure holes by hand. A post-hole-type auger, a digging bar, a 5-ft shovel, a 9-ft shovel, and a 9-ft spoon should be available. The power-digging equipment should be inspected before the auger is operated. The lineman should be sure that all personnel maintain adequate clearance to prevent clothing from being caught by the auger or other mechanical equipment.

When pole holes have been dug along paved streets, near farm buildings, or on any premises used by the public and poles are not set at once, the holes should be barricaded. If a hole is to be left overnight without setting the pole, the hole should be completely covered with a plank, and dirt, and the barricade should be lighted.

Pole-Hole Power Borer Pole holes are dug with a power earth borer. This greatly reduces the physical effort required and speeds up the work. A power borer is very similar to the familiar wood brace except that it is much larger and is power driven; Fig. 11-8 shows a typical power borer mounted on the rear of a truck body. The borer can be tilted down while moving.

When a hole is to be dug, the truck is driven over the pole location and placed so that the borer is directly over the pole stake. The truck outriggers are lowered to stabilize the vehicle. The table supporting the earth-borer motor and mechanism is moved, or rotated, to place the point of the auger over the stake when it is lowered (Fig. 11-9). The boring equipment must be positioned to maintain proper alignment while the pole hole is dug. Then the operator lowers the auger (Fig.

Fig. 11-8   Power pole-hole borer mounted on rear of truck and tilted forward into traveling position. Power for auger is obtained from special engine. *(Courtesy Iowa Illinois Gas and Electric Co.)*

11-10) while it rotates until it has taken a full "bite." Then the auger is raised until it is above the ground and made to spin rapidly (Figs. 11-11 to 11-13) thereby throwing the dirt around and away from the hole. Then the auger is lowered again, and another bite is taken. Again the auger is raised and made to spin fast. This is repeated until the desired depth has been reached.

The competed hole will have a ridge of loose dirt completely around it. The loose dirt should be moved away from the edge of the hole to prevent it from falling into the hole. All loose dirt should be removed from the bottom of the hole by use of the hand tools. A post-hole digger, a long-handled shovel, and a spoon are usually required (see Figs. 11-15 and 11-16).

Holes of different diameter are obtained by the use of augers of different diameter. Augers having different diameters can be mounted on the digging bar. The change from one to the other can be made in a few minutes. The 9-in size, the smallest, is generally used for digging guy-anchor holes. The 16-, 24-, 36-, and 48-in sizes are carried on the truck and are used for the different sizes of poles that must be set.

Pole-Hole Derrick Auger   Derrick-mounted power pole-hole augers (Fig. 11-14) are com-

Fig. 11-9   Pole-hole power borer in position to start digging. Truck outriggers are down and point of auger is touching ground at location established by a stake. *(Courtesy Iowa Illinois Gas and Electric Co.)*

Fig. 11-10   Auger lowered into hole for another bite of ground. Note loose dirt around hole. *(Courtesy Iowa Illinois Gas and Electric Co.)*

monly used for digging distribution-pole holes. This type of power hole digger increases the flexibility of the distribution-line crew in utilizing its vehicle and equipment efficiently. The procedures and precautions are similar to those described for the pole-hole power-boring equipment. The derrick-mounted auger can be used in locations several feet from the truck by extending the derrick, or boom, on which it is mounted. The auger is driven by the hydraulic system that operates the boom. The point of the auger is lowered to the ground at the pole location, usually identified by a stake. It is important to locate the auger accurately. The lineman must be sure the boom will move easily to keep the pole hole plumb and prevent damage to the equipment. The auger should be started to rotate slowly to prevent tilting.

The auger rotation should be stopped when the auger worm is full of dirt to avoid damage to the equipment. The rotation of the auger must be stopped before the auger worm gets stuck. The

Fig. 11-11   Auger raised with full bite of ground. *(Courtesy Iowa Illinois Gas and Electric Co.)*

Fig. 11-12 Auger is being rotated at high speed to throw dirt from auger and away from hole. *(Courtesy Iowa Illinois Gas and Electric Co.)*

dirt is removed from the hole by raising the derrick and reversing the rotation. If the derrick will not raise the auger, reverse the rotation of the unit to free it from the earth. A short-handled shovel can be used to remove the dirt from the auger.

Pole Hole Digging by Hand   Digging holes with shovel and spoon is resorted to when it is impossible to use a power digger. The shovel is a straight-bladed tool built like a spade (Fig. 11-15) but the spoon has the blade at an angle (Fig. 11-16). The shovel is used to loosen the soil, and the spoon is used to remove soil from the hole.

An experienced hole digger will dig the hole largely without the aid of the spoon under ordinary conditions of soil and moisture. A certain amount of practice is required to do this. One must know just how to strike the soil with the shovel so that the soil will break loose from its position and yet stick to the blade of the shovel. In case the soil is very dry, the shovel can be used to loosen the soil and the spoon used to lift the loose soil from the hole.

Fig. 11-13 Auger has been rotated at high speed and dirt has been thrown from auger and away from hole. *(Courtesy Iowa Illinois Gas and Electric Co.)*

Fig. 11-14 Drilling hole for pole with power auger mounted on boom of truck. *(Courtesy Pitman Manufacturing Co.)*

Sometimes a bar must be used to break the earth loose if there is frost, rock, or hard clay. One end of the bar should be sharpened to a blunt chisel point and the other end to a round point similar to the sharpened end of a pencil. Such a digging bar is shown in Fig. 11-17.

The usual procedure to be followed is to outline the hole around the stake. The diameter of this circle should be at least 6 in greater than the diameter of the pole at the butt. For the first 2 or 3 ft of digging, it is advisable to use a short-handled shovel (Fig. 11-18), as it is more convenient to hold. The spoon will not be necessary as the earth can be lifted out with the shovel. For the remainder of the hole the long-handled shovel and spoon are used in the usual manner. Where

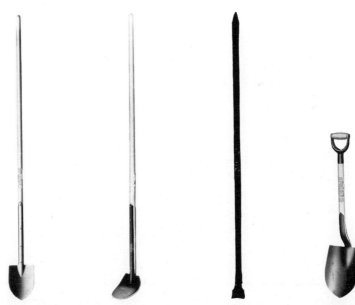

Fig. 11-15 Straight shovel used to lessen soil in digging a pole hole. *(Courtesy Leach Co.)*

Fig. 11-16 Spoon used to lift soil out of a pole hole while digging. *(Courtesy Leach Co.)*

Fig. 11-17 Digging bar used in hard or stony soil. *(Courtesy Leach Co.)*

Fig. 11-18 Short D-handled shovel often used in digging the first few feet of a pole hole. *(Courtesy Leach Co.)*

Fig. 11-19 Lineman using bottomless drum to dig a pole hole in sand or swampy soil. The drum is forced down as the hole is dug, thereby preventing the sides of the hole from caving in. Drum also helps to hold pole erect because of greater bearing surface. (*Courtesy Electric Lineman, Series 100.*)

quicksand or swamp soil is encountered and cave-ins occur, a large bottomless barrel is lowered into the hole as it is dug. As the hole is dug deeper, the barrel or drum is forced down, thus preventing the side of the hole from falling in (Fig. 11-19). The drum will also provide a greater bearing surface for the pole when it is set and tamped in the inside of the drum.

Setting Poles with Derrick   The most common method of setting poles is with a truck derrick. The truck should be equipped with a derrick (boom), winch line, and pole grabber (claws). Hand tools required include a cant hook, pike poles, and rope lines. Protective equipment required includes rubber gloves, hard hats, safety glasses, and polyethylene pole covers, or epoxy-rod pole guards.

The line truck should be positioned appropriately to dig the hole, if a boom-mounted auger is to be used, and to erect the pole (Fig. 11-20). The crew should be informed how the work is to be

Fig. 11-20 Line truck with poles to be set on trailer behind truck. Crew leader is holding tailboard conference to discuss the hazards and to give directions for completing the job. (*Courtesy Iowa Illinois Gas and Electric Co.*)

Fig. 11-21   Pole is insulated with polyethylene pole covers designed for use when setting a pole near 13.2Y/ 7.62-kV three-phase circuit. Winch line is attached to pole so that the pole is a little butt heavy. *(Courtesy Iowa Illinois Gas and Electric Co.)*

performed. The hazards should be identified so that all personnel will understand the safe way of performing the work. The truck outriggers must be positioned in the proper manner to stabilize the vehicle when the pole is raised. Wood poles are heavy and could tip a line truck over if proper precautions are not taken. The weights of common sizes and types of wood poles are itemized in Table 11-2.

If the pole is to be set near an energized line, the pole should be covered with protective equipment with proper insulation for the voltage of the energized circuit. The winch line is attached to the pole slightly above the midpoint so that the butt end outweighs the top end (Fig. 11-21). The truck derrick is used to move the pole from the pole trailer to the proper position for setting it in the hole. The linemen may have to balance the pole and guide it to the proper position (Fig. 11-22). The linemen should not stand or pass under the pole while it is suspended by the winch cable and derrick. The pole is gradually raised until it is in a near vertical position and the butt end is clear of the ground. Linemen wearing rubber gloves guide the pole as it is raised (Fig. 11-23).

The crew leader, or foreman, directs the operation using standard hand signals for line work. As the pole reaches vertical position, the pole claws are used to grab the pole and stabilize it (Fig. 11-24). When the pole claws have grabbed the pole, the derrick and winch line are manipulated

Fig. 11-22   Lineman guiding and balancing pole while it is supported by derrick and winch line. *(Courtesy Iowa Illinois Gas and Electric Co.)*

Table 11-2   Weights of Wood Poles That Are Used Extensively

Length of pole (ft)	Species*	Class 1H	1	2	4	5
		Average weight, lb per pole				
30	WC	1320	880	750	540	440
	DF	1760	1110	930	690	600
	SP	1884	1279	1082	784	660
	WL	1515	955	800	595	515
35	WC	1585	1055	880	660	570
	DF	2070	1435	1260	875	770
	SP	2223	1568	1343	1004	862
	WL	1780	1235	1085	750	660
40	WC	1760	1320	1145	790	705
	DF	2500	1760	1560	1120	920
	SP	2585	1884	1623	1219	1059
	WL	2150	1515	1340	965	790
45	WC	2025	1585	1365	1010	880
	DF	2860	2070	1845	1350	1125
	SP	2993	2223	1911	1444	1274
	WL	2460	1780	1585	1160	965
50	WC	2290	1760	1585	1230	1145
	DF	3360	2500	2150	1600	1300
	SP	3451	2585	2214	1687	1494
	WL	2890	2150	1850	1375	1120
55	WC	2815	2025	1760	1410	1410
	DF	3835	2860	2475	1815	1540
	SP	4015	2993	2567	1934	1719
	WL	3300	2460	2130	1560	1325
60	WC	3170	2290	1935	1670	
	DF	4340	3360	2820	2040	1740
	SP	4620	3451	2943	2186	1953
	WL		2890	2425	1755	
65	WC	3695	2815	2200	1935	
	DF	4800	3835	3250	2340	2015
	SP	5198	4015	3341	2457	2237
	WL		3300	2795	2010	
70	WC	4400	3170	2640	2290	
	DF	5240	4340	3640	2590	2240
	SP	6400	4620	3781	2732	2488
75	WC	4840	3695	3170	2640	
	DF	5780	4800	4050	2925	
	SP	7200	5198	4235	3021	
80	WC	5810	4400	3695	3080	
	DF	6300	5240	4400	3200	
	SP	8140	6400	5170	3615	
85	WC	6750	4840	3960		
	DF	6935	5780	4930		
	SP		7200	5745		
90	WC	7500	5810	4930		
	DF	7600	6300	5400		
	SP		8140	6405		
95	WC		6750	5950		
	DF		6935	5985		
100	WC		7500	6550		
	DF		7600	6700		

SOURCE: Commonwealth Edison Co.
*WC = western red cedar; DF = Douglas fir; WL = western larch; SP = southern pine.

Fig. 11-23 Linemen, wearing rubber gloves and protectors, guide pole as it is raised. *(Courtesy Iowa Illinois Gas and Electric Co.)*

Fig. 11-24 Pole claws are used to grab and stabilize pole. *(Courtesy Iowa Illinois Gas and Electric Co.)*

Fig. 11-25 Crew leader uses standard hand signal to direct linemen operating derrick to set pole in hole. *(Courtesy Iowa Illinois Gas and Electric Co.)*

as directed by the crew leader to set the pole in the hole (Fig. 11-25). If the pole is to be set in an existing energized line, it may be necessary to hold the energized conductor apart to have adequate clearance. This can be done with hot-line tools and rope lines, as shown in Fig. 11-26.

After the pole is set in the hole and properly aligned in accordance with the specifications, the hole around the pole is backfilled and tamped (Figs. 11-27 and 11-28). After the pole is set and

Fig. 11-26 Energized conductor held in the clear with roller link-stick hot-line tools and rope lines. *(Courtesy A. B. Chance Co.)*

Fig. 11-27 Lineman and groundman backfilling and tamping dirt in hole around pole. *(Courtesy Iowa Illinois Gas and Electric Co.)*

Fig. 11-28 Hydraulic-powered tamper packing dirt around pole to keep pole in place. *(Courtesy Pitman Manufacturing Co.)*

Fig. 11-29 Lineman in bucket mounted on top of boom attaching equipment and conductor to pole after it is set. Note protective equipment on the lineman and the energized conductor.

secured by backfilling and tamping, linemen remove the temporary protective covering used to set the pole and install the equipment on the pole necessary to complete the project (Fig. 11-29).

Transmission structures are usually assembled on the ground complete with insulators, hardware, conductor stringing blocks, and small rope lines (Fig. 11-30). The structures are set in place with caterpillar-mounted power derricks or mobile cranes (Fig. 11-31).

Setting Poles with Helicopter Erecting single-pole lines and transmission structures using conventional construction methods in mountainous, swampy, muddy, or snow and ice conditions is difficult, slow, and very costly (Fig. 11-32). Single poles are often transported by helicopter to

Fig. 11-30 Linemen assembling H-frame transmission-line structure on the ground to replace structure destroyed by ice storm. Note damaged structure and conductors laying on the ground. *(Courtesy Nebraska Public Power District.)*

Fig. 11-31   Raising assembled H frame for 132,000-volt line with caterpillar-mounted power boom. Note use of lifting yoke to equalize lift. *(Courtesy American Electric Power System.)*

Fig. 11-32   Linemen setting poles singly for 161-kV transmission-line structure in rough terrain. *(Courtesy L. E. Myers Co.)*

Fig. 11-33 Setting pole by helicopter. The fully framed pole, weighing about 1100 lb is delivered by a Sikorsky S-58 helicopter and set into hole. Only one lineman is needed to guide pole into hole. A 4-mile average round trip over a 1500-ft elevation change took only 8½ min. *(Courtesy Pacific Gas and Electric Co.)*

mountainous locations where land transportation is difficult or practically impossible. The poles are normally airlifted with crossarm, insulators, and hardware in place (Fig. 11-33). When the poles arrive at the site, they can usually be lowered into the pole hole directly. Instead of unloading the pole, the helicopter hovers over the assigned pole hole and slowly lowers the pole into it. One or two crewmen are usually needed to guide the butt of the pole into the hole.

Transmission structures can be assembled on the ground at a convenient marshaling yard (Figs. 11-34 and 11-35).

Fig. 11-34 Two-pole transmission structure assembled on airport concrete apron. Poles and timbers for additional structures are located near the structure. *(Courtesy L. E. Myers Co.)*

Fig. 11-35 Assembled two-pole transmission structures laid out on airport concrete apron in rotation so that when the helicopter picks them up they are headed in the proper direction. *(Courtesy L. E. Myers Co.)*

Fig. 11-36 Setting H frame by helicopter. Two linemen guide each pole into hole. *(Courtesy American Electric Power System.)*

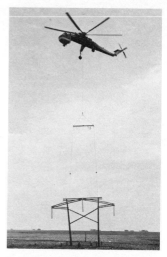

Fig. 11-37 Helicopter use to transport and erect K-frame transmission structure to rebuild 345-kV transmission line destroyed by ice storm. *(Courtesy Nebraska Public Power District.)*

The helicopter hovers over the assembled transmission structure at the marshaling yard while the ground crew attaches the structure to the cables extended from a winch installed on the helicopter. Communications are executed using two-way radios and standard hand signals for line work. The helicopter airlifts the structure to the site where it is to be installed. Figure 11-36 shows an H-frame structure being lowered into its holes. Figure 11-37 shows a K-frame transmission structure being set in predrilled holes after being transported 15 miles by helicopter. The helicopter pilot is directed from the ground by two-way radio. Sometimes considerable saving in time and money can be realized by the use of a helicopter for transport and for erection.

Fig. 11-38 Pike pole used in raising pole beyond the reach of the crew. Note steel spike on end of pole. *(Courtesy Leach Co.)*

Table 11-3   Average Size of Crew Required to Raise Poles of Different Lengths

Pole length, ft	Size of crew	Number of pikers	Number of jennymen	Number of men at butt
25	5	3	1	1
30	6	4	1	1
35	7	5	1	1
40	8	6	1	1
45	10	8	1	1
50	10	8	1	1

Setting Poles Manually   The piking method is the oldest method of raising poles. It is a manual method. It gets its name from the tool called a "pike pole" (Fig. 11-38) employed after the pole is lifted beyond the reach of the crew raising the pole. The piking method is used only where a truck with a derrick cannot be brought in. The use of a power boom is a faster and more economical method as only a few workers are needed.

The size of the piking crew depends upon the length and the weight of the pole to be raised. It varies from 5 for a 25-ft pole to 10 for a 50-ft pole, as shown in Table 11-3. The size of the crew required is not absolutely fixed, but the numbers indicated in Table 11-3 are the size of crews generally employed. As will be noted, one of the crew is the jennyman, one is stationed at the butt of the pole near the hole, and the balance of the crew are called the "pikers."

The first step in the procedure of raising a pole by the piking method is to lay the butt end of the pole over the hole against a bump board or bar, as shown in Fig. 11-39. The use of the board or bar protects the walls of the hole and prevents them from being caved in by the butt of the pole as the pole is raised. In the second step (Fig. 11-40), the upper end of the pole is raised and placed on the pole support. Pole supports are made in various forms, two of which are shown in Fig. 11-41. The main duty of the man at the butt is to keep the pole from rolling. This is done by means of a cant hook, illustrated in Fig. 11-42. In the third step (Fig. 11-43), the men stand side by side on both sides at the top end of the pole. They then push toward each other and up by use of their arms, the jennyman catching the weight between lifts. In this manner they move along the pole until the pole is high enough to permit the use of pikes. The fourth step (Fig. 11-44) is to punch the pikes into the pole and prepare to raise the pole. As the pole is raised, the man carries the jenny forward always ready to support the pole if need be. The raising continues until "high pike" is called by one of the men. This means that his pike is too high to be effective if raised any

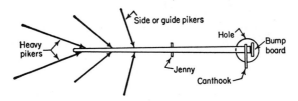

Fig. 11-39   Plan view of pole-piking method of pole raising.

Fig. 11-40   Second step in raising a pole. Pole is being lifted from ground into jenny.

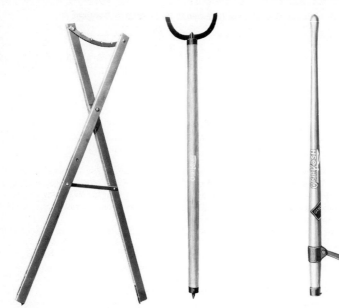

Fig. 11-41 Two forms of pole supports. The one on the left shown open is called the "jenny," and the one on the right is known as the "mule." *(Courtesy Leach Co.)*

Fig. 11-42 Cant hook used by man at butt of pole to keep pole from rolling and to turn pole if necessary. *(Courtesy Leach Co.)*

Fig. 11-43 Third step in raising a pole. Pikers have raised pole onto jenny and are now raising pole as high as they can without use of pike poles.

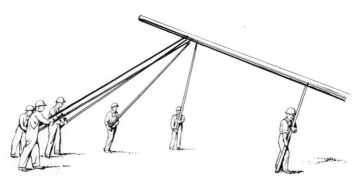

Fig. 11-44 Fourth step in raising pole. Pikers have placed pike poles against pole and are raising it into vertical position. Note how pikers are spread out.

further. The low man brings his pike down first, because he does not have to lower his pike through the rest of the pikes, but has a clear path. The other men follow in order until all the pikes are lowered. The pole is raised in this manner until it drops into the hole. Figure 11-45 shows the pole almost raised and ready to slide into the hole. This picture shows very clearly the duty of each one of the crew. The five pikers raise the pole, the jennyman keeps the jenny always snugly under the pole, and the buttman guides the bottom of the pole and keeps the pole from rolling.

It will be noted that the pikers are not directly underneath the pole. At least one should be well out on each side to guide the pole as well as to help lift it.

Another thing worth noting is the manner of holding the pike pole after the pole is partly raised. It is held in the palm of the hand with the other hand underneath, both arms extended downward. If the pike pole is held in this manner, the worker is in a comfortable position, and if the weight of the pole should be suddenly thrown on the pike pole, he is in a good position for grounding the pike pole or moving into the clear. If the pole should be a very tall pole, the pike pole can be supported on the shoulder as shown in Fig. 11-45. The pike still rests in the palm of the piker's hand. In this way the piker can exert a strong push on the pike.

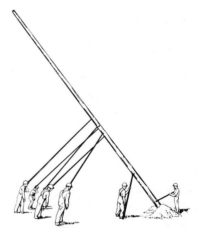

Fig. 11-45   Last step in raising a pole. All pikes have been lowered. Pole is about to slide into hole. Note jennyman and buttman with cant hook. Also note manner of pushing on pike poles.

If the pole to be raised is a large heavy pole and the crew is small, it is well to "trench" the hole, that is, to cut a ditch back from the hole. The pole is then placed in this trench. This allows the pole to begin to slide into the hole earlier than if the pole lay flush on the ground. Furthermore, it allows the weight of the lower end of the pole to balance a portion of the weight of the pole above the point of the trench on which the pole is resting.

Facing the Poles   "Facing" the poles means turning the poles after they have been raised so that the crossarms will be on the proper side. The side of the pole on which the gains are cut is known as the "face" of the pole. Facing the pole therefore means turning the pole until the gains are on the proper side of the pole. In general, poles should be set so that the ridge of the pole roof is in line with the lead. The ridge of a junction pole should be placed in line with the main lead. The ridge of a guy stub pole should be in line with the guy.

On straight lines it is customary to set adjacent poles with the gains facing in opposite directions (Fig. 11-46). This system is sometimes called "gain to gain" or "back to back." It provides for the maximum strength in the line.

Crossarms on poles on each side of the center of a curve should face the center of the curve (Fig. 11-47). The pole at the center of the curve is often equipped with double arms.

Poles next to angles should face the angle, as shown in Fig. 11-48. In case of large angles the angle pole is generally double armed or dead-ended.

Poles next to corners should face the corner (Fig. 11-49). In case of city distribution, the arms on the corner pole are generally double arms as illustrated.

Poles on steep grades should face up the grade as shown in Fig. 11-50.

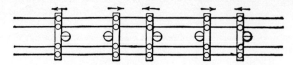

Fig. 11-46 Facing crossarms on straight lines. Every second pole has crossarm facing in the same direction.

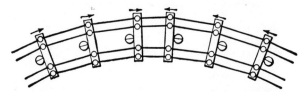

Fig. 11-47 Facing crossarms on curve. Note that the three poles on each side of the center of the curve face the curve.

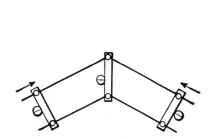

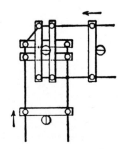

Fig. 11-48 Facing crossarms at an angle. Adjacent poles face the angle.

Fig. 11-49 Facing crossarms at a corner. Adjacent poles face the corner.

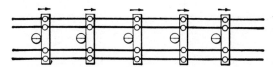

Fig. 11-50 Facing poles on steep grade. All poles face upgrade.

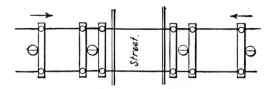

Fig. 11-51 Facing poles at a crossing. Adjacent poles face the crossing.

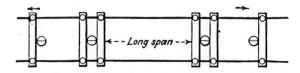

Fig. 11-52 Facing poles on long spans. Adjacent poles face away from long span.

Poles next to street crossings should face the street (Fig. 11-51). The poles at the intersections are generally double armed as illustrated.

Poles next to poles supporting long spans should face away from the long span, thereby being in a better position to carry part of the load (Fig. 11-52). The poles on each side of a long span are generally of special construction.

Poles immediately preceding end poles shall face toward the end of the line (Fig. 11-53). The

Fig. 11-53 Facing poles at terminals. Adjacent poles face terminal.

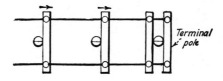

Fig. 11-54 Crew backfilling and tamping. Note that crew consists of one man at shovel and two tampers. Also note pike poles still in position holding pole in alignment. The dark lower portion of the pole is the creosoted butt. *(Courtesy Wisconsin Electric Power Co.)*

Fig. 11-55 Installing a creosoted block to make the pole self-sustaining. This pole happens to be at a slight angle in the line. It was set against the side of the hole on the "strain" side to give it good solid ground backing. Then No. 2 crushed stone was placed at the bottom of the hole opposite to the strain. And finally the 3- by 12- by 30 in creosoted block was placed against the pole on the "strain" side about 6 in below the ground. A pole set in this manner can take some side strain without being guyed. The figure shows one shoveler, two tampers, one man placing the creosoted block, and the crew foreman directing operations. Note that all men, including the foreman, wear hard plastic hats for protection against head injuries. *(Courtesy Wisconsin Electric Power Co.)*

Fig. 11-56 Wood and steel tamping bars used in pole setting. *(Courtesy Leach Co.)*

end pole, being a terminal pole, is usually of special construction. Sometimes it is well to have the last two poles face the terminal pole.

Poles set at line terminals, curves, corners, and other points of abnormal stress should be given a slight rake against the direction of the pull. This should be sufficient to allow for the change in pole position caused by the continued pull of the load and the normal creepage of the anchor.

Backfilling and Tamping Pole Hole   After the pole is lined in, the hole is backfilled and tamped (Figs. 11-54 and 11-55).

Tamping of the dirt around the pole is completed with a hydraulic-powered tamper (Fig. 11-28) or with a tamping bar (Fig. 11-56).

Too much stress cannot be laid on proper tamping, as a poorly tamped hole will not hold the pole in alignment. The earth should be backfilled slowly and each layer thoroughly tamped until the tamp makes a solid sound as the earth is struck. In general, the tamping should be done so thoroughly that no dirt need be hauled away.

If small stones or gravel are readily available, these should be used in backfilling as a better foundation can thus be obtained. Care should be taken that plenty of earth is tamped in to fill the spaces between the stones. The backfill should be piled well up around the base of the pole to allow for settling. An examination should be made 1 or 2 months later and backfilling added if it has settled below the ground level.

Section 12

Guying Poles

A guy is a brace or cable fastened to the pole to strengthen it and keep it in position. Guys are used wherever the wires tend to pull the pole out of its normal position and to sustain the line during the abnormal loads caused by sleet, wind, and cold. Guys counteract the unbalanced forces imposed on the poles by dead-ending conductors, changing conductor sizes, types, and tensions, or by angles in the transmission or distribution line. The guy should be considered as counteracting the horizontal component of the forces with the pole or supporting structure as a strut resisting the vertical component of the forces.

Anchor or Down Guys An "anchor" or "down guy" consists of a wire running from the attachment near the top of the pole to a rod and anchor installed in the ground (Fig. 12-1). This type of a guy is preferable if field conditions permit its installation since it transfers the unbalanced force on a pole or structure to the earth without intermediate supports.

An anchor or down guy used at the ends of pole lines in order to counterbalance the pull of the line conductors is called a "terminal guy" (Fig. 12-2). All corners in the line are considered as dead ends. They should be guyed the same as terminal poles except that there will be two guys, one for the pull of the conductors in each direction (Fig. 12-3).

Where the line makes an angle, a side pull is produced on the pole. Side guys should be installed to balance the side pull (Figs. 12-4 and 12-5). Figure 12-6 shows side guys installed in the line as it changes from one side of the highway to the other.

Where a branch line takes off from the main line, an unbalanced side pull is produced. A side guy should be placed on the pole directly opposite to the pull of the branch line. Anchor or down guys are installed at regular intervals in transmission lines that extend long distances in one direction to protect the line from excessive damage as a result of broken conductors (Fig. 12-7). Guys installed to protect the facilities and limit the damage if a conductor breaks are called "line guys" or "storm guys."

An anchor guy with a horizontal strut at a height above the sidewalk to clear the pedestrians on the sidewalk is referred to as a "sidewalk guy." Figure 12-8 illustrates two sidewalk guys installed at one pole to serve as branch guys.

Span Guy A span or overhead guy consists of a guy wire installed from the top of a pole to the top of an adjacent pole to remove the strain from the line conductors. The overhead or span guy transfers the strain on a pole to another structure. This may be to another line pole or to a stub pole on which there is no energized equipment. A span guy is always installed to extend from the strain pole to the same or lower level on the next line pole.

Head Guy A guy wire running from the top of a pole to a point below the top of the adjacent pole is called a "head guy" (Fig. 12-9). Lines on steep hills are normally constructed with head guys to counteract the downhill strain of the line.

Electric lines crossing railroad tracks must be reinforced by the use of double arms and head guys as shown in Fig. 12-10. These guys run from the top of the pole adjacent to the track to a point about halfway down on the next pole.

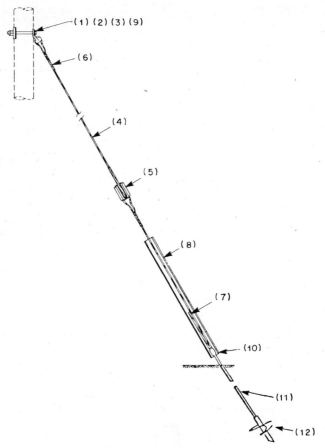

Fig. 12-1   Anchor or down guy assembly. (1) Galvanized machine bolt with nut; (2) locknut; (3) square curved galvanized washer; (4) galvanized steel guy wire; (5) porcelain guy strain insulator; (6) prefabricated guy dead-end grip; (7) prefabricated guy dead-end grip; (8) plastic guy guard; (9) angle thimbleye; (10) eyenut; (11) steel anchor rod; (12) power-installed screw achor.

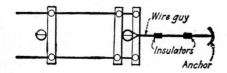

Fig. 12-2   Wire guy installed on terminal or end pole.

Fig. 12-3 Terminal guys installed at corner of H-frame transmission line. Entire pull of line conductors has to be carried by two sets of three guys. Note wood insulators inserted in down guy wires. *(Courtesy Joslyn Mfg. and Supply Co.)*

Fig. 12-4 Guy installed at angle in line. The guy opposes the side pull of the line conductors.

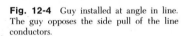
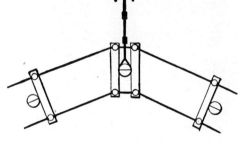

Fig. 12-5 Side guys counterbalancing the side pull of the line conductors due to an angle in the line. Note use of wood strain insulators in each guy. *(Courtesy Iowa-Illinois Gas and Electric Co.)*

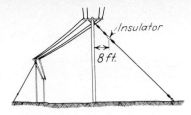

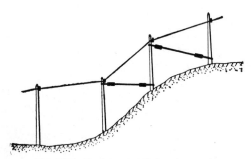

Fig. 12-6 Down guys installed in line to balance side pulls caused by angles in the line.

Fig. 12-7 Line guys installed on three-phase H-frame transmission line. Note that poles are guyed in both directions, primarily installed to protect line from broken conductors. *(Courtesy Hubbard and Co.)*

Fig. 12-8 Sidewalk guys used to counteract the pull of branch wires crossing street.

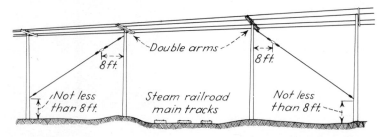

Fig. 12-9 Head guys installed on steep grade.

Fig. 12-10 Line reinforcement by use of double arms and head guys at a railroad crossing.

Arm Guys   A guy wire running from one side of a crossarm to the next pole is called an "arm guy." Arm guys are used to counteract the forces on crossarms which have more wires dead-ended on one side than on the other (Fig. 12-11).

Stub Guys   A guy wire installed between a line pole and a stub pole on which there is no energized equipment is called a "stub guy" (Fig. 12-12). An anchor or down guy is used to secure the stub pole. This type of guy is often installed to obtain adequate clearance for guy wires extend-

Fig. 12-11   Arm guy. Guy opposes unbalanced pull on crossarm.

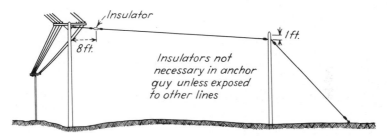

Fig. 12-12   Use of pole stub to secure clearance over highway. Pole stub is then guyed in usual manner.

Fig. 12-13   Stub guy used to obtain clearance of guy wire over highway. *(Courtesy Copperweld Steel Co.)*

ing across streets or highways (Fig. 12-13). When transmission or distribution lines parallel to streets or roads must be guyed toward the street or road, the necessary clearance can be obtained by the use of a pole stub on the opposite side.

Push Guys   A pole used as a brace to a line pole is often referred to as a "push guy" (Fig. 12-14). A push brace or guy is used where it is impossible to use anchor or down guys. When it is impossible to obtain sufficient right of way for a down guy, the push brace can usually be installed (Fig. 12-15).

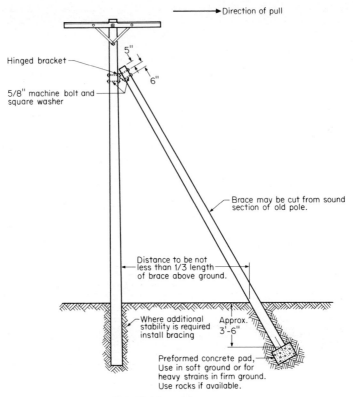

Direction of pull

Hinged bracket

5"

6"

5/8" machine bolt and
square washer

Brace may be cut from sound
section of old pole.

Distance to be not
less than 1/3 length
of brace above ground.

Where additional
stability is required
install bracing

Approx.
3'-6"

Preformed concrete pad,
Use in soft ground or for
heavy strains in firm ground.
Use rocks if available.

Fig. 12-14   Push brace or guy.

Fig. 12-15   Single-pole brace used in place of stub guy.
(Courtesy Line Material Industries.)

Table 12-1   Guy Strand Sizes and Strengths

Number of wires in strand	Diameter, in	Minimum breaking strength of strand, lb				
		Utilities grade	Common grade	Siemens-Martin grade	High-strength grade	Extra high-strength grade
7	⁹⁄₃₂	4,600	2,570	4,250	6,400	8,950
7	⁵⁄₁₆	6,000	3,200	5,350	8,000	11,200
7	⅜	11,500	4,250	6,950	10,800	15,400
7	⁷⁄₁₆	18,500	5,700	9,350	14,500	20,800
7	½	25,000	7,400	12,100	18,800	26,900
7	⁹⁄₁₆	—	9,600	15,700	24,500	35,000
7	⅝	—	11,600	19,100	29,600	42,400
19	½	—	—	12,700	19,100	26,700
19	⁹⁄₁₆	—	—	16,100	24,100	33,700
19	⅝	—	—	18,100	28,100	40,200
19	¾	—	—	26,200	40,800	58,300
19	⅞	—	—	35,900	55,800	79,700
19	1	—	—	47,000	73,200	104,500
37	1	—	—	46,200	71,900	102,700
37	1⅛	—	—	58,900	91,600	130,800
37	1¼	—	—	73,000	113,600	162,200

Guy Wire The wire or cable normally used in a down guy is seven-strand galvanized steel wire or seven-strand alumoweld wire (Fig. 12-16). Alumoweld wire consists of steel wire strands coated with a layer of aluminum to prevent corrosion. Guy wire is used in various sizes with diameters from ¼ to 1¼ in. The breaking or ultimate strength of the various types of guy wire are given in Table 12-1. Alumoweld guy cable, seven-strand No. 8 wire, has a minimum breaking strength of 15,930 lb, and seven-strand No. 5 wire has a minimum breaking strength of 27,030 lb.

Anchors The log anchor is the oldest type anchor used to counteract the unbalanced forces on electric transmission and distribution line structures. The log anchor is frequently referred to as a "dead man." This nickname comes from the grave-like dimensions of the excavation necessary to install the log anchor. The cost of the excavation necessary to install the log anchor in the ground limits its use. Figures 12-17 and 12-18 illustrate the component parts of a log anchor assembly.

Manufactured anchors are easier, quicker, and less expensive to install than log anchors. The common types of manufactured anchors consist of the expansion anchor (Fig. 12-19), the screw-type anchor (Fig. 12-20), and the "never-creep" or plate-type anchor (Fig. 12-21).

Log or manufactured anchors cannot be installed to secure a down guy assembly in an area where rock formations are close to the surface of the ground. If rock is located close to the surface, expanding rock-type anchors are used (Fig. 12-22). Rock anchors, properly installed, will develop holding power equal to the full strength of the anchor rod.

A log "deadman" anchor 8 ft long with a diameter of 9 in installed 6 ft deep in loose or sandy-type soil, with an angle of pull for the guy wire and rod assembly equal to 45 degrees, should have a holding power of approximately 28,000 lb. This holding power can be determined by calculating the weight of the soil that must be lifted to extract the log from the earth. Loose sandy-gravel-type soil weighs approximately 100 lb per cubic foot (ft^3). The volume of the soil of the fill above the anchor, and in line with the pull, is approximately 280 ft^3. Tests performed by the American Electric Power Co. on this type of anchor, when installed in wet loam and clay-type

Fig. 12-16 Seven-strand steel guy cable. (*Courtesy Joslyn Mfg. and Supply Co.*)

Fig. 12-17 Guy anchor rod used with log anchor. (*Courtesy Joslyn Mfg. and Supply Co.*)

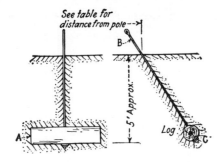

Size of pole, ft	Minimum distance from pole, ft	Material		
		Loc.	No.	Description
30 and 40	15	A	1	12″ × 6′—0″log
45 and 50	20	B	1	Anchor rod
55 and 60	25	C	1	Anchor-rod washer

Fig. 12-18 Views showing log anchor installed. Tables list materials required as well as distance from pole at which anchor should be placed for various heights of poles.

soil, determined that a ¾-in anchor rod failed with a pull of 27,000 lb without disturbing the log in the earth. The holding strength of the different types of manufactured anchors are itemized in Table 12-2.

Anchor Rods   The anchor rod serves as the connecting link between the anchor and the guy cable (Fig. 12-17). The rod must have an ultimate strength equal to, or greater than, that required by the down-guy assembly. The strength of commonly used anchor rods is itemized in Table 12-3. Anchor rods vary in diameter from ½ to 1¼ in and in length from 3½ to 12 ft. Power installed ⅝ in D anchor rods are normally 6 to 8 ft long. If a greater length is needed, a 3½-ft rod can be coupled to the longer rod to reach a greater depth (Fig. 12-23).

(a)   (b)

Fig. 12-19 Three-way-type Everstick anchor. (a) Expanded position and (b) closed position. The anchor is inserted in the hole in the closed position. Then the top of the anchor is struck several blows with a ram to cause the leaves to expand into solid undisturbed earth at the bottom of the hole. (*Courtesy The Everstick Anchor Co.*)

Table 12-2   Holding Power of Commonly Used Manufactured-Type Anchors

Anchor		Holding strength, lb	
Type	Size in	Poor soil	Average soil
Expansion	8	10,000	17,000
Expansion	10	12,000	21,000
Expansion	12	16,000	26,500
Screw or Helix°	8	6,000	15,000
Screw or Helix	11⅝₆	9,500	15,000
Screw or Helix	32	10,000	23,000
Screw or Helix	34	15,500	27,000
Never-Creep		21,000	34,000
Cross-Plate		18,000	30,000

°Screw or Helix anchors power installed.

Fig. 12-20   Screw-type anchor. *(Courtesy Hubbard and Co.)*

Fig. 12-21   View showing "never-creep" or plate-type anchor installed. *(Courtesy A. B. Chance Co.)*

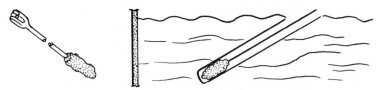

Fig. 12-22   Rock anchor expanded in hole drilled in the rock. The greater the strain, the more firmly it is wedged in the rock.

Table 12-3   Anchor Rod Strength

| Anchor rod nominal diameter, in | Ultimate tensile strength, psi | Yield point, psi | Loads required to produce these stresses | | | |
| | | | Full rod section | | Threaded section | |
			Ultimate load lb	Yield load, lb	Ultimate load lb	Yield load, lb
½	70,000	46,500	13,700	9,120	12,750	8,450
⅝	70,000	46,500	20,400	13,500	18,700	12,400
¾	70,000	46,500	29,600	19,600	27,100	18,000
1	70,000	46,500	53,200	35,300	42,400	28,100
1 high strength	90,000	59,500	68,200	45,100	54,500	36,000
1¼	70,000	46,500	85,800	57,000	67,800	45,000
1¼ high strength	90,000	59,500	110,000	73,000	87,200	57,600

Figure 12-24 illustrates a helix anchor. Several helixes are stacked on a 1½-in sq steel shaft. Each helix acts essentially as a separate anchor for increased holding power. Anchor rods and extension rods for power-installed helix anchors are fabricated with a connecting device designed to fit over the 1½-in sq drive hub (Fig. 12-25). Figure 12-26 illustrates a power-installed helix swamp anchor. Extra-heavy galvanized steel pipe is used as an anchor rod to obtain the proper depth. The pipe is manufactured in standard 21-ft lengths. The pipe is cut to the desired length or joined together with couplings, if necessary. Commonly used sizes are 1½-in diameter pipe with a 10-in swamp anchor and 2-in diameter pipe with a 15-in swamp anchor. Galvanized steel eyenuts are installed on power-installed screw anchor rods to provide a connecting device for the guy-strand cable (Fig. 12-27).

Guy Insulators   The National Electrical Safety Code requires that ungrounded guys which are attached to supporting structures that carry open-supply conductors of more than 300 volts,

Fig. 12-23   Galvanized steel coupling for power-installed screw anchor rods.

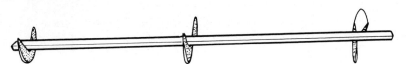

Fig. 12-24   Helix anchor designed for heavy-guy loading.

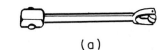

(a)

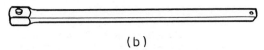

(b)

Fig. 12-25   Helix anchor guy rod (a) and extension (b).

or guys that are exposed to such conductors, must be insulated. Otherwise, the guys must be effectively grounded. It may be desirable to install insulators in guys that are effectively grounded to maintain adequate electrical clearances and provide safe working space for the lineman. Linemen must recognize the grounded guy. If work is to be done on a grounded guy assembly, electrical continuity must be maintained. The guy should be treated like any other grounded device. Proper protective equipment must be installed on the guy in the area of the work while the lineman is working on or near energized conductors. Ungrounded guys are insulated with porcelain, wood, fiberglass, or other material of suitable mechanical and electrical properties.

A porcelain guy insulator usually consists of a porcelain piece pierced with two holes, at right angles to each other, through which the guy wires are looped (Fig. 12-28). Since this puts the porcelain between the loops under compression, it makes for great strength.

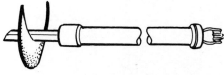

Fig. 12-26 Swamp anchor to be used in swampy areas where depth is necessary.

Fig. 12-27 Lineman installing threaded eyenut on anchor rod. *(Courtesy A. B. Chance Co.)*

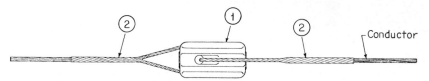

Fig. 12-28 Porcelain guy insulator. (1) Insulator porcelain guy strain; (2) guy grip.

When higher values of insulation are needed in a guy, wood or fiberglass strain insulators are used. Long wood strain insulators require long arcing horns (as shown in Figs. 12-3, 12-5, and 12-29) to bypass lightning strokes around the wood. Fiberglass strain insulators (Figs. 12-30 and 12-31), on the contrary, are impervious to moisture. They can therefore withstand a direct stroke of lightning to the pole structure without bursting. For the same reason, they do not require arcing horns to bypass the stroke. The fiberglass insulator, therefore, takes up less air space in the vicinity of the pole.

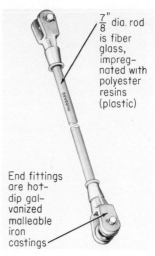

$\frac{7}{8}$" dia. rod
is fiber
glass,
impreg-
nated with
polyester
resins
(plastic)

End fittings
are hot-
dip gal-
vanized
malleable
iron
castings

Fig. 12-29 Wood strain insulators equipped with arcing horns inserted in down guys. *(Courtesy Copperweld Steel Co.)*

Fig. 12-30 Fiberglass strain insulator. Fiberglass rod is impregnated with polyester resins. *(Courtesy Hubbard and Co.)*

Fig. 12-31 Epoxiglas strain insulator installed in down guys. Note absence of arcing horns. *(Courtesy A. B. Chance Co.)*

Guy Size Determination[*]   Factors to be considered in guying pole lines are the weight of the conductor or cable, the size and weight of crossarms and insulators, wind pressure on poles and conductors, strains due to the contour of the earth, line curvatures, pole heights, and dead-end loads, plus the vertical load due to sleet and ice. To reduce unbalanced stresses to a minimum, correct angling and positioning of guy wires is essential. Where obstructions make it impossible to locate a single guy in line with the load or pull, two or more guys can be installed with their resultant guying effect in line with the load.

Where lines make an abrupt change in direction, the guy anchor is normally placed so that it bisects the angle formed by the two conductors as shown by anchor "A" in Fig. 12-32. Under ·

[*]Material under this heading is from *Encyclopedia of Anchoring.* Copyright A. B. Chance Co. Used by permission.

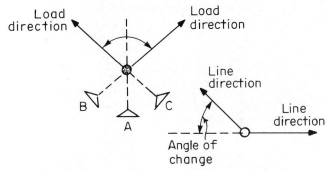

Fig. 12-32   Proper location of guys and line angle as illustrated.

heavy load conditions, it may be necessary to use two anchors, each dead-ending a leg of the line load as shown by anchors "B" and "C." Long straight spans require occasional side and end guys to compensate for heavy icing and crosswind on conductors and poles.

These, and all other factors that might make it advisable to use guys, should be carefully considered in initial designs for line construction.

To compute the load on the guy, the line load must first be determined. When the line is dead-ended, the line load can be calculated by multiplying the ultimate breaking strength of the conductor used (S) times the number conductors (N).

For example, if three 1/0 ACSR conductors are dead-ended on a pole, the line load will be 12,840 lb:

$$S \times N = \text{line load}$$
$$4280 \times 3 = 12,840$$

The ultimate breaking strength of the conductor used is found for the conductor size as shown in Tables 12-4 through 12-10.

To determine the line load to be guyed on a single anchor where the line changes direction, multiply the ultimate breaking strength of the conductor used (S) times the number of conductors used (N) times the multiplication factor for the angle change of line direction (M) as listed in Table 12-11. For example, if the pole to be guyed carries three 1/0 ACSR conductors with an angle change of 90°, the line load will be 18,156 lb:

$$S \times N \times M = \text{line load}$$
$$4280 \times 3 \times 1.414 = 18,156$$

If separate anchors are installed, each in line with a leg of the line, consider the legs as dead ends.

The manner of measuring the height and lead of a guy on sloping ground and of a stub pole guy is illustrated in Fig. 12-33. On sloping ground the lead is the horizontal distance from the guy to the pole, and the height is the distance above this horizontal line, as shown in Fig. 12-33a. On a pole stub guy the lead is the horizontal distance from the pole stub to line pole, and the height

Table 12-4   Hard Drawn Copper Wire, Solid

Size, AWG	Diameter, in	Breaking strength, lb
10	.102	529
8	.129	826
6	.162	1,280
4	.204	1,970
2	.258	3,003
1	.289	3,688
1/0	.325	4,517
2/0	.365	5,519
3/0	.410	6,722
4/0	.460	8,143

Table 12-5   Stranded Hard-Drawn Copper Wire

Size, AWG or cir. mils	No. of strands	Bore diameter, in	Breaking strength, lb
6	7	.184	1,288
4	7	.232	1,938
2	7	.292	3,045
1	7	.328	3,804
1/0	7	.368	4,752
2/0	7	.414	5,926
3/0	7	.464	7,366
4/0	7	.522	9,154
4/0	19	.528	9,617
250	19	.574	11,360
300	19	.628	13,510
350	19	.679	15,590
400	19	.726	17,810
450	19	.770	19,750
500	19	.811	21,950
500	37	.813	22,500
600	37	.891	27,020
700	61	.964	31,820
750	61	.998	34,090
800	61	1.031	36,360
900	61	1.094	40,520
1,000	61	1.152	45,030
1,250	91	1.289	56,280
1,500	91	1.412	67,540
1,750	127	1.526	78,800
2,000	127	1.632	90,050
2,500	127	1.824	111,300
3,000	169	1.998	134,400

Table 12-6   Copperweld Copper, 3-Wire

	Type A	
Conductor	Cable diameter, in	Breaking load, lb
2A	.366	5,876
3A	.326	4,810
4A	.290	3,938
5A	.258	3,193
5D	.310	6,035
6A	.230	2,585
6D	.276	4,942
7A	.223	2,754
7D	.246	4,022
8A	.199	2,233
8D	.219	3,256
9½D	.174	1,743

Table 12-7 Copperweld Stranded

Size		Diameter, in	Breaking strength, lb	
			High-strength	Extra-high-strength
¹³⁄₁₆	(19 No. 6)	.810	45,830	55,530
²¹⁄₃₂	(19 No. 8)	.642	31,040	37,690
⅝	(7 No. 4)	.613	24,780	29,430
½	(7 No. 6)	.486	16,890	20,460
⅜	(7 No. 8)	.385	11,440	13,890
⁵⁄₁₆	(7 No. 10)	.306	7,758	9,196
	3 No. 6	.349	7,639	9,754
	3 No. 8	.277	5,174	6,282
	3 No. 10	.220	3,509	4,160
	3 No. 12	.174		

Table 12-8 Aluminum Stranded

Circular mils or AWG	No. of strands	Diameter, in	Ultimate strength, lb
6	7	.184	528
4	7	.232	826
3	7	.260	1,022
2	7	.292	1,266
1	7	.328	1,537
1/0	7	.368	1,865
2/0	7	.414	2,350
3/0	7	.464	2,845
4/0	7	.522	3,590
266800	7	.586	4,525
266800	19	.593	4,800
336400	19	.666	5,940
397500	19	.724	6,880
477000	19	.793	8,090
477000	37	.795	8,600
556500	19	.856	9,440
556500	37	.858	9,830
636000	37	.918	11,240
715500	37	.974	12,640
715500	61	.975	13,150
795000	37	1.026	13,770

Table 12-9 Aluminum Stranded

Circular mils or AWG	No. of strands	Diameter, in	Ultimate strength, lb
795000	61	1.028	14,330
874500	37	1.077	14,830
874500	61	1.078	15,760
954000	37	1.124	16,180
954000	61	1.126	16,860
1033500	37	1.170	17,530
1033500	61	1.172	18,260
1113000	61	1.216	19,660
1272000	61	1.300	22,000
1431000	61	1.379	24,300
1590000	61	1.424	27,000
1590000	91	1.454	28,100

Table 12-10   ACSR

Size, cir. mils or AWG	No. of strands	Diameter, in	Ultimate strength, lb
6	6× 1	.198	1,170
6	6× 1	.223	1,490
4	6× 1	.250	1,830
4	7× 1	.257	2,288
3	6× 1	.281	2,250
2	6× 1	.316	2,790
2	7× 1	.325	3,525
1	6× 1	.355	3,480
1/0	6× 1	.398	4,280
2/0	6× 1	.447	5,345
3/0	6× 1	.502	6,675
4/0	6× 1	.563	8,420
266800	18× 1	.609	7,100
266800	6× 7	.633	9,645
266800	26× 7	.642	11,250
300000	26× 7	.680	12,650
336400	18× 1	.684	8,950
336400	26× 7	.721	14,050
336400	30× 7	.741	17,040
397500	18× 1	.743	10,400
397500	26× 7	.783	16,190
397500	30× 7	.806	19,980
477000	18× 1	.814	12,300
477000	24× 7	.846	17,200
477000	26× 7	.858	19,430
477000	30× 7	.883	23,300
556500	26× 7	.914	19,850
556500	26× 7	.927	22,400
556500	30× 7	.953	27,200
605000	24× 7	.953	21,500
605000	26× 7	.966	24,100
605000	30×19	.994	30,000
636000	24× 7	.977	22,600
636000	26× 7	.990	25,000
636000	30×19	1.019	31,500
666600	24× 7	1.000	23,700
715500	54× 7	1.036	26,300
715500	26× 7	1.051	28,100
715000	30×19	1.081	34,600
795000	54× 7	1.093	28,500
795000	26× 7	1.108	31,200
795000	30×19	1.140	38,400
874500	54× 7	1.146	31,400
900000	54× 7	1.162	32,300
954000	54× 7	1.196	34,200
1033500	54× 7	1.246	37,100
1113000	54×19	1.292	40,200
1272000	54×19	1.382	44,800
1431000	54×19	1.465	50,400
1590000	54×19	1.545	56,000

Table 12-11   Line-Angle Change Multiplication Factor

Angle change of line direction, °	Multiplication factor, M
15	0.262
30	0.518
45	0,766
60	1.000
75	1.218
90	1.414

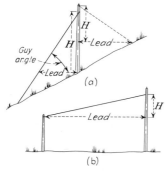

Fig. 12-33 Methods of measuring height and lead to determine guy angle. (*a*) Sloping ground; (*b*) stub guy.

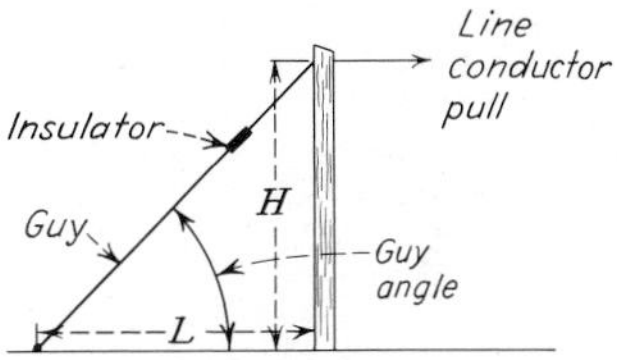

Fig. 12-34 Sketch of guy installation. *H* is height of guy attachment to pole. *L* is lead or distance from pole to anchor rod.

is the vertical distance above this horizontal line, as shown in Fig. 12-33*b*. The tension in the guy wire for a given line-conductor load depends on the distance the anchor is installed from the base of the pole or the guy angle (Fig. 12-34).

After the line load is known, the chart in Fig. 12-35 is used as a quick reference for determining the load on the guy at different angles of pull. To use the chart, determine the line load, and using the figures across the top of the chart, follow the vertical and curved line until it intersects the line indicating the angle of the proposed guy. From this point, follow the horizontal line across the chart to the right-hand side. The figure at this point will be the guy load in pounds. For example, assuming a line load of 10,000 lb and a guy angle of 60°, follow the line at 10,000 down the chart, following its curve, until it intersects the line indicating the 60° angle. Reading across the chart to the right, the figure obtained is 20,000 lb, which is the guy load. An alternative method is to multiply the factor at the bottom of the chart by the line load. For example, the multiplication factor for a 60° angle is 2.00 and the guy load is 10,000 lb; 2.00 multiplied by 10,000 lb equals 20,000 lb.

In using breaking strength of conductor, it should be considered that a conductor properly sagged (according to the N.E.S. Code) will not exceed 60 percent of its breaking strength when fully loaded. This automatically allows a safety factor of 1⅔. However, additional safety factors will be required on important crossings especially over highways, railroads, or rivers, where safety factors of 2 and 3 are generally used. After the guy load has been found, select an anchor with holding power in the soil class, allowing for the desired safety factor. The anchor, rod, and guy strand must be selected with proper size and strength to coordinate with the anchor maintaining the desired safety factor (see Tables 12-1 through 12-3).

Guy Construction The installation of a guy divides itself naturally into the five steps outlined:

1. Digging in the anchor
2. Inserting the strain insulator in the guy wire
3. Fastening the guy wire to the pole
4. Tightening the guy wire and fastening to the anchor
5. Mounting the guy-wire guard

Digging in the Anchor   The manner of digging in the guy anchor depends on the type of anchor. See Fig. 12-36 for various types of guy anchors in general use. The screw-type anchor is screwed into the earth in the manner shown in Fig. 12-37.

The patented "never-creep" anchor is installed in the manner shown in Fig. 12-38. For the driving of the rod a special maul fitted with oil-soaked hickory inserts (Fig. 12-39) is used. The inserts prevent damage to the anchor rod. For hanging of the plate an installing bar fitted with a special hook (Fig. 12-40) is employed. The same bar is also used for tamping and for retrieving the anchor in case the guy is no longer needed. The Everstick anchor (Fig. 12-19) is installed by inserting the anchor into the bottom of the hole and then causing the blades to expand into the

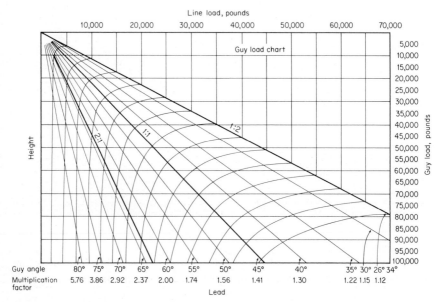

Fig. 12-35   Chart for determining guy load if line load and guy angle is known.

solid ground (Fig. 12-41). The deadman, which is a log anchor, has to be dug in. First a hole is dug, and then the anchor is placed in the hole (Fig. 12-42). The hole is backfilled and thoroughly tamped. The deadman is usually cut from a broken pole or from one that is too short. It should be treated with a preserving compound to make its life greater than that of the pole it is holding in place.

The anchor is an important part of a line, for if the anchor fails, the guy fails, and if the guy fails, the pole fails, and if the pole fails, the corner fails, and if the corner fails, the line fails. In other words, the chain is no stronger than its weakest link, and the line is no stronger than its weakest angle or corner. The importance of proper guying has led to the expression, "A line well guyed is half built, but a line poorly guyed is never built."

Digging the Anchor Hole with a Power Digger   Most anchor holes are dug with a power digger. Only isolated holes or holes in locations inaccessible to a digger are dug by hand. The usual diameter of a hole is 9 in. The length of the hole should be such, as a general rule, that the anchor itself will be not less than 6 ft below the surface of the ground. Figure 12-43 illustrates the use of a power digger in digging an anchor hole in a crowded place. This digger is the same digger that is used in digging pole holes. The 9-in auger replaces the larger size needed for pole holes.

Power-Drive Screw Anchor   Screw anchors may also be installed by use of a power drive. A special tubular wrench (see Fig. 12-44) slipped over the anchor rod transmits the torque from the power drive directly to the hub of the anchor wing. Since no torque is transmitted through the anchor rod, the rod has to be only heavy enough to withstand the pull of the guy. Moreover,

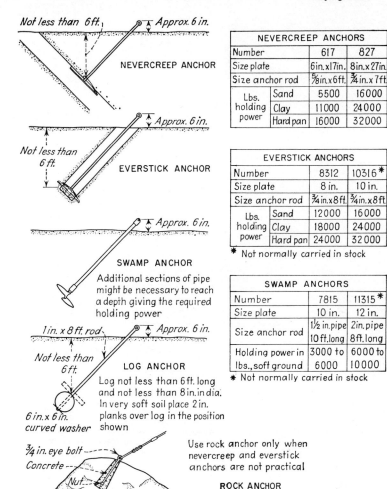

NEVERCREEP ANCHOR

Not less than 6ft.   Approx. 6 in.

NEVERCREEP ANCHORS		
Number	617	827
Size plate	6in.x17in.	8in.x27in.
Size anchor rod	⅝in.x6ft.	¾in.x7ft
Lbs. holding power	Sand	5500	16000
	Clay	11000	24000
	Hard pan	16000	32000

Approx. 6 in.

Not less than 6 ft.

EVERSTICK ANCHOR

EVERSTICK ANCHORS		
Number	8312	10316＊
Size plate	8 in.	10 in.
Size anchor rod	¾in.x8ft.	¾in.x8ft.
Lbs. holding power	Sand	12000	16000
	Clay	18000	24000
	Hard pan	24000	32000

＊ Not normally carried in stock

Approx. 6 in.

SWAMP ANCHOR

Additional sections of pipe might be necessary to reach a depth giving the required holding power

SWAMP ANCHORS		
Number	7815	11315＊
Size plate	10 in.	12 in.
Size anchor rod	1½ in.pipe 10ft.long	2in.pipe 8ft.long
Holding power in lbs., soft ground	3000 to 6000	6000 to 10000

＊ Not normally carried in stock

1in. x 8ft. rod   Approx. 6 in.

Not less than 6 ft.

LOG ANCHOR

6 in.x 6 in. curved washer

Log not less than 6ft. long and not less than 8in.in dia. In very soft soil place 2 in. planks over log in the position shown

¾ in. eye bolt
Concrete
Nut

Use rock anchor only when nevercreep and everstick anchors are not practical

ROCK ANCHOR

Fig. 12-36 Various types of guy anchors in general use.

a   b

Fig. 12-37 Steps in installation of screw-type anchor. (a) Digging small hole with bar; (b) screwing in anchor. (*Courtesy Hubbard and Co.*)

(a) Bore the hole

(b) Drive the rod

(c) Hang the plate

(d) Tighten the guy

(e) Fill the hole

Fig. 12-38 Steps in the installation of the "never-creep" anchor. Locate the spot desired for the anchor rod, measure from that point back from the pole the length of the rod, and start the hole at this point. A boring-machine auger of almost any size can be used. (*a*) Bore the hole as nearly at right angles to the line of strain as conditions will allow. Then proceed with steps (*b*), (*c*), (*d*), and (*e*) as shown. (*Courtesy A. B. Chance Co.*)

Fig. 12-39 Driving the rod for a "never-creep" anchor. Lineman is using a special maul fitted with hickory inserts to prevent damage to the rod. *(Courtesy A. B. Chance Co.)*

Fig. 12-40 Lineman lowering the "never-creep" anchor plate with special installing bar. Bar has hook which engages plate. The same bar can also be used to tamp soil into hole and to retrieve plate if necessary. *(Courtesy A. B. Chance Co.)*

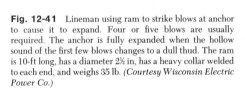

Fig. 12-41 Lineman using ram to strike blows at anchor to cause it to expand. Four or five blows are usually required. The anchor is fully expanded when the hollow sound of the first few blows changes to a dull thud. The ram is 10-ft long, has a diameter 2½ in, has a heavy collar welded to each end, and weighs 35 lb. *(Courtesy Wisconsin Electric Power Co.)*

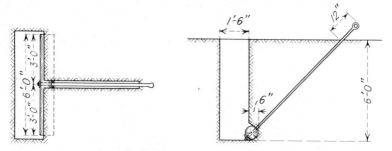

Fig. 12-42 Single-rod log anchor assembly.

a larger diameter wing can be readily installed in this manner, thus providing greater holding power.

The steps in the installation procedure are as follows:

1. Remove the eye from the anchor rod.

2. Slide a tubular wrench over the anchor rod to engage the square hub shank of the helix. The wrench is held in position by spring-loaded dogs and a hex nut (see Fig. 12-45).

3. Attach the wrench to the driving bar of the power digger (Fig. 12-46).

4. Drive the anchor down with the power digger, which feeds as well as screws the anchor into the earth. The feed should match the natural penetration of the helix so that a minimum of earth is disturbed. Driving at proper rpm and feed eliminates churning the earth, which often results from hand turning, as the anchor then must be drawn down by the ability of the soil to resist crumbling. Churning is similar to stripping a thread (see Figs. 12-47 and 12-48).

5. Release the spring-loaded dogs and remove the wrench from the anchor rod (Fig. 12-49).

6. Replace the eye on the anchor rod (Fig. 12-50).

Besides eliminating the hard work of installation when done manually, the use of power saves time. The average time required for actual installation is about 5 min.

Fig. 12-43 Digging an anchor hole with a diameter of 9 in with a power borer in a crowded place. The procedure is the same as that used in digging a pole hole. The hole is just 2 ft away from the building on the left and runs diagonally under a large tank. *(Courtesy Wisconsin Electric Power Co.)*

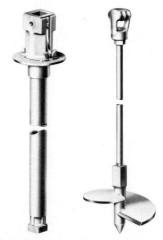

Fig. 12-44 Special tubular wrench used with power drive to rotate screw anchor. Turning torque is applied on square hub shank next to helix. *(Courtesy Hubbard and Co.)*

Fig. 12-45 Tubular wrench is slid over anchor rod. Square hub of anchor is engaged by square opening in wrench. Spring-loaded "dogs" on wrench slip over hex nut to lock in position. *(Courtesy A. B. Chance Co.)*

Fig. 12-46 Wrench being fastened to driving bar of power drive. A bolt securely locks wrench in place. *(Courtesy A. B. Chance Co.)*

Fig. 12-47 Installing screw anchor by means of power drive. Power is used to rotate the screw as well as feed anchor into earth. The anchor is driven into the ground until locking "dogs" on the wrench are just above ground level. *(Courtesy A. B. Chance Co.)*

Fig. 12-48 Anchor almost completely in place. Note ratchet on driving bar which is used to advance the anchor according to the pitch of the helix. *(Courtesy A. B. Chance Co.)*

Guy Assembly The tools needed to assemble a guy include belt cutters, wrenches, pliers, hoist, wire grips, and pulling eye for preformed grips. A typical guy assembly is illustrated in Fig. 12-51. The material required to connect the anchor rod to the pole includes guy wire, hardware, insulators, and preformed guy-wire grips. The lineman assembling a guy must determine the length of the guy wire needed to cut the wire accordingly. All guy wire, whether galvanized, copperweld, or aluminoweld, has steel core strands, the wire is springy and difficult to handle. Before a guy wire is cut, it is good practice to wrap tape around the wire on each side of the place where the cut is to be made. This will keep the strands from fraying or unlaying after the cut is made. When the guy wire is cut, it is well to hold both sides of the cut firmly so that the free ends will not fly into one's face or into a live conductor. A good procedure is to place both feet on the guy wire and make the cut between them. After the cut has been made, each end can then be carefully released.

When the guy wire has been cut to the proper length the insulators should be installed in accordance with the specifications using prefabricated guy-wire grips (Fig. 12-52) or three-bolt clamps (Fig. 12-53). An insulator properly assembled in series with the guy strand is shown in Fig. 12-28.

Fig. 12-49 Releasing wrench from anchor by lifting the "dogs" clear of the hex nut on the anchor. The wrench can then be pulled free and uncoupled from the drive shaft. *(Courtesy A. B. Chance Co.)*

Fig. 12-50 Replacing nut on end of anchor rod. Anchor is now ready for hooking up guy wire. *(Courtesy A. B. Chance Co.)*

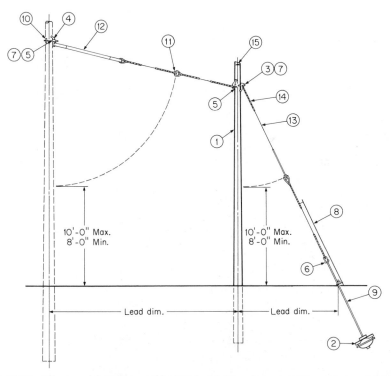

Fig. 12-51 Stub guy and down guy assembly. (1) Pole stub; (2) anchor, patent; (3) guy attachment, thimble eye; (4) guy attachment, eye type; (5) bolt, standard machine, with standard nut; (6) clamp, guy bond, for ⅝-in rod, twin eye; (7) locknut, ⅝-in galvanized; (8) protector, guy; (9) rod, twin eye, ⅝-in × 9 ft, with 2 standard nuts; (10) washer, 4-in × 4-in square, curved; (11) insulator, strain, porcelain; (12) insulator, strain, fiberglass guy; (13) wire, guy; (14) grip, guy wire; (15) pole topper.

The guy wire is attached to the pole by the lineman using the proper procedures. Figure 12-54 illustrates guys secured with three-pole clamps and prefabricated grips. The guy wire can be fastened to the pole by wrapping the turns of the wire around the pole and clamping the free end of the wire. Strain plates are used with guy hooks to provide a bearing surface for the guy strand to prevent damage to the wood fibers of the pole. The guy hooks are used to keep the guy strand from slipping down on the pole (Fig. 12-55). The guy strands are clamped together with three-bolt clamps. The end of the guy strand extends approximately 1 ft beyond the clamps and is clipped to the main guy wire.

Fig. 12-52 Prefabricated guy grip partially installed on end of guy strand. *(Courtesy Preformed Line Products Co.)*

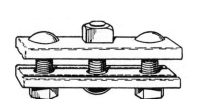

Fig. 12-53 Three-bolt guy-wire clamp. (Courtesy Hubbard and Co.)

Fig. 12-54 Lineman has secured span guy with three-bolt clamps and down guy with prefabricated grip. Thimble eye nuts are installed on through bolt and pole is protected with curved washers. *(Courtesy Copperweld Steel Co.)*

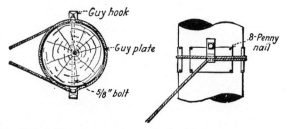

Fig. 12-55 Method of fastening guy wire to pole. Sketch shows use of guy hooks and plates.

Fig. 12-56 Lineman pulling up on anchor guy. Note the Coffing hoist or chain jack and the two "come-alongs" being used. It replaces and is easier to use than the ordinary block and tackle. It is operated with a back-and-forth pump-handle motion of the handle. The main advantage is that it locks itself in place. It will not slack off when the hand is taken off the handle. *(Courtesy Wisconsin Electric Power Co.)*

Guy wires should be placed on the poles and "pulled" before the line conductors are placed on the pole. If the line wires should be placed first, the poles would be pulled out of position.

A guy wire is pulled up or tightened from the ground by means of a hoist and approved grips, as shown in Figs. 12-56 and 12-57. These grips (Fig. 12-58) are also called "come-alongs." Come-alongs or wire eccentrics will grip a straight wire at any point without slipping. The guy is drawn up until the pole is pulled over slightly toward the guy.

In threading the guy wire through the anchor rod, a thimble should be used to protect the guy wire from a sharp bend. This is shown in Fig. 12-59.

Prefabricated guy dead-end grips can be used to terminate the guy strand to the eye on the anchor rod (see Fig. 12-60). The guy strand is pulled taut with a hoist by the lineman, then the

Fig. 12-57 Cut-a-way view of "never-creep" anchor installation, showing lineman tightening up on guy wire with hoist attached to guy wire with come-alongs. *(Courtesy A. B. Chance Co.)*

Fig. 12-58 Improved "Chicago" steel wire grip or come-along. *(Courtesy Mathias Klein and Sons.)*

Fig. 12-59 Lineman tightening bolts on anchor guy clamp. Enough end is left on the guy wire beyond the clamp for the attachment of a come-along for later retightening. *(Courtesy Wisconsin Electric Power Co.)*

Fig. 12-60 Prefabricated guy dead end partially installed. The legs of the grip must be completely wrapped around the guy strand and ends must be snapped in place. *(Courtesy Preformed Line Products Co.)*

Fig. 12-61 Lineman tightening a pole or head guy with a block and tackle. *(Courtesy Wisconsin Electric Power Co.)*

Fig. 12-62 Showing use of come-along, dynamometer, and chain jack in pulling down guy to proper tension. *(Courtesy Coffing Hoist Co.)*

guy strand is wrapped with tape and cut to the proper length. The prefabricated grip is threaded through the eye on the anchor rod and the legs of the grip are wrapped around the guy strand. The hoist is removed and the ends of the legs of the grip are snapped into position completing the installation.

Figure 12-61 shows the manner of pulling up on a head guy. It should be noted that the lineman on the pole does not do any direct pulling on the rope. This should always be done by someone on the ground.

The lineman is pulling on the fall line to his right with his left hand and on the same rope on the other side of the sheave to his left with his right hand. His right and left hands are pulling in

Fig. 12-63 Typical metal guy guard or protector made of galvanized steel. *(Courtesy Hubbard and Co.)*

Fig. 12-64 Lineman installing plastic guy guard. *(Courtesy Preformed Line Products Co.)*

Fig. 12-65 Vertical or "sidewalk" guy completely installed with strain insulator and guy guard. This type of guy is used when the anchor must be set within 5 ft of the pole. The horizontal strut shown consists of a 2-in galvanized-iron pipe and should have a length equal to the distance between the anchor and the pole. This type of guy is often called a sidewalk guy because the pole can be set on one side of a sidewalk and the anchor on the other. (*Courtesy Wisconsin Electric Power Co.*)

opposite directions; hence no extra strain is placed on his "hooks" and there is less danger of breakouts. A "foul" was thrown into the set of blocks before the operation was started. This can be seen in Fig. 12-61 just above the lineman's left forearm. This enables the lineman to cinch the blocks to hold the strain. When the lineman cannot pull the guy wire up tight enough by himself, the helper on the ground also pulls on the fall line. In this case the fall line is passed over the lineman's safety strap and the helper sets himself about one-half span away before he starts to pull. In this case, however, the "foul" cannot be used to cinch the strain. The helper must keep pulling on the rope until the lineman fastens the guy with a clamp.

The purpose of this pole guy is to take up the unbalanced pull on the adjacent pole by a primary line which "dead-ends" on that pole. The anchor guy shown to the left of the pole in the figure will balance the pull of the pole guy and the secondary main shown, after it has been pulled up to sag. The lineman is wearing a hard plastic hat and rubber gloves.

Figure 12-62 shows the use of a dynamometer in pulling a down guy to the proper tension.

Guys should not be installed where the guys will interfere with traffic. Figure 12-63 shows a typical metal guy protector. The ground end of anchor guys exposed to pedestrian traffic must be provided with a substantial and conspicuous marker not less than 8-ft long. Figure 12-64 shows a lineman installing a plastic guard. The guard should be light in color so that it will be visible at night. The guard should extend from a point near the ground to 8-ft above the ground. A guard installed on a vertical guy is shown in Fig. 12-65.

An insulator is a material that prevents the flow of an electric current and can be used to support electrical conductors. The function of an insulator is to separate the line conductors from the pole or tower. Insulators are fabricated from porcelain, glass, and fiberglass treated with epoxy resins. Rubberlike compounds are applied to fiberglass rods treated with epoxy resins to fabricate suspension and post-type insulators.

Porcelain insulators are manufactured from clay. Special clays are selected and mixed mechanically until a plasticlike compound is obtained. The plasticlike compound is placed in molds to form the insulators. The insulators are dried in an oven. When the clay is partially dry, the mold is removed and the drying process is completed. When the insulator is dry, it is dipped in a glazing solution and fired in a kiln. The glaze colors the insulator and provides a glossy surface. This makes the insulator surface self-cleaning.

Large porcelain insulators are made up of several shapes cemented together. Care must be taken when cementing the insulators together to prevent a chemical reaction on the metal parts causing cement growth. Cement growth can cause stresses on the porcelain great enough to crack the porcelain.

Glass insulators are made from sand, soda ash, and lime. The materials are properly mixed and melted in an oven until a clear liquid plasticlike material is produced. The plasticlike compound is placed in a mold and allowed to cool. After the glass insulator is cool, it is placed in an annealing oven.

Pin-Type Insulators   The pin insulator gets its name from the fact that it is supported on a pin. The pin holds the insulator, and the insulator has the conductor tied to it. Figure 13-1 shows a typical pin insulator in position.

Pin insulators are made of either glass or porcelain. The glass insulator is always one solid piece of glass, that is, it is a one-piece insulator, as Fig. 13-1 shows. The porcelain insulator (Fig. 13-2) is also a one-piece insulator when used on low-voltage lines but consists of two, three, or four layers cemented together to form a rigid unit when used on higher voltage lines. It is usually one piece for voltages below 23,000 volts, two piece for voltages from 23,000 to 46,000 volts, three piece for voltages from 46,000 to 69,000 volts, and four piece for voltages from 69,000 to 88,000 volts. The use of several layers helps to spill the rain and provides a long, dry arc-over path. These layers flare out at the bottom into a bell shape. Sometimes these layers are called "petticoats" because of their appearance, and pin insulators of this multilayer construction are sometimes called "petticoat insulators." Figure 13-3 shows a typical multilayer petticoat insulator.

Pin insulators are seldom used on transmission lines having voltages above 44,000 volts, although some 88,000-volt lines using pin insulators are in operation. The glass pin insulator is principally used on low-voltage circuits. The porcelain pin insulator is used on secondary mains and services, as well as primary mains, feeders, and transmission lines.

The smaller insulators are threaded to fit on a 1-in pin, and the larger units are threaded to fit on a 1⅜-in pin. When insulators are arranged to be mounted on steel pins, a thimble, threaded to fit the pin, is cemented into the porcelain.

Fig. 13-1 Pin insulator mounted in position. Typical low-voltage side-groove glass-pin insulator. Note use of steel insulator pin. *(Courtesy Kimble Glass Co.)*

Fig. 13-2 Porcelain-top groove pin insulators mounted in position on double crossarm. Note use of steel insulator pins and crossarm spreader bolt. *(Courtesy Locke Dept., General Electric Co.)*

Insulator Pins   The function of an insulator pin is to hold the insulator mounted on it in a vertical position. Insulator pins are made of wood or metal. Wooden pins are usually made of locust (see Fig. 13-4). Locust is durable and retains its strength longer than other wood. Iron and steel pins are used wherever the pins must be extra long, because of high voltage, and wherever the tension on the conductor is large. One make is arranged to encircle the crossarm as a clamp pin, the clamp being held by bolts (Fig. 13-5). In many cases a steel rod is used as the base in order to permit the use of a ⅝- or ¾-in hole instead of the 1½-in hole required by a wooden pin. This saves some of the strength of the arm.

The standard wood pin for low-voltage circuits is 9-in long. The diameter of the cylindrical portion that fits into the pinhole is 1¹⁄₁₆ in. Above this part is a shoulder with a diameter of 1¾ in, and from the shoulder the pin tapers gradually to a diameter of 1 in at the point where the thread begins. The details of this pin are shown in Fig. 13-6. Over 2 in of the top is threaded to hold the insulator.

Steel pins are in general use. An all steel pin is shown in detail in Fig. 13-7. It has a broad base which rests squarely on the crossarm as shown in Fig. 13-8. The pin is usually provided with lead threads which prevent any localized pressure on the insulator.

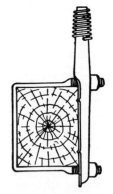

Fig. 13-3 A 23,000-volt two-layer porcelain-pin insulator. *(Courtesy Ohio Brass Co.)*

Fig. 13-4 Standard low-voltage wood insulator pin. *(Courtesy Joslyn Mfg. and Supply Co.)*

Fig. 13-5 Steel clamp pin. *(Courtesy Hubbard and Co.)*

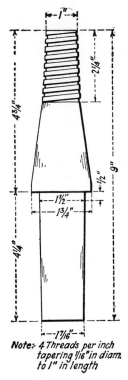

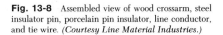

Note:- 4 Threads per inch
tapering 1/16" in diam.
to 1" in length

Fig. 13-6   Standard wood pin.

Fig. 13-7   Steel pin provided with lead threads. *(Courtesy Hubbard and Co.)*

Fig. 13-8   Assembled view of wood crossarm, steel insulator pin, porcelain pin insulator, line conductor, and tie wire. *(Courtesy Line Material Industries.)*

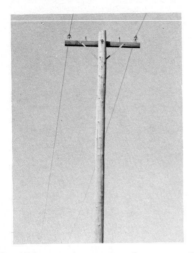

Fig. 13-9 View showing a straight run in a pole line and illustrating single-crossarm construction. Alternate arms should face in same direction. *(Courtesy Iowa-Illinois Gas and Electric Co.)*

Fig. 13-10 View illustrating a three-phase buck-arm assembly. Since the line turns a corner, lines are dead-ended on double-arm construction. Pole is guyed in both directions to balance pull of dead-ended lines. Down guys make angle of about 45° and are equipped with strain insulators and guy guards. *(Courtesy American Electric Power System.)*

The spacing of the pins is generally suited to the voltage of the circuit. In addition, the spacing should provide sufficient working space for the lineman. For general distribution work the spacing is 14½ in between centers. Wider spacings are more common on four-pin than on six-pin cross-arms. The two pins next to the pole are usually spaced 30 in apart. This should allow space for the lineman to climb through to work on the upper arms. The end pins are generally spaced 4 in from the end of the arm.

Fig. 13-11 Double-arm dead-end construction at corner pole. Note use of down guys to counterbalance pull of line conductors. *(Courtesy Iowa-Illinois Gas and Electric Co.)*

Fig. 13-12 Alley-arm brace used with side arm to clear obstructions. Note pole step on brace for use by lineman. *(Courtesy Iowa-Illinois Gas and Electric Co.)*

Wooden pins are merely driven into the hole, and then a sixpenny nail is driven through the crossarm from the side to hold the pin in place. It is well not to drive the nail entirely flush with the wood. Enough of the head should project so that the lineman can catch hold of it with the pliers in case it is necessary to replace the pin.

Metal pins which clamp the crossarm are mounted by clamping them on the arm at the desired spacings. Since these pins do not require pin holes, their use avoids weakening the crossarm.

Crossarms   Pin-type insulators are supported by crossarms. *Single arms* are used on straight lines where no excessive strain needs to be provided for, as in Fig. 13-9. As already mentioned, every other crossarm faces in the same direction.

Double arms should be used at line terminals, at corners, at angles, or at other points where there is an excessive strain (Fig. 13-10). Where lines cross telephone circuits or railroad crossings, double arms should also be used, as more than ordinary safety is required at such points. When two or more transformers are mounted on the same pole, double arms, as a rule, are used for their support.

Fig. 13-13   Standard flat-strap crossarm brace. *(Courtesy Hubbard and Co.)*

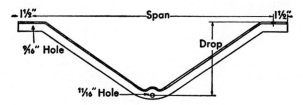

Fig. 13-14   A V-shaped angle-iron crossarm brace. *(Courtesy Joslyn Mfg. and Supply Co.)*

Buck arms are used at corners and at points where branch circuits are taken off at right angles to the main line (Fig. 13-11).

Side arms are used in alleys or other locations where it is necessary to clear buildings, etc. (Fig. 13-12).

Crossarm braces are used to give strength and rigidity to the crossarm. Metal crossarm braces are made of either flat bar or light angle iron. The size used varies with the size of the arm and the weight of the conductors. The usual flat-strap brace for ordinary distribution work (Fig. 13-13) is 38 in long and ¼ by 1¼ in. One end is attached to the crossarm by means of a carriage bolt and the other to the pole by means of a lag screw. One brace extends to each side of the arm, as shown in Fig. 13-9. Angle-iron braces are made in one piece and bent into the shape of a V, as shown in Fig. 13-14. These braces are fitted to the bottom of the crossarm instead of the side as is the flat type.

Wooden-crossarm braces are used extensively for medium voltages. Their use increases the insulation of the line. A set of wooden braces is illustrated in Fig. 13-15, and their use is illustrated in Figs. 13-16 and 13-17.

Where crossarms extend to one side of the pole only, as is sometimes necessary in alley construction, special braces are required. The brace must be longer and more rigid. A single brace made of angle iron 1¾ by 1¾ by ¾₆ in, and 5 to 7 ft long is generally used for distribution lines, and is shown in Figs. 13-12 and 13-18. The brace is provided with a step near the middle for use of the lineman.

A variety of bolts and screws are used in line construction, and only the common ones will be illustrated. Figure 13-19 shows a typical crossarm through bolt used for fastening the crossarm to the pole. Its common size is ⅝ by 12 to 16 in. A bolt for attaching two crossarms, one on each side of the pole, is shown in Fig. 13-20. Figure 13-21 illustrates a carriage bolt used for attaching the brace to the crossarm. The typical bolt for this purpose has a diameter of ⅜ in and is 4, 4½, or 5 in long. Figure 13-22 shows a lag screw used to fasten the crossarm braces to the pole. The rec-

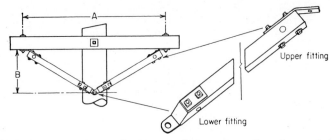

Fig. 13-15   Wooden crossarm brace used to increase insulation properties of line. *(Courtesy Joslyn Mfg. and Supply Co.)*

ommended size is ½ by 4 in. Lag screws are supposed to be screwed into place either into a small bored hole or after being started by hammering. They should be screwed into place to keep the threads from injuring the fibers of the wood.

The complete assembly of the pieces of pole hardware described above is shown in Fig. 13-23. This figure also lists the standard parts and illustrates their correct use.

Crossarms may be mounted before or after the pole is erected; the practice varies. If the pole is located where it is accessible from a bucket truck and if the pole has not been preframed and drilled, the lineman frames and drills the pole working from the insulated bucket (Fig. 13-24). The pin-type insulators are installed on the crossarm by a groundman or lineman working on the ground (Fig. 13-25). The groundman hands the crossarm, complete with pins, insulators, and crossarm braces, to the lineman in the bucket of the truck (Fig. 13-26). The lineman installs the crossarm on the pole using a through bolt while working from insulated bucket on truck (Fig. 13-27).

In case the crossarm is mounted after the pole is in place and the pole is located in an area inaccessible for a bucket truck, the lineman climbs the pole, carrying his small tools (pliers, knife, and connectors), hand line, hammer, and lag wrench, and fastens the pulley of the hand line. The crossarm is then pulled up by the groundman as illustrated in Fig. 13-28. Various methods of tying the hand line to the crossarm are shown in Fig. 13-29.

The crossarm is attached to the pole (Fig. 13-30) by means of a ⅝-in galvanized machine bolt driven from the back of the pole. A 2¼-in square washer is placed under the head and the nut of the bolt, and then the bolt is drawn up.

Fig. 13-16   Use of wooden crossarm braces on distribution line. Use of wood pole, wooden crossarm, and wooden braces provides considerable insulation to line in addition to that provided by the pin insulators. *(Courtesy Iowa-Illinois Gas and Electric Co.)*

Fig. 13-17   Wooden braces used on double arms of low-voltage transmission line. *(Courtesy Iowa-Illinois Gas and Electric Co.)*

After the through bolt is in place, the crossarm braces are attached. It is necessary first to sight the crossarm, that is, make sure it is horizontal. In case more conductors are to be carried on one side than on the other, that side can be higher, as it will settle later. The braces are usually 1¼ by ¼ by 28 in. Sometimes the braces are bolted to the crossarm before the arm is pulled up. The end of the brace attached to the crossarm is held in place by means of a ⅜- by 4-in galvanized carriage bolt, and the ends at the pole are fastened by means of a ½- by 5-in lag screw. The braces are attached on the back of the crossarm (see Fig. 13-31). When the standard 28-in braces are used, the hole in the crossarm should be located 19 in from the middle of the crossarm.

Fig. 13-18   Alley-arm brace. *(Courtesy Hubbard and Co.)*

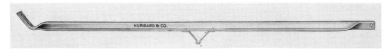

Fig. 13-19   Crossarm bolt. *(Courtesy Hubbard and Co.)*

Fig. 13-20   Bolt for double crossarms. *(Courtesy Hubbard and Co.)*

Fig. 13-21   Carriage bolt for fastening brace to crossarm. *(Courtesy Hubbard and Co.)*

Fig. 13-22   Lag screw for fastening crossarm brace to pole. *(Courtesy Hubbard and Co.)*

When double arms are placed on a pole, the arms are bolted together on each end and held a given distance apart by means of spreaders, which fill in the space between the arms and make a solid structure. The second arm should be a completely equipped arm of the same size and pin spacing as the first arm. However, no gain is cut for the second arm.

The appearance of the pole line can be improved by the use of armless-type construction. Pin-type insulators can be installed on epoxy rods to provide proper clearances without the use of crossarms. The epoxy rods are available in single, double, and triple assemblies (Fig. 13-32).

Post-Type Insulators   Post-type insulators are used on distribution, subtransmission, and transmission lines and are installed on wood, concrete, and steel poles. The line-post insulators are manufactured for vertical or horizontal mounting. The line-post insulators are usually manufactured as one-piece solid porcelain units (Fig. 13-33) or fiberglass epoxy-covered rods with metal end fittings and rubber weather sheds (Fig. 13-34). In some instances the fiberglass rod is combined with a porcelain insulator unit (Fig. 13-35). The insulators are fabricated with a mounting base for curved or flat surfaces, and the top is designed for tying the conductor to the insulator or fitted with a clamp designed to hold the conductor.

Line-post insulators designed for vertical mounting are mounted on crossarms (Fig. 13-36). This type of construction is often used for long span rural distribution circuits. Figure 13-37 illustrates

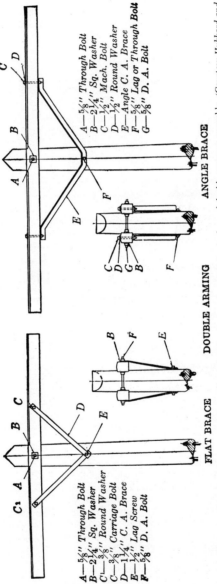

Fig. 13-23 Sketches showing standard sizes and use of bolts, lags, and washers in single- and double-crossarm assembly. (*Courtesy Hubbard and Co.*)

ANGLE BRACE

A—5/8'' Through Bolt
B—2 1/4'' Sq. Washer
C—1/2'' Mach. Bolt
D—1/2'' Round Washer
E—Angle C. A. Brace
F—5/8'' Lag or Through Bolt
G—5/8'' D. A. Bolt

DOUBLE ARMING

FLAT BRACE

A—5/8'' Through Bolt
B—2 1/4'' Sq. Washer
C—3/8'' Round Washer
D—3/8'' Carriage Bolt
D—1 1/4'' C. A. Brace
E—1/2'' Lag Screw
F—5/8'' D. A. Bolt

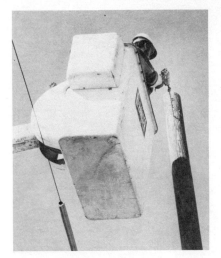

Fig. 13-24   Lineman drills hole in pole using power drill operated by truck hydraulic system. *(Courtesy Iowa-Illinois Gas and Electric Co.)*

Fig. 13-25   Groundman installing pin-type insulator with steel pin on crossarm. *(Courtesy Iowa-Illinois Gas and Electric Co.)*

Fig. 13-26   Lineman in bucket of truck takes crossarm completely assembled from groundman in preparation to raise it to proper height for installation. *(Courtesy Iowa-Illinois Gas and Electric Co.)*

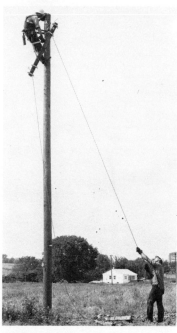

Fig. 13-27 Lineman completing installation of crossarm. Pin-type insulator for center-phase conductor is supported by special pole-top steel pin fabricated so it can be bolted to pole. (*Courtesy Iowa-Illinois Gas and Electric Co.*)

Fig. 13-28 Groundman hoisting crossarm to lineman on pole. Note that groundman stands to one side so that he is out of danger in case any materials or tools should drop. The lineman has just taken hold of the crossarm and is about to remove the hand-line hitch from its upper end. The groundman is holding the hand line taut to take up the weight of the crossarm. (*Courtesy Wisconsin Electric Power Co.*)

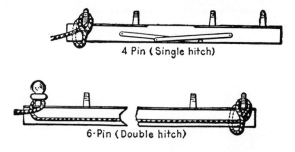

4 Pin (Single hitch)

6-Pin (Double hitch)

Transmission Arm

This hitch is put on close to end of arm so that the lineman does not have to hold the weight of the arm any longer than necessary. After he removes the top hitch, the groundman pulls the arm up the rest of the way.

Fig. 13-29 Methods of tying hand line to crossarm for purpose of raising.

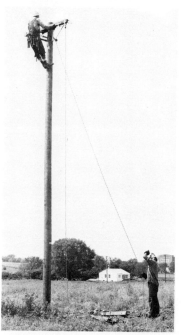

Fig. 13-30 Lineman placing the crossarm on the through bolt. Notice the "assist" given to the lineman by the helper. The hand line is still attached to the trailing end of the crossarm, and the helper is keeping it taut to take up some of the weight. The through bolt was put into the pole previously with the threaded end toward the lineman. The bolt has a diameter of ⅝ in, and the hole in the pole has a diameter of ¹¹⁄₁₆ in. If the bolt fits loosely, the lineman usually bends it slightly near its head end by hitting it with a hammer when it is part way in. This jams the bolt in the hole so that it will not push out when the lineman pushes the crossarm on. On secondary transmission-line construction where through-bolt strength is very important, the through bolt is installed so that the head end is toward the lineman. This places the weaker threaded end in the pole opposite the crossarm where the strain is least. *(Courtesy Wisconsin Electric Power Co.)*

Fig. 13-31 Lineman fastening crossarm braces to the pole with a ½- by 5-in lag bolt. The lag bolt has a "drive" thread so that the edge of the thread toward the end of the bolt is beveled. The bolt is not driven in full length. It is left about ¼ in off tight position. It is finally seated by three or four turns with a lag wrench. Note the working position of the lineman. He has set himself so that he can get a good two-handed free swing at the lag. His safety strap is just long enough to place him the correct distance away. He does not have to "reach"; neither does he have to "choke up." *(Courtesy Wisconsin Electric Power Co.)*

a distribution circuit constructed with porcelain horizontal line-type post insulators. Armless construction using post-type insulators permits the construction of subtransmission and transmission lines on narrow rights of way and along city streets. A 69-kV subtransmission line constructed with porcelain line-post insulators with conductor clamp top is pictured in Fig. 13-38. A 230-kV transmission line constructed with steel poles and line-post-type insulators is illustrated in Fig. 13-39.

Post-type insulators have been used to construct circuits that operate at voltages through 345 kV. The higher transmission voltages require more than one segment of porcelain to be bolted together (Fig. 13-40). Lines constructed with post-type insulators are clean looking and esthetically acceptable to the public (Fig. 13-39).

Fig. 13-32 Pin-type insulators mounted on epoxy rod single and bi-unit assembly to eliminate use of crossarm. Eyes on bi-unit assembly are for mounting stringing equipment. *(Courtesy A. B. Chance Co.)*

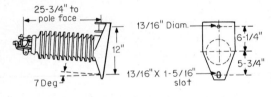

Characteristics				
Typical application			kV	69
Flash-over voltage	Impulse critical 1.2 x 50 mU-sec wave	Positive	kV	330
		Negative	kV	425
	60 Hz	Dry	kV	200
		Wet	kV	180
Standard glaze color			Sky tone	
Radio influence voltage	Test voltage to ground		kV	44
	Maximum RIV at 1000 kHz		Micro-volts	200
Leakage distance			Inches	53
Dry arcing distance			Inches	19-1/4
Cantilever			Pounds	2800

Fig. 13-33 An 88-kV porcelain horizontal post insulator.

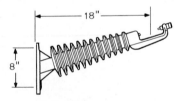

Mechanical characteristics (ultimate ratings)	
Vertical cantilever, lbs	1,500
Longitudinal cantilever, lbs	1,000
Tensile, lbs	4,000
Compression, lbs	1,500
Electrical characteristics	
60 Hz flashover, dry, kV	140
60 Hz flashover, wet, kV	120
Impulse flashover, positive, kV	210
Impulse flashover, negative, kV	320

Fig. 13-34 A 15 kV Epoxilator tapered skirt standoff insulator. *(Courtesy A. B. Chance Co.)*

Suspension Insulators   The suspension insulator, as its name implies, is suspended from the crossarm and has the line conductor fastened to the lower end (Fig. 13-41). The suspension-type insulator was developed when voltages were increased above 44,000 volts. At higher voltages the pin insulator becomes quite heavy, and it is difficult to obtain sufficient mechanical strength

Fig. 13-35 Fiberglass rod and pin insulator used to support top conductor and fiberglass rod and porcelain assemblies to support side conductors. This arrangement permitted increasing the voltage on the circuit from 4.16 to 13.2 kV without replacing pole. *(Courtesy A. B. Chance Co.)*

Fig. 13-36 Post-type insulator in use in vertical position. Note special supporting hardware used in place of conventional pin. *(Coutesy Lapp Insulator Co.)*

Fig. 13-37 A 13.2Y/7.6-kV three-phase four-wire distribution circuit constructed with sky-gray colored line-post insulators to enhance the appearance. Phase conductors are 477-mcm ACSR. Neutral conductor, 4/0 AWG ACSR is supported by porcelain spool-type insulator fastened to pole with a metal bracket.

Fig. 13-38 Post-type insulator in use in horizontal position. Note narrow right of way required.

Fig. 13-39 A 230-kV transmission line located along a city street. Line uses porcelain post-type insulators for the transmission circuit and the underbuilt distribution circuit on steel poles.

in the pin to support the insulator. The suspension-type insulator is manufactured from porcelain (Fig. 13-42), glass (Fig. 13-43), and epoxy glass and rubber materials (Fig. 13-44). Transmission (Fig. 13-41), subtransmission (Fig. 13-45), and distribution (Fig. 13-44) circuits are constructed with suspension insulators. Economics, voltage of the circuit, width of right of way, size of the conductor, length of the span, and clearances required all enter into the decision regarding the type and material of the insulators to be used.

The porcelain insulator unit consists essentially of two metal pieces insulated from one another with porcelain. First the porcelain is cemented into the metal cap or the top of the unit, and then the metal pin on the bottom of the unit is cemented into the porcelain. A typical cemented-type suspension insulator is shown in Fig. 13-46.

The diameter of the porcelain disks varies from 3¾ to 11½ in or more. The 10-in disk is perhaps the most common size. The number of units to use in a string depends largely on the voltage of the line (Fig. 13-47). Other factors, such as climate, type of construction, and degree of reliability required also enter into consideration. Table 13-1 gives a general idea of the usual number of units employed for the various standard transmission voltages.

Strain insulators are used where a pull must be carried as well as insulation provided (Fig. 13-48). Such places occur wherever a line is dead-ended—at corners, at sharp curves, or at extra long spans, as at river crossings or mountainous country. In such places the insulator must not only be a good insulator electrically, but it must also have sufficient mechanical strength to counterbalance the forces due to the tension of the line conductors.

Table 13-1   Approximate Number of Disks Required in Suspension String for Various Line Voltages

Line voltage, volts	Number of suspension units required in string	Line voltage, volts	Number of suspension units required in string
13,200	2	138,000	8, 9, or 10
23,000	2 or 3	154,000	9, 10, or 11
34,500	2 or 3	230,000	12 to 16
69,000	4 or 5	345,000	15 to 20
88,000	5 or 6	500,000	23 to 28
110,000	6, 7, or 8	765,000	32 to 36

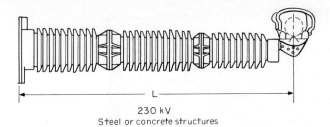

230 kV
Steel or concrete structures

138 kV
Steel or concrete structures

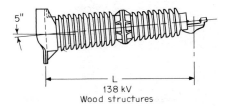

5"

138 kV
Wood structures

Line Voltage	Lapp Catalog no.	Unit length (L), inches	Ultimate strength	
			Cantilever, lb	Tension lb
138 kV	70148	53.0	2800	5,000
	79135	58.0	4000	10,000
	301543	56.75	2800	5,000
	305071	60.50	4000	10,000
230 kV	301372	101.50	3500	10,000
	90974	89.0	5800	10,000

Fig. 13-40 Porcelain horizontal line-post insulators. *(Courtesy Lapp Insulator Co.)*

Fig. 13-41 Single-Circuit 345,000-volt three-phase high-voltage transmission line. Note the long string of insulators needed to insulate the conductors. Also note two ground wires at extreme top of the tower. These are used to shield the line from lightning.

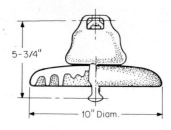

Characteristics				
ANSI class (Std. C.29.2-1971)				52.8
Flash-over voltage	Impulse critical 1.2 x 50 mu-sec wave	Positive	kV	125
		Negative	kV	130
	60 Hz	Dry	kV	80
		Wet	kV	50
Low-frequency puncture voltage			kV	110
Radio influence voltage	Test voltage to ground		kV	10
	Maximum RIV at 1000 kHz		Micro-volts	50
Leakage distance			Inches	13
Dry arcing distance			Inches	7.75
Section length			Inches	5.75
Porcelain disc diameter			Inches	10
Strength rating	M & E rating		Pounds	40,000
	ANSI M & E category		Pounds	36,000
Recommended maximum sustained load			Pounds	20,000
Impact strength			Inch Pounds	90
Color				Skytone

Fig. 13-42 Porcelain suspension insulator strength 36,000 lb, ball-and-socket type.

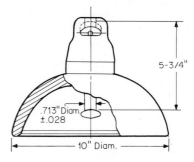

Fig. 13-43 Toughened glass suspension insulator, strength 30,000 lb, ball-and-socket type.

Characteristics				
ANSI class (Std. C29.2-1962)				52-5
Flash-over voltage	Impulse	Positive	kV	125
		Negative	kV	125
	Low frequency	Dry	kV	70
		Wet	kV	55
Low-frequency puncture voltage			kV	130
Radio influence voltage	Test voltage to ground		kV	10
	Maximum RIV at 1000 kHz		Micro-volts	50
Leakage distance			Inches	11-1/2
	Impact strength		Inches Pounds	400
	Routine proof test		Pounds	15,000
Strength ratings	M & E rating		Pounds	30,000
	Maximum sustained load		Pounds	16,500

Fig. 13-44 Epoxy glass rod with rubber weather sheds suspension insulators used on vertical corner of a distribution circuit. *(Courtesy A. B. Chance Co.)*

Fig. 13-45 A 69-kV subtransmission circuit with porcelain suspension insulators. Wishbone wooden crossarms provide triangular spacing. *(Courtesy Iowa-Illinois Gas and Electric Co.)*

Fig. 13-46 Sectional view of cemented-type ball-and-socket porcelain suspension insulator. *(Courtesy Lapp Insulator Co.)*

Fig. 13-47 A string of eight suspension insulator units suspended from a crossarm. Note conductor clamp at bottom of string. *(Courtesy Locke Dept., General Electric Co.)*

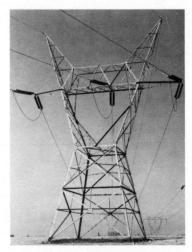

Fig. 13-48 Distribution circuit with small-diameter porcelain suspension insulators used to dead-end the conductors. The insulators are in series with the conductors and must be designed to withstand the conductor strain. *(Courtesy A. B. Chance Co.)*

Fig. 13-49 Two strings of suspension insulators used in multiple to serve as strain insulators at 230-kV dead-end tower. *(Courtesy Locke Dept., General Electric Co.)*

Fig. 13-50 Linemen dead-ending conductors on 345kV EHV line with two porcelain insulator strings in parallel. Insulators are heavy and difficult to handle. *(Courtesy L. E. Myers Co.)*

Fig. 13-51 Lineman installing Ohio Brass Co. Hi-Lite insulators on steel pole before pole is erected with the aid of crane. Note the insulator is light enough to be handled easily by the lineman. *(Courtesy Iowa-Illinois Gas and Electric Co.)*

Fig. 13-52 Crane is being used to set a steel pole on a concrete foundation with synthetic-material insulators in place for a double-circuit 161-kV transmission circuit. *(Courtesy Iowa-Illinois Gas and Electric Co.)*

Fig. 13-53 Lineman securing a spacer cable to a porcelain spool insulator with a prefabricated tie. The spacer cable is insulated but not shielded. The porcelain spool insulators separate the cables from each other and from the pole or ground. *(Courtesy A. B. Chance Co.)*

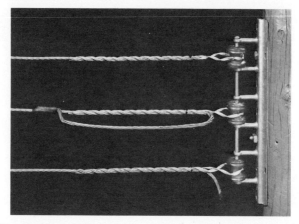

Fig. 13-54 Open-wire bare secondary wires supported by porcelain spool insulators on a galvanized steel rack. Wires are secured to insulators with preformed distribution grips. *(Courtesy Preformed Line Products Co.)*

Strain insulators are built in the same way as suspension insulators except that they are made stronger mechanically. Furthermore, when a single string is not able to withstand the pull, two or more strings are arranged in multiple (Fig. 13-49).

Glass suspension insulators have been used on the circuits in many cases. Glass is used more commonly in countries other than the United States. Glass insulators can be used interchangeably with porcelain insulators wherever their strength and voltage characteristics are adequate.

The development of extra-high-voltage (EHV) bundle-conductor transmission circuits resulted in heavy and cumbersome installations when porcelain suspension insulators were used (Fig. 13-50). To solve that problem, the Ohio Brass Co. developed the Hi-Lite suspension insulator, which has a long epoxy-fiberglass rod of extremely high mechanical and dielectric strength with weather sheds of rubber. A heavy silicone grease is used as an interface between the rod and weather sheds. Forged-steel end attachments are swaged to the rod under pressure. The insulators are manufactured with maximum design tension ratings of 10,000, 20,000, 40,000, and 80,000 lb. A single Hi-Lite suspension insulator can replace four parallel strings of porcelain or glass suspension insulators. The synthetic-material insulators weigh only 5 to 10 percent of the weight of the sus-

Fig. 13-55 Lineman securing bare neutral conductor to porcelain spool insulator fastened to pole with a galvanized steel clevis by a through bolt. Conductor is being secured with a preformed grip. *(Courtesy Preformed Line Products Co.)*

pension insulators they replace (Fig. 13-51). A 500-kV insulator weighs 40 lb, and a 765-kV insulator weighs 65 lb. The synthetic-material insulators are less susceptible to mechanical damage. Structures can be assembled on the ground and erected without danger of destroying the insulation (Fig. 13-52). The Hi-Lite insulators have a high ratio of strength to weight, resistance to vandalism (gunfire), a life expectancy equal to that of porcelain, freedom from tracking and deterioration caused by surface leakage currents in contaminated environments, and a low cost. The synthetic-material insulators have been used experimentally on 1200-kV alternating-current test lines.

Low-Voltage Insulators   Porcelain spool insulators are used for spacer-cable supports (Fig. 13-53) and secondary-voltage circuits operating at voltages below 600 volts. Open-wire secondaries of bare conductor or conductor covered with weather-proof material are usually supported by porcelain spool insulators mounted on secondary steel racks (Fig. 13-54).

Porcelain spool insulators supported by steel clevises are often used to support neutral conductors for three-phase, four-wire high-voltage distribution circuits or the bare neutral conductor of insulated twisted-multicable secondary circuits (Fig. 13-55). Porcelain insulators reinforced by a one-piece pressed-steel housing with attached galvanized steel screws, often referred to as "house knobs," are available for supporting low-voltage circuits.

Strain insulators used in guy wires are described and illustrated in Sec. 12 "Guying Poles."

Line Conductors

The wires and cables over which electrical energy is transmitted are made of copper, aluminum, steel, or a combination of copper and steel or aluminum and steel. Some insulated cables use sodium for a conductor. A conductor is a material that readily permits the flow of an electric current. Most metals can be used for conductors in an electric circuit. Materials, other than those mentioned, that conduct electricity are not generally used to make wires and cables because of economic or physical reasons. Gold, silver, platinum, nickel, zinc, tungsten, molybdenum, boron, cobalt, cadmium, berylium, magnesium, silicon, and the like, may be used for special wire applications or combined with other materials to improve their characteristics for forming wires.

Copper Conductors   Copper is a commonly used line conductor. It conducts electric current very readily, ranking next to silver. It is very plentiful in nature, and, therefore, its cost is comparatively low. It can be easily spliced.

Three kinds of copper wire are in use: hard-drawn copper, medium-hard-drawn copper, and annealed copper, also called "soft drawn." Copper wire is hard drawn as it comes from the drawing die. To obtain soft or annealed copper wire, the hard-drawn wire is heated to a red heat to soften it.

For overhead line purposes, hard-drawn copper wire is preferable on account of its greater strength. Annealing, or softening it, reduces the tensile strength of the wire from about 55,000 to 35,000 lb/in². Because of this, it is not good practice to have any soldered splices when using hard-drawn wire, as the soldering anneals the wire near the joint, thereby reducing its strength. Joints in hard-drawn wire should, therefore, be made with splicing sleeves. Annealed or soft-drawn copper wire is used for ground wires and special applications where it is necessary to bend and shape the conductor. Medium-hard drawn copper is used for distribution especially for wire sizes smaller than No. 2. Soft copper wire has a yield point less than one-half that of medium-hard copper and hence stretches permanently with a correspondingly lighter loading of ice and wind.

Aluminum Conductors   Aluminum is widely used for distribution and transmission-line conductors. Its conductivity, however, is only about two-thirds that of copper. Compared with a copper wire of the same physical size, aluminum wire only has 60 percent of the conductivity, only 45 percent of the tensile strength, and only 33 percent of the weight. To have the same conductivity, therefore, the aluminum wire must be 100/60 = 1.66 times as large as the copper wire in cross section. An aluminum wire of this size will have 75 percent of the tensile strength and 55 percent of the weight of the equivalent copper conductor.

In Fig. 14-1 are shown cross sections of copper, aluminum, and aluminum-conductor steel-reinforced (ACSR) conductors which have equal conductivity or, what amounts to the same thing, equal current-carrying capacities.

The tensile strength of aluminum is only about one-half that of hard-drawn copper, namely, 27,000 lb/in². But since the cross section of an aluminum wire must be about twice that of copper to have the same conductivity, the actual breaking strength of the aluminum conductor is about the same as that of copper. In spite of being twice as large in section, the weight per foot is still only 55 percent as great, as pointed out above. This fact makes aluminum conductors preferred in many cases, because the lighter weight permits longer spans and, therefore, fewer towers and insulators. This is a decided advantage where long spans are common. The larger size also helps to hold the corona loss down.

When an aluminum conductor is stranded, the central strand is often made of steel, which serves to reinforce the cable. Such reinforcement gives great strength for the weight of conductor. Reinforced aluminum cable called ACSR (aluminum-conductor steel-reinforced) is therefore especially suited for long spans.

Steel Conductors   Steel wire is used to a limited extent where very cheap construction is desired. Steel wire, because of its high tensile strength of 160,000 lb/in², permits relatively long spans, therefore requires few supports. Bare steel wire, however, rusts rapidly and is therefore very short-lived. Steel is also a poor conductor compared with copper, being only about 10 to 15 percent as good. Galvanized steel conductors are commonly used for shield or static wires on transmission and subtransmission lines. Most guy wires are galvanized steel stranded cables.

COPPER   ALUMINUM   A.C.S.R.

Fig. 14-1 Relative cross sections of copper, aluminum, and ACSR conductors having equal current-carrying capacities.

Fig. 14-2 Stranded copperweld conductor. *(Courtesy Copperweld Steel Co.)*

Copperweld Steel Conductors   The disadvantages of short life and low conductivity of steel led to the development of the copperweld steel conductor. In this conductor a coating of copper is securely welded to the outside of the steel wire. The copper acts as a protective coating to the steel wire, thus giving the conductor the same life as if it were made of solid copper. At the same time, the layer of copper greatly increases the conductivity of the steel conductor, while the steel gives it great strength. This combination produces a very satisfactory yet inexpensive line conductor. Its chief field of application is in rural lines, for guy wires, and for overhead ground wires.

The conductivity of copperweld conductors can be raised to any desired percentage, depending

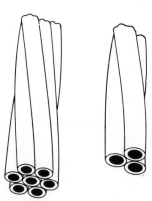

Fig. 14-3 Alumoweld guy strand. Each strand is covered with a welded layer of aluminum to protect the steel core from rusting. *(Courtesy Copperweld Steel Co.)*

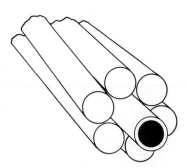

Fig. 14-4 ACSR cable with steel strand covered with layer of welded aluminum. The layer of aluminum on the steel strand prevents electrolytic corrosion between strands, prevents rusting of the steel strand, and increases the conductivity of the steel strand by a substantial amount. *(Courtesy Copperweld Steel Co.)*

on the thickness of the copper layer. The usual values of conductivity of wires as manufactured are 30 and 40 percent. Figure 14-2 shows a stranded copperweld conductor.

Alumoweld Steel Conductors   In like manner steel can be covered with aluminum to prevent the steel from rusting as well as to improve its conductivity. Figure 14-3 shows such an aluminum-covered stranded steel cable used for guys. By its use the guys can be expected to last as long as modern treated wooden poles.

Figure 14-4 shows an ACSR conductor in which the steel strand is covered with a welded layer of aluminum. The layer of aluminum increases the conductivity of the steel strand in addition to preventing it from rusting. All the strands of the cable, the alumoweld strand and the aluminum strands thus have the same life.

Conductor Classes   Conductors are classified as solid or stranded. A solid conductor, as the name implies, is a single conductor of solid circular section. A stranded conductor is composed of a group of wires made into a single conductor. A stranded conductor is used where the solid conductor is too large and not flexible enough to be handled readily. Large solid conductors are also easily injured by bending. The size No. 0 wire is the approximate dividing line between solid and stranded conductors. The strands in the stranded conductor are usually arranged in concentric

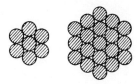

Fig. 14-5   Sketches showing arrangement of wires in 7- and 19-strand conductors. The wires are arranged in concentric layers about a central core.

Fig. 14-6   Typical stranded cable. *(Courtesy Western Electric Co.)*

Fig. 14-7   Solid conductor covered with two layers of braided cotton and saturated with a weatherproofing compound.

layers about a central core (Fig. 14-5). The smallest number of wires in a stranded conductor is three. The next number of strands is 7, then 19, 37, 61, 91, 126, etc. Figure 14-6 illustrates a typical stranded conductor. Both copper and aluminum are thus stranded.

Conductor Covering   Conductors on overhead transmission lines are bare conductors; that is, they are not covered. Conductors on overhead distribution circuits, below 5000 volts, are usually covered. The common covering is triple-braid weatherproof cotton, neoprene, or polythene. The triple-braid cotton is saturated with a black moistureproof compound. Figure 14-7 shows a conductor and two layers of braid. This covering, of course, is not sufficient to withstand the voltage at which the line is operating, and the conductors must, therefore, also be mounted on insulators. In fact, the wires should always be treated as though they were bare.

Wire Sizes   Wire sizes are ordinarily expressed by numbers. There are, however, several different numbering methods, so that in specifying a wire size by number it is also necessary to state which wire gauge or numbering method is used. The most used wire gauge in the United States is the so-called "Brown and Sharpe" gauge, also called "American Wire Gauge."

American Wire Gauge   The American standard wire gauge is shown in Fig. 14-8. This cut is full size, so that the widths of the openings on the rim of the gauge correspond to the diameters of the wires whose numbers stand opposite the openings. The actual size of these wire numbers is perhaps still better illustrated by Table 14-1. This table gives, for wire sizes from No. 0000 to No. 8, the diameter of the wire in inches and a full size end and side view of the wire corresponding to each number of the gauge. This table should make it easy to gain an idea of the actual physical size of the wires commonly used in distribution work. It will be noted that the greater the number of the wire, the smaller the wire is. That came about by numbering the wire by the

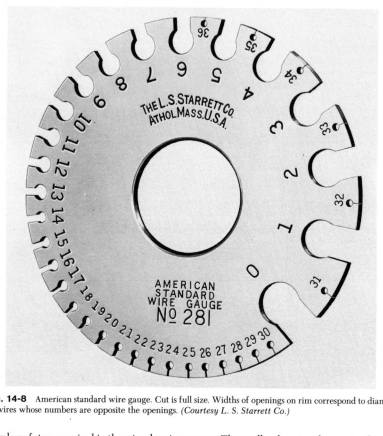

Fig. 14-8 American standard wire gauge. Cut is full size. Widths of openings on rim correspond to diameters of wires whose numbers are opposite the openings. *(Courtesy L. S. Starrett Co.)*

number of steps required in the wire-drawing process. The smaller the wire, the greater the number of steps required to draw it.

Descriptive details for solid copper wire are itemized in Table 14-2. It can be noted in Table 14-2 that with every increase of three in the gauge number, the cross section and weight are halved and the resistance is doubled. An increase of ten in the gauge number increases the resistance ten times and cuts the weight and cross section to one-tenth. Thus, for example, the No. 5 wire is three sizes smaller than the No. 2 wire. Its cross section is one-half that of the No. 2, for 33,090 is approximately one-half of 66,360, and its resistance is twice that of the No. 2, for 0.3260 is twice 0.1625. Furthermore, the resistance of the No. 12 is ten times that of the No. 2 for 1.65 is nearly ten times 0.1625. This general relation between gauge numbers is well to remember. Properties of solid aluminum wire are listed in Table 14-3.

Circular Mil The circular mil is the unit of cross-sectional area customarily used in designating the area of wires. A circular mil is the area contained in a circle having a diameter of $\frac{1}{1000}$ in (Fig. 14-9). A mil is a thousandth of an inch. The use of the circular mil as the unit of area came about because of the custom of using a square as the unit of area in measuring the areas of squares or rectangles. Naturally, when the areas of circles had to be measured a circle having unit diameter suggested itself as the natural unit of measurement.

Table 14-1   American Wire Sizes (Bare Conductor)

Gage number	Diameter, in.	Full size, end view	Full size, side view
8	0.1285		
7	0.1443		
6	0.162		
5	0.1819		
4	0.2043		
3	0.2294		
2	0.2576		
1	0.2893		
0	0.3249		
00	0.3648		
000	0.4096		
0000	0.460		

Table 14-2   Copper Wire—Weight, Breaking Strength, DC Resistance
(Based on ASTM Specifications B1, B2, B3)

Size, AWG	Diameter, in.	Area		Weight		Hard		Medium		Soft	
		Cir mils	Sq in.	Lb per 1,000 ft	Lb per mile	Breaking strength, minimum, lb	D-c resistance at 20 C (68 F) maximum,† ohms per 1,000 ft	Breaking strength, minimum, lb	D-c resistance at 20 C (68 F) maximum,† ohms per 1,000 ft	Breaking strength, maximum,‡ lb	D-c resistance at 20 C (68 F) maximum,† ohms per 1,000 ft
4/0	0.4600	211,600	0.1662	640.5	3382	8143	0.05045	6980	0.05019	5983	0.04901
3/0	0.4096	167,800	0.1318	507.8	2681	6720	0.06362	5666	0.06330	4744	0.06182
2/0	0.3648	133,100	0.1045	402.8	2127	5519	0.08021	4599	0.07980	3763	0.07793
1/0	0.3249	105,600	0.08291	319.5	1687	4518	0.1022	3731	0.1016	2985	0.09825
1	0.2893	83,690	0.06573	253.3	1338	3688	0.1289	3024	0.1282	2432	0.1239
2	0.2576	66,360	0.05212	200.9	1061	3002	0.1625	2450	0.1617	1928	0.1563
3	0.2294	52,620	0.04133	159.3	841.1	2439	0.2050	1984	0.2039	1529	0.1971
4	0.2043	41,740	0.03278	126.3	667.1	1970	0.2584	1584	0.2571	1213	0.2485
5	0.1819	33,090	0.02599	100.2	528.8	1590	0.3260	1265	0.3243	961.5	0.3135
6	0.1620	26,240	0.02061	79.44	419.4	1280	0.4110	1010	0.4088	762.6	0.3952
7	0.1443	20,820	0.01635	63.03	332.8	1030	0.5180	806.7	0.5153	605.1	0.4981
8	0.1285	16,510	0.01297	49.98	263.9	826.1	0.6532	644.0	0.6498	479.8	0.6281
9	0.1144	13,090	0.01028	39.61	209.2	660.9	0.8241	513.9	0.8199	380.3	0.7925
10	0.1019	10,380	0.008155	31.43	166.0	529.3	1.039	410.5	1.033	314.0	0.9988
11	0.0907	8,230	0.00646	24.9	131	423	1.31	327	1.30	249	1.26
12	0.0808	6,530	0.00513	19.8	104	337	1.65	262	1.64	197	1.59
13	0.0720	5,180	0.00407	15.7	82.9	268	2.08	209	2.07	157	2.00
14	0.0641	4,110	0.00323	12.4	65.7	214	2.63	167	2.61	124	2.52
15	0.0571	3,260	0.00256	9.87	52.1	170	3.31	133	3.29	98.6	3.18
16	0.0508	2,580	0.00203	7.81	41.2	135	4.18	106	4.16	78.0	4.02

source: *Standard Handbook for Electrical Engineers*, Donald G. Fink, Editor-In-Chief, H. Wayne Beaty, Associate Editor. Copyright 1978 by McGraw-Hill, Inc. Used with permission of McGraw-Hill Book Company.
†Based on nominal diameter and ASTM resistivities.
‡No requirements for tensile strength are specified in ASTM B3. Values given here based on Anaconda data.

The circular mils of cross section in a wire of any diameter are obtained by multiplying the diameter in thousandths of an inch by itself. Thus the number of circular mils cross section in the No. 10 wire is $102 \times 102 = 10,404$. The circular mils cross section of the No. 2 wire is $257.6 \times 257.6 = 66,370$.

Table 14-3   Aluminum Wire—Dimensions, Weight, DC Resistance
(Based on ASTM Specifications B230, B262, and B323)

Conductor size, AWG	Diam. at 20 C (68 F), mils	Area at 20 C (68 F)		D-c resistance at 20 C (68 F),* ohms per 1,000 ft	Weight at 20 C (68 F),† lb		Length at 20 C (68 F), ft per ohm
		Cir mils	Sq in.		Per 1,000 ft	Per ohm	
2	257.6	66,360	0.05212	0.2562	61.07	238.4	3903
3	229.4	52,620	0.04133	0.3231	48.43	149.9	3095
4	204.3	41,740	0.03278	0.4074	38.41	94.30	2455
5	181.9	33,090	0.02599	0.5139	30.45	59.26	1946
6	162.0	26,240	0.02061	0.6479	24.15	37.28	1544
7	144.3	20,820	0.01635	0.8165	19.16	23.47	1225
8	128.5	16,510	0.01297	1.030	15.20	14.76	971.2
9	114.4	13,090	0.01028	1.299	12.04	9.272	769.7
10	101.9	10,380	0.008155	1.637	9.556	5.836	610.7
11	90.7	8,230	0.00646	2.07	7.57	3.66	484
12	80.8	6,530	0.00513	2.60	6.01	2.31	384
13	72.0	5,180	0.00407	3.28	4.77	1.45	305
14	64.1	4,110	0.00323	4.14	3.78	0.914	242
15	57.1	3,260	0.00256	5.21	3.00	0.575	192
16	50.8	2,580	0.00203	6.59	2.38	0.361	152
17	45.3	2,050	0.00161	8.29	1.89	0.228	121
18	40.3	1,620	0.00128	10.5	1.49	0.143	95.5
19	35.9	1,290	0.00101	13.2	1.19	0.0899	75.8
20	32.0	1,020	0.000804	16.6	0.942	0.0568	60.2
21	28.5	812	0.000638	20.9	0.748	0.0357	47.8
22	25.3	640	0.000503	26.6	0.589	0.0222	37.6
23	22.6	511	0.000401	33.3	0.470	0.0141	30.0
24	20.1	404	0.000317	42.1	0.372	0.00884	23.8
25	17.9	320	0.000252	53.1	0.295	0.00556	18.8
26	15.9	253	0.000199	67.3	0.233	0.00346	14.9
27	14.2	202	0.000158	84.3	0.186	0.00220	11.9
28	12.6	159	0.000125	107	0.146	0.00136	9.34
29	11.3	128	0.000100	133	0.118	0.000883	7.51
30	10.0	100	0.0000785	170	0.0920	0.000541	5.88

SOURCE: *Standard Handbook for Electrical Engineers*, Donald G. Fink, Editor-In-Chief, H. Wayne Beaty, Associate Editor. Copyright 1978 by McGraw-Hill, Inc. Used with permission of McGraw-Hill Book Company.
* Conductivity = 61.0% IACS.
† Density = 2.703 g per cu cm (0.09765 lb per cu in.).

If one can remember the relations pointed out above and the data for the No. 10 copper wire, one can build up the remainder of the table any time. The approximate data for the No. 10 copper wire are easily remembered because its resistance per 1000 ft is 1 ohm, its diameter is $\frac{1}{10}$ in, and its circular-mil cross section is 10,000. Its weight per 1000 ft is 31.4 lb. With these figures as a basis, one can calculate the data for the Nos. 7, 4, 1, 000, and 13, 16, 19, etc., wire numbers by using the factors 2 and ½, remembering that for every increase in wire number of three sizes, the cross section and weight are halved and the resistance is doubled. In other words, the larger the number of the wire, the smaller the wire.

To obtain the data for wires one number larger, the factors are 1.25 and 0.80, and for two numbers larger they are 1.60 and 0.625. Thus the data for the No. 11 wire are obtained by mul-

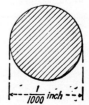

Fig. 14-9   Area of circle whose diameter is $\frac{1}{1000}$ in is 1 cir mil.

tiplying the resistance of the No. 10 wire by 1.25, that is, 1.0 × 1.25 equals 1.25, and dividing the cross-sectional area of 10,000 by 1.25, which equals 8000. In the same way, the resistance of the No. 12 wire is obtained by multiplying, 1.0 × 1.6 = 1.6; and the cross section is obtained by dividing, 10,000 ÷ 1.6 = 6250. The data for the No. 13 wire would be obtained by using the factor 2 as explained before, and the value for the No. 14, which is one number larger than the No. 13, would be obtained by using the factor 1.25 again on the values obtained for the No. 13 wire. The results obtained in this way are, of course, approximate, but they are close enough to give one a fair idea of the wire.

Table 14-4   Aluminum Cable—Stranding Classes, Uses
(ASTM Specification B231)

Construction	Class	Application
Concentric lay	AA	For bare conductors usually used in overhead lines
	A	For conductors to be covered with weather-resistant (weatherproof), slow-burning materials and for bare conductors where greater flexibility than is afforded by Class AA is required. Conductors intended for further fabrication into tree wire or to be insulated and laid helically with or around aluminum or ACSR messengers shall be regarded as Class A conductors with respect to direction of lay only
	B	For conductors to be insulated with various materials such as rubber, paper, varnished cloth, etc., and for the conductors indicated under Class A where greater flexibility is required
	C, D	For conductors where greater flexibility is required than is provided by Class B conductors

SOURCE: *Standard Handbook for Electrical Engineers*, Donald G. Fink, Editor-In-Chief, H. Wayne Beaty, Associate Editor. Copyright 1978 by McGraw Hill, Inc. Used with permission of McGraw-Hill Book Company.

Stranded Conductor Data   The classes of concentric lay stranded cable for various applications are given in Table 14-4. Descriptive data for copper stranded conductors is itemized in Table 14-5. Table 14-6 provides data for copper-clad steel-copper conductors.

Table 14-7 lists properties of aluminum-stranded conductors. Aluminum-steel-reinforced cable (ACSR) descriptive data is itemized in Table 14-8. Galvanized-steel cable properties are listed in Table 14-9.

Conductor Selection   The size of a line conductor depends on a number of factors. The most important of these are the line voltage, the amount of power to be transmitted, and mechanical strength. The line voltage is a very important factor, for it will be recalled that for a given amount of power the greater the voltage, the less will the current be. That is the reason for using as high a line voltage as possible. The current in turn determines the size of the conductor.

Equally important is the magnitude of the load to be transmitted. For a given voltage, the larger the load, the larger the current. That a conductor must also have sufficient mechanical strength to carry its own weight and any load due to sleet or wind is obvious if the line is to give reliable and uninterrupted service. Other factors are length of line, power factor of load, length of span, etc. Table 14-10 gives the current-carrying capacities of copper and aluminum wires for the wire sizes used frequently in outdoor circuits. These values of current were obtained by experiment and are of such magnitude that the heat produced by them in the conductor will bring it up to a temperature of 75°C (167°F). The air is assumed to have a temperature of 25°C, and a steady wind of 1.4 miles per hour (2 ft per sec) is blowing over the conductors. The frequency is 60 Hz. If ability to carry current were the only requirement, one could easily pick from this table the right size of conductor to be used for any given current.

While the values of current shown for the different size conductors could be carried in emergency operation, they are far too high for normal loading. Normal values would be approximately one-half of the values shown. Furthermore, current-carrying capacity is not the only consideration in determining conductor size. Percent line loss and voltage regulation are equally important considerations.

If no other factor would need to be considered, a good rule to follow for copper conductors is to allow 1000 cir mils per amp. For example, a current of 300 amp would thus require a copper conductor having 300,000 cir mils of area.

The conductor, besides requiring adequate current-carrying capacity, must be of sufficient size and strength to support itself and any additional load due to ice, sleet, and wind. The weight of the conductor for a given span can be easily computed, but the magnitude of the additional load due to sleet and wind cannot be accurately determined. For this purpose the Bureau of Standards of the United States government has divided the entire area of the United States into so-called "loading districts," as shown in Fig. 14-10. The three loading districts are known as "heavy," "medium," and "light."

Table 14-5 Copper Cable, Classes AA, A, A, B—Weight, Breaking Strength, DC Resistance
(ASTM Specifications B1, B2, B3, B8)

Conductor size, Mcm or Awg	No. of wires (ASTM stranding class)	Wire diameter, in.	Conductor diameter, in.	Conductor area, sq in.	Conductor weight, lb Per 1000 ft	Conductor weight, lb Per mile	Hard Breaking strength, minimum,* lb	Hard D-c resistance at 20 C (68 F), ohms per 1000 ft	Medium Breaking strength, minimum,* lb	Medium D-c resistance at 20 C (68 F), ohms per 1000 ft	Soft Breaking strength, maximum,† lb	Soft D-c resistance at 20 C (68 F), ohms per 1000 ft
5000	169 (A)	0.1720	2.580	3.927	15,890	83,910	216,300	0.002265	172,000	0.002253	145,300	0.002178
5000	217 (B)	0.1518	2.581	3.927	15,890	83,910	219,500	0.002265	173,200	0.002253	145,300	0.002178
4500	169 (A)	0.1632	2.448	3.534	14,300	75,520	197,200	0.002517	154,800	0.002504	130,800	0.002420
4500	217 (B)	0.1440	2.448	3.534	14,300	75,520	200,400	0.002517	156,900	0.002504	130,800	0.002420
4000	169 (A)	0.1538	2.307	3.142	12,590	66,490	175,600	0.002804	138,500	0.002790	116,200	0.002697
4000	217 (B)	0.1358	2.309	3.142	12,590	66,490	178,100	0.002804	139,500	0.002790	116,200	0.002697
3500	127 (A)	0.1660	2.158	2.749	11,020	58,180	153,400	0.003205	120,400	0.003188	101,700	0.003082
3500	169 (B)	0.1439	2.158	2.749	11,020	58,180	155,900	0.003205	122,000	0.003188	101,700	0.003082
3000	127 (A)	0.1537	1.998	2.356	9,353	49,390	131,700	0.003703	103,900	0.003684	87,180	0.003561
3000	169 (B)	0.1332	1.998	2.356	9,353	49,390	134,400	0.003703	104,600	0.003684	87,180	0.003561
2500	91 (A)	0.1657	1.823	1.963	7,794	41,150	109,600	0.004444	85,990	0.004421	72,650	0.004273
2500	127 (B)	0.1403	1.824	1.963	7,794	41,150	111,300	0.004444	87,170	0.004421	72,650	0.004273
2000	91 (A)	0.1482	1.630	1.571	6,175	32,600	87,790	0.005501	69,270	0.005472	58,120	0.005289
2000	127 (B)	0.1255	1.632	1.571	6,175	32,600	90,050	0.005501	70,210	0.005472	58,120	0.005289
1750	91 (A)	0.1387	1.526	1.374	5,403	28,530	77,930	0.006286	61,020	0.006254	50,850	0.006045
1750	127 (B)	0.1174	1.526	1.374	5,403	28,530	78,800	0.006286	61,430	0.006254	50,850	0.006045
1500	61 (A)	0.1568	1.411	1.178	4,631	24,450	65,840	0.007334	51,950	0.007296	43,590	0.007052
1500	91 (B)	0.1284	1.412	1.178	4,631	24,450	67,540	0.007334	52,650	0.007296	43,590	0.007052
1250	61 (A)	0.1431	1.288	0.9817	3,859	20,380	55,670	0.008801	43,590	0.008755	36,320	0.008463
1250	91 (B)	0.1172	1.289	0.9817	3,859	20,380	56,280	0.008801	43,880	0.008755	36,320	0.008463
1000	37 (AA)	0.1644	1.151	0.7854	3,088	16,300	43,830	0.01100	34,400	0.01094	29,060	0.01058
1000	61 (A-B)	0.1280	1.152	0.7854	3,088	16,300	45,030	0.01100	35,100	0.01094	29,060	0.01058
900	37 (AA)	0.1560	1.092	0.7069	2,779	14,670	39,510	0.01222	31,170	0.01216	26,150	0.01175
900	61 (A-B)	0.1215	1.094	0.7069	2,779	14,670	40,520	0.01222	31,590	0.01216	26,150	0.01175
850	37 (AA)	0.1516	1.061	0.6676	2,624	13,860	37,310	0.01294	29,440	0.01288	24,700	0.01245
850	61 (A-B)	0.1180	1.062	0.6676	2,624	13,860	38,270	0.01294	29,840	0.01288	24,700	0.01245
800	37 (AA)	0.1470	1.029	0.6283	2,470	13,040	35,120	0.01375	27,710	0.01368	23,250	0.01322
800	61 (A-B)	0.1145	1.031	0.6283	2,470	13,040	36,360	0.01375	28,270	0.01368	23,250	0.01322
750	37 (AA)	0.1424	0.997	0.5890	2,316	12,230	33,400	0.01467	26,150	0.01459	21,790	0.01410
750	61 (A-B)	0.1109	0.998	0.5890	2,316	12,230	34,090	0.01467	26,510	0.01459	21,790	0.01410

Table 14-5   Copper Cable, Classes AA, A, B—Weight, Breaking Strength, DC Resistance   (Continued)

Conductor size, Mcm or Awg	No. of wires (ASTM stranding class)	Wire diameter, in.	Conductor diameter, in.	Conductor area, sq in.	Conductor weight, lb Per 1000 ft	Per mile	Hard Breaking strength, minimum,* lb	Hard D-c resistance at 20 C (68 F), ohms per 1000 ft	Medium Breaking strength, minimum,* lb	Medium D-c resistance at 20 C (68 F), ohms per 1000 ft	Soft Breaking strength, maximum,† lb	Soft D-c resistance at 20 C (68 F), ohms per 1000 ft
700	37 (AA)	0.1375	0.963	0.5498	2,161	11,410	31,170	0.01572	24,410	0.01563	20,340	0.01511
700	61 (A-B)	0.1071	0.964	0.5498	2,161	11,410	31,820	0.01572	24,740	0.01563	20,340	0.01511
650	37 (AA)	0.1325	0.928	0.5105	2,007	10,600	29,130	0.01692	22,670	0.01684	18,890	0.01627
650	61 (A-B)	0.1032	0.929	0.5105	2,007	10,600	29,770	0.01692	22,970	0.01684	18,890	0.01627
600	37 (AA-A)	0.1273	0.891	0.4712	1,853	9,781	27,020	0.01834	21,060	0.01824	17,440	0.01763
600	61 (B)	0.0992	0.893	0.4712	1,853	9,781	27,530	0.01834	21,350	0.01824	18,140	0.01763
550	37 (AA-A)	0.1219	0.853	0.4320	1,698	8,966	24,760	0.02000	19,310	0.01990	15,980	0.01923
550	61 (B)	0.0950	0.855	0.4320	1,698	8,966	25,230	0.02000	19,570	0.01990	16,630	0.01923
500	19 (AA)	0.1622	0.811	0.3927	1,544	8,151	21,950	0.02200	17,320	0.02189	14,530	0.02116
500	37 (A-B)	0.1162	0.813	0.3927	1,544	8,151	22,510	0.02200	17,550	0.02189	14,530	0.02116
450	19 (AA)	0.1539	0.770	0.3534	1,389	7,336	19,750	0.02445	15,590	0.02432	13,080	0.02351
450	37 (A-B)	0.1103	0.772	0.3534	1,389	7,336	20,450	0.02445	15,900	0.02432	13,080	0.02351
400	19 (AA-A)	0.1451	0.726	0.3142	1,235	6,521	17,810	0.02750	13,950	0.02736	11,620	0.02645
400	37 (B)	0.1040	0.728	0.3142	1,235	6,521	18,320	0.02750	14,140	0.02736	11,620	0.02645
350	12 (AA)	0.1708	0.710	0.2749	1,081	5,706	15,140	0.03143	12,040	0.03127	10,170	0.03022
350	19 (A)	0.1357	0.679	0.2749	1,081	5,706	15,590	0.03143	12,200	0.03127	10,170	0.03022
350	37 (B)	0.0973	0.681	0.2749	1,081	5,706	16,060	0.03143	12,450	0.03127	10,580	0.03022
300	12 (AA)	0.1581	0.657	0.2356	926.3	4,891	13,170	0.03667	10,390	0.03648	8,718	0.03526
300	19 (A)	0.1257	0.629	0.2356	926.3	4,891	13,510	0.03667	10,530	0.03648	8,718	0.03526
300	37 (B)	0.0900	0.630	0.2356	926.3	4,891	13,870	0.03667	10,740	0.03648	9,071	0.03526
250	12 (AA)	0.1443	0.600	0.1963	771.9	4,076	11,130	0.04400	8,717	0.04378	7,265	0.04231
250	19 (A)	0.1147	0.574	0.1963	771.9	4,076	11,360	0.04400	8,836	0.04378	7,265	0.04231
250	37 (B)	0.0822	0.575	0.1963	771.9	4,076	8,952	0.04400	8,952	0.04378	7,559	0.04231
4/0	7 (AA-A)	0.1739	0.522	0.1662	653.3	3,450	9,154	0.05199	7,278	0.05172	6,149	0.04999
4/0	12 —	0.1328	0.552	0.1662	653.3	3,450	9,483	0.05199	7,378	0.05172	6,149	0.04999
4/0	19 (B)	0.1055	0.528	0.1662	653.3	3,450	9,617	0.05199	7,479	0.05172	6,149	0.04999
3/0	7 (AA-A)	0.1548	0.464	0.1318	518.1	2,736	7,366	0.06556	5,812	0.06522	4,876	0.06304
3/0	12 (B)	0.1183	0.492	0.1318	518.1	2,736	7,556	0.06556	5,890	0.06522	4,876	0.06304
2/0	19 (AA-A)	0.0940	0.470	0.1318	518.1	2,736	7,698	0.06556	5,970	0.06522	5,074	0.06304
2/0	12 —	0.1379	0.414	0.1045	410.9	2,169	5,926	0.08267	4,640	0.08224	3,867	0.07949
2/0	19 (B)	0.1053	0.438	0.1045	410.9	2,169	6,048	0.08267	4,703	0.08224	3,867	0.07949
1/0	7 (AA-A)	0.0837	0.419	0.1045	410.9	2,169	6,152	0.08267	4,765	0.08224	4,024	0.07949
1/0	12 (B)	0.1228	0.368	0.08289	325.8	1,720	4,752	0.1042	3,705	0.1037	3,067	0.1002
1/0	19 (B)	0.0938	0.390	0.08289	325.8	1,720	4,841	0.1042	3,755	0.1037	3,191	0.1002
1/0	—	0.0745	0.373	0.08289	325.8	1,720	4,901	0.1042	3,805	0.1037	3,191	0.1002

Conductor size, Mcm or Awg	No. of wires (ASTM stranding class)	Wire diameter, in.	Conductor diameter, in.	Conductor area, sq in.	Conductor weight, lb		Hard		Medium		Soft	
					Per 1000 ft	Per mile	Breaking strength, minimum,* lb	D-c resistance at 20 C (68 F), ohms per 1000 ft	Breaking strength, minimum,* lb	D-c resistance at 20 C (68 F), ohms per 1000 ft	Breaking strength, maximum,† lb	D-c resistance at 20 C (68 F), ohms per 1000 ft
1	3 (AA)	0.1670	0.360	0.06573	255.9	1,351	3,621	0.1302	2,879	0.1295	2,432	0.1252
1	7 (A)	0.1093	0.328	0.06573	258.4	1,364	3,804	0.1314	2,958	0.1308	2,432	0.1264
1	19 (B)	0.0664	0.332	0.06573	258.4	1,364	3,899	0.1314	3,037	0.1308	2,531	0.1264
2	3 (AA)	0.1487	0.320	0.05213	202.9	1,071	2,913	0.1641	2,299	0.1633	1,929	0.1578
2	7 (A-B)	0.0974	0.292	0.05213	204.9	1,082	3,045	0.1657	2,361	0.1649	2,007	0.1594
3	3 (AA)	0.1325	0.285	0.04134	160.9	849.6	2,359	0.2070	1,835	0.2059	1,530	0.1990
3	7 (A-B)	0.0867	0.260	0.04134	162.5	858.0	2,433	0.2090	1,885	0.2079	1,592	0.2010
4	3 (AA)	0.1180	0.254	0.03278	127.6	673.8	1,879	0.2610	1,465	0.2596	1,213	0.2509
4	7 (A-B)	0.0772	0.232	0.03278	128.9	680.5	1,938	0.2636	1,505	0.2622	1,262	0.2534
5	7 (A-B)	0.0688	0.206	0.02600	102.2	539.6	1,542	0.3323	1,201	0.3306	1,001	0.3196
6	7 (B)	0.0612	0.184	0.02062	81.05	427.9	1,288	0.4191	958.6	0.4169	793.8	0.4030
7	7 (B)	0.0545	0.164	0.01635	64.28	339.4	977.1	0.5284	765.2	0.5257	629.5	0.5081
8	7 (B)	0.0486	0.146	0.01297	50.97	269.1	777.2	0.6663	610.7	0.6629	499.2	0.6408
9	7 (B)	0.0432	0.130	0.01028	40.42	213.4	618.2	0.8402	487.4	0.8359	395.9	0.8080
10	7 (B)	0.0385	0.116	0.008155	32.06	169.3	491.7	1.060	388.9	1.054	314.0	1.019
12	7 (B)	0.0305	0.0915	0.005129	20.16	106.5	311.1	1.685	247.7	1.676	197.5	1.620
14	7 (B)	0.0242	0.0726	0.003225	12.68	66.95	197.1	2.679	157.7	2.665	124.2	2.576
16	7 (B)	0.0192	0.0576	0.002028	7.974	42.10	124.7	4.259	100.4	4.237	81.14	4.096
18	7 (B)	0.0152	0.0456	0.001276	5.015	26.48	78.99	6.773	63.91	6.738	51.03	6.513
20	7 (B)	0.0121	0.0363	0.0008023	3.154	16.65	50.04	10.77	40.67	10.71	32.09	10.36
ASTM Designation				B8			B1 & B8		B2 & B8		B3 & B8	

*No. 10 AWG and smaller, based on Anaconda data.

†No requirements for tensile strength are specified in ASTM B3. Values given here based on Anaconda data.

Weight and Resistance

Stranding class	Conductor size, Mcm or Awg	Increment of resistance and weight, %
AA........	{ 4–1	
	1/0–1000	1
A, B, C, D......	2000 and under	2
	Over 2000–3000	3
	Over 3000–4000	4
	Over 4000–5000	5

Resistance
(ASTM requirements)

Temper	Conductivity at 20 C (68 F), IACS, %	Resistivity at 20 C (68 F), ohms (mile, lb)
Hard......	96.16	910.15
Medium......	96.66	905.44
Soft......	100	875.20

The resistance values in this table are trade maximums and are higher than the average values for commercial cable.

SOURCE: *Standard Handbook for Electrical Engineers*, Donald G. Fink, Editor-In-Chief, H. Wayne Beaty, Associate Editor. Copyright 1978 by McGraw-Hill, Inc. Used with permission of McGraw-Hill Book Company.

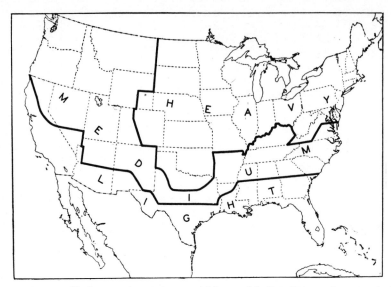

Fig. 14-10   General loading map showing the territorial division of the United States with respect to loading of overhead lines. *(Taken from National Electrical Safety Code, 6th ed.)*

Table 14-6   Copper-clad Steel-Copper Cable—Weight, Breaking Strength, DC Resistance
(ASTM Specification B229)

Hard-drawn copper equivalent,* Mcm or AWG	Conductor type	Conductor stranding				Conductor diam., in	Conductor area, in²	Conductor weight, lb		Breaking strength, min., lb	D-c resistance at 20°C (68°F), Ω/1,000 ft
		EHS 30% copper-clad wires		Hard-drawn copper wires				Per 1,000 ft	Per mi		
		No.	Diam., in	No.	Diam, in						
350	E	7	0.1576	12	0.1576	0.788	0.3706	1403	7409	32,420	0.03143
350	EK	4	0.1470	15	0.1470	0.735	0.3225	1238	6536	23,850	0.03143
300	E	7	0.1459	12	0.1459	0.729	0.3177	1203	6351	27,770	0.03667
300	EK	4	0.1361	15	0.1361	0.680	0.2764	1061	5602	20,960	0.03667
250	E	7	0.1332	12	0.1332	0.666	0.2648	1002	5292	23,920	0.04400
250	EK	4	0.1242	15	0.1242	0.621	0.2302	884.2	4669	17,840	0.04400
4/0	E	7	0.1225	12	0.1225	0.613	0.2239	848.3	4479	20,730	0.05199
4/0	EK	4	0.1143	15	0.1143	0.571	0.1950	748.4	3951	15,370	0.05199
4/0	F	1	0.1833	6	0.1833	0.550	0.1847	710.2	3750	12,290	0.05199
3/0	E	7	0.1091	12	0.1091	0.545	0.1776	672.7	3552	16,800	0.06556
3/0	EK	4	0.1018	15	0.1018	0.509	0.1546	593.5	3134	12,370	0.06556
3/0	F	1	0.1632	6	0.1632	0.490	0.1464	563.2	2974	9,980	0.06556
2/0	F	1	0.1454	6	0.1454	0.436	0.1162	446.8	2359	8,094	0.08265
1/0	F	1	0.1294	6	0.1294	0.388	0.09206	354.1	1870	6,536	0.1043
1	F	1	0.1153	6	0.1153	0.346	0.07309	280.9	1483	5,266	0.1315
2†	A	1	0.1699	2	0.1699	0.366	0.06801	256.8	1356	5,876	0.1658
2	F	1	0.1026	6	0.1026	0.308	0.05787	222.8	1176	4,233	0.1658
4†	A	1	0.1347	2	0.1347	0.290	0.04275	161.5	852	3,938	0.2636
6†	A	1	0.1068	2	0.1068	0.230	0.02688	101.6	536.3	2,585	0.4150
8†	A	1	0.1127	2	0.07969	0.199	0.01995	74.27	392.2	2,233	0.6598

SOURCE: *Standard Handbook for Electrical Engineers*, Donald G. Fink, Editor-In-Chief, H. Wayne Beaty, Associate Editor. Copyright 1978 by McGraw-Hill, Inc. Used with permission of McGraw-Hill Book Company.
* Area of hard-drawn copper cable having the same d-c resistance as that of the composite cable.
† Sizes commonly used for rural distribution.

Table 14-7 Aluminum Conductor—Physical Characteristics EC-H19 Classes AA and A

Cable code word	Conductor size cmils or AWG	Conductor size in²	Current-carrying capacity* A	Stranding Class	Stranding No. and diam of wires, in	Conductor diam, in	Rated strength, lb	Nominal weight, lbf Per 1000 ft	Nominal weight, lbf Per mile
Peachbell	6	0.0206	95	A	7 × 0.0612	0.184	563	24.6	130
Rose	4	0.0328	130	A	7 × 0.0772	0.232	881	39.2	207
Iris	2	0.0522	175	AA, A	7 × 0.0974	0.292	1,350	62.3	329
Pansy	1	0.0657	200	AA, A	7 × 0.1093	0.328	1,640	78.5	414
Poppy	1/0	0.0829	235	AA, A	7 × 0.1228	0.368	1,990	99.1	523
Aster	2/0	0.1045	270	AA, A	7 × 0.1379	0.414	2,510	124.9	659
Phlox	3/0	0.1317	315	AA, A	7 × 0.1548	0.464	3,040	157.5	832
Oxlip	4/0	0.1663	365	AA, A	7 × 0.1739	0.522	3,830	198.7	1,049
Sneezewort	250,000	0.1964	405	AA	7 × 0.1890	0.567	4,520	234.7	1,239
Valerian	250,000	0.1963	405	A	19 × 0.1147	0.574	4,660	234.6	1,239
Daisy	266,800	0.2097	420	AA	7 × 0.1953	0.586	4,830	250.6	1,323
Laurel	266,800	0.2095	425	A	19 × 0.1185	0.593	4,970	250.4	1,322
Peony	300,000	0.2358	455	A	19 × 0.1257	0.629	5,480	281.8	1,488
Tulip	336,400	0.2644	495	A	19 × 0.1331	0.666	6,150	316.0	1,668
Daffodil	350,000	0.2748	506	A	19 × 0.1357	0.679	6,390	328.4	1,734
Canna	397,500	0.3124	550	AA, A	19 × 0.1447	0.724	7,110	373.4	1,972
Goldentuft	450,000	0.3534	545	AA	19 × 0.1539	0.770	7,890	422.4	2,230
Cosmos	477,000	0.3744	615	AA	19 × 0.1584	0.793	8,360	447.5	2,363
Syringa	477,000	0.3743	615	A	37 × 0.1135	0.795	8,690	447.4	2,362
Zinnia	500,000	0.3926	635	AA	19 × 0.1622	0.811	8,760	469.2	2,477
Hyacinth	500,000	0.3924	635	A	37 × 0.1162	0.813	9,110	469.0	2,476
Dahlia	556,500	0.4369	680	AA	19 × 0.1711	0.856	9,750	522.1	2,757
Mistletoe	556,500	0.4368	680	AA, A	37 × 0.1226	0.858	9,940	522.0	2,756
Meadowsweet	600,000	0.4709	715	AA, A	37 × 0.1273	0.891	10,700	562.8	2,972
Orchid	636,000	0.4995	745	AA, A	37 × 0.1311	0.918	11,400	596.9	3,152
Heuchera	650,000	0.5102	755	AA	37 × 0.1325	0.928	11,600	609.8	3,220
Verbena	700,000	0.5494	790	AA	37 × 0.1375	0.963	12,500	656.6	3,467
Flag	700,000	0.5495	790	A	61 × 0.1071	0.964	12,500	656.8	3,468
Violet	715,500	0.5622	800	AA	37 × 0.1391	0.974	12,800	672.0	3,548
Nasturtium	715,500	0.5619	800	A	61 × 0.1083	0.975	13,100	671.6	3,546
Petunia	750,000	0.5892	825	AA	37 × 0.1424	0.997	13,100	704.3	3,719
Cattail	750,000	0.5892	825	A	61 × 0.1109	0.998	13,500	704.2	3,718
Arbutus	795,000	0.6245	855	AA	37 × 0.1466	1.026	13,900	746.4	3,941
Lilac	795,000	0.6248	855	A	61 × 0.1142	1.028	14,300	746.7	3,943

Table 14-7 Aluminum Conductor—Physical Characteristics EC-H19 Classes AA and A (Continued)

Cable code word	Conductor size		Current-carrying capacity* A	Stranding		Conductor diam, in	Rated strength, lb	Nominal weight, lb†	
	cmils or AWG	in²		Class	No. and diam of wires, in			Per 1000 ft	Per mile
Cockscomb	900,000	0.7072	925	AA	37 × 0.1560	1.092	15,400	845.2	4,463
Snapdragon	900,000	0.7072	925	A	61 × 0.1215	1.094	15,900	845.3	4,463
Magnolia	954,000	0.7495	960	AA	37 × 0.1606	1.124	16,400	895.8	4,730
Goldenrod	954,000	0.7498	960	A	61 × 0.1251	1.126	16,900	896.1	4,731
Hawkweed	1,000,000	0.7854	990	AA	37 × 0.1644	1.151	17,200	938.7	4,956
Camellia	1,000,000	0.7849	990	A	61 × 0.1280	1.152	17,700	938.2	4,954
Bluebell	1,033,500	0.8124	1015	AA	37 × 0.1672	1.170	17,700	970.9	5,126
Larkspur	1,033,500	0.8122	1015	A	61 × 0.1302	1.172	18,300	970.6	5,125
Marigold	1,113,000	0.8744	1040	AA, A	61 × 0.1351	1.216	19,700	1,045	5,518
Hawthorn	1,192,500	0.9363	1085	AA, A	61 × 0.1398	1.258	21,100	1,119	5,908
Narcissus	1,272,000	0.999	1130	AA, A	61 × 0.1444	1.300	22,000	1,194	6,304
Columbine	1,351,500	1.062	1175	AA, A	61 × 0.1489	1.340	23,400	1,269	6,700
Carnation	1,431,000	1.124	1220	AA, A	61 × 0.1532	1.379	24,300	1,344	7,096
Galdiolus	1,510,500	1.187	1265	AA, A	61 × 0.1574	1.417	25,600	1,419	7,492
Coreopsis	1,590,000	1.250	1305	AA	61 × 0.1615	1.454	27,000	1,493	7,883
Jessamine	1,750,000	1.375	1385	AA	61 × 0.1694	1.525	29,700	1,643	8,675
Cowslip	2,000,000	1.570	1500	A	91 × 0.1482	1.630	34,200	1,876	9,911
Sagebrush	2,250,000	1.766	1600	A	91 × 0.1572	1.729	37,700	2,132	11,257
Lupine	2,500,000	1.962	1700	A	91 × 0.1657	1.823	41,800	2,368	12,503
Bitterroot	2,750,000	2.159	1795	A	91 × 0.1738	1.912	46,100	2,606	13,760
Trillium	3,000,000	2.356	1885	A	127 × 0.1537	1.996	50,300	2,844	15,016
Bluebonnet	3,500,000	2.749	2035	A	127 × 0.1660	2.158	58,700	3,350	17,688

SOURCE: *Standard Handbook for Electrical Engineers*, Donald G. Fink, Editor-In-Chief, H. Wayne Beaty, Associate Editor. Copyright 1978 by McGraw-Hill, Inc. Used with permission of McGraw-Hill Book Company.

Class of stranding. The class of stranding must be specified on all orders. Class AA stranding is usually specified for bare conductors used on overhead lines. Class A stranding is usually specified for conductors to be covered with weather-resistant (weatherproof) materials and for bare conductors where greater flexibility than afforded by Class AA is required.

Lay. The direction of lay of the outside layer of wires with Class AA and Class A stranding will be right hand unless otherwise specified.

*Ampacity for conductor temperature rise of 40°C over 40°C ambient with a 2 ft/s crosswind and an emissivity factor of 0.5 without sun.

†Nominal conductor weights are based on ASTM standard stranding increments. Actual weights will vary with lay lengths. Invoicing will be based on actual weights.

Table 14-8 ACSR Conductor—Physical Characteristics

Code word	ACSR Cross section Aluminum, cmils or AWG	Aluminum in²	Total in²	Current-carrying capacity,* A	Stranding No. and diam of strand, in Aluminum	Steel	Diameter, in Complete cond.	Steel core	Nominal weight, lbf Total	Per 1000 ft Al	Steel	Rated strength, lb Standard weight coating	Zinc-coated core Class B coating	Class C coating	Aluminum-coated core
Turkey	6	0.0206	0.0240	95	6 × 0.0661	1 × 0.0661	0.198	0.0661	36.1	24.5	11.6	1,190	1,160	1,120	1,120
Swan	4	0.0328	0.0382	130	6 × 0.0834	1 × 0.0834	0.250	0.0834	57.4	39.0	18.4	1,860	1,810	1,760	1,760
Swanate	4	0.0328	0.0411	130	7 × 0.0772	1 × 0.1029	0.257	0.1029	67.0	39.0	28.0	2,360	2,280	2,200	2,160
Sparrow	2	0.0522	0.0608	175	6 × 0.1052	1 × 0.1052	0.316	0.1052	91.3	62.0	29.3	2,850	2,760	2,680	2,640
Sparate	2	0.0522	0.0654	175	7 × 0.0974	1 × 0.1299	0.325	0.1299	106.7	62.0	44.7	3,640	3,510	3,390	3,260
Robin	1	0.0657	0.0767	200	6 × 0.1181	1 × 0.1181	0.355	0.1182	115.1	78.2	36.9	3,550	3,450	3,340	3,290
Raven	1/0	0.0830	0.0968	230	6 × 0.1327	1 × 0.1327	0.398	0.1327	145.3	98.7	46.6	4,380	4,250	4,120	3,980
Quail	2/0	0.1046	0.1221	265	6 × 0.1490	1 × 0.1490	0.447	0.1490	183.2	124.4	58.8	5,310	5,130	5,050	4,720
Pigeon	3/0	0.1317	0.1537	310	6 × 0.1672	1 × 0.1672	0.502	0.1672	230.8	156.7	74.1	6,620	6,410	6,300	5,880
Penguin	4/0	0.1662	0.1939	350	6 × 0.1878	1 × 0.1878	0.563	0.1878	291.1	197.7	93.4	8,350	8,080	7,950	7,420
Waxwing	266,800	0.2094	0.2210	430	18 × 0.1217	1 × 0.1217	0.609	0.1217	289.5	250.3	39.2	6,880	6,770	6,650	6,540
Owl	266,800	0.2096	0.2368	410	6 × 0.2109	7 × 0.0703	0.633	0.2109	342.4	250.5	91.9	9,680	9,420	9,160	9,160
Partridge	266,800	0.2095	0.2436	440	26 × 0.1013	7 × 0.0788	0.642	0.2364	367.3	251.7	115.6	11,300	11,000	10,600	10,640
Merlin	336,400	0.2642	0.2789	500	18 × 0.1367	1 × 0.1367	0.684	0.1367	365.2	315.7	49.5	8,680	8,540	8,400	8,260
Linnet	336,400	0.2640	0.3070	510	26 × 0.1137	7 × 0.0884	0.720	0.2652	462.5	317.0	145.5	14,100	13,700	13,300	13,300
Oriole	336,400	0.2642	0.3259	515	30 × 0.1059	7 × 0.1059	0.741	0.3177	527.1	318.1	209.0	17,300	16,700	16,200	15,900
Chickadee	397,500	0.3121	0.3295	555	18 × 0.1486	1 × 0.1486	0.743	0.1486	431.6	373.1	58.5	9,940	9,780	9,690	9,530
Brant	397,500	0.3122	0.3527	565	24 × 0.1287	7 × 0.0858	0.772	0.2574	512.1	375.0	137.1	14,600	14,300	13,900	13,900
Ibis	397,500	0.3119	0.3627	570	26 × 0.1236	7 × 0.0961	0.783	0.2883	546.6	374.7	171.9	16,300	15,800	15,300	15,100
Lark	397,500	0.3121	0.3849	575	30 × 0.1151	7 × 0.1151	0.806	0.3453	622.7	375.8	246.9	20,300	19,600	18,900	18,600
Pelican	477,000	0.3747	0.3955	625	18 × 0.1628	1 × 0.1628	0.814	0.1628	518.0	447.8	70.2	11,800	11,600	11,500	11,100
Flicker	477,000	0.3747	0.4233	635	24 × 0.1410	1 × 0.1628	0.846	0.2820	614.6	450.1	164.5	17,200	16,700	16,200	16,000

Table 14-8 ACSR Conductor—Physical Characteristics (Continued)

Code word	ACSR Cross section — Aluminum, cmils or AWG	Aluminum, in²	Total, in²	Current-carrying capacity,* A	Stranding No. and diam of strand, in — Aluminum	Steel	Diameter, in — Complete cond.	Steel core	Nominal weight, lb† Per 1000 ft — Total	Al	Steel	Rated strength, lb — Standard weight coating	Zinc-coated core Class B coating	Class C coating	Aluminum-coated core
Hawk	477,000	0.3744	0.4354	640	26 × 0.1354	7 × 0.1053	0.858	0.3159	656.0	449.6	206.4	19,500	18,900	18,400	18,100
Hen	477,000	0.3747	0.4621	645	30 × 0.1261	7 × 0.1261	0.883	0.3783	747.4	451.1	296.3	23,800	23,000	22,100	21,300
Osprey	556,500	0.4369	0.4612	690	18 × 0.1758	1 × 0.1758	0.879	0.1758	604.1	522.2	81.9	13,700	13,500	13,400	12,900
Parakeet	556,500	0.4372	0.4938	700	24 × 0.1523	7 × 0.1015	0.914	0.3045	716.9	525.1	191.8	19,800	19,300	18,700	18,500
Dove	556,500	0.4371	0.5083	710	26 × 0.1463	7 × 0.1138	0.927	0.3414	766.0	524.9	241.1	22,600	21,900	21,200	20,900
Eagle	556,500	0.4371	0.5391	710	30 × 0.1362	7 × 0.1362	0.953	0.4086	871.9	526.2	345.7	27,800	26,800	25,800	24,800
Peacock	605,000	0.4753	0.5370	740	24 × 0.1588	7 × 0.1059	0.953	0.318	779.7	570.9	208.8	21,600	21,000	20,400	20,100
Squab	605,000	0.4749	0.5522	745	26 × 0.1525	7 × 0.1186	0.966	0.356	832.3	570.4	261.9	24,300	23,600	22,800	22,500
Teal	605,000	0.4751	0.5834	750	30 × 0.1420	19 × 0.0852	0.994	0.426	939.5	572.0	367.5	30,000	29,000	28,000	28,000
Swift	636,000	0.4994	0.5133	745	36 × 0.1329	1 × 0.1329	0.930	0.1329	643.7	596.9	46.8	13,800	13,600	13,500	13,400
Kingbird	636,000	0.4997	0.5275	750	18 × 0.1880	1 × 0.1880	0.940	0.1880	690.8	597.2	93.6	15,700	15,400	15,300	14,800
Rook	636,000	0.4996	0.5643	765	24 × 0.1628	7 × 0.1085	0.977	0.326	819.2	600.0	219.2	22,600	22,000	21,400	21,100
Grosbeak	636,000	0.4995	0.5808	775	26 × 0.1564	7 × 0.1216	0.990	0.365	875.2	599.9	275.3	25,200	24,400	23,600	22,900
Egret	636,000	0.4995	0.6135	775	30 × 0.1456	19 × 0.0874	1.019	0.437	988.2	601.4	386.8	31,500	30,500	29,400	29,400
Scoter	653,900	0.5136	0.5321	760	18 × 0.1906	3 × 0.0885	0.953	0.1906	676.2	613.8	62.4	14,800	14,700	14,500	14,500
Flamingo	666,600	0.5238	0.5917	790	24 × 0.1667	7 × 0.1111	1.000	0.333	858.9	629.1	229.8	23,700	23,100	22,400	22,100
Gannet	666,600	0.5234	0.6086	795	26 × 0.1601	7 × 0.1245	1.014	0.373	917.3	628.7	288.6	26,400	25,600	24,800	24,000
Starling	715,500	0.5620	0.6535	835	26 × 0.1659	7 × 0.1290	1.051	0.387	984.8	675.0	309.8	28,400	27,500	26,600	25,700
Redwing	715,500	0.5617	0.6896	840	30 × 0.1544	19 × 0.0926	1.081	0.463	1110	676	434	34,600	33,400	32,700	31,600
Coot	795,000	0.6243	0.6416	860	36 × 0.1486	1 × 0.1486	1.040	0.1486	804.7	746.2	58.5	16,800	16,600	16,500	16,300
Tern	795,000	0.6242	0.6674	875	45 × 0.1329	7 × 0.0886	1.063	0.266	895.8	749.7	146.1	22,100	21,700	21,200	21,200
Cuckoo	795,000	0.6244	0.7053	885	24 × 0.1820	7 × 0.1213	1.092	0.364	1024	750	274	27,900	27,100	26,400	25,600
Condor	795,000	0.6240	0.7049	885	54 × 0.1213	7 × 0.1213	1.093	0.364	1024	750	274	28,200	27,400	26,600	25,800
Drake	795,000	0.6247	0.7264	890	26 × 0.1749	7 × 0.1360	1.108	0.408	1094	750	344	31,500	30,500	29,600	28,600
Mallard	795,000	0.6245	0.7669	900	30 × 0.1628	19 × 0.0977	1.140	0.489	1235	752	483	38,400	37,100	35,800	35,100
Ruddy	900,000	0.7066	0.7555	945	45 × 0.1414	7 × 0.0943	1.131	0.283	1015	849	166	24,400	24,000	23,500	23,300
Canary	900,000	0.7068	0.7984	955	54 × 0.1291	7 × 0.1291	1.162	0.387	1159	849	310	31,900	31,000	30,200	29,300

SOURCE: *Standard Handbook for Electrical Engineers*, Donald G. Fink, Editor-In-Chief, H. Wayne Beaty, Associate Editor. Copyright 1978 by McGraw-Hill, Inc. Used with permission of McGraw-Hill Book Company.

*Ampacity for conductor temperature rise of 40°C over 40°C ambient with a 2 ft/s crosswind and an emissivity factor of 0.5 without sun.

†Nominal conductor weights are based on ASTM standard stranding increments. Actual weights will vary within standard tolerances for wire diameters and lay lengths. Invoicing will be based on actual weights.

Table 14-9 Galvanized-Steel Strand—Dimensions, Weight, Breaking Strength
(ASTM Specifications A363, A475)

Strand diameter, in. Nominal	Actual	Stranding No. of wires	Diameter of coated wires, in.	Strand area, sq in.	Strand weight, lb per 1000 ft	Utilities grade* 1	2	3	4	Common	Siemens-Martin	High strength	Extra-high strength
1¼	1.253	37	0.179	0.9311	3248	....	....	....	......	44,600	73,000	113,600	162,200
1⅛	1.127	37	0.161	0.7533	2691	....	....	....	......	36,000	58,900	91,600	130,800
1	1.001	37	0.143	0.5942	2057	....	....	....	......	28,300	46,200	71,900	102,700
1	1.000	19	0.200	0.5969	2073	....	....	....	......	28,700	47,000	73,200	104,500
⅞	0.885	19	0.177	0.4675	1581	....	....	....	......	21,900	35,900	55,800	79,700
¾	0.750	19	0.150	0.3358	1155	....	....	....	......	16,000	26,200	40,800	58,300
⅝	0.625	19	0.125	0.2332	796	....	....	....	......	11,000	18,100	28,100	40,200
⅝	0.621	7	0.207	0.2356	813	....	....	....	......	11,600	19,100	29,600	42,400
9⁄16	0.565	19	0.113	0.1905	637	....	....	....	......	9,640	16,100	24,100	33,700
9⁄16	0.564	7	0.188	0.1943	671	....	....	....	......	9,600	15,700	24,500	35,000
½	0.500	19	0.100	0.1492	504	....	....	....	......	7,620	12,700	19,100	26,700
½	0.495	7	0.165	0.1497	517	....	....	....	25,000	7,400	12,100	**18,800**	26,900
7⁄16	0.435	7	0.145	0.1156	399	....	....	....	18,000	5,700	9,350	**14,500**	20,800
⅜	0.360	7	0.120	0.07917	273	....	....	....	11,500	4,250	6,950	10,800	15,400
⅜	0.356	3	0.165	0.06415	220.3	....	....	8500					
5⁄16	0.327	7	0.109	0.06532	225	6000							
5⁄16	0.312	7	0.104	0.05946	205	....	....	....	......	3,200	5,350	**8,000**	11,200
5⁄16	0.312	3	0.145	0.04954	170.6	....	....	6500					
9⁄32	0.279	7	0.093	0.04755	164	4600	....	....	......	2,570	4,250	6,400	8,950
¼	0.240	7	0.080	0.03519	121	....	....	....	......	**1,900**	**3,150**	4,750	6,650
¼	0.259	3	0.120	0.03393	116.7	....	3150	4500					
7⁄32	0.216	7	0.072	0.02850	98.3	....	....	....	......	1,540	2,560	3,850	5,400
3⁄16	0.195	7	0.065	0.02323	80.3	2400							
3⁄16	0.186	7	0.062	0.02113	72.9	....	....	....	......	**1,150**	1,900	2,850	3,990
5⁄32	0.156	7	0.052	0.01487	51.3	....	....	....	......	870	1,470	2,140	2,940
⅛	0.123	7	0.041	0.00924	31.8	....	....	....	......	540	910	1,330	1,830
Elongation in 24 in.:						% 10	8	5	4	10	8	5	4

SOURCE: *Standard Handbook for Electrical Engineers*, Donald G. Fink, Editor-In-Chief, H. Wayne Beaty, Associate Editor. Copyright 1978 by McGraw-Hill, Inc. Used with permission of McGraw-Hill Book Company.
* Used principally by communication and power and light industries.
NOTE: Sizes and grades in bold-faced type are those most commonly used and readily available.

Table 14-10 Current-Carrying Capacities of Copper and Aluminum Conductors When Used Outdoors

Copper Circular mils or AWG	Amperes	Aluminum Circular mils or AWG	Amperes
		3,364,000	2150
		3,156,000	1580
		1,780,000	1460
1,000,000	1300	1,590,000	1380
900,000	1220	1,431,000	1300
800,000	1130	1,272,000	1210
700,000	1040	1,113,000	1110
600,000	940	954,000	1000
500,000	840	795,000	897
400,000	730	636,000	776
300,000	610	477,000	646
No. 4/0	480	336,400	514
No. 3/0	420	266,800	443
No. 2/0	360	No. 4/0	381
No. 1/0	310	No. 3/0	328
No. 1	270	No. 2/0	283
No. 2	230	No. 1/0	244
No. 3	200	No. 1	209
No. 4	180	No. 2	180

In the heavy loading district, the resultant loading on the conductor is assumed to be that due to the weight of the conductor plus the added weight of a layer of ice with a radial thickness of ½ in, combined with a transverse horizontal wind pressure of 4 lb/ft² on the projected area of the ice-covered conductor at 0°F.

In the medium loading districts the loading is assumed to be that caused by ¼ in of radial thickness of ice combined with a transverse wind pressure of 4 lb/ft² at 15°F.

In the light loading district the loading assumed is that produced without any ice on the conductor but with a wind pressure of 9 lb/ft² at 30°F.

To these resultant values a constant in pounds per foot as shown in the table is added. As the map shows, the light loading district consists of the extreme South and a part of the Southwest; the medium loading district covers a band through the South and the entire West; and the heavy loading district includes the north central and northeastern states. Figure 14-11 shows conductors covered with sleet and ice after one of the storms in the midwest.

Fig. 14-11 Distribution circuit conductors covered with ice elliptical in shape with a vertical axis of 3 in and a horizontal axis of 5 in. *(Courtesy L. E. Myers Co.)*

The following is the statement on conductor loading as specified by the National Electrical Safety Code:

The loading on conductors shall be assumed to be the resultant loading per foot equivalent to the vertical load per foot of the conductor, ice-covered where specified, combined with the transverse loading per foot due to a transverse horizontal wind pressure upon the projected area of the conductor, ice-covered where specified, to which equivalent resultant shall be added a constant. Table 14-11 gives the values for ice, wind, temperature, and constants which shall be used to determine the conductor loading.

In addition to the ice load, the side push due to the wind on the conductor and ice must be taken into account, as mentioned above. After taking everything that causes tension on the conductor into account, namely, the weight of the conductor itself, the weight of the probable ice load, the transverse force due to probable high winds, and the shortening of the conductor and consequent increase of tension due to extreme cold, the engineer multiplies the resultant pull by 2 and selects the conductor that will stand that force before breaking. This gives a factor of safety of 2 under the assumed conditions.

Table 14-11   Conductor Loading Table

	Loading district		
	Heavy	Medium	Light
Radial thickness of ice, in	0.50	0.25	0
Horizontal wind pressure, lb per sq ft	4	4	9
Temperature, °F	0	+15	+30
Constant to be added to the resultant for the following types of conductors, lb per ft:			
Bare:			
Copper, steel, copper alloy, copper-covered steel, and combinations thereof	0.29	0.19	0.05
Bare aluminum (with or without steel reinforcement)	0.31	0.22	0.05

Fig. 14-12 Conductors of a 161 kV transmission line with ice measuring 5 in. in diameter. The devastation resulted from the ice and wind conditions. *(Courtesy L. E. Myers Co.)*

If the ice and wind exceed the design conditions sufficiently, the transmission line will fail structurally. Figure 14-12 shows ice conditions on transmission line conductors that greatly exceeds the conditions for which it was designed. A snow storm in Nebraska on March 29, 1976, destroyed 60 miles of 345 kV transmission lines when the diameter of snow accumulation on the conductors measured 6 in and tornado-like winds reached a velocity of 70 miles per hour (Fig. 14-13).

The electrical losses of a transmission line are equal to the current in amperes squared multiplied by the resistance of the conductors.

$$\text{Losses (watts)} = I^2 \text{ (amperes)} \times R \text{ (ohms)}$$

The current required for a given quantity of power decreases as the voltage increases.

$$P \text{ (watts)} = I \text{ (amperes)} \times V \text{ (volts)}$$

Fig. 14-13 Damaged 345 kV transmission line resulting from severe snow and wind conditions. *(Courtesy Nebraska Public Power District.)*

Fig. 14-14   Linemen installing 4-conductor bundle for center phase of 765 kV ac transmission line. Conduc-tor is 954 mcm ACSR. *(Courtesy American Electric Power Service Corp.)*

It is much more effective to reduce the current than to decrease the resistance of the conductors to permit the transmission of energy economically. If the losses are too great, the conductor will overheat and fail mechanically.

The need to transmit large quantities of electrical energy over circuits for long distances has led to the extensive use of extra high voltage (EHV) and ultra high voltage (UHV) transmission lines. The selection of conductors for EHV and UHV transmission lines is influenced by line losses, current carrying capacity, conductor material costs, and environmental considerations.

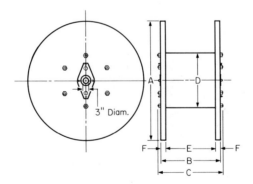

Approximate dimensions, inches							Approximate weight, pounds	
	Diameter							
A	B	C	D	E	F	Lagging thickness	With lags	Without lags
69	43½	46	30	38	2¾	1¾	900	—
69	32½	35	30	27	2¾	1¾	720	—
65	32½	35	30	27	2¾	1¾	650	—
58	32½	35	30	27	2¾	1¾	540	—
52	31½	33	23½	26	2¾	1¾	420	—
46	31½	33	23½	26	2¾	1¾	380	—
44	22½	24½	22½	18	2¼	1¾	205	140
40	22½	24½	22½	18	2¼	1¾	185	125
36	22½	24½	22½	18	2¼	—	—	100

Fig. 14-15   Reel for all-aluminum and ACSR cables.

Aluminum conductor with steel reinforcement is normally selected for transmission line conductors because of the good current-carrying capacity and minimum conductor and structure cost. Steel reinforcement of the aluminum conductor does not affect the electrical characteristics of the conductor appreciably. The ACSR conductor will permit higher current densities without loss of the strength since the sag of the conductors is limited by the steel strands as the aluminum expands. Limiting the sag of the aluminum conductor permits the use of longer spans, and shorter and fewer supporting structures.

Environmental considerations involve visual impact and voltage gradient. Steel reinforcement of the conductor results in shorter structures, or fewer structures, or both, thus improving the appearance of the facilities. The voltage gradient is an electrical characteristic determined by the physical size, arrangement, and surface condition of the conductors. The voltage gradient determines corona loss, radio and TV interference, and audible noise level. The voltage gradient should be minimized. This is accomplished by increasing phase-to-phase spacing and the size and/or number of conductors per phase. The increasing costs of installing generation equipment requires the use of larger conductors to reduce the cost of supplying the losses. This has resulted in the selection of conductors with high ampere capacity, lower voltage gradients, and higher supporting structure costs (Fig. 14-14).

The proper conductor selection is a compromise between economics associated with the cost of electric generation, conductor material, and supporting structures; mechanical strength requirements; current-carrying capacity required; and environmental impact.

Conductor Reels   All types and sizes of wire are shipped on reels normally. Limited amounts of small wire can be shipped in coils wrapped with paper. Figure 14-15 illustrates reel construction and size for all-aluminum and ACSR wire.

Section **15**

Distribution
Transformers

The purpose of a distribution transformer is to reduce the primary voltage of the electric distribution system to the utilization voltage serving the customer. A distribution transformer is a static device constructed with two or more windings used to transfer alternating-current electric power by electromagnetic induction from one circuit to another at the same frequency but with different values of voltage and current.

Figure 15-1 shows distribution transformers in stock at an electric utility company service building. The distribution transformers available for use for various applications, as shown, include pole-type, (Figures 15-2 and 15-3), padmounted (Fig. 15-4), vault or network type (Fig. 15-5), submersible (Fig. 15-6), and direct-buried (illustrated in Sec. 34).

Fig. 15-1   Electric utility distribution transformer storage yard. Forklift trucks are used to load transformers on line trucks. Storage area is covered with concrete to provide accessibility and protect transformers.

Fig. 15-2 Typical pole-type distribution transformer bolted directly to the pole. The unit is equipped with a lightning arrester, a low-voltage circuit breaker, and an overload warning light. *(Courtesy General Electric Co.)*

Fig. 15-3 Cutaway view of internal components and connections of a self-protected distribution transformer. *(Courtesy General Electric Co.)*

The distribution transformer in Fig. 15-2 is self-protected, equipped with a lightning arrester, secondary breaker, and warning light. The primary bushing is connected to the single-phase primary on the pole-top insulator utilized in armless-type construction. The transformer tank is grounded and connected to the secondary neutral wire.

The cutaway view in Fig. 15-3 shows the internal components of a self-protected pole-type transformer. The transformer contains a core and coils, a primary fuse mounted on the bottom of the primary bushing, a secondary terminal block, and a low-voltage breaker.

Padmounted transformers are used with underground systems. Three-phase padmounted trans-

Fig. 15-4 Three-phase padmounted transformer installed in shopping center parking lot. Transformer reduces voltage from 13,200 GRD Y/7620 to 208Y/120 volts. Primary and secondary terminals of transformer are connected to underground cables.

Fig. 15-5 Vault-type distribution transformer 1000 kVA, three-phase 13,200 GRD Y/7620 to 208Y/120 volts. High-voltage terminals H_1, H_2, and H_3 are designed for elbow-type cable connections, and low-voltage terminals X_1, X_2, and X_3 are spade-type for bolted connections to cable lugs. *(Courtesy Westinghouse Electric Corp.)*

Fig. 15-6 Submersible distribution transformer strapped to a pallet for shipment. Cables are for low-voltage connections. High-voltage connections are made with elbow cable connectors to bushing wells shown on top of transformer. *(Courtesy Westinghouse Electric Corp.)*

formers are used for commercial installations, as illustrated in Fig. 15-4, and single-phase pad-mounted transformers are used for underground residential installations, as described and shown in Section 34. Vault-type distribution transformers are installed for commercial customers where adequate space is not available for padmounted transformers. The vault-type transformer (Fig. 15-5) may be installed in vaults under sidewalks or in buildings. They are often used in network areas.

Submersible single-phase distribution transformers (Fig. 15-6) are used in some underground systems installed in residential areas, as described in Section 34. Corrosion problems and the high cost of installation have minimized the use of submersible transformers.

Distribution Transformer Operation   The basic principle of operation of a transformer is explained in Section 1. A schematic drawing of a single-phase distribution transformer appears in Fig. 15-7. The single-phase distribution transformer consists of a primary winding and a secondary winding wound on a laminated steel core. If the load is disconnected from the secondary winding of the transformer and a high voltage is applied to the primary winding of the transformer, a magnetizing current will flow in the primary winding. If we assume the resistance of the primary winding is small, which is usually true, this current is limited by the counter voltage of self-induction induced in the highly inductive primary winding. The windings of the transformer are constructed with sufficient turns in each winding to limit the no-load or exciting current and produce a counter voltage approximately equal to the applied voltage. The exciting current magnetizes, or produces, a magnetic flux in the steel transformer core. The magnetic flux

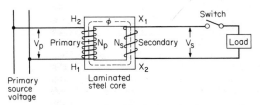

Fig. 15-7   Schematic drawing of single-phase distribution transformer.

reverses each half cycle as a result of the alternating voltage applied to the primary winding. The magnetic flux cuts the turns of the primary and secondary windings. This action induces a counter voltage in the primary winding and produces a voltage in the secondary winding. The voltage induced in each turn of the primary and secondary winding coils will be approximately equal, and the voltage induced in each winding will be equal to the voltage per turn multiplied by the number of turns.

$$\text{Voltage } V = \text{Volts per turn} \times \text{number of turns } N$$
$$\text{Voltage primary winding} = V_p$$
$$\text{Voltage secondary winding} = V_s$$
$$\text{Volts per turn} = V_t$$
$$\text{Number of turns of primary winding} = N_p$$
$$\text{Number of turns of secondary winding} = N_s$$

Therefore, the voltage of the primary winding is equal to the applied voltage and equal to the number of turns in the primary winding multiplied by the volts per turn

$$V_p = V_t \times N_p$$

Likewise the voltage of the secondary winding will equal the volts per turn multiplied by the number of turns in the secondary winding.

$$V_s = V_t \times N_s$$

If we use basic algebraic operations and divide both sides of an equation by the same number, we can obtain the following equation:

$$\frac{V_p}{N_p} = \frac{V_t \times N_p}{N_p}$$

$$\frac{V_p}{N_p} = \frac{V_t \times N_{\not{p}}}{N_{\not{p}}}$$

$$\frac{V_p}{N_p} = V_t$$

and

$$\frac{V_s}{N_s} = \frac{V_t \times N_s}{N_s}$$

$$\frac{V_s}{N_s} = \frac{V_t \times N_{\not{s}}}{N_{\not{s}}}$$

$$\frac{V_s}{N_s} = V_t$$

Since quantities equal to the same quantity are equal,

$$\frac{V_p}{N_p} = \frac{V_s}{N_s}$$

Since both quantities are equal to V_t. If we multiply both sides of the equation by the same quantity N_p, we obtain

$$\frac{V_p}{N_p} \times N_p = N_p \times \frac{V_s}{N_s}$$

or

$$V_p = \frac{N_p}{N_s} \times V_s$$

likewise

$$V_s = \frac{N_s}{N_p} \times V_p$$

Therefore, if we know the source voltage, or primary voltage, and the number of turns in the primary and secondary windings, we can calculate the secondary voltage. For example, if a distribution transformer has a primary winding with 7620 turns and a secondary winding of 120 turns and the source voltage equals 7620 volts, the secondary voltage equals 120 volts.

$$V_s = V_p \times \frac{N_s}{N_p}$$

$$V_s = 7620 \times \frac{120}{7620} \text{ volts}$$

$$V_s = \cancel{7620} \times \frac{120}{\cancel{7620}} \text{ volts}$$

$$V_s = 120 \text{ volts}$$

When a load is connected to the secondary winding of the transformer, a current I_s will flow in the secondary winding of the transformer. This current is equal to the secondary voltage divided by the impedance of the load Z.

$$I_s = \frac{E_s}{Z}$$

Lenz's law determined that any current that flows as a result of an induced voltage will flow in a direction to oppose the action that causes the voltage to be induced. Therefore, the secondary current, or load current, in the distribution transformer secondary winding will flow in a direction to reduce the magnetizing action of the exciting current or the primary current. This action reduces the magnetic flux in the laminated steel core momentarily and, therefore, reduces the counter voltage in the primary winding. Reducing the counter voltage in the primary winding causes a larger primary current to flow, restoring the magnetic flux in the core to its original value. The losses in a distribution transformer are small, and therefore, for all practical purposes, the volt-amperes of the load, or secondary winding, equal the volt-amperes of the source, or primary winding.

$$V_p \times I_p = V_s \times I_s$$

If we divide both sides of this equation by the same quanity V_p, we obtain

$$\frac{V_p \times I_p}{V_p} = \frac{V_s \times I_s}{V_p}$$

$$\frac{\cancel{V_p} \times I_p}{\cancel{V_p}} = \frac{V_s \times I_s}{V_p}$$

or

$$I_p = \frac{V_s}{V_p} \times I_s$$

Therefore, if the primary source voltage of a transformer is 7620 volts, the secondary voltage is 120 volts, and the secondary current, as a result of the load impedance, is 10 amps, the primary current is 0.16 amps.

$$I_p = \frac{120}{7620} \times 10$$

$$I_p = 0.16 \text{ amps}$$

A three-phase transformer is basically three single-phase transformers in a single tank in most cases. The windings of the three transformers are normally wound on a single multilegged laminated steel core.

Distribution Transformer Construction   Two types of construction are outstanding: the *core type* and the *shell type*. In the core type the core is in the form of a rectangular frame with the coils placed on the two vertical sides (Fig. 15-8). The coils are cylindrical and relatively long. They are divided, part of each primary and secondary being on each of the two vertical legs. In the shell type (Fig. 15-9), the core surrounds the coils, instead of the coils surrounding the core.

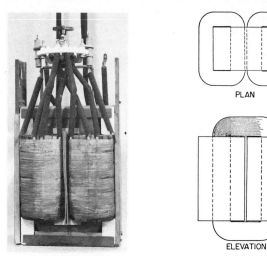

Fig. 15-8   Views illustrating core-type transformer construction. *(Courtesy Westinghouse Electric Corp.)*

The coils in the shell type are generally flat instead of cylindrical, and the primary and secondary coils are alternated (Fig. 15-10). The core and coil assembly of the distribution transformer is completely immersed in insulating oil or other insulating liquids to keep the operating temperatures low and to provide additional insulation for the windings.

Distribution Transformer Ratings   The capacity of a distribution transformer is determined by the amount of current it can carry continuously at rated voltage without exceeding the design temperature. The transformers are rated in kilovolt-amperes (kVA) since the capacity is limited by the load current which is proportional to the kVA regardless of power factor. The standard kVA ratings are itemized in Table 15-1.

Distribution Transformer Connections   Single-phase distribution transformers are manufactured with one or two primary bushings. The single-primary-bushing transformers can only be used on grounded wye systems. The two-primary-bushing transformers can be used on three-wire-delta systems or four-wire-wye systems if they are properly connected. Figure 15-11 illustrates schematically the connections of a single-phase transformer to a three-phase 2400-volt

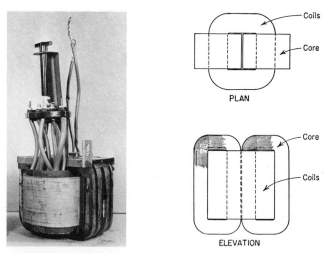

Fig. 15-9   Views illustrating shell-type transformer construction. *(Courtesy Westinghouse Electric Corp.)*

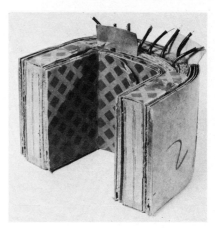

Fig. 15-10 Windings of a shell-type distribution transformer cut to show cross section of winding conductors. High-voltage winding is located between the secondary winding on the inner and outer surfaces. The high-voltage winding has small conductors and a large number of turns. The secondary winding has large rectangular-shaped conductors and a small number of turns. *(Courtesy Westinghouse Electric Corp.)*

three-wire ungrounded delta primary-voltage system to obtain 120-volt single-phase two-wire secondary service. The connections for similar systems operating at other primary distribution voltages such as 4800, 7200, 13,200, and 34,400 would be identical.

Figure 15-12 illustrates the proper connections for a single-phase transformer to a three-phase three-wire ungrounded delta primary-voltage system to obtain 120/240-volt single-phase three-wire service. Normally the wire connected to the center bushing will be connected to ground. Grounding the wire connecting to the center bushing limits the voltage above ground to 120 volts even though the wires connecting to the outside secondary bushings have 240 volts between them.

Figure 15-13 illustrates schematically the single-phase distribution transformer connections to a three-phase four-wire-wye grounded neutral primary system rated 4160Y/2400 volts to obtain 120/240-volt single-phase secondary service. The three-phase four-wire-wye grounded neutral system has voltages between phases equal to the phase or line to neutral voltage multiplied by 1.73. In Fig. 15-13 the primary system line to neutral voltage is 2400 volts and the voltage between phases is 1.73 × 2400 or 4160 volts. This system is designated as a 4160Y/2400-volt system. Other standard three-phase four-wire-wye grounded neutral primary system voltages are 8320Y/4800, 12,470Y/7200, 13,200Y/7620, 13,800Y/7970, 20,780Y/12,000, 22,860Y/13,200, 24,940Y/14,400, and 34,500Y/19,920.

Table 15-1   Standard Ratings of Distribution Transformers
(in kilovolt-amperes, kVA)

Overhead-type		Pad-mounted type	
Single-phase	Three-phase	Single-phase	Three-phase
5	15	25	75
10	30	37½	112½
15	45	50	150
25	75	75	225
37½	112½	100	300
50	150	167	500
75	225		750
100	300		1000
167	500		1500
250			2000
333			2500
500			

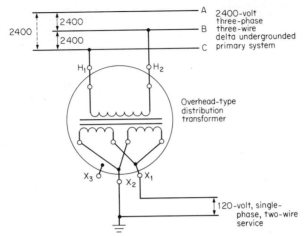

Fig. 15-11 Single-phase overhead or pole-type distribution transformer connection for 120-volt two-wire secondary service. Transformer secondary coils are connected in parallel.

The transformer connections to obtain 120-, 120/240-, or 240-volt secondary service are normally completed inside the tank of the transformer (Fig. 15-14). The transformer name plate provides information necessary to complete the connections. The voltages should be measured when the transformer is energized to be sure the connections are correct before the load is connected to the transformer.

Single-phase pole or overhead-type transformers can be used to obtain three-phase secondary service (Fig. 15-15). There are four normal connections: the delta-delta (Δ-Δ), the wye-wye (Y-Y), the delta-wye (Δ-Y), and the wye-delta (Y-Δ). Figure 15-16 illustrates the proper connections for three single-phase distribution transformers connected to a three-phase three-wire, ungrounded delta primary-voltage system to obtain three-phase three-wire delta secondary service. The illustration is for a 2400-volt primary system and 240-volt secondary service. Other similar systems operating at other primary voltages would be connected the same.

Single-phase transformers with secondary windings constructed for voltages or 240/480 could

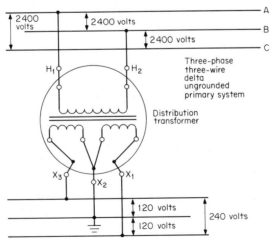

Fig. 15-12 Single-phase distribution transformer connected to give 120/240-volt three-wire single-phase service. Transformer secondary coils are connected in series.

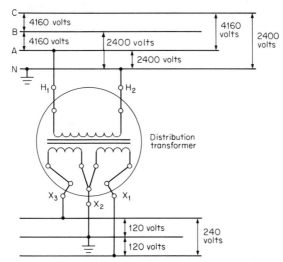

Fig. 15-13   Single-phase distribution transformer connected to provide 120/240-volt three-wire single-phase service. Primary winding is connected line to neutral or ground.

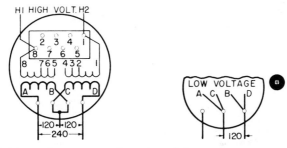

Fig. 15-14   The left view shows connections for 120/240-volt three-wire service. For 240-volt two-wire service, the middle lead is not brought out. For 120-volt two-wire service, the connections are made as shown in right view. All connections are made inside tank. *(Courtesy Westinghouse Electric Corp.)*

Fig. 15-15   Cluster-mounted bank of transformers bolted directly to pole and protected by lightning arresters and fuses. *(Courtesy S. & C. Electric Co.)*

Three-phase three-wire delta primary system

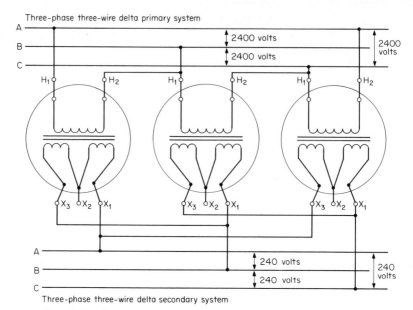

Fig. 15-16   Three single-phase distribution transformers connected delta-delta (Δ-Δ).

Three-phase four-wire grounded-neutral primary system

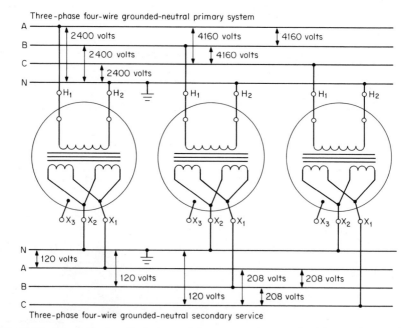

Fig. 15-17   Three single-phase distribution transformers connected wye-wye (Y-Y). Proper voltage windings would be used with other primary voltages and secondary windings constructed for 277 volts would provide 480Y/277-volt secondary service.

be used to obtain 480-volt three-phase secondary service. Figure 15-17 illustrates the proper connections for three single-phase distribution transformers to a three-phase four-wire grounded neutral wye primary-voltage system to obtain 208Y/120- or 480Y/277-volt secondary service. Figure 15-18 illustrates the proper connections for three single-phase distribution transformers to a three-phase three-wire ungrounded neutral delta primary-voltage system to obtain 208Y/120 or 480Y/277-volt secondary service.

Transformers must be constructed with the proper windings for the primary-voltage system and the desired secondary voltage. Properly manufactured distribution transformers can be connected wye-delta if it is desired to obtain three-wire three-phase secondary voltages from a three-phase four-wire grounded neutral wye primary-voltage system. The standard secondary system voltages are 208Y/120, 240, 480Y/277, and 480 volts.

If one of the transformers from a delta-connected bank is removed, the remaining two are said to be "open-delta connected." This can be done, and the remaining two transformers will still transform the voltages in all three phases and supply power to all three phases of the secondary

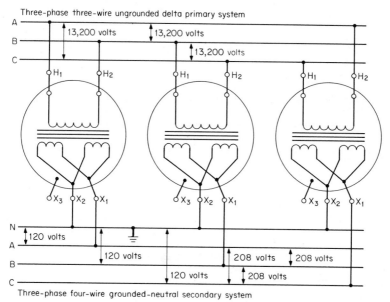

Three-phase three-wire ungrounded delta primary system

Three-phase four-wire grounded-neutral secondary system

Fig. 15-18 Three single-phase distribution transformers connected delta-wye (Δ-Y). Windings must be rated for the primary system and the desired secondary service voltage.

mains. The connections for a 2400-volt circuit are shown in Fig. 15-19. The capacity of the two is now, however, only 58 percent instead of 66⅔ percent of what it would appear to be with two transformers remaining.

The open-delta connection is often used where an increase in load is anticipated. The third unit is added when the load grows to the point at which it exceeds the capacity of the two transformers. Furthermore, if one transformer of a three-phase bank should become defective, the defective transformer can be removed, and the remaining two transformers continue to render partial service.

Distribution Transformer Polarity In connecting transformers in parallel or in a three-phase bank, it is important to know the polarity of the transformer terminals or leads. In building transformers in the factory, the ends of the windings can be so connected to the leads extending out through the case that the direction of the flow of current in the secondary terminal with respect to the corresponding primary terminal is in the same direction or in the opposite direction. When the current flows in the same direction in the two adjacent primary and secondary terminals, the polarity of the transformer is said to be "subtractive," and when the current flows in opposite directions, the polarity is said to be "additive."

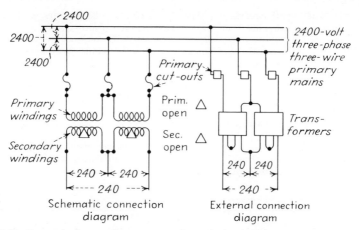

Fig. 15-19 Two single-phase transformers connected into a bank to transform power from 2400 volts three-phase to 240 volts three-phase. Primaries and secondaries are open-delta connected.

Polarity may be further explained as follows: Imagine a single-phase transformer having two high-voltage and two low-voltage external terminals. Connect one high-voltage terminal to the adjacent low-voltage terminal, and apply voltage across the two high-voltage terminals. Then if the voltage across the unconnected high-voltage and low-voltage terminals is less than the voltage applied across the high-voltage terminals, the polarity is *subtractive;* while if it is greater than the voltage applied across the high-voltage terminals, the polarity is *additive* (see Fig. 15-20).

From the foregoing, it is apparent that, when the voltage indicated on the voltmeter is greater than the impressed voltage, it must be the sum of the primary and the secondary voltages, and the sense of the two windings must be in opposite directions, as in Fig. 15-21. Likewise, when the voltage read on the voltmeter is less than the impressed voltage, the voltage must be the difference, as is illustrated in Fig. 15-22. When the terminal markings are arranged in the same numerical order, H_1H_2 and X_1X_2 or H_2H_1 and X_2X_1, on each side of the transformer, the polarity of each winding is the same (subtractive). If either is in reverse order, H_2H_1 and X_1X_2 or H_1H_2 and X_2X_1, their polarities are opposite (additive). The name plate of a transformer should always indicate the polarity of the transformer to which it is attached.

Additive polarity is standard for all single-phase distribution transformers in sizes of 200 kVA and below, having high-voltage ratings of 9000 volts and below.

Subtractive polarity is standard for all single-phase distribution transformers in sizes above 200 kVA irrespective of voltage rating.

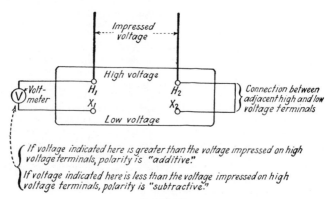

Fig. 15-20

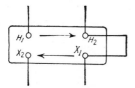

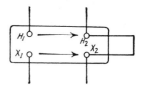

Fig. 15-21 Sketch showing polarity markings and directions of voltages when polarity is additive.

Fig. 15-22 Sketch showing polarity markings and directions of voltages when polarity is subtractive.

Subtractive polarity is standard for all single-phase transformers in sizes of 200 kVA and below having high-voltage ratings above 9000 volts.

Standard low-voltage terminal designations are illustrated in Fig. 15-23.

Distribution Transformer Tapped Windings Windings of distribution transformers can be manufactured with various taps to broaden the range of primary voltages suitable for a given installation. The standard taps available are on the primary winding of the transformer and are designed to provide variations in the number of winding turns in steps of 2½ percent of the rated voltage. The distribution transformers may be designed with four 2½ percent above rated voltage, two 2½ percent above and two 2½ percent below rated voltage, or four 2½ percent taps below rated voltage. A standard distribution transformer with a primary winding rated 2400 volts and supplied with two 2½ percent taps above and two 2½ percent taps below rated voltage would be constructed with winding taps rated 2520, 2460, 2340, and 2280 volts. Tap changers on distribution transformers are designed to be operated while the transformer is deenergized. The operating handle (Fig. 15-24) can be located outside of the tank or inside the tank above the oil level. The transformer nameplate provides information concerning the taps available on the distribution transformer.

Distribution Transformer Standards The standards are published by American National Standards Institute (ANSI). ANSI Standard C57.12.20 covers "Requirements for Overhead-Type Distribution Transformers, 67,000 Volts and Below; 500 kVA and Smaller." ANSI Standard C57.12.25 covers "Requirements for Pad Mounted Compartmental-Type Single Phase Distribution Transformers with Separable Insulated High-Voltage Connectors, High-Voltage, 24,940 GRD Y/14,400 Volts and Below; Low-Voltage, 240/120; 167 kVA and Smaller." ANSI Standard C57.12.26 covers "Requirements for Pad Mounted Compartmental-Type Self-Cooled, Three-Phase Distribution Transformers for Use with Separable Insulated High-Voltage Connectors, High-Voltage, 24,940 GRD Y/14,400 Volts and Below; 2,500 kVA and Smaller."

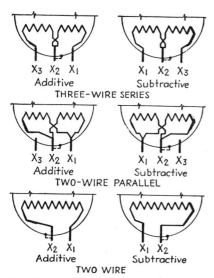

Fig. 15-23 Standard low-voltage connections for distribution transformer.

Fig. 15-24 No-load tap changer for distribution transformer. *(Courtesy Westinghouse Electric Corp.)*

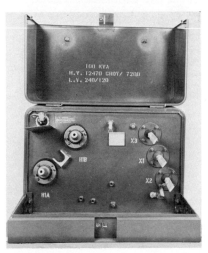

Fig. 15-25 Single-phase, pad-mounted distribution transformer. *(Courtesy A. B. Chance Co.)*

Fig. 15-26 Single-phase, pad-mounted distribution transformer with cover open. *(Courtesy Westinghouse Electric Corp.)*

A typical specification for a self-protected single-phase overhead distribution transformer of the kVA size needed (Figs. 15-2 and 15-3) would read as follows:

Distribution transformers, 13,200 Grd.Y/7620-120/240-volts, 95-kV BIL, self-protected with primary weak link fuse and secondary breaker, 1 phase, OISC, no taps, 60 Hz, 1 lightning arrester, 1 primary bushing, ground terminal pad, no ground gap, with standard accessories, overhead type with insulated cover for squirrel protection, 65 degree Celsius rise rating, units to be sky gray in color, with Qualitrol No. 202-2 Automatic Pressure Relief Device or approved equivalent.

The primary bushing shall be supplied with birdguard and eyebolt connector equipped with an insulated handknob tightening device to accept $\frac{5}{16}''$ diameter knurled stud terminal with birdguard in place.

The arrester shall be a 9/10 kV, direct connected, lead type, distribution class lightning arrester equipped with ground lead disconnector, Joslyn Type "QL" or approved equivalent. The line lead shall not be connected to the primary bushing. The lead shall be insulated and of standard length for connection to the line side of a current limiting fuse.

The transformers' electrical characteristics and mechanical features shall be in accordance with the latest revision of ANSI Standard C57.12.20.

A typical specification for a single-phase pad-mounted distribution transformer of the kVA size needed (Figs. 15-25 and 15-26) would read as follows:

Low profile, dead front, pad-mounted transformer, primary to be 13,200 Grd.Y/7620-volts, secondary 120/240-volts, 95-kV BIL, 1 phase, 60 Hz, 65 degree C rise, OISC, and no taps on primary. Units to be equipped with two primary bushing wells in accordance with the latest revision of ANSI Standard C119.2. The bushing wells to accommodate an 8.3/15-kV Grd.Y load break bushings and elbows for sectionalizing cable in a loop feed primary system. The bushing wells to be equipped with 8.3/15-kv Grd.Y load break bushing inserts General Electric 9U02AAA001, Elastimold 1601A3R, RTE 2604797B01, or approved equivalent. An RTE Bay-O-Net, or approved equivalent, overload sensing primary fuse for coordinated overload and secondary fault protection replaceable externally, coordinated with and in series with an 8.3-kV partial range current limiting fuse. The current limiting fuse shall be coordinated with the overload sensing fuse to prevent the current limiting fuse from blowing on secondary short circuits or overloads. The current limiting fuse shall be installed inside the transformer tank. Secondary terminals to be equipped with standard threaded stud only. Unit to be equipped with secondary ground terminal, primary ground terminal, tank ground terminal, parking stand bracket for primary elbow, provisions for padlocking and all standard accessories including a Qualitrol No. 202-2 Automatic Pressure Relief Device or approved equivalent. The transformer electrical characteristics and mechanical features shall be in accordance with the latest revision of ANSI Standard C57.12.25. Transformers shall be furnished with Type 2 interchangeability dimensions as per above ANSI Standard.

A typical specification for a three-phase pad-mounted distribution transformer available in kVA ratings of 75, 112½, 150, 225, 300, 500, 750, and 1,000 (Fig. 15-4) would read as follows:

Three phase, pad-mounted transformer, primary 13,200 Grd.Y/7620-volts, with standard taps, secondary 208 Grd.Y/120-volts, 95-kV BIL, 65 degree C rise, 60 Hz, OISC, primary and secondary windings to be wound on a five legged core and connected Grd.Y-Grd.Y. Primary compartment to have six 200-ampere bushing wells to accommodate 8.3/15-kV Grd.Y load break bushings and elbows for sectionalizing cable in a loop feed primary system. The bushing wells shall be supplied in accordance with the latest revision of ANSI Standard C119.2. The bushing wells to be equipped with 8.3/15-kV Grd.Y load break bushing inserts General Electric 9U02AAA001, Elastimold 1601A3R, RTE 2604797B01, or approved equivalent. The primary compartment shall be of dead front design and shall be equipped with parking stands for the primary elbows. An RTE Bay-O-Net, or approved equivalent, overload sensing primary fuse per phase for coordinated overload and secondary fault protection, replaceable externally, coordinated with and in series with a 8.3-kV partial range current limiting fuse. The current limiting fuse shall be coordinated with the overload sensing fuse to prevent the current limiting fuse from blowing on secondary short circuits or overloads. Each current limiting fuse and the overload sensing fuse shall be capable of withstanding 15-kV system recovery voltage across an open fuse. The current limiting fuse shall be installed inside the transformer tank. Secondary terminals are to be equipped with standard spade terminals. Tap changer to be in the high voltage compartment. Transformer to be equipped with a secondary ground terminal, primary ground terminal, tank ground terminal, with provisions for padlocking and all standard accessories, including an oil sampling device, oil drain valve and Qualitrol No. 202-2 Automatic Pressure Relief Device or approved equivalent. The transformer electrical characteristics and mechanical features shall be in accordance with the latest revision of ANSI Standard C57.12.26. The primary bushing wells and parking stands shall be located as shown in "Fig. 6" specific dimensions for type "B" transformers in the above ANSI Standard. The high and low voltage segments shall be compartmentalized by a metal barrier. An oil drip shield shall be installed directly below RTE Bay-O-Net fuse to prevent oil from dripping on primary elbows or bushings during removal of fuse.

Lightning Protection

Lightning Everyone has seen lightning flashes (Fig. 16-1). These flashes are currents of electricity flowing from one cloud to another or between cloud and earth. They always flow by way of the path of least resistance, just as water takes the path of least resistance when it flows down the hillside. Such paths down the hillside are not generally straight but, on the contrary, are very crooked. The same is true of the path taken by a bolt of lightning; this fact explains the zigzag path of the flash. Trees, high buildings, towers, transmission lines, and poles are paths of less resistance than air, and therefore lightning often strikes them and flows through them to ground.

Overhead lines also have charges of electricity induced on them when charged clouds pass over them. Such charges are released when the cloud overhead is discharged as by a stroke to another cloud or to ground. When the charge is released, the voltage of the line is suddenly raised for a brief period of time. This abrupt rise of voltage is likely to be sufficient to cause insulators to flash over or to break down the insulation on the windings of transformers or other apparatus connected to the line.

Lightning-producing storms are very common. It is estimated that 44,000 thunderstorms rage daily over the earth, producing 100 flashes of lightning every second. Figure 16-2 shows graphically the approximate number of lightning-producing storm days in various parts of the United States during the year. In some areas the number is as low as 5 and in other parts as high as 90. Thunderstorms bring rain, strong winds, tornados, and lightning. All of these can cause major damage to electric facilities.

The induced and direct lightning strokes cause abnormal voltages on the electric circuits. Overhead ground wires called static wires, lightning arresters, and protector tubes are used to prevent or eliminate the voltage surges from the circuits.

Ground Wire A ground wire, or static wire, on an overhead line may be a galvanized steel, copper-covered steel, or aluminum-covered steel cable strung along the line the same as other line conductors, but it is mounted above the other conductors (see Figs. 16-3 and 16-4).

The ground wire is not a part of any electrical circuit, however, but instead is connected to the earth at frequent intervals, at least every fifth pole, on pole lines. On steel tower lines it is grounded to each tower, by means of the metal clamp with which it is fastened. The ground or earth potential is thereby virtually brought above the transmission line and, hence, the stress on the line and transformer insulation due to lightning is greatly reduced.

A properly installed ground wire is highly effective, but its effectiveness depends on low ground resistance. Good ground connections should therefore be made at frequent intervals along the line.

Lightning Arresters American National Standards, "Surge Arresters for Alternating-Current Power Circuits" (ANSI C62.1) and "Guide for Application of Valve-Type Lightning Arresters for Alternating-Current Systems" (ANSI C62.2), provide specifications for lightning arresters and proper methods to apply them. A lightning arrester is a device that prevents high voltages, which

Fig. 16-1 Lightning flashes, illustrating zigzag path. Overhead distribution and transmission lines are in constant danger of being damaged by lightning. *(Courtesy Line Material Industries.)*

would damage equipment, from building up on the circuit by providing a low-impedance path to ground for the lightning or transient currents, and then restores normal circuit conditions.

Lightning arresters serve the same purpose on a line as a safety valve on a steam boiler. A safety valve on a boiler relieves a high pressure by blowing off steam until the high pressure is reduced to normal. When the pressure is down to normal, the safety valve closes again and is ready for the next abnormal condition. Such is also the operation of a lightning arrester. When a high voltage, greater than the normal line voltage, exists on the line, the arrester immediately furnishes a path to ground, and thus drains off the excess voltage. Furthermore, when the excess voltage is relieved, the action of the arrester must prevent any further flow of power current. The function of a lightning arrester is, therefore, twofold, first to provide a point in the circuit at which the

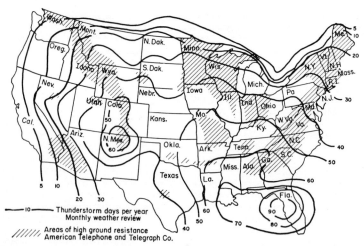

Fig. 16-2 Number of electrical storm days per year. Lines are marked with the average number of lightning-producing storm days occurring in each area per year. *(Courtesy American Telephone and Telegraph Co.)*

lightning impulse can pass to earth without injury to line insulators, transformers, or other connected equipment and second to prevent any follow-up power current from flowing to ground.

Elementary Lightning Arrester   The elementary form of an arrester is a simple horn gap connected in series with a resistance, as shown in Fig. 16-5. A horn gap is the space that is formed between the two sides of a V when the lower point of the V is removed. One reason for providing the gap is to prevent current leakage when the voltage on the line is normal. If there were no lightning arrester protecting the motor shown in Fig. 16-5, the path of least resistance would likely be through the windings of the machine, to its frame, and then to ground. This, of course, would damage the motor by puncturing its insulation. If a lightning arrester is connected from the line

Fig. 16-3 Subtransmission line with 13,200Y/ 7620-volt distribution feeder, underbuild protected with ground wire supported on pole top. Ground wire is grounded at frequent intervals.

Fig. 16-4 A 69,000-volt transmission line protected with overhead ground wire. Note use of Stockbridge dampers on conductors to reduce vibration. *(Courtesy Iowa-Illinois Gas and Electric Co.)*

to the earth, however, it is likely that the lightning would find it easier to flow to ground by way of the arrester and thus leave the motor undamaged. While the high-voltage charge is being drained off, the current flow to ground is limited by the resistance in series with the horn gap. When the discharge has passed, the heat of the arc between the V gap creates an upward draft of air, which tends to carry the arc upward with it until it is stretched out to such a length that it breaks. In addition the magnetic field on the inside of the V helps to force the arc upward. Thus it is clear that all this may be done without interrupting the load circuit.

It should be pointed out, however, that this elementary form of arrester is rarely used in practice. The general principle of operation, however, is correct.

Valve-Type Lightning Arrester   Four classes of valve type lightning arresters are manufactured for alternating-current systems: *secondary* (Fig. 16-6), *distribution* (Fig. 16-7), *intermediate* (Fig. 16-8), and *station* (Fig. 16-9). The four classes of lightning arresters have similar components, but are manufactured with different characteristics. Secondary arresters are manufactured with voltage ratings of 175 and 650 volts. Distribution arresters are manufactured with voltage ratings of 1, 3, 6, 9, 10, 12, 15, 18, 21, 25, 27, and 30 kV. Intermediate arresters are manufactured with voltage ratings of 3 through 120 kV. Station arresters are manufactured with voltage ratings of 3 kV and above. The four different classifications of arresters have different sparkover voltages, current discharge capabilities, and maximum surge discharge voltages. The cost of the lightning arresters increases as the capacities are increased.

Secondary arresters are used on service and other low-voltage alternating-current circuits. Distribution arresters are used on primary distribution systems to protect insulators, distribution transformers, and other equipment. Intermediate-type lightning arresters are often used on substation

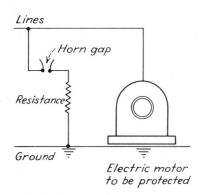

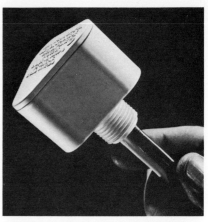

Fig. 16-5 Elementary lightning arrester consisting of horn gap and resistance.

Fig. 16-6 Secondary valve-type lightning arrester. Arrester can be connected to threewire, 120/240-volt service at service entrance or mounted on lighting distribution panel board and connected to main cable terminators. *(Courtesy Joslyn Manufacturing and Supply Co.)*

exit cables and other locations on the distribution system needing a high level of lightning and surge protection. Station-type lightning arresters are used in substations and generating stations to provide a high level of surge protection for the major pieces of equipment. Surge voltages can be generated by operating switches in the electric transmission system as well as by lightning.

The valve-type arrester consists of gaps in a sealed porcelain casing, grading resistors, or capacitors, to control the sparkover voltages of the gaps, valve blocks consisting of a nonlinear resistor that pass large surge currents and limit the power follow current, and a porcelain housing (Fig. 16-10).

When a surge voltage of a high magnitude develops on a circuit with a valve-type lightning arrester, the series gaps in the arrester will spark over permitting the nonlinear resistance valve

Fig. 16-7 Distribution valve-type lightning arrester rated 9–10 kV. *(Courtesy Westinghouse Electric Corp.)*

Fig. 16-8 Intermediate valve-type lightning arresters rated 39 kV installed on a nominal 34.5-kV three-phase circuit. *(Courtesy Ohio Brass Co.)*

block to pass the surge current to ground. The surge voltage will be limited by the voltage drop across the arrester to ground ($V = I \times R$). The resistance of the valve blocks while conducting the large surge current is very low keeping the voltage across the lightning arrester low. This keeps the voltage across equipment in parallel with the lightning arrester below the value that would damage the insulation of the equipment. After the surge current has been bypassed to ground, power follow current will flow through the lightning arrester to ground. The nonlinear resistor valve blocks develop a high resistance to the 60-Hz (hertz) current preventing a short

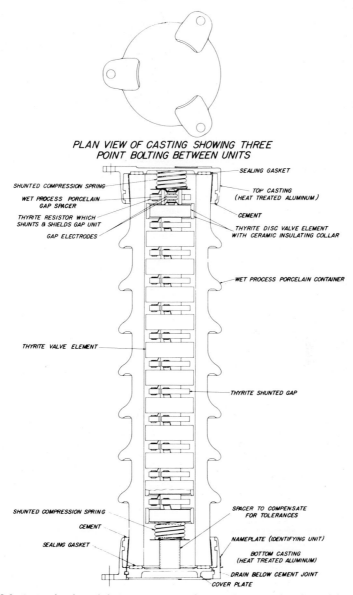

PLAN VIEW OF CASTING SHOWING THREE
POINT BOLTING BETWEEN UNITS

SEALING GASKET
SHUNTED COMPRESSION SPRING
WET PROCESS PORCELAIN
GAP SPACER
TOP CASTING
(HEAT TREATED ALUMINUM)
THYRITE RESISTOR WHICH
SHUNTS & SHIELDS GAP UNIT
CEMENT
GAP ELECTRODES
THYRITE DISC VALVE ELEMENT
WITH CERAMIC INSULATING COLLAR
WET PROCESS PORCELAIN CONTAINER
THYRITE VALVE ELEMENT
THYRITE SHUNTED GAP
SHUNTED COMPRESSION SPRING
SPACER TO COMPENSATE
FOR TOLERANCES
CEMENT
NAMEPLATE (IDENTIFYING UNIT)
SEALING GASKET
BOTTOM CASTING
(HEAT TREATED ALUMINUM)
DRAIN BELOW CEMENT JOINT
COVER PLATE

Fig. 16-9 Station class thyrite lightning-arrester unit with section cut away to show thyrite disks and series-gap unit. *(Courtesy General Electric Co.)*

Fig. 16-10 Internal view of distribution class valve-type lightning arrester showing gap structures and nonlinear resistor valve blocks. *(Courtesy Westinghouse Electric Corp.)*

circuit to ground on the line. When the instantaneous value of the 60-Hz power follow current goes through zero, the arrester gaps interrupt the current, returning the circuit conditions to normal.

Application of lightning arresters requires knowledge of the maximum phase-to-ground system voltage, maximum discharge voltage of the arrester, and basic impulse insulation level (BIL) of the equipment to be protected. Basic impulse insulation level is the crest value of voltage of a standard full impulse voltage wave the insulation will withstand. The lightning arrester must be selected so that it will not spark over as a result of maximum system voltage and have a maximum surge discharge voltage approximately 20 percent less than the basic impulse insulation of the equipment to be protected. Valve-type lightning arrester typical descriptive data is given in Table 16-1 for distribution class, in Table 16-2 for intermediate class, and in Table 16-3 for station class.

Researchers are currently developing valve-type arresters without gaps. The new arresters use a ceramic material, sintered at high temperature, containing zinc oxide and metal oxide additives. The new nonlinear valve-block material replaces the widely used silicone carbide valve blocks. The voltage characteristics of the zinc oxide valve blocks are flat and stable enough to permit elimination of the gap structure.

Isolation Lightning Arrester Distribution class lightning arresters' failure rate is high enough to make it desirable to provide them with an isolating device. Isolation is accomplished with an

Table 16-1   Typical Application Data for Distribution Class, Valve-Type Lightning Arresters
(Electrical Characteristics)

Arrester rating max line to ground kV rms	Minimum 60-Hz sparkover kV rms	Maximum 1½ × 40 wave kV crest	Maximum impulse sparkover ANSI front of wave kV crest, maximum	Discharge voltage in kV crest for discharge currents of 8 × 20 microsecond wave shape with following maximum crest amplitudes				
				1500 amp kV crest, maximum	5000 amp kV crest, maximum	10,000 amp kV crest, maximum	20,000 amp kV crest, maximum	65,000 amp kV crest, maximum
3	6	14	17	10	12.4	13.8	15.5	19
6	11	26	33	19	23	26	29	35
9	18	38	47	30	36.5	41	46	56
10	18	38	48	30	36.5	41	46	56
12	23.5	49	60	38	46	52	58	71
15	27	50	75	45	55	62	70	85
18	33	58	90	54	66	74	83	103

SOURCE: Westinghouse Electric Corp.

Table 16-2 Typical Application Data for Intermediate Class, Valve-Type Lightning Arresters
(Electrical Characteristics)

Arrester rating kV rms maximum valve-off or maximum reseal rating	Maximum circuit voltage phase to phase kV rms		Maximum front-of-wave impulse sparkover, kV crest	Maximum 100% impulse sparkover 1.2 × 50 wave, kV crest	Minimum 60-Hz sparkover, kV rms	Maximum discharge voltage with discharge current, 8 × 20 wave, kV crest				
	Ungrounded neutral 100% arrester	Effectively grounded neutral 80% arrester				1.5 kA	3 kA	5 kA	10 kA	20 kA
3	3	3.75	11	11	4.5	5.2	6	6.6	7.5	8.7
4.5	4.5	5.63	16	15	6.8	7.8	9	9.9	11.3	13.1
6	6	7.50	21	19	9	10.4	11.9	13.2	15	17.4
7.5	7.5	9.38	26	23.5	11.3	13	14.9	16.5	18.8	21.8
9	9	11.25	31	27.5	13.5	15.6	17.9	19.8	22.5	26.1
10	10	12.50	35	31	15	17.5	20.0	22.0	25.0	29.0
12	12	15.00	40	35.5	18	20.8	23.8	26.4	30	34.8
15	15	18.75	50	43.5	22.5	25.9	29.7	32.9	37.5	43.5
18	18	22.50	59	51.5	27	31.1	35.7	39.5	45	52.2
21	21	26.25	68	59	31.5	36.3	41.6	46.1	52.5	60.9
24	24	30.00	78	67	36	41.5	47.6	52.7	60.0	69.6
27	27	33.75	88	75	40.5	46.7	53.5	59.2	67.5	78.3
30	30	37.50	97	81	45	51.8	59.4	65.8	75	87
36	36	45.00	116	95	54	62.2	71.3	79	90	104.4
39	39	48.75	126	102	58.5	67.4	77.3	85.5	97.5	113.1
45	45	56.25	144	116	67.5	77.7	89.1	98.7	112.5	130.5
48	48	60.00	154	123	72	82.9	95.1	105.3	120	139.2
60	60	75.00	190	153	90	103.6	119	131.2	150	174
72	72	90.00	228	180	108	124.3	142.6	158	180	208.8
78	78	97.50	245	195	117	134.7	154.5	171	185	226.2
84	84	105.00	262	209	126	145	166.4	184.2	210	243.6
90	90	112.50	282	223	135	155.4	178.2	197.3	225	261
96	96	120.00	300	236	144	165.7	190.1	210.5	240	278.4
108	108	135.00	335	263	162	186.5	213.9	237	270	313.1
120	120	150.00	370	290	180	207.2	238	263	300	347.9

SOURCE: Westinghouse Electric Corp.

Table 16-3 Typical Application Data for Station Class, Valve-Type Lightning Arresters
(Electrical Characteristics)

Arrester rating, maximum permissible line-to-ground kV rms (maximum valve off or maximum reseal rating)	Minimum 60-Hz sparkover, kV rms	Maximum switching surge sparkover, kV crest	Maximum 100% 1½ × 40 impulse sparkover, kV crest	Maximum impulse sparkover ASA front-of-wave, kV crest	Maximum discharge voltage for discharge currents of 8 × 20 microsecond wave shape with the following crest amplitudes					
					1500 amperes, kV crest	3000 amperes, kV crest	5000 amperes, kV crest	10,000 amperes, kV crest	20,000 amperes, kV crest	40,000 amperes, kV crest
3	5	8	8	12	7	8	8.5	9	10	11.5
6	11	17	17	24	15	16	17	19	20	23
9	16	25	24	35	21	23	24	26	28	31.5
12	22	34	32	45	28	30	32	35	38	42
15	27	42	40	55	35	38	40	44	47	52.5
21	36	56	55	72	47	52	55	60	65	73
24	45	70	65	90	56	61	65	71	76	84
30	54	85	80	105	69	75	80	87	94	105
36	66	104	96	125	83	90	96	105	113	126
39	72	113	104	130	90	98	104	114	123	137
48	90	141	130	155	112	122	130	142	153	168
60	108	169	160	190	137	150	160	174	189	210
72	132	206	195	230	167	184	195	212	230	252
90	160	242	228	271	206	226	240	262	283	315
96	175	257	247	294	222	242	258	280	304	336
108	195	294	266	332	244	264	282	316	333	378
120	220	323	304	370	275	301	320	350	378	420

SOURCE: Westinghouse Electric Corp.

external air gap (Fig. 16-11) or a ground lead isolating device (Fig. 16-12). The external air gap is designed to prevent sparkover by normal system voltage if the lightning arrester fails. The defective arrester will fail to interrupt power follow current after an operation as a result of a lightning surge causing the substation circuit breaker, or other circuit protective device, to operate. When the circuit breaker is reclosed, the external gap will isolate the defective arrester and permit normal operation of the distribution circuit. The external gap on a lightning arrester has

Fig. 16-11   Distribution class valve-type lightning arrester with external gap structure. *(Courtesy Westinghouse Electric Corp.)*

Fig. 16-12   Distribution class valve-type lightning arrester with a ground-lead isolating device. Isolating device is the dark plastic projection on the bottom of the arrester with the ground-lead connection. *(Courtesy Westinghouse Electric Corp.)*

the disadvantage of increasing the sparkover voltage of the arrester but fails to provide visible evidence of lightning arrester failure. The ground lead isolating device is a heat-sensitive device that separates the ground lead from a defective lightning arrester. The heat sensing device detonates an explosive rivet when power follow current is not interrupted successfully. The detonation breaks the isolator housing and separates the ground lead from the arrester housing. The power follow current may continue to flow until it is interrupted by operation of the substation circuit breaker or other circuit protective device. When the circuit is reenergized, the lightning arrester will be isolated from ground permitting normal operation. The lineman patrolling the circuit will see the separated ground lead informing him that the lightning arrester is defective and should be replaced.

Expulsion-Type Arrester   As in other types of arresters, the expulsion-type arrester consists essentially of a spark gap connected in series with a so-called characteristic or arrester element (see Fig. 16-13). In the expulsion arrester the characteristic element consists of a hollow fiber tube which serves as the power arc quenching element (Fig. 16-14).

The spark gap is set to spark over at a predetermined value greater than line voltage. With the spark gap so set, the circuit through the arrester is kept open under normal conditions. Under abnormal conditions, it sparks over and closes the circuit through the arrester and allows the lightning surge current to drain off to ground. When the follow-up power current flows through the arcing chamber, it produces intense heat which causes cool nonconducting gases to be given off the inner walls of the fiber tube. These gases form so rapidly that they produce an explosive blast which expels the gases through the open lower end. In so doing, the conducting gas is replaced

Fig. 16-13 External view of expulsion-type arrester. *(Courtesy James R. Kearney Corp.)*

Fig. 16-14 Cutaway view of assembled expulsion arrester showing open series gap and arcing chamber. *(Courtesy Westinghouse Electric Corp.)*

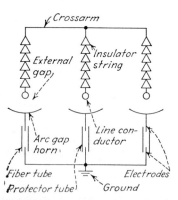

Fig. 16-15 Diagrammatic representation of protector-tube installation on a three-phase tower line.

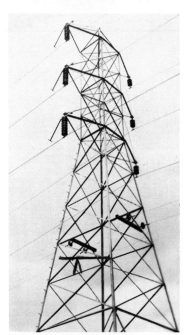

Fig. 16-16 Installation of expulsion protective gaps on a 115,000-volt three-phase transmission line. Tubes are mounted on one side of insulator string instead of below conductor. *(Courtesy General Electric Co.)*

with nonconducting gas which reduces the follow-up current flow to such a low value that it goes out at the next current zero.

Protector Tubes   Another device that is used to protect overhead transmission lines from lightning discharges is the protector tube. Basically, the protector tube consists of a fiber tube with an electrode in each end. It is so designed that the impulse voltage breakdown through the tube is lower than that of the line insulation to be protected. On a transmission tower the tube is often mounted below the conductor, as shown in Fig. 16-15. The upper electrode is connected to an arc-shaped horn located the proper distance below the conductor. The arc-shaped horn maintains a constant length of external gap between the upper electrode and the line conductor even though the insulator string swings from side to side. The lower electrode is solidly grounded. An installation in which the protector tubes hang downward beside the insulator strings as shown in Fig. 16-16.

When lightning strikes the line, the external and internal series gaps break down instead of flashing over the insulator string because the tube has the lower breakdown voltage. After breakdown of the gaps the power follow current volatilizes a small layer of the fiber wall and the gas given off mixes in the arc to help deionize the space between the electrodes. A pressure is built up in the tube, and the hot gases are discharged through the lower electrode, which is hollow. If the deionizing action is sufficiently strong and if the voltage does not build up too rapidly across the tube, the arc will go out at a current zero and will not be reestablished. While the tube is discharging it is a good conductor, but after the arc is extinguished it becomes a good insulator again.

Section **17**

Fuses

Section **17**

Fuses

Fuses are relatively inexpensive protective devices connected into circuits to protect the circuit and any connected apparatus from damage due to overload or short circuit. A fuse is an intentionally weakened spot in an electrical circuit. It usually consists of a short piece of wire made of lead or more often of an alloy of lead and tin which will melt at low temperature. When the current through such a metal becomes excessive, the resistance offered by the metal to the flow of current develops enough heat to melt the metal, thereby opening the circuit before the abnormal current can damage the circuit or any connected apparatus.

Fuses are usually enclosed to prevent the molten metal from flying and doing damage or causing a fire. Enclosing the fuse also aids in quenching the arc. The melting of the fuse is often accompanied by a puff of smoke of vaporized metal. This action is sometimes referred to as the "blowing" of the fuse.

Fuses can be broadly classified into low- and high-voltage fuses. The low voltage fuses are of the plug or cartridge type. High-voltage fuses commonly used on an electric distribution system are the expulsion, open-link, current-limiting, liquid, and boric acid types.

Fig. 17-1 Plug fuse screwed in place in socket. Fusible element is contained in an insulated cup. Cup has transparent cover to permit inspection of fusible element.

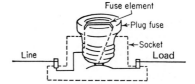

Plug-Type Fuses The low-voltage plug-type fuse can be used for residential services and for ordinary lighting branch circuits on panel boards. The fuse consists of a small cup of solid insulating material within which the fusible wire connects the center contact with the outer metal screw shell, as shown in Fig. 17-1. The plug is the same size as the ordinary incandescent-lamp base.

Cartridge Fuse Cartridge fuses are used on circuits rated 600 volts and below. The fuses are manufactured differently for use on 240- and 480-volt circuits. A fuse of the proper voltage and current rating should be used for the circuit application. In the cartridge fuse the fusible element is contained within a completely enclosed insulating tube. This tube is called the "cartridge," because of its resemblance to an ordinary shotgun cartridge. Cartridge fuses in which the element is not replaceable after the fuse has blown are called "nonrenewable" fuses. Cartridge fuses in which the fusible element may be readily replaced by the user with suitable renewal elements supplied by the manufacturer are called "renewable" fuses (Fig. 17-2). The cartridge or tube may thus be used repeatedly with new renewal elements. In this way, a considerable saving is effected, especially in the larger fuse sizes.

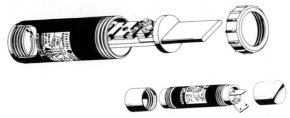

Fig. 17-2 Renewable cartridge-type fuse. Only the blade shown in the inside of the tube need be replaced in case the fuse blows. *(Courtesy General Electric Co.)*

Distribution Cutouts   A distribution cutout provides a high-voltage mounting for the fuse element used to protect the distribution system or the equipment connected to it (Fig. 17-3). Distribution cutouts are used with installations of transformers, capacitors, cable circuits, and sectionalizing points on overhead circuits (Fig. 17-4). Enclosed, open, and open-link cutouts are used for different distribution circuit applications. All the cutouts use an expulsion fuse. An expulsion fuse operates to isolate a fault or overload from a circuit as a result of the arc from the fault current eroding the tube of the fuse holder, producing a gas that blasts the arc out through the fuse tube vent or vents (Fig. 17-5).

The mechanical differences between enclosed, open, and open-link cutouts are in their external appearance and method of operation. Enclosed cutouts have terminals, fuse clips, and fuse holders mounted completely within an insulating enclosure. Open cutouts have these parts completely exposed as the type name indicates. Open-link cutouts have no integral fuse holder; the arc confining tube for the cutouts is incorporated in the fuse link.

The construction of the cutout fuse holder can provide for non-drop-out, or drop-out operation. Some of the fuses are manufactured to provide indication if the fuse is blown; other fuses may have an expendable cap.

Enclosed Distribution Cutout   An enclosed distribution fuse cutout is one in which the fuse clips and fuse holder are mounted completely within an enclosure. A typical enclosed cutout as shown in Fig. 17-6 has a porcelain housing and a hinged door supporting the fuse holder. The fuse holder is a hollow vulcanized-fiber expulsion tube. The fuse link is placed inside the tube and connects with the upper and lower line terminals when the door is closed. When the fuse blows or melts because of excessive current passing through it, the resultant arc attacks the walls of the fiber tube, producing a gas which blows out the arc. The melting of the fusible element of some cutouts causes the door to drop open, signaling to the lineman that the fuse has blown.

Fig. 17-3 Distribution cutout with expulsion fuse mounted. Fuse is in the closed position completing circuit between top and bottom terminals. *(Courtesy A. B. Chance Co.)*

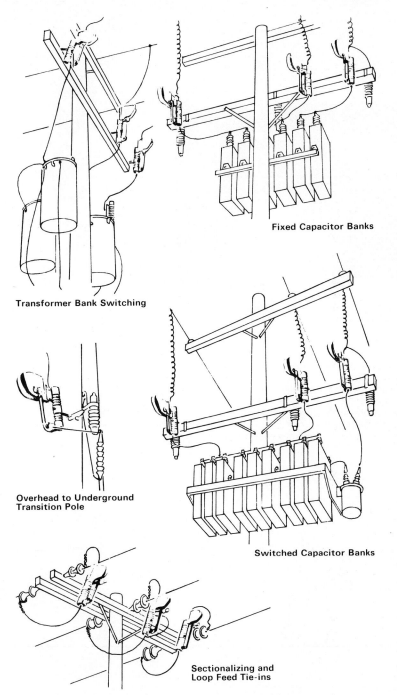

Fixed Capacitor Banks

Transformer Bank Switching

**Overhead to Underground
Transition Pole**

Switched Capacitor Banks

**Sectionalizing and
Loop Feed Tie-ins**

Fig. 17-4   Distribution cutouts installed in electric distribution circuits. Cutouts are equipped with arc chute
to permit switching equipment on and off under load. *(Courtesy Westinghouse Electric Corp.)*

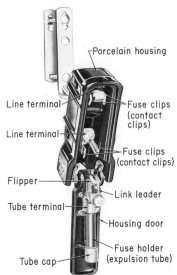

Fig. 17-5 Night photograph of the blowing of an expulsion cutout fuse. Note violent gas blast out of lower end of fiber tube. *(Courtesy S. & C. Electric Co.)*

Fig. 17-6 Enclosed primary cutout showing door and cartridge assembly. *(Courtesy Line Material Industries.)*

Each time the fuse blows, a small amount of the vulcanized fiber of the expulsion tube is eroded away. The larger the value of the current interrupted, the more material is consumed. In general, a hundred or more operations of average current values can be successfully performed before the cutout fails.

Enclosed cutouts can also be arranged to indicate when the fuse link has blown by dropping the fuse holder. Figure 17-7 shows an indicating enclosed cutout after the fuse link has melted and the tube has become extended below the housing. The enclosed cutout is designed and man-

Fig. 17-7 Vertical drop-out fuse cutout indicating that fuse link has melted. *(Courtesy Line Material Industries.)*

Fig. 17-8 Open distribution cutout with arc chute designed for load break operation. *(Courtesy Westinghouse Electric Corp.)*

ufactured for operation on distribution circuits of 7200 volts and below. The standard current ratings of the cutouts are 50, 100, or 200 amps.

Open Distribution Cutout   Open-type cutouts are similar to the enclosed types except that the housing is omitted (Fig. 17-8). The open-type cutout is designed and manufactured for all distribution system voltages. The open-type is made for 100- or 200-amp operation. Some cutouts can be uprated from 100 to 200 amps by using a fuse tube rated for 200-amp operation.

Open-Link Distribution Cutout   This type of cutout differs from the open cutout in that it does not employ the fiber expulsion tube. The fuse link is supported by spring terminal contacts as shown in Fig. 17-9. An arc-confining tube surrounds the fusible element of the link. During

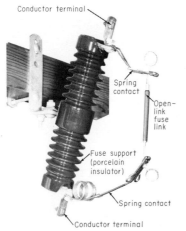

Fig. 17-9 Open-link primary cutout showing spring terminal contacts and fuse link enclosed in an arc-confining tube. During fault clearing, the spring contacts provide link separation and arc stretching. *(Courtesy S. & C. Electric Co.)*

Fig. 17-10 Open-link primary cutout after link has melted and lower spring contact arm has separated the lower fuse link terminal, thus stretching the arc and extinguishing it. *(Courtesy S. & C. Electric Corp.)*

fault clearing, the spring contacts provide link separation and arc stretching. The arc-confining tube furnishes the necessary expulsion action for circuit interruption.

As shown in Fig. 17-9, the fuse link is held in tension between the two contact arms. During fault current operation the upper arm remains stationary while the lower arm swings outward and downward. Thus, when the fusible element melts, the lower fuse link terminal is drawn instantly through and away from the sleeve which surrounds it, the parts coming to rest as shown in Fig. 17-10. The contact arms retain the fuse link cables after operation as protection against fires.

To re-fuse an open-link cutout, a new fuse link is hung on a hook stick by its top ring. The bottom cable of the link is slid along either side of the lower arm into one of its contacts. In like manner the top cable of the link is slid into one of the upper arm contacts, all as shown in Fig. 17-11.

An exchange of fuse links can be accomplished without interrupting the service by inserting a second link into the unused lower and upper contacts before removing the first link.

Primary Fuse Links   The basic primary fuse link consists of four parts, namely, the button, shank, fusible element, and leader, all as shown in Fig. 17-12. The button is the upper terminal of the fuse link, and the leader is the lower terminal. Fuse links intended to be handled with hot-line tools have pull rings at the top and bottom for the insertion of hooks when the link is placed in a fuse cutout (see Fig. 17-13). Links rated 100 amp and below have fiber tubes enclosing the fusible element for protection from damage during handling and to stabilize the melting time by enclosing the element in a uniform volume of air.

Standards specify the size of fuse holder in which the link must fit freely. Thus, links rated 1 to 50 amp must fit into a ⁵⁄₁₆-in-diameter holder, 60- to 100-amp links must fit into a ⁷⁄₁₆-in-diameter holder, and 125- to 200-amp links must fit into a ¾-in holder. Links must also withstand a 10-lb pull while carrying no load current but are generally given a 25-lb test.

Fusible elements are made in four designs, namely,

1. High surge dual element for links rated 1 to 8 amp
2. Wire element for links rated 5 to 20 amp
3. Diecast tin element for links rated 25 to 100 amp
4. Formed strip element for links rated over 100 amp

These four types are illustrated in Fig. 17-14a to e, respectively. Link designs are further distinguished as strain or spring type. The fusible element of a strain-type link incorporates a high-resistance strain wire for strengthening purposes, while the fusible element of a spring-type link

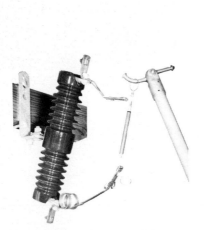

Fig. 17-11 Re-fusing open-link primary cutout by use of insulated hook stick. Link is first slid into lower contact, then spring arm is raised until upper end of link can be slid into upper link contact. Exchange of fuse links is accomplished without interruption to service by inserting a second link in the unused contacts before removing the first link. *(Courtesy S. & C. Electric Co.)*

Fig. 17-12 Typical primary cutout fuse link showing component parts, button, shank, fusible element, and leader. *(Courtesy James R. Kearney Corp.)*

Fig. 17-13 Open primary cutout fuse link provided with pull rings for handling with hot-line tools. Hooks are inserted in rings for placement in cutout. Cutaway shows fusible element inside arc-confining tube. *(Courtesy Line Material Industries.)*

has in addition to the strain wire a pretensioned spring to stretch the arc, thereby aiding circuit clearing after the fusible element has melted.

The high-surge-element link of fuses rated 1 to 8 amp (Fig. 17-14a) is a spring-type link having two heater wires in series with a reinforced eutectic alloy solder joint. The solder joint is the central component of the element. One wire is fastened to the shank, and the other wire is fastened to the leader.

Links rated 5 to 20 amp (Fig. 17-14b) are also of the spring type consisting of a fusible tin wire and a separate strain wire. Both wires are soldered to the shank and leader.

Links rated 25 to 100 amp (Fig. 17-14c and d) are strain type with diecast elements, whereas links rated over 100 amp (Fig. 17-14e) are strain type with formed strip elements. The strip element is lap soldered to the shank and the leader, and the strain wire is securely fastened to the shank and leader.

Fuse Link Operation   When a fault occurs, the fusible element is melted by the excessive fault current. Immediately following, the strain wire is heated and melted owing to its high resistance. An arc now forms across the severed fusible element. This arc is a conducting path of ionized particles consisting of metallic ions of the melted element and wire and ions of trapped air. The arc heats the fiber of the tube wall, thereby generating deionizing gases. When the cur-

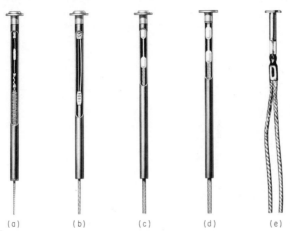

Fig. 17-14   Primary cutout fuse links having different fusible elements. (*a*) High-surge dual element (1 to 8 amp); (*b*) wire element (5 to 20 amp); (*c*) diecast tin element (25 to 50 amp); (*d*) diecast tin element (65 to 100 amp); (*e*) formed strip element (140 to 200 amp). (*Courtesy Line Material Industries.*)

rent passes through its next zero value as it changes direction of flow, the arc is momentarily interrupted (Fig. 17-15). As the current increases in value again, the arc attempts to reestablish itself. This restrike, however, is prevented by the presence of the deionizing gases. The increasing pressure of these gases expels the arc-supporting ions from the tube, which prevents any further restrike. The lower fuse terminal is usually blown out of the cutout tube. To facilitate this the lower connection is made with a flexible pigtail.

Time-Current Curves   A time-current curve is a curve that is plotted between the magnitude of a fault current and the time required for the fuse link to open the circuit. It is obvious that the greater the current, the faster the fuse melts and the shorter the time required for it to blow. Figure 17-16 is a typical time-current curve for a 10K fuse link (normal current rating 10 amp). Thus, a current of 10 amps would not cause the fuse to blow; 20 amps, which is twice normal, would require 300 sec or 5 min; 30 amp would require 3 sec; and 100 amp would require about 0.15 sec. These time-current curves are useful in providing required coordination of fuses, reclosers, and sectionalizers.

Links are further divided into two types, fast and slow, designated as K (kwick) and T (tardy), respectively. Links of the same current rating have the same 300- or 600-sec points but have slightly different time-current curves, as illustrated in Fig. 17-17 for a 15-amp link. The 15K link is faster than the 15T link on the higher current end of the curves by an amount of time equal to the vertical difference between the two curves. Thus, for an overcurrent of 100 amp the 15K fuse will operate in 0.5 sec while the 15T fuse will require 1.5 sec, a difference of 1.0 sec.

Recommended Size of Primary Fuse   Table 17-1 gives the recommended size of primary fuse to use with different transformer voltage and kilovolt-ampere ratings. The table also gives the normal full-load primary current rating of the transformer.

Fig. 17-15   Diagram of voltages, curent, and timing reference recorded with an oscillograph to show fuse operation.

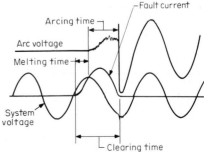

Table 17-1   Suggested Transformer Fusing Schedule (Protection Between 200 and 300 percent rated load)*
(Link sizes are for EEI-NEMA K or T fuse links except for the H links noted.)

Transformer size, kva	2,400 Δ Figures 1 and 2 Rated amp	2,400 Δ Figures 1 and 2 Link rating	2,400 Δ Figure 3 Rated amp	2,400 Δ Figure 3 Link rating	4,160Y/2,400 Figures 4, 5, and 6 Rated amp	4,160Y/2,400 Figures 4, 5, and 6 Link rating	4,800 Δ Figures 1 and 2 Rated amp	4,800 Δ Figures 1 and 2 Link rating	4,800 Δ Figure 3 Rated amp	4,800 Δ Figure 3 Link rating	8,320Y/4,800 Figures 4, 5, and 6 Rated amp	8,320Y/4,800 Figures 4, 5, and 6 Link rating
3	1.25	2H	2.16	3H	1.25	2H	0.625	1H†	1.08	1H	0.625	1H†
5	2.08	3H	3.61	5H	2.08	3H	1.042	1H	1.805	3H	1.042	1H
10	4.17	6	7.22	10	4.17	6	2.083	3H	3.61	5H	2.083	3H
15	6.25	8	10.8	12	6.25	8	3.125	5H	5.42	6	3.125	5H
25	10.42	12	18.05	25	10.42	12	5.21	6	9.01	12	5.21	6
37.5	15.63	20	27.05	30	15.63	20	7.81	10	13.5	15	7.81	10
50	20.8	25	36.1	50	20.8	25	10.42	12	18.05	25	10.42	12
75	31.25	40	54.2	65	31.25	40	15.63	20	27.05	30	15.63	20
100	41.67	50	72.2	80	41.67	50	20.83	25	36.1	50	20.83	25
167	69.4	80	119.0	140	69.4	80	34.7	40	60.1	80	34.7	40
250	104.2	140	180.5	200	104.2	140	52.1	60	90.1	100	52.1	60
333	138.8	140	238.0		138.8	140	69.4	80	120.1	140	69.4	80
500	208.3	200	361.0		208.3	200	104.2	140	180.5	200	104.2	140

* Reprinted with permission from "Distribution System Protection and Apparatus Co-ordination," published by Line Material Industries.

† Since this is the smallest link available and does not protect for 300 per cent load, secondary protection is desirable.

Table 17-1   Suggested Transformer Fusing Schedule (Protection Between 200 and 300 percent rated load) (Continued)

| Transformer size, kva | 7,200 Δ | | | | 12,470Y/7,200 | | 13,200Y/7,620 | | 12,000 Δ | | | |
| | Figures 1 and 2 | | Figure 3 | | Figures 4, 5, and 6 | | Figures 4, 5, and 6 | | Figures 1 and 2 | | Figure 3 | |
	Rated amp	Link rating	Rated amp	Link rating	Rated amp	Link rating	Rated amp	Link rating	Rated amp	Link rating	Rated amp	Link rating
3	0.416	1H†	0.722	1H†	0.416	1H†	0.394	1H†	0.250	1H†	0.432	1H†
5	0.694	1H†	1.201	1H	0.694	1H†	0.656	1H†	0.417	1H†	0.722	1H†
10	1.389	2H	2.4	5H	1.389	2H	1.312	2H	0.833	1H†	1.44	2H
15	2.083	3H	3.61	5H	2.083	3H	1.97	3H	1.25	1H	2.16	3H
25	3.47	5H	5.94	8	3.47	5H	3.28	5H	2.083	3H	3.61	5H
37.5	5.21	6	9.01	12	5.21	6	4.92	6	3.125	5H	5.42	6
50	6.49	8	12.01	15	6.94	8	6.56	8	4.17	6	7.22	10
75	10.42	12	18.05	25	10.42	12	9.84	12	6.25	8	10.8	12
100	13.89	15	24.0	30	13.89	15	13.12	15	8.33	10	14.44	15
167	23.2	30	40.1	50	23.2	30	21.8	25	13.87	15	23.8	30
250	34.73	40	59.4	80	34.73	40	32.8	40	20.83	25	36.1	50
333	46.3	50	80.2	100	46.3	50	43.7	50	27.75	30	47.5	65
500	69.4	80	120.1	140	69.4	80	65.6	80	41.67	50	72.2	80

Table 17-1 Suggested Transformer Fusing Schedule (Protection Between 200 and 300 percent rated load) (Continued)

Transformer size, kva	13,200 Δ				14,400 Δ				24,900Y/14,400	
	Figures 1 and 2		Figure 3		Figures 1 and 2		Figure 3		Figures 4, 5, and 6	
	Rated amp	Link rating	Rated amp	Link rating	Rated amp	Link rating	Rated amp	Link rating	Rated amp	Link rating
3	0.227	1H†	0.394	1H†	0.208	1H†	0.361	1H†	0.208	1H†
5	0.379	1H†	0.656	1H†	0.347	1H†	0.594	1H†	0.374	1H†
10	0.757	1H†	1.312	2H	0.694	1H†	1.20	2H	0.694	1H†
15	1.14	1H	1.97	3H	1.04	1H	1.80	3H	1.04	1H
25	1.89	3H	3.28	5H	1.74	2H	3.01	5H	1.74	2H
37.5	2.84	5H	4.92	6	2.61	3H	4.52	6	2.61	3H
50	3.79	6	6.56	8	3.47	5H	5.94	8	3.47	5H
75	5.68	6	9.84	12	5.21	6	9.01	12	5.21	6
100	7.57	8	13.12	15	6.94	8	12.01	15	6.94	8
167	12.62	15	21.8	25	11.6	12	20.1	25	11.6	12
250	18.94	25	32.8	40	17.4	20	30.1	40	17.4	20
333	25.23	30	43.7	50	23.1	30	40.0	50	23.1	30
500	37.88	50	65.6	80	34.7	40	60.0	80	34.7	40

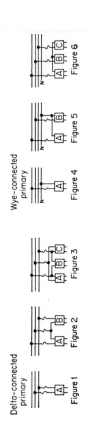

Delta-connected primary

Figure 1

Figure 2

Figure 3

Wye-connected primary

Figure 4

Figure 5

Figure 6

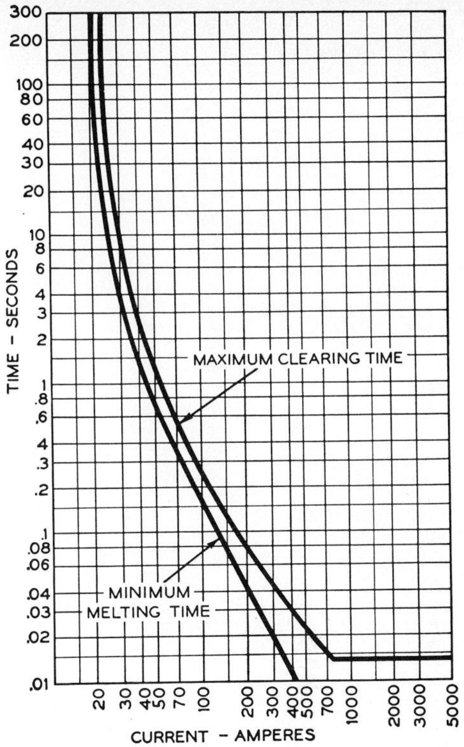

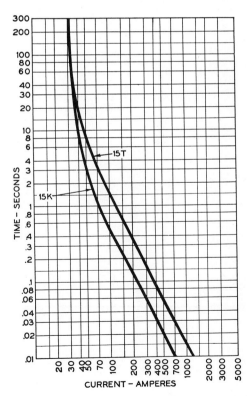

Fig. 17-16 Typical time-current curves for 10K fuse link. Fault currents are plotted along abscissa, and time to blow along the ordinate. *(Courtesy Line Material Industries.)*

Fig. 17-17 Curves showing melting times of fast (K) links and slow (T) links having the same 15-amp rating. *(Courtesy Line Material Industries.)*

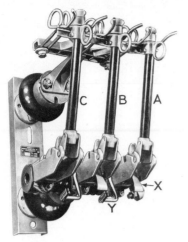

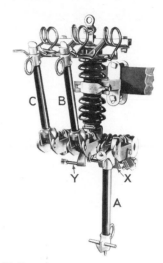

Fig. 17-18 All tubes ready for operation. *(Courtesy Railway and Industrial Engineering Co.)*

Fig. 17-19 Tube A open; tube B in circuit. *(Courtesy Railway and Industrial Engineering Co.)*

It is general practice not to protect distribution transformers against small overloads. To do so would cause unnecessary blowing of the fuses and frequent interruption of the service, both of which are undesirable. It is therefore customary to provide fuses which have a higher current rating than the current rating of the transformer. This is clearly shown by a comparison of the normal current-rating column with the fuse-rating column in Table 17-1.

Repeater Cutout Fuses   A repeater fuse consists of two or three primary cutout fuses so mounted and arranged that when one fuse blows it drops out and the second fuse is cut in. If the fault is of a temporary nature, it will be cleared and service will be restored. If the fault is permanent, the second fuse will blow and trip out, and the third fuse is cut in. This continues until all the fuses have blown or until the fault is cleared. Figures 17-18 to 17-21 illustrate the operation of the repeater fuse or "repeater cutout" as it is sometimes called. Figure 17-18 shows all fuses loaded and in place. All fuse tubes are connected to the line side, but only tube A is connected to the load. In Fig. 17-19 a fault has occurred and opened tube A. In dropping open, the tube forced

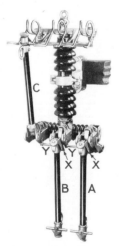

Fig. 17-20 Tubes A and B open; tube C in circuit. *(Courtesy Railway and Industrial Engineering Co.)*

Fig. 17-21 All tubes open; line disconnected. *(Courtesy Railway and Industrial Engineering Co.)*

trip lever X back to the position shown and a transfer arm Y has placed tube B in the line. If the trouble has cleared, service will be restored. Tube A in the down position indicates to the service-man that it has operated and the fuse element should be replaced. The tubes should be removed with a switch stick and reloaded on the ground away from all danger of contact with live circuits. After reloading, tube A is replaced in the mounting and closed. In closing the tube onto the line, lever X is automatically reset to its original position and tube A again carries the load. In case of a more serious fault, tubes A and B may both blow in turn, operating trip lever X and transfer arms Y, causing C to carry the load (Fig. 17-20). Both blown tubes would then have to be removed, reloaded on the ground, and replaced and closed. In Fig. 17-21 a permanent fault has

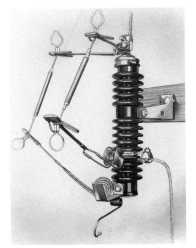

Fig. 17-22 Two-shot open-link primary cutout repeater fuse. (*Courtesy James R. Kearney Corp.*)

Fig. 17-23 Two-shot cutout repeater fuse. (*Courtesy James R. Kearney Corp.*)

occurred and blown and operated all the tubes. After the fault has been cleared, all tubes are reloaded on the ground and replaced in the mounting.

Repeater cutout fuses having only two fuses are shown in Figs. 17-22 and 17-23. Their method of operation is similar to the three-tube repeater.

Liquid Fuse   This fuse consists of a glass tube of proper length and diameter, at one end of which a fusible element is mounted. The fusible element is held in tension by means of a helically coiled spring secured at the bottom end, the fusible element being attached to the upper end. The tube is filled with an arc-quenching liquid like oil which has a high dielectric strength. When fault current melts the fusible element, the spring contracts very suddenly and opens a gap in the tube proportionate to the voltage rating of the fuse. As the arc is drawn down into the liquid, the liquid extinguishes or quenches it. See Fig. 17-24.

The fuse should be mounted in a vertical position and is usually held in spring clips, so that it is readily removable with the use of fuse-handling tools.

In case of an excessive short-circuit current, the arc produced may cause enough heat to vapor-ize some of the liquid within the fuse. A vent cap is provided which yields to relieve the internal pressure.

Boric Acid Fuse   The interrupting medium in this fuse is compressed boric acid. It is pressed into cylindrical blocks with a hole in the center of the block forming the bore. When an arc is drawn in the boric acid tube, the generated gas consists principally of steam formed from the water of crystallization of the boric acid. Steam is less readily ionized than organic gases and thus helps to maintain high dielectric strength around the fuse under operating conditions. There is no flame discharge when the fuse operates. (See Figs. 17-25 and 17-26).

Because steam can be condensed, a copper-mesh condenser can be provided for indoor installations.

Fig. 17-24 Two views of liquid fuse showing (right) fuse before being blown, (left) fuse after being blown. *(Courtesy S. & C. Electrc Co.)*

Fig. 17-25 External view of power fuse of boric acid type in position in fuse mounting. Fuse drops out when blown. Fuse is rated 69 kV, 300 amp continuous, with an interrupting rating of 2,000,000 kVA. *(Courtesy S. & C. Electric Co.)*

To replace a blown fuse, the fittings are removed from the blown fuse tube and clamped on a new fuse unit.

Boric acid fuses are made in both fixed and drop-out forms. Figure 17-27 illustrates an outdoor drop-out switch-hook-operated type. Boric acid fuses are used in pad-mounted switchgear to isolate cable and equipment faults on underground circuits (Fig. 17-28). The operating characteristics of the boric acid fuse meet the requirements necessary to maintain the dielectric strength of the atmosphere in metalclad switchgear.

Current-Limiting Fuse   The current-limiting fuse (CLF) is a nonexpulsion fuse consisting of a silver fuse element wound on a central core enclosed in a tube filled with high purity silica sand. Under fault conditions the silver element melts and establishes an arc. The fuse element is burned back by the arc. The heat of the arc melts the adjacent sand to produce a glass-like substance known as a fulgurite. The rapid heat absorption of the sand, together with the rapid burn back of the fuse element acts to insert a high resistance into the arc path that prevents the current from reaching its natural crest. This produces the fuse's current-limiting action and lowers the energy let-through during faults. In addition, the current-limiting fuse has an interrupting rating in the 50,000-amp symmetric range and a less than ½ cycle clearing time. The current-limiting fuse is produced in two versions: the backup or partial range and the general purpose or full range.

The backup CLF is constructed as described with the central core generally being of ceramic construction. No gas is produced during fault clearing; therefore, the fuse can be hermetically sealed making it suitable for submersion in oil filled apparatus. This type of fuse is not capable of interrupting low-magnitude fault currents. It will only interrupt a range of fault currents from stated minimum interrupting rating (usually 3 to 7 times the continuous current rating) to the maximum interrupting rating. Currents below this range but above the continuous current rating of the fuse will melt the fuse element but will not have enough energy to cause interruption. These currents must be interrupted by a series-connected interrupting device (Fig. 17-29). If the fuse is exposed to low-fault currents, the continued arcing will cause the fuse to burn up or explode.

The general purpose CLF adds modifications to the basic CLF design to make the fuse capable of clearing any current greater than the continuous rating and less than the maximum interrupting rating of the fuse (Fig. 17-30). In this fuse, part of the fuse element is constructed from a solder alloy with a melting temperature ⅓ of the 960° melting temperature of the silver portion of the fuse element. The alloy portion is known as the "M" spot. The central core of the fuse is constructed from a material which evolves gas when heated by element melting. The lowered melting temperature and the gas evolving core make the general purpose CLF capable of clearing the lower range of currents that the backup CLF cannot clear. High-current operation is the same.

Current-limiting fuses are expensive compared to the cost of expulsion fuses. The use of the partial range current-limiting fuse in series with an expulsion fuse minimizes the operating costs since the expulsion fuse will operate for the most common faults that occur most frequently.

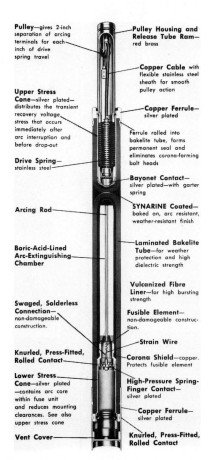

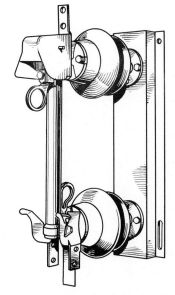

Fig. 17-26 Internal view of boric acid power fuse showing construction details. Note boric acid-lined arc-extinguishing chamber. (*Courtesy S. & C. Electric Co.*)

Fig. 17-27 Boric acid fuse designed to drop out when blown. Fuse is operated with customary switch hook. (*Courtesy Westinghouse Electric Corp.*)

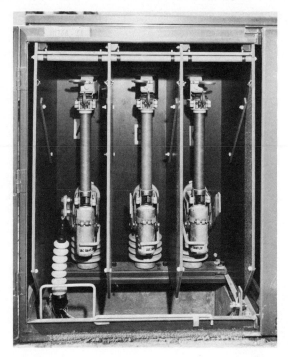

Fig. 17-28 Pad-mounted switchgear with boric acid fuses installed for a cable underground circuit. *(Courtesy S. & C. Electric Co.)*

Fig. 17-29 Partial range current-limiting fuse is installed on top of distribution cutout with expulsion fuse. Current-limiting fuse will operate for high-fault-current insulation failures in distribution transformer. Operating characteristics of current-limiting fuse will prevent destruction of transformer and lid from blowing off if a fault develops inside the transformer tank. Expulsion fuse will operate for low-primary current or secondary faults. *(Courtesy A. B. Chance Co.)*

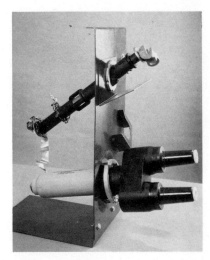

Fig. 17-30 General purpose or full range current-limiting fuse mounted on bushing of distribution transformer. Fuse will operate for all faults and is designed to prevent destruction of transformer and lid of transformer from blowing off if a short circuit develops inside transformer tank. *(Courtesy Westinghouse Electric Corp.)*

Fig. 17-31 Partial range current-limiting fuse connected in series with draw-out type expulsion fuse. Both fuses would be installed in tank of pad-mounted transformer or switching equipment. Draw-out feature facilitates replacement of expulsion fuse which is designed to operate for most common electrical faults. *(Courtesy A. B. Chance Co.)*

Figure 17-31 illustrates an expulsion fuse in series with a partial range CLF mounted on a metal sheet as it would be installed in a pad-mounted transformer or pad-mounted switchgear used with underground circuits.

Protective Overcurrent Coordination The coordination of overcurrent protective devices involves their selection and use in such a manner that temporary faults will be quickly removed and permanent faults will be restricted to the smallest section of the system possible.

The locations of the protective devices are known as coordinating points. Coordinating points are usually established at the substation, at positions along the feeder, in branch lines off the feeder, and on the primary side of distribution transformers. In the single-line diagram of Fig. 17-32 representing an elementary distribution system, points A, B, C, E, F, H, and I are coordinating points. Each device must be selected so that it can carry the normal line or load current and will respond properly to an excessive current. The device nearest the power source is known as the protecting device, and the one next to it the backup or protected device. Thus, device E is the protecting device, device C being the protected device. However, with respect to device A, C is the protecting device, A being the protected device. In each case the protecting device must clear a permanent fault before the protected device does. Therefore, the permanent fault shown

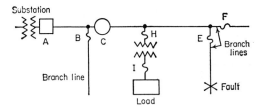

Fig. 17-32 One-line diagram of an elementary distribution system. Overcurrent protective devices are shown as A, B, C, E, F, H, and I. Fault is shown as X. Device E should disconnect branch line and permit remainder of system to operate.

on the branch line should be cleared by device E. After device E has isolated the faulted branch, normal load current will continue to flow in the remainder of the system.

Additional detailed information on fuses can be obtained from the American National Standards ANSI C37.40, "Service Conditions and Definitions for Distribution Cutouts and Fuse Links, Secondary Fuses, Distribution Enclosed Single-Pole Air Switches, Power Fuses, Fuse Disconnecting Switches, and Accessories"; ANSI C37.42, "Specifications for Distribution Enclosed, Open and Open-Link Cutouts"; ANSI C37.43 "Specifications for Distribution Fuse Links for Use in Distribution Enclosed, Open and Open-Link Cutouts"; ANSI C37.46, "Specifications for Power Fuses and Fuse Disconnecting Switches"; and ANSI C37.47, "Specifications for Distribution Fuse Disconnecting Switches, Fuse Supports and Current-Limiting Fuses."

A switch is used to disconnect or close circuits that may be energized. If the circuit conducts current when it is operated, special devices need to be installed on the switch contacts to interrupt or establish the current flow. High-voltage switches are operated remotely with a mechanism or a hot-line tool, called a switch stick.

Switches may be divided into three general classes, namely,

1. Air switches
   a. Air circuit breaker
   b. Air break
   c. Disconnect
2. Oil switches
   a. Oil circuit breaker
   b. Oil circuit recloser
      (1) Sectionalizer
3. Vacuum switches
   a. Vacuum circuit breaker
   b. Vacuum recloser

Air Switches   As their names imply, air switches are switches whose contacts are opened in air, while oil switches are switches whose contacts are opened under oil, and vacuum switches are switches whose contacts are opened in a vacuum. Air switches are further classified as air circuit breakers, air-break switches, and disconnects. Air-break switches are almost always manually operated. Oil switches and vacuum switches are almost always designed for automatic operation, that is, they will open automatically on overload or short circuit and can also be made to reclose automatically one or more times.

Air Circuit Breaker   A device used to complete, maintain, and interrupt currents flowing in circuits under normal or faulted conditions is called a circuit breaker. The circuit breaker has a mechanism that mechanically, hydraulically, or pneumatically operates the circuit-breaker contacts. Insulating oil, air, compressed air, vacuum, or sulfur hexafluoride gas is used as an arc-interrupting medium and a dielectric to insulate the contacts after the arc is interrupted.

When a circuit carrying considerable current must be opened while the load is on, switches, called "circuit breakers," are employed. If it is desired to open the circuit automatically during overload or short circuit, the circuit breaker can be equipped with a tripping mechanism to accomplish this. Circuit breakers are, therefore, used where control of the circuit as well as protection from overload, short circuit, etc., is desired as at generating stations and substations.

Most of the different types of circuit breakers are illustrated in Section 3 "Substations."

Air-Break Switch   The air-break switch can have both blade and stationary contacts equipped with arcing horns as shown in Fig. 18-1. These horns are pieces of metal between which the arc forms when a circuit carrying current is opened. As the switch opens, these horns are spread farther and farther apart, thereby lengthening the arc until it finally breaks. That such arcs do form is convincingly shown in Fig. 18-2, which shows the arcs formed in air when a three-phase gang-operated air-break switch is opened at the source end of a 50-mile 132,000-volt transmission line. The line was without load, but 20 amp of charging current was flowing into the line. The arcs measured 123 ft in length, while the spacing of the phases was only 16 ft. Favorable oblique winds kept the arcs from shorting across phases. Wind also has a cooling effect which aids extinction.

Air-break switches are usually mounted on substation structures or on poles and are operated from the ground. They are operated by means of a rod mounted on the supporting structure (see

Fig. 18-1   Single pole of high-voltage horn-gap air-break switch. Switch is in closed position. Switch is opened by tilting middle insulator to left. Operating bar is shown below. Long metal rods constitute horn gap. *(Courtesy James R. Kearney Corp.)*

Fig. 18-2   Arcs formed as a three-phase gang-operated switch is opened at the source end of a 132-kV 50-mile unloaded transmission line. Arcs measured 123 ft in length. Spacing between phases was 16 ft. Charging current interrupted was 20 amp in each line. *(Courtesy Hubbard and Co.)*

Fig. 18-3). In a three-phase circuit all three switches, one in each phase, are opened and closed together as a "gang," as the system is called.

Disconnect Switch   A disconnect switch is an air-break switch not equipped with arcing horns or other load-break devices. It therefore cannot be used to open circuits while current is flowing. The disconnect switch cannot be opened until the circuit in which it is connected is interrupted by some other means, such as an oil switch. If a disconnect switch should be opened while current is flowing in the line, an arc would likely be drawn between the blade and stationary contact which might easily jump across to the other conductor or to some grounded metal and cause a short. The hot arc would also melt part of the metal, thereby damaging the switch. Disconnect switches are operated in the same manner as air-break switches or by use of a switch hook as shown in Fig. 18-4.

Disconnect switches are used to isolate a dead section of line from a live line, a dead branch line from a feeder, a dead feeder from a substation, or a dead substation from a transmission line. In every case the purpose is merely to isolate a line which is already dead from a line or piece of apparatus for the purpose of making the disconnected line or apparatus dead electrically, thus making it safe to make repairs, tests, inspections, or changes after being connected to ground.

Switches for underground distribution circuits are usually installed in pad-mounted switchgear (Fig. 18-5). The switches are operated with the cabinet doors closed to provide protection for the lineman or cableman.

Customers with essential loads may require dual electrical sources with automatic transfer from one source to the other if one circuit is taken out of service.

Pad-mounted metal enclosed switchgear is built to house equipment to switch underground circuits (Fig. 18-6). Switchgear is equipped with switches in series with each underground source circuit and a bus-tie switch. These switches are power operated and controlled by an automatic control device for source transfer. The underground load feeder circuits are connected to the switchgear bus through power fuses and air-break switches with interrupter contacts to permit

Fig. 18-3 Gang-operated three-pole air-break switch with arc interrupter in closed position. *(Courtesy S. & C. Electric Co.)*

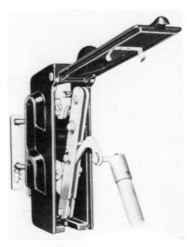

Fig. 18-4 A 7500-volt porcelain-housed primary disconnect. Knife blade is hinged on special insulator so that when switch is opened the knife blade is absolutely dead. Note use of special switch hook for operating the switch. *(Courtesy Line Material Industries.)*

energized operation while carrying load. The bus-tie switch will permit the load to be served from either source. The power source circuits are monitored by an automatic control device which in turn controls and sequences the power operation of the incoming circuits and bus-tie switches.

Upon loss of service from either of the two power sources—for a preset time sufficient to confirm permanent loss of service—the interrupter switch associated with the deenergized source automatically opens and the bus-tie switch closes to restore service to the affected feeders. When the normal power source is returned to service, the affected feeder bus section is retransferred to its normal power source, either automatically or manually depending on the operating mode which has been selected. Over-current relays are used to block closing of the bus-tie switch in the event of a bus fault. Paralleling of power sources is precluded by mechanical and electrical interlocks.

Oil Switch An oil switch is a high-voltage switch whose contacts are opened and closed in oil. The switch is actually immersed in an oil bath, contained in a steel tank as shown in Fig. 18-7. The reason for placing high-voltage switches in oil is that the oil may help to break the circuit when the switch is opened. With high voltages, a separation of the switch contacts does not always

Fig. 18-5 Pad-mounted 15-kV metal-clad switchgear with three-pole switches for 600-amp three-phase circuits. Switches are equipped with interrupting devices to permit operation while energized carrying load. Cabinet has fused taps for two three-phase underground circuits on opposite side. *(Courtesy S. & C. Electric Co.)*

break the current flow, because an electric arc forms between the contacts. If the contacts are opened in oil, however, the oil will help to quench the arc. Oil is an insulator and, therefore, helps to quench the arc between the contacts. Furthermore, if an arc should form in the oil, it will evaporate part of the oil, because of its high temperature, and will, therefore, partially fill the interruptors surrounding the switch contacts with vaporized oil. This vapor develops a pressure in the interruptors which assists in quenching or breaking the arc by elongating the arc.

The three lines of a three-phase circuit can be opened and closed by a single oil switch. If the voltage is not extremely high, the three poles of the switch are generally in the same tank (Fig. 18-8), but if the voltage of the line is high, the three poles of the switch are placed in separate oil

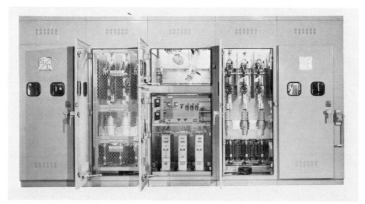

Fig. 18-6 Pad-mounted, outdoor, metal-enclosed switchgear rated 13.8 kV, 600 amperes, with dual electric source circuits equipped with power operated switches to provide for automatic transfer from one source to the other. *(Courtesy S. & C. Electric Co.)*

Fig. 18-7 Exposed view of contacts of a three-phase oil switch. These contacts are immersed in oil, which insulates the phases and assists in quenching the arcs formed when the switch contacts are opened. *(Courtesy Westinghouse Electric Corp.)*

tanks. The poles are placed in separate tanks to make it impossible for an arc to form between any two lines when the switch is opened or closed. An arc between lines would be a short circuit across the line and would probably blow up the tank

When an oil switch is to open the circuit automatically because of overload or short circuit, it is provided with a trip coil. This trip coil consists of a coil of wire and a movable plunger. In low-voltage circuits carrying small currents, this coil is connected in series with the line. When the current exceeds its permissible value, the coil pulls up its plunger. The plunger trips the mechanism, and a spring opens the switch suddenly.

In high-voltage circuits or in circuits carrying large currents, a current transformer is connected into the line and the secondary leads from this transformer supply the current to the trip coil of the oil switch (see Fig. 18-9). Since there is a fixed ratio between primary and secondary currents of the current transformer, the coil can be adjusted to trip at any predetermined value of current in the line.

The use of the current transformer on such circuits serves the double purpose of providing a small current for operating the tripping coil and of insulating the coil from the high voltage of the line.

Fig. 18-8 Three-phase oil circuit recloser for use on circuits operating at 15 kV or below. Operating mechanism can be operated with a switch stick to switch circuits on and off while energized carrying load. *(Courtesy McGraw-Edison Co.)*

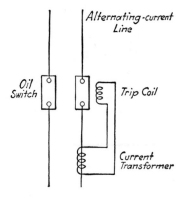

Fig. 18-9 Current transformer used to supply current to trip coil of oil switch.

Oil Circuit Recloser An oil circuit recloser is a type of oil switch designed to interrupt and reclose an alternating-current circuit automatically (Fig. 18-10). It can be made to repeat this cycle several times. Reclosers are designed for use on single-phase circuits (Fig. 18-11) or on three-phase circuits (Fig. 18-12).

A recloser opens the circuit in case of fault as would a fuse or circuit breaker. The recloser, however, recloses the circuit after a predetermined time (for hydraulically controlled reclosers about 2 sec). If the fault persists, the recloser operates a predetermined number of times (1 to 4) and "locks out," after which it must be manually reset before it can be closed again. If, however, the fault was of a temporary nature and cleared before "lock out" the recloser would reset itself and be ready for another full sequence of operations.

Temporary faults arise from wires swinging together when improperly sagged, from tree branches falling into the line, or from lightning surges causing temporary flashover of line insulators.

A recloser is unlike a fuse link because it distinguishes a temporary from a permanent fault. A fuse link interrupts temporary and permanent faults alike. Reclosers give temporary faults repeated chances, usually four, to clear or be cleared by a subordinate device like a fuse or sec-

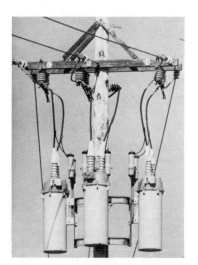

Fig. 18-10 Three single-phase oil circuit reclosers installed in 13,200Y/7620-volt three-phase line. Reclosers have a current rating of 560 amp and interrupting rating of 8000 amp. Also note installation of lightning arresters. (*Courtesy McGraw Edison Co.*)

Fig. 18-11 Single-phase oil circuit recloser removed from oil tank. Recloser shown has a maximum voltage rating of 14.4 kV and a maximum current-interrupting rating of 1250 amp.

Fig. 18-12 Three-phase recloser removed from oil tank. Recloser shown is of the hydraulically controlled type. Recloser has a maximum voltage rating of 14.4 kV and a current-interrupting rating of 4000 amp. *(Courtesy Line Material Industries.)*

tionalizer. If the fault is not cleared after four operations, the recloser recognizes it as a permanent fault and operates to lock out and leave the line open.

A recloser can be magnetically operated by a solenoid connected in series with the line. Minimum trip current is usually twice the normal load current rating of the recloser coil. The operations are performed by a hydraulic mechanism and a mechanical linkage system. When the fault current reaches twice normal line current, the increased magnetic field pulls the plunger down into the coil. As the plunger moves downward, the lower end trips the contact assembly to open the contacts and break the circuit. As soon as the contacts are open, there is no more current in the coil to hold them open, so a spring closes the mechanism and reenergizes the line.

This and succeeding operations are depicted on the oscillogram in Fig. 18-13 as it cycles through to lockout. Note that the line is held open for approximately 60 cycles between reclosures. This provides time for sectionalizers to operate while no current is flowing in the line. During lockout the contacts are held open until the recloser is reset manually. If a temporary fault clears before the recloser locks out, all mechanical operations cease and the recloser becomes ready to cycle over again when the next fault occurs.

Examination of the oscillogram (Fig. 18-13) also shows that the first two openings of the recloser are faster than the last two. The first two openings occur in about 2 Hz each, whereas the last two occur in about 5 Hz each. This is arranged so that in case the fault takes place on a branch line or tap protected with a fuse cutout, the recloser will try to remove the fault before the fuse has

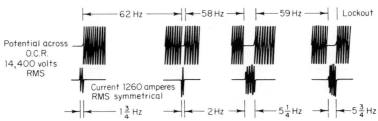

Fig. 18-13 Oscillogram of single-phase recloser operation interrupting 1260 amp at 14.4 kV. Recloser is set for two fast reclosures of approximately 2 Hz and two delayed reclosures of approximately 6 Hz. *(Courtesy Line Material Industries.)*

time to melt. If the recloser does not clear the fault after two trials, it will give the fuse time to blow on the longer 5-Hz openings and disconnect the faulted branch. The fact that the fault did not clear with two openings indicates that it was of a permanent nature and could be removed only by disconnecting the defective branch line.

The objective of recloser and fuse coordination is to eliminate outages due to temporary faults, and in the case of permanent faults, to restrict outages to a minimum number of customers for the shortest possible time. This requires satisfying the following conditions whenever possible:

1. Fuses should be protected by a recloser fast trip to clear transient faults.

2. Reclosers and/or fuses in series should be sized such that the protecting device operates before the protected device over the entire range of fault currents.

3. Fuses should be sized as large as fault current and source side protective devices will permit in order to avoid blowing fuses on overload.

In determining time-current characteristic (TCC) curves, tests are conducted at 25°C. ambient without any preloading. In accordance with NEMA Standards, minimum melting curves are

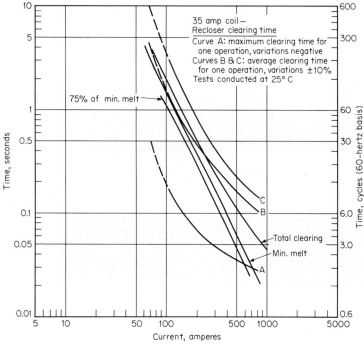

Fig. 18-14   Coordination curves for 35-amp recloser and 15T fuse link.

determined by taking average test values minus the manufacturer's tolerance, while total clearing curves represent average melting time plus the manufacturer's tolerance plus the arcing time.

Since fuse links are thermal devices, it should be evident that preloading and/or ambient temperatures higher than 25°C. will result in links blowing faster than indicated on the published curves. To account for operating variables, it is recommended that 75 percent of the minimum melting curve be used in coordination work. It is generally agreed that for tin fuse links, this 25 percent margin consists of 10 percent for preloading, 5 percent for extraordinary ambient temperatures, and 10 percent to prevent damaging the fuse link. The concept of damaging a fuse link provides an explanation for fuses blowing on "nice, sunny days for no apparent reason." Fuse link damage is caused by a fault which lasts long enough for the link to reach its melting temperature, but not long enough for the necessary heat of fusion to be added to completely melt the link. When the link cools, the physical dimensions have changed and consequently the time-current characteristics have been altered.

Recloser curves are plotted to average values with manufacturing tolerances resulting in variations of ± 10 percent in current or time, whichever is greater. The hydraulic fluid used provides consistent timing for ambient temperatures above 0°C. Below this, operations may be slightly slower, but coordination with fuses will be maintained since they are also affected by the low ambient temperature.

To examine fuse-recloser coordination, assume the recloser is the source side or protected device and the fuse is the load side or protecting device. The curve shown in Fig. 18-14 will illustrate coordination between a 35A Kyle Type "H" recloser and a "15T" fuse link. Four operations to lockout may be selected using a maximum of two curves. If fast tripping is desired, select at least one operation on the "A" curve with the remainder on the "B" or "C" curve. The "C" curve is chosen since it allows more room for coordination. To determine the number of fast trips, the following should be considered. First, the purpose of a fast trip is to prevent transient faults from becoming permanent ones and also to prevent fuse blowing due to transient faults. Since the momentary interruptions resulting from using two fast trips should not be objectionable, it should be used to provide better fuse protection. By comparing the maximum clearing time for a recloser fast trip to 75 percent of the fuse melting time, it is evident that a 35A Type "H" recloser can protect a "15T" fuse with one fast trip for faults up to 550A. For two fast trips it would obviously be too conservative to compare twice the maximum clearing time with 75 percent of the fuse minimum melting time since the reclosing interval has a cooling effect on the fuse. For a reclosing

Fig. 18-15   Typical external view of single-phase sectionalizer. *(Courtesy Line Material Industries.)*

interval of 1–2 sec, the cooling factor can be handled by comparing twice the average clearing time for a fast trip to 75 percent of the fuse minimum melting time. Thus, for two fast trips, a 35A Type "H" recloser can protect a "15T" fuse for faults up to 350A.

Thus, the fault duty at the fuse location determines the degree of protection afforded by the recloser fast trips.

Sectionalizer   A sectionalizer is another type of oil switch designed to isolate faults on distribution circuits in conjunction with reclosers. Sectionalizers are usually installed on taps or branches off main lines. While in appearance (Fig. 18-15) a sectionalizer is similar to a recloser, it should not be confused with it because it does not interrupt a fault current. In fact, a sectionalizer waits until the recloser has opened the line and then sectionalizes the faulty line while the line is still open and no current is flowing. It will be remembered from Fig. 18-13 that the recloser holds the line open for about 1-sec intervals (60 Hz) between reclosures. It is during these periods that the sectionalizer functions.

When a fault occurs behind the sectionalizer, the recloser will operate. If the fault is of a permanent nature, the sectionalizer will count the number of operations of the recloser and trip and lock itself out after a predetermined number, usually three, of recloser operations. The recloser continues on its fourth operation and restores service up to the sectionalizer. A sectionalizer must therefore always be backed up by a recloser of proper size.

Fig. 18-16 Internal view of single-phase section-alizer shown in Fig. 18-15. *(Courtesy Line Material Industries.)*

Fig. 18-17 Three-phase vacuum recloser installed to protect an electric distribution circuit from failure of an underground cable tap. The distribution circuit operates at 13,200Y/7620 volts. *(Courtesy McGraw-Edison Co.)*

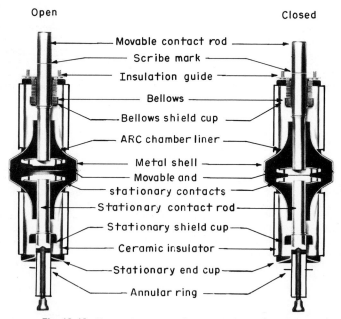

Open   Closed

Movable contact rod

Scribe mark

Insulation guide

Bellows

Bellows shield cup

ARC chamber liner

Metal shell

Movable and
stationary contacts

Stationary contact rod

Stationary shield cup

Ceramic insulator

Stationary end cup

Annular ring

Fig. 18-18 Vacuum interrupter. *(Courtesy McGraw-Edison Co.)*

Fig. 18-19   Electronic controls for recloser. Operating characteristics can be selected by switches and interchangeable components on front of panel. *(Courtesy McGraw-Edison Co.)*

The sectionalizer illustrated in Fig. 18-16 consists of a set of contacts, a spring-controlled trip, and a solenoid operating coil also connected in series with the line. All are immersed in a bath of insulating oil and housed in a tank.

Vacuum Reclosers   Vacuum reclosers operate in the same manner as oil circuit reclosers and can be used in the electric distribution system in place of oil circuit reclosers. A vacuum recloser installation is illustrated in Fig. 18-17. Vacuum medium has an outstanding dielectric strength and develops rapid recovery of the dielectric strength in the arc path between the separating contacts following a current zero (Fig. 18-18). The vacuum interruptor limits the arc time to a short period, usually interrupting the current at the first current zero. The contact travel is short, minimizing the energy requirements of the operating mechanism.

The vacuum reclosers will operate for long periods of time without maintenance. Contact erosion occurs each time the recloser operates. The amount of contact erosion is directly related to the current magnitude when the operations occur. When the contacts have eroded approximately ⅛ in, the vacuum interrupter should be replaced. The recloser mechanism should be kept clean and checked for proper operation at regular intervals. The insulators should be kept clean and inspected for damage when other routine maintenance is performed.

Electronic Controlled Reclosers   In the hydraulically controlled three-phase reclosers it is normally necessary to un-tank the unit to change the characteristics, such as number of fast and retarded trips, total number of operations, time-current trip characteristics, or the reclose times. However, in the electronically controlled reclosers, the changing of all of the functions of the control is accomplished by easy replacement of minimum trip resistors and time-current plugs, as shown in Fig. 18-19. A much finer and repeatable control function is obtained since the temperature of the oil does not have to be taken into account.

Section **19**

Voltage Regulators

Voltage at the customer's premises must be maintained within a range that will permit the equipment to operate properly. Voltage regulators are used to vary the alternating-current supply or source voltage to the customer by the proper amount to keep the voltage within the limits desired. Standards specify that 120-volt alternating-current nominal service must be maintained between 114 and 126 volts.

Tap-Changing Transformers Distribution substation power transformers may be equipped with tap-changing underload (TCUL) equipment (Fig. 19-1). The distribution substation transformers are normally three-phase with delta-connected primary windings and wye-connected secondary windings to provide a source for three-phase four-wire grounded-neutral alternating-current distribution feeder circuits (Fig. 19-2). The windings of the transformer illustrated are constructed to reduce the subtransmission voltage to the distribution system primary voltage level. The three distribution voltage windings of the transformer have taps near the neutral, or common connection. The taps in each winding connect to switch contacts making it possible to change the number of turns in the transformer secondary windings to vary the substation distribution feeder bus voltage as desired.

The tap-changer switches in the low-voltage windings are constructed to permit their operation underload without interrupting the circuits. The high-voltage tap changer is designed for deenergized operation. The tap changer is set to the proper position before the transformer is energized to coordinate with the subtransmission system voltage level at the substation. The rated voltage for each tap in the low-voltage winding is tabulated in the chart shown in Fig. 19-2. The ratio between high- and low-voltage windings is determined by the rated voltage of the taps the transformer operates at. For example, if the high-voltage winding is set on no-load tap position 3, rated 69,000 volts, and the low-voltage winding is set on underload tap position lower 4, rated 13,455 volts, the ratio of the windings will be 69,000/13,455, or 5.13/1. If the subtransmission voltage is 68,500 volts, the distribution primary voltage will be 68,500/5.13, or 13,353Y/7710 volts for the transformer tap position stated.

The transformer tap changer can be controlled manually by an operator using control switches or automatically by voltage-sensitive devices. The automatic regulating equipment can adjust the primary distribution voltage for various distribution feeder line loadings and associated voltage drop with the aid of line-drop compensator circuits. Voltage regulation and the control circuits for a transformer tap changer are described in other sections. If the distribution primary feeder circuits are short, the substation power transformer tap-changing equipment may be adequate to regulate the voltage. Power transformers with tap-changing underload equipment normally have a capacity to vary the voltage ±10 percent from rated voltage.

Feeder Voltage Regulators The function of a feeder voltage regulator is to maintain constant voltage on an alternating-current primary distribution feeder circuit with variations in load. The feeder voltage regulator can be three-phase (Fig. 19-3), or single phase (Fig. 19-4). If the

Fig. 19-1 Distribution substation transformer with tap-changing underload (TCUL) equipment. Tap-changing switches are located in upper compartment on side of transformer and control equipment is located in lower compartment.

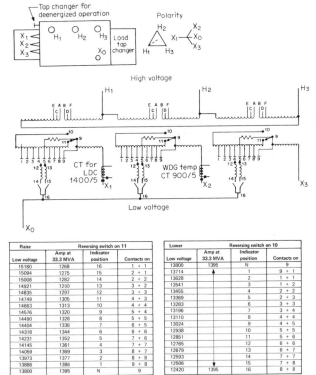

Raise	Reversing switch on 11			Lower	Reversing switch on 10		
Low voltage	Amp at 33.3 MVA	Indicator position	Contacts on	Low voltage	Amp at 33.3 MVA	Indicator position	Contacts on
15180	1268	16	1 + 1	13800	1395	N	9
15094	1275	15	2 + 1	13714		1	9 + 1
15008	1282	14	2 + 2	13628		2	1 + 1
14921	1290	13	3 + 2	13541		3	1 + 2
14835	1297	12	3 + 3	13455		4	2 + 2
14749	1305	11	4 + 3	13369		5	2 + 3
14663	1313	10	4 + 4	13283		6	3 + 3
14576	1320	9	5 + 4	13196		7	3 + 4
14490	1328	8	5 + 5	13110		8	4 + 4
14404	1336	7	6 + 5	13024		9	4 + 5
14318	1344	6	6 + 6	12938		10	5 + 5
14231	1352	5	7 + 6	12851		11	5 + 6
14145	1361	4	7 + 7	12765		12	6 + 6
14059	1369	3	8 + 7	12679		13	6 + 7
13973	1377	2	8 + 8	12593		14	7 + 7
13886	1386	1	9 + 8	12506		15	7 + 8
13800	1395	N	9	12420	1395	16	8 + 8

Fig. 19-2 Schematic diagram of three-phase distribution substation power transformer windings. Transformer is rated 20/26.7/33.3 mVA high-voltage winding 69,000 Δ volts, low-voltage winding 13,800Y/7970 volts.

Fig. 19-3 Three-phase step-type primary distribution feeder circuit voltage regulator installed in a distribution substation. Regulator is rated 13,800Y/7970 volts. Switches in structure permit regulator to be bypassed and isolated.

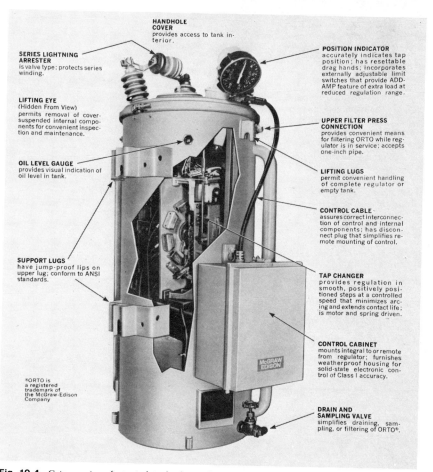

HANDHOLE COVER
provides access to tank interior.

SERIES LIGHTNING ARRESTER
is valve type; protects series winding.

LIFTING EYE
(Hidden From View)
permits removal of cover-suspended internal components for convenient inspection and maintenance.

OIL LEVEL GAUGE
provides visual indication of oil level in tank.

SUPPORT LUGS
have jump-proof lips on upper lug; conform to ANSI standards.

*ORTO is
a registered
trademark of
the McGraw-Edison
Company

POSITION INDICATOR
accurately indicates tap position; has resettable drag hands; incorporates externally adjustable limit switches that provide ADD-AMP feature of extra load at reduced regulation range.

UPPER FILTER PRESS CONNECTION
provides convenient means for filtering ORTO while regulator is in service; accepts one-inch pipe.

LIFTING LUGS
permit convenient handling of complete regulator or empty tank.

CONTROL CABLE
assures correct interconnection of control and internal components; has disconnect plug that simplifies remote mounting of control.

TAP CHANGER
provides regulation in smooth, positively positioned steps at a controlled speed that minimizes arcing and extends contact life; is motor and spring driven.

CONTROL CABINET
mounts integral to or remote from regulator; furnishes weatherproof housing for solid-state electronic control of Class I accuracy.

DRAIN AND SAMPLING VALVE
simplifies draining, sampling, or filtering of ORTO*.

Fig. 19-4 Cutaway view of a typical single-phase pole-type distribution feeder voltage regulator. Regulator is rated 167 kVA. (*Courtesy McGraw-Edison Co.*)

Fig. 19-5 Three-phase step-type primary distribution feeder circuit voltage regulator installed in a long circuit remote from a substation.

Fig. 19-6 Single-phase pole-type distribution feeder voltage regulator installed in distribution circuit.

substation power transformer is not manufactured with tap-changing underload equipment, or the primary distribution feeder circuit is long and heavily loaded, a voltage regulator will be installed to maintain proper voltage (Fig. 19-5).

A three-phase voltage regulator basically consists of three single-phase regulators installed in a single tank. Three single-phase regulators are often used to control the voltage on a three-phase distribution circuit. One single-phase voltage regulator will be used on long single-phase taps to the main three-phase distribution circuit (Fig. 19-6).

The single-phase feeder voltage regulator, or step voltage regulator, is essentially a transformer having a low-voltage secondary winding connected in series with the line and so arranged that the number of turns in the winding can be varied. The voltage is changed by changing the number of turns in the secondary winding. This is accomplished by means of a rotary tap-changing switch. A schematic diagram of the connections is shown in Fig. 19-7. The primary winding is connected across the line, while the secondary winding with its taps is connected in series with the line.

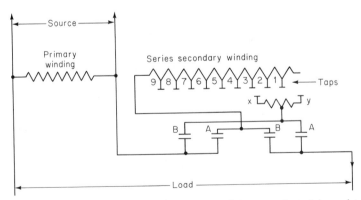

Fig. 19-7 Connection diagram of step voltage regulator. A and B are reversing switches, and 1 to 9 are transfer switches which contact X and Y. When A switches are closed, the line voltage is raised, and when B switches are closed, the line voltage is lowered.

When the A switches are closed, the series winding will extend the primary winding and add to the line voltage in an amount depending on which transfer switches are contacted by X and Y. When the B switches are closed, the voltage in the series winding will be opposite in direction to the primary and so reduce the load voltage.

Regulators generally provide for a variation in load voltage from 10 percent below to 10 percent above normal line voltage. This is usually accomplished in 32 steps of ⅝ percent each. The tap-changing switch is motor driven and is immersed in oil. The reversing switch provides the 10

Fig. 19-8   Pad-mounted primary distribution feeder voltage regulator with compartment door open. High-voltage cables connect to the regulator with elbow connectors. Voltage-sensitive control device case is open displaying voltage control dials. *(Courtesy Allis Chalmers Manufacturing Co.)*

percent regulation on either side of the neutral position. The use of vacuum switches when changing taps on the voltage regulators has reduced maintenance. A tap changer, using a vacuum interrupter, can operate a million times while carrying large current before it is necessary to complete major maintenance. Single-phase step voltage regulators are manufactured in pad-mounted tanks with provision for cable elbow connectors providing dead front construction (Fig. 19-8).

The control transformer illustrated in the schematic diagram (Fig. 19-9) provides a voltage source for the regulator control circuits. The control transformer is an autotransformer that supplies half voltages above and below ground to the control circuits (Fig. 19-10).

Two diode rectifiers are connected to this center tapped supply: one supplies power to the circuit, and the other supplies a sensing voltage through a resistive divider.

Reference voltages are generated from a Zener reference and a resistive divider. These are applied to one input each of two detectors (Comparators). The difference in the two reference voltages is the basis of the bandwidth of the control. The sensed voltage is filtered and applied to the other input of both detectors.

One output of each of the detectors goes low if the sensed voltage is out of band on the low side. This initiates the timer and turns on the OUT-BAND light-emitting diode (LED). The other output of the two detectors goes high when the voltage is out of band on the high side. This is ANDED with the output of the timer which goes high after the predetermined delay and the associated Triac fires.

When the control selector switch is in the automatic position, the tap changer-motor (Fig. 19-11) returns are connected to the control output Triacs. When the selector switch is in either the manual raise or manual lower position, the tap-changer motor is disconnected from the automatic

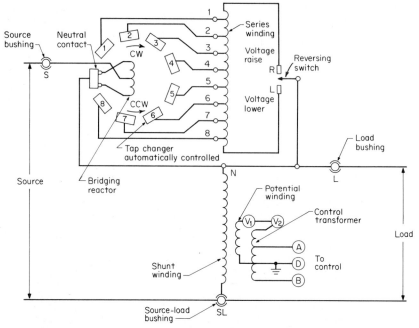

Fig. 19-9   Single-phase regulator power circuit schematic diagram. *(Courtesy McGraw-Edison Co.)*

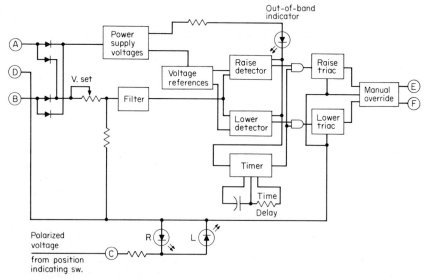

Fig. 19-10   Regulator control circuit schematic diagram. *(Courtesy McGraw-Edison Co.)*

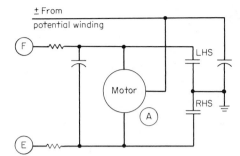

Fig. 19-11 Regulator motor circuit schematic diagram. *(Courtesy McGraw-Edison Co.)*

circuit and the proper motor circuit is connected to ground to drive the tap changer in the direction selected.

In automatic operation, the voltage from the potential winding is compared with the voltage-level setting on the control. If the voltage falls outside the band, the time-delay sequence is initiated. At the end of the delay time (30 sec is standard), the proper Triac is fired, driving the tap changer in the direction that will correct the voltage. The block diagram of the control (Fig. 19-10) in conjunction with the motor circuit diagram (Fig. 19-11) shows this sequence of operation.

Time delay, which can be altered by changing a resistor, is set by an R/C timing circuit. A capacitive discharge circuit rapidly discharges the capacitor if the input to the timer goes high, resetting the timer. Thus, the timer is quickly reset when voltage goes back in band. Time delay must be continuous without resetting before a tap change is accomplished.

A holding switch—either LHS or RHS (Fig. 19-11)—seals in shortly after the motor starts to run, thus assuring the completion of the tap change. Resistors in the Triac return from the motor limit Triac current to an acceptable level under all conditions. The Triacs only have to initiate a tap change, at which time the motor torque requirement of the tap changer is low. The holding switch takes over almost immediately after the start of a tap change and carries the motor current for the higher torque required to complete the operation.

A pair of microswitches (Fig. 19-12) are actuated by the tap changer reversing switch. When the reversing switch actuating segment is centered (the tap changer is in the neutral position), the potential voltage is fed through a series connection of the normally closed contacts of the microswitches to the neutral lamp mounted on the tank wall. As the tap changer moves off the neutral position, the reversing switch segment actuates one of the microswitches (which microswitch depends on whether the segment is moving in a lower or raise direction). On one side, the supply is half-wave rectified by a diode to form positive pulses and, on the other side, to form negative pulses. These positive and negative pulses are fed through a current-limiting resistor to a pair of parallel reverse-connected LEDs: Positive pulses actuate one; negative pulses actuate the other. Thus, one LED—designated L—remains on as long as the tap changer is in the lower position. The other LED—designated R—remains on as long as the tap changer is in the raise position.

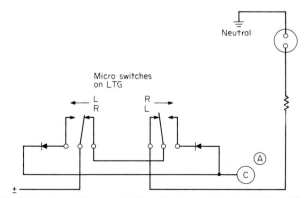

Fig. 19-12 Regulator indicator circuit schematic diagram. *(Courtesy McGraw-Edison Co.)*

Fig. 19-13 Substation-type induction feeder-voltage regulator. *(Courtesy Westinghouse Electric Corp.)*

Induction Voltage Regulator   An induction feeder-voltage regulator is essentially a voltage transformer (Fig. 19-13). The primary or shunt winding, which is also the high-voltage winding, is connected across the line in the same way as in an ordinary transformer (see Fig. 19-14). The secondary or series winding is the low-voltage winding and is connected in series with the line. The shunt winding is arranged so that it can be rotated on an axis. The series winding is stationary. Figure 19-15 shows the windings in their relative positions.

The voltage induced in the series winding depends upon the position of the shunt winding. When the shunt winding is turned in one direction, the voltage induced in the series winding is large and is in such a direction as to add to the voltage of the line, thus raising the voltage of the feeder. When the shunt winding is turned in the opposite direction, the induced voltage in the series winding has the same value, but this time the voltage will be in such a direction as to reduce the feeder voltage. Any intermediate position of the shunt winding between these limits will increase or decrease the feeder voltage by a correspondingly smaller amount. The curve shown in Fig. 19-16 gives the values of voltage above and below 2300 volts for different positions of the primary coil. It will be noted that the regulator can boost the feeder voltage to 2600 volts or reduce it to 2000 volts. This amounts to a boost and a reduction, respectively, of 300 volts. In case of a large load on the feeder, the coil will be rotated in the direction to boost the feeder voltage, thereby bringing it up to normal. In case of light load or no load, the coil will be rotated in the opposite direction. The voltage induced in the secondary will then buck the line voltage and thus keep it down to its normal value.

The shunt coil may be rotated by hand or automatically by means of a motor. The latter method is practically the only method used. In the automatic control the regulator is provided with a

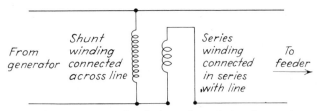

Fig. 19-14   Connections of single-phase induction feeder-voltage regulator.

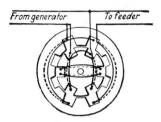

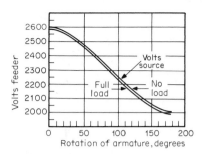

Fig. 19-15 Arrangement of shunt and series windings in a single-phase induction voltage regulator. *(Courtesy General Electric Co.)*

Fig. 19-16 Curves showing boosting and lowering of feeder voltage by induction voltage regulator. *(Courtesy General Electric Co.)*

contact-making relay, a double-pole double-throw switch, and a motor mounted on the regulator and geared to the axis of the rotating coil. This equipment is shown in Fig. 19-17.

In general, the operation is as follows: When the voltage of the feeder is normal, the plunger of the contact-making relay is in its middle position, the switch in the motor line is open, and the regulator is neither bucking nor boosting the voltage. When the voltage of the feeder drops, the relay plunger drops. This closes the switch in the motor circuit to make the motor revolve in a direction to boost the voltage. The motor revolves until the voltage is raised to the value which will lift the relay plunger to its medium position. If the voltage should rise above normal, the relay plunger would rise above its medium position, thereby closing a contact above. Closing this contact closes the motor switch in the opposite direction, thereby reversing the direction of the

Fig. 19-17 Single-phase automatic induction-voltage regulator for indoor and outdoor service, with control cabinet door open to show accessibility of all control devices. *(Courtesy General Electric Co.)*

Fig. 19-18 Voltage regulator by-pass switch mounted on substation steel structure above voltage regulator. Interrupter on switch is designed to deenergize the magnetizing current. Voltage regulator must be in neutral position to operate switch with circuit energized.

motor. The regulator will now buck the voltage and lower it to normal value. In every case, the shunt winding is revolved until the voltage is brought to its normal value.

Voltage Regulator By-Pass Switches   By-pass switches for a voltage regulator are shown in Fig. 19-18. The voltage regulator must be in the neutral position to operate the by-pass switch with the circuit energized. When the voltage regulator is in the neutral position the series winding output voltage is zero. When the lineman operates the by-pass switch to deenergize the regulator switch, the switch contacts short circuit the series winding of the voltage regulator to maintain the continuity of the feeder circuit and then in sequence open the contacts in series with the primary winding interrupting the magnetizing current. If the switches are operated without the regulator in the neutral position, zero voltage output from the series winding, a short circuit will develop creating a dangerous condition.

Transmission Tower Erection

The general procedure for erecting transmission towers is similar to that for poles, but towers present more problems. The towers require foundations; they are higher and heavier and are therefore more difficult to erect; the conductors are larger and the spans longer, making wire stringing a more difficult job.

Only the general procedure of the construction process will be given. Illustrations will be used to show the various steps. The discussions will be limited to a description of the operations shown in the illustrations.

The order of the operations in tower erection may be briefly outlined as follows:

1. Clearing right of way
2. Installing tower footings
3. Grounding tower base
4. Erecting transmission towers
5. Insulator installation

These operations will be illustrated in the pages that follow, and comments will be made on the operations performed.

Clearing Right of Way The right-of-way must be cleared of all obstructions that will interfere with the operation of the electric transmission line. In scenic and residential areas, clearing of natural vegetation must be limited. Trees, shrubs, grass, and topsoil, which are not cleared, should be protected from damage during the construction of the tower line. At road crossings, or other special locations of high visibility, right-of-way strips through forest and timber areas should be cleared with varying alignment to comport with the topography of the terrain. Where rights-of-way enter dense timber from a meadow or other clearing, trees should be feathered in at the entrance of the timber for a distance of 150 to 200 yards. Small trees and plants should be used for transition from natural ground cover to larger areas (Fig. 20-1).

A strip of land must be cleared on each side of the centerline of the transmission line by cutting and/or trimming the trees and brush. All "danger trees," trees considered to be hazardous to the transmission line, must be removed (Fig. 20-2). All trees, brush, stumps, and other inflammable material, except grass and weeds, must be removed from the right-of-way (Fig. 20-3). All trees and brush shold be cut 3-in or less from the ground line so that the passage of trucks and tractors will not be hindered. The trees and brush cut must be disposed of by chipping and spreading, burning, or hauling away. Disposal of the debris by burning, or otherwise, must be accomplished in accordance with state and local laws and regulations without creating a hazard or a nuisance. The right-of-way should be treated with a chemical spray to retard the growth of brush or trees that could endanger the operation of the transmission line (Fig. 20-4).

 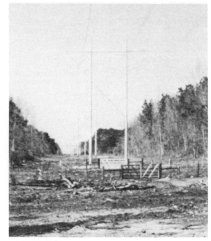

Fig. 20-1 Right-of-way cleared for transmission line. The contour of the land and the environment must be taken into consideration by the use of selective clearing techniques. *(Courtesy Asplundh Tree Expert Co.)*

Fig. 20-2 High-voltage alternating-current transmission line with right-of-way cleared. *(Courtesy A. B. Chance Co.)*

Installing Tower Footings   Tower sites must be properly graded in accordance with the specifications (Fig. 20-5). Usually the slope of the grade must not be greater than 3:1. All topsoil should be removed prior to grading the tower location. After the tower construction has been completed and the footings backfilled, the topsoil should be replaced. Any excess graded material must be removed from the right-of-way.

The excavations must be made according to the dimensions and depths shown on the drawings allowing sufficient space for proper construction of forms and installation of caissons as required (Figs. 20-6 through 20-12).

Excavations may require the use of equipment such as a wellpoint for dewatering the excavation. The excavations must be adequately braced and shored to guard against movement or settlement of adjacent structures, utilities, roadways, or railroad facilities.

Reinforcing bars and embedded stub angles must be installed properly. The reinforcing steel must be properly placed and firmly wired before the concrete pouring is started (Fig. 20-13). Exposed reinforcement steel for bonding future construction must be protected from corrosion.

Fig. 20-3   Trees and brush cut with power saws are pushed into windrows on edge of right-of-way by tractor equipped with bulldozer rake. *(Courtesy Union Electric Co.)*

Fig. 20-4 Right-of-way is sprayed with chemicals to control unwanted woody growth. Chemicals and methods must be tailored to the season of the year and the varying conditions of the area to meet ecological needs. *(Courtesy Asplundh Tree Expert Co.)*

Reinforcing steel ties, or spacers, should be 1 in or more from the finished surface of the exterior or exposed portion of the footings. All accessories in contact with the formwork, or the soil, must be galvinized steel, plastic, or some other corrosion-resistant material.

Piles must be driven to the depth specified in the foundation drawing. Driving must be done with fixed leads which will hold the pile firmly in position and alignment. Suitable anvils should be used to prevent excessive damage to the piles. Driving of the piles must be continuous without intermission until the pile has been driven to the specified depth. The tops of the piles must be cut off true and level at the proper elevation.

Concrete for the footings must be thoroughly mixed and in a uniform workable state when placed. Concrete must be placed before the initial set has occurred which can be accomplished by pouring the concrete in the forms within 1¼ hours or before the drum of agitating transport equipment has revolved 300 revolutions (Fig. 20-14). Before placing concrete, the earth foundation must be compact and granular soil should be thoroughly moistened by intermittent sprinkling. The surface, however, must not be muddy or frozen when the concrete is poured. Rock surfaces on which concrete is to be placed must be thoroughly cleaned of dirt, debris, and disin-

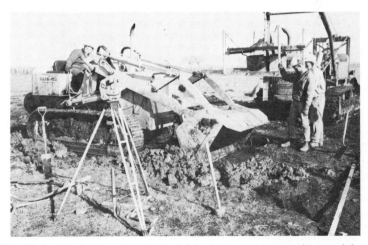

Fig. 20-5 Workmen grading tower site with a track driven tractor. Note transit used to properly locate footing. *(Courtesy L. E. Myers Co.)*

tegrated material by a high-velocity jet of air or water. All standing water should be removed from the depressions in the area where the concrete is to be placed. The steel reinforcing rods, imbedded steel, and forms must be cleaned of all loose rust scale, paint, mud, dirt, or dried mortar. The forms should be oiled before the concrete is poured.

The concrete must be conveyed with the proper consistency to the place of final deposit as rapidly as practicable by methods which will prevent segregation, loss of ingredients, rehandling, or premature coating of reinforcing steel or forms (Fig. 20-15). The concrete must be consolidated by use of high-frequency internal vibrations and hand spading and rodding while it is poured. The tops of all foundations are troweled smooth and all irregular projections should be removed when the forms have been removed. Any exposed reinforcing steel, broken corners, or edges should be cleaned and painted with an epoxy resin bonding compound (Fig. 20-16). The concrete foundations must be properly cured. Curing of the concrete usually requires seven consecutive days, or more, with the temperature of the air in contact with the concrete at 50°F or greater,

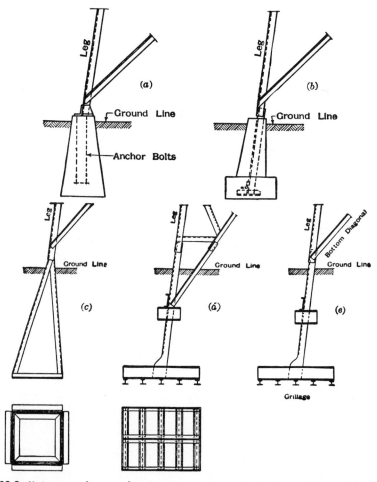

Fig. 20-6 Various types of tower anchors. Since towers are not massive, footings must be provided to anchor them firmly in the ground. Anchors support the tower and prevent it from blowing over when the lines are subjected to sleet and wind. The anchors shown in (a) and (b) consist of tapering masses of concrete in which anchor bolts or stub angles are embedded. The anchors shown in (c), (d), and (e) are merely extensions of the tower structure embedded in the earth. These extensions are enlarged so that a large bearing surface is presented. *(Courtesy American Bridge Co.)*

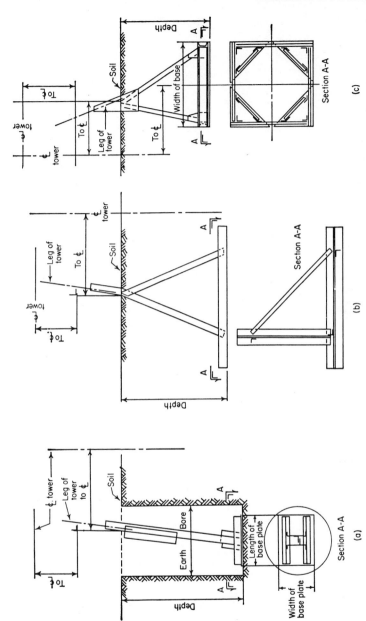

Fig. 20-7 Tower anchoring systems. (a) Single-post earth anchor, used with tangent and small-angle towers in line and at sites where earth augering equipment can economically and feasibly be employed. (b) Triped earth anchor, used for normal soil conditions with line and angle towers. (c) Quadruped earth anchor, used with dead-end, line, and heavy-angle towers in normal soil conditions. The soil-bearing capacity of this anchor is doubled when it is encased in a concrete footing. (*Courtesy Blaw-Knox Co.*)

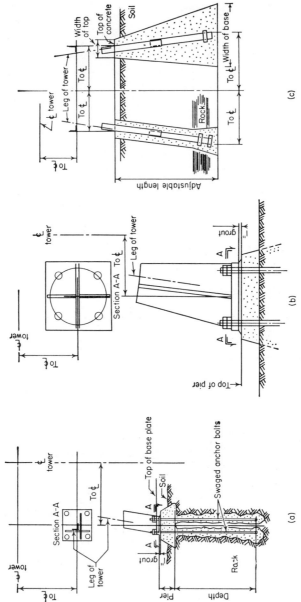

Fig. 20-8 Tower anchoring systems. (*a*) Rock anchor, used where welded base assembly is preferred and at sites where elevation of subsurface rock is known before erection. Insertion of the anchor's narrow shaft in drilled rock eliminates rock removal by blasting or other costly means. Swedged anchor bolts in the shaft increase the bond between steel and grout. (*b*) Welded base assembly, an easy-to-erect anchor for use where structural loads require the strength of a welded unit in good soil conditions. (*c*) Adjustable rock anchor, concrete anchor, used for applications where economical concrete anchorage is desirable and where rock may be encountered. (*Courtesy Blaw-Knox Co.*)

Fig. 20-9 Holes being dug for a 132,000-volt transmission-line tower. Note the "back-hoe" caterpillar power excavator being used. A clamshell digger could be used to equal advantage. The holes are about 15 ft square and 9 ft deep. In favorable soil such a hole can be dug in about ½ hr. This particular footing is being constructed for a 50-ft four-legged tower. (Distance from ground surface to lowest crossarm is 50 ft.) Three holes have been completed while the fourth is about half done. This tower is located at the dead end of the line adjacent to the bulk substation it will feed. The legs or anchors of this tower will be embedded in huge blocks of concrete to sustain the unbalanced pull which the line wires place on a dead-end tower. Note that one concrete form is already in place in the hole adjacent to the power digger. A second one is being assembled in the right background. These forms are bolted together. They are unbolted and removed after the concrete has hardened. *(Courtesy Wisconsin Electric Power Co.)*

Fig. 20-10 Excavating footing for steel tower anchor on 345,000-volt line by use of "back-hoe" caterpillar power excavator. Towers on this line will be 148 ft high and will carry six 1.6-in-diameter ACSR conductors hung from 18 suspension insulators. The average span will be 1200 ft. *(Courtesy American Electric Power System.)*

Fig. 20-11   Steel towers using footings of the grillage type require large footing holes. An extra-large auger is therefore needed. The foundation digger shown employs an 84-in-diameter auger bit. The foundation digger is mounted on a 45,000-lb GVW 6- by 6-ft truck. This equipment can handle up to 96-in (8-ft) augers and place over 15 tons of hydraulically powered thrust (weight of truck rear) on it. Hydraulic stabilizers at the rear of truck level and steady the rig during digging. The controls are hydraulic. *(Courtesy Idaho Power Co.)*

Fig. 20-12   Lining up tower footings with a template. The template, which has the same dimensions as the lower end of the tower, must be used to give correct spacing and slope to the anchors which will be embedded in the concrete. The anchors are bolted to this template and are held in their proper position until the concrete is poured and has hardened. Complete template and all four concrete forms are in place all ready for the concrete. A tower anchor or stub is bolted to each corner of the template. Note the ends of the template sides extending beyond the limits of the tower footings at the corners. This extra length is used for lining up towers with a larger base than the one shown. *(Courtesy Wisconsin Electric Power Co.)*

Fig. 20-13 Reinforcing steel in place and wired. Concrete footing is partially poured.

Fig. 20-14 Concrete is being poured into forms for transmission-line tower footing from transport agitating equipment. *(Courtesy L. E. Myers Co.)*

Fig. 20-15   Helicopter delivering concrete to hopper. A ⅛ yd³ 2400-lb bucket was used for delivery. Unloading was done by suspending bucket over hopper and releasing concrete with a ground line. Several lifts per tower were needed for the footings. *(Courtesy Southern California Edison Co.)*

before the steel for the tower is attached. Backfilling of dirt around the completed tower footing should be completed as soon as practicable after the concrete work is finished. Care should be taken to exclude large lumps of dirt from the backfill. The backfilling should be completed evenly on all sides and properly tamped.

Grounding Tower Base   All steel towers are grounded, usually by means of ground rods. Some tower bases are actually extended to buried steel footers which ground them, while others are connected to buried copper or steel wires. The usual ground rod is a ¾-in-diameter 8-ft. long

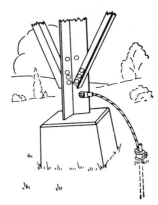

Fig. 20-16   Finished concrete foundation with bottom section of steel-transmission-line pole set in position.

Fig. 20-17   Ground rod connected to tower leg. *(Courtesy Copperweld Steel Co.)*

Fig. 20-18 Delivering members of steel transmission tower by helicopter. Made-up bundles were tightly banded for slinging below helicopter. All loads were carried horizontally on a single hook having pilot-controlled mechanical and electrical releases and an automatic underload release. *(Courtesy Southern California Edison Co.)*

copperweld rod. The rods are driven into the earth near the tower footing, one rod near each leg. Each tower leg is connected to its own ground rod. If the tower is located near a substation, the four ground rods are often connected together with a heavy copper wire which runs around the tower in a shallow trench. The actual ground connection to the tower is made at one leg only. This connection consists of a heavy clamp bolted to the tower leg above the ground surface where it can be removed for testing purposes if need be (Fig. 20-17). Often the tower ground is also connected to the substation ground by means of the same heavy copper connecting wire.

Erecting Transmission Towers   The linemen should exercise care in unloading, delivery, handling, and erecting towers so as not to damage the finish (Fig. 20-18). Tower members stored pending erection should be sorted and neatly piled and supported by suitable blocking (Fig. 20-19). Tower members must not be used as unloading or loading skids. Steel erection drawings and

Fig. 20-19   Horizontal assembly of the base of a high-voltage transmission-line tower. The first side or panel of the tower has been assembled. Note one anchor or stub in the center foreground. With this type of assembly and modern methods of tower erection, the tower can be assembled on level land, even though it is some distance from the tower footing. Later a crane will pick up the completed tower, carry it to its location, and set it on its foundation. *(Courtesy Wisconsin Electric Power Co.)*

Fig. 20-20 Linemen are using flatbed truck with A frame to assemble section of transmission-line tower for a 345-kV alternating current circuit. *(Courtesy L. E. Myers Co.)*

structure lists provide the guides necessary to assemble the towers. The towers may be erected by assembling them in sections on the ground and hoisting the successive sections into place (Figs. 20-20 through 20-27). If the towers are erected by assembling in sections, 50 percent, or more, of the bolts must be in place on each section before starting another section.

The "piece-by-piece" method of tower erection is similar to that used in erecting steel buildings or other permanent structures. The method is used when the towers are large and heavy and when the ground is rough.

In assembling a tower in place, light members are often simply lifted into place. Sometimes one of the corner legs is used for raising the other members. For heavier towers, a small boom is rigged on one of the tower legs for hoisting purposes. The usual procedure followed is illustrated in Figs. 20-28 through 20-39.

When the tower assembly is complete all bolts must be drawn up tight to specified torque. Palnuts should be placed on all bolts and tightened with a special palnut wrench.

The development of the helicopter has made possible its use for transport of poles, tower, and line materials for erection of poles and towers. It is especially suited for transport if the terrain is rugged, if access roads are not available, or if the right-of-way has not been cleared. Figures 20-40 to 20-47 illustrate the helicopter's suitability in the transport and erection of towers in sections, or V-type transmission line towers fully assembled.

Fig. 20-21 A 132,000-volt transmission-line tower subassembly. This is the top of the tower, here shown, being assembled separately. It will be bolted to the bottom part of the tower after it has been assembled. Note steel in the right background being sorted and placed for the assembly of the bottom part of the tower. *(Courtesy Wisconsin Electric Power Co.)*

Fig. 20-22 Bottom part of a 132,000-volt transmission-line tower in process of assembly. Note the crane which is used to hoist the separate parts of the tower to workmen on the partial assembly. The workmen fit each part to its proper place and bolt it securely. *(Courtesy Wisconsin Electric Power Co.)*

Fig. 20-23 A 132,000-volt transmission-line tower almost completely assembled. Note that the bottom part is being bolted to the subassembly of the top part. This tower is almost complete and ready for erection. The dark square sections of steel shown in the top assembly are the mounts for the insulator strings which carry the line conductors. *(Courtesy Wisconsin Electric Power Co.)*

Fig. 20-24   A completely assembled 132,000-volt transmission-line tower ready for erection. The crane is about to pick the tower off the ground and carry it to its footing. This is a far cry from the old gin-pole method of erection, where the tower had to be hinged to its footings while being erected. With this method of erection, a level piece of ground can be picked for the tower assembly even though it is a short distance away from the tower base. *(Courtesy Wisconsin Electric Power Co.)*

Fig. 20-25   A 132,000-volt transmission-line tower being set on its footings. The first leg is in its approximately correct position. From this point on, the crane operator must operate his crane to get the other three legs correctly placed. After that, it is just a matter of bolting the tower legs to their footings. *(Courtesy Wisconsin Electric Power Co.)*

Fig. 20-26 Erecting lofty tower by use of two crawler cranes. Tower was first completely assembled on ground. Then two cranes double-team to erect tower by the "up-ending" method. *(Courtesy Bethlehem Steel Co.)*

Fig. 20-27 A 132,000-volt transmission-line tower being bolted to its anchors or stubs after it has been placed in position. The workmen are lining up bolt holes by means of drift pins so that the bolts can be inserted readily and the tower secured to its footing. *(Courtesy Wisconsin Electric Power Co.)*

Fig. 20-28 Beginning the erection of a 45-ft corner tower by the "piecemeal" method. *(Courtesy Central Illinois Light Co.)*

Fig. 20-29 Gin pole placed on one of the corner legs for raising parts for the second section of the frame. *(Courtesy Central Illinois Light Co.)*

Fig. 20-30 Leg members for second section in place. *(Courtesy Central Illinois Light Co.)*

Fig. 20-31 Struts on four sides of second section in place. *(Courtesy Central Illinois Light Co.)*

Fig. 20-32 Bottom arm of tower in place. Gin pole ready to put up first leg for third section. *(Courtesy Central Illinois Light Co.)*

Fig. 20-33 First leg of third section in place. *(Courtesy Central Illinois Light Co.)*

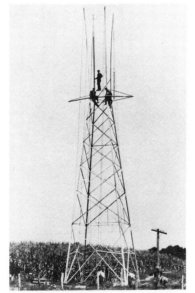

Fig. 20-34 All leg members for third section in place. Ready for struts of third section. *(Courtesy Central Illinois Light Co.)*

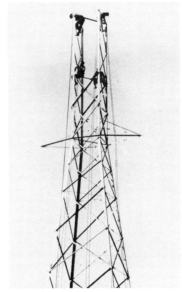

Fig. 20-35 Ready for second crossarm. *(Courtesy Central Illinois Light Co.)*

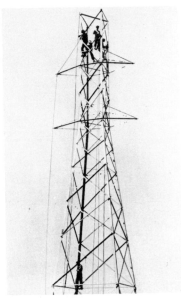

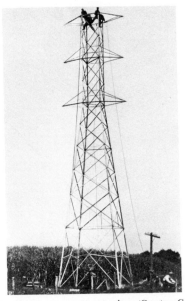

Fig. 20-36 Second crossarm in position. *(Courtesy Central Illinois Light Co.)*

Fig. 20-37 Top crossarm in place. *(Courtesy Central Illinois Light Co.)*

Fig. 20-38 Gin pole fastened to one leg of tower. Tower is being erected piece by piece and bolted into place. Tower base is being completed. *(Courtesy American Electric Power System.)*

Fig. 20-39 Use of gin pole in completing the erection of the top of the tower by the piecemeal method. Gin pole is moved up as the tower grows. Assembly of one set of crossarms is under way. *(Courtesy American Electric Power System.)*

Fig. 20-40 Helicopter used to erect transmission tower in sections. Bottom section 40 ft high and weighing 2000 lb was lifted onto the stubs. Upper section was lifted in place by use of guide shoes on upper and lower assemblies. Sikorsky S-58 helicopter has a lifting capability of about 4000 lb. This varies with altitude and temperature. *(Courtesy Southern California Edison Co.)*

Fig. 20-41 Helicopter lowering a steel tower section into position where it will be bolted into place by waiting riggers. Conventional assembly methods were used to raise the tower to a height of about 200 ft. The helicopter served to add the remaining sections to a height of 306 ft. *(Courtesy Sikorsky Aircraft.)*

Fig. 20-42 Helicopter lifting an aluminum V-shaped two-legged tower in the marshaling yard where it was fabricated to transport it to the line location. Towers are 50 to 95 ft high and weigh upward of 3900 lb. *(Courtesy American Electric Power System.)*

Fig. 20-43 Helicopter transporting completely assembled tower including guy wires to line location. *(Courtesy American Electric Power System.)*

Fig. 20-44 Helicopter lowering V-shaped aluminum tower onto its single-pin foundation. Helicopter
hovers overhead while tower is being securely guyed.
(Courtesy American Electric Power System.)

Fig. 20-45 While the helicopter holds the tower in
place, members of the ground crew grab each of four
guy lines, race to their preset respective anchor points,
and secure the tower. *(Courtesy American Electric
Power System.)*

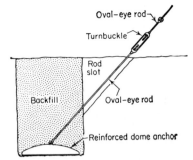

Fig. 20-46 Ground crew connecting and tightening the tower guys. As soon as the guy lines are pulled
taut, the helicopter disengages and returns to the marshaling yard for another tower. *(Courtesy American
Electric Power System.)*

Fig. 20-47 Typical anchor assembly used to
anchor the guys on the V-shaped two-legged aluminum towers. A single center compression foundation,
1 by 2 ft, supports the tower. The anchors range in
size from 4 to 6 ft in diameter. *(Courtesy A. B.
Chance Co.)*

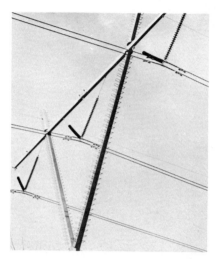

Fig. 20-48 Steel pole H frame structure. Note linemen working from fiberglass ladder connecting conductors to insulators. *(Courtesy A. B. Chance Co.)*

Fig. 20-49 Dreyfuss-type Y-shaped steel-pole structure used for 500-kV ac transmission line. Note steel wings welded to tower members for linemen to use for climbing tower. *(Courtesy A. B. Chance Co.)*

Steel pole H-frame structures (Fig. 20-48), Y-type steel structures (Fig. 20-49), and single steel pole structures (Fig. 20-50) are used to improve the esthetic effects and minimize the right-of-way required. Steel poles eliminate the need for down guys. Single steel poles can be used to support transmission-line circuits along city streets. The steel poles are fabricated in a factory and shipped to the site in sections. The foundations are constructed in the same manner as those described for towers. Linemen assemble the steel-pole towers in the field and erect the structures with the use of cranes and mechanized equipment (Figs. 20-51 through 20-57). Some steel transmission-line poles are fabricated with telescoping construction. The steel poles are shipped to the site, bolted to the concrete foundation, and raised by pumping the interior of the pole full of

Fig. 20-50 Single steel-pole structure on edge of golf course along railroad right-of-way. Structures support double circuit 161-kV ac transmission. *(Courtesy A. B. Chance Co.)*

Fig. 20-51 Single steel-pole transmission structure assembled on ground prior to installation. Note brackets on side of pole to facilitate the installation of steps for climbing.

Fig. 20-52 Bottom section of steel pole has been installed and bolted to concrete foundation. Crane is in position to lift top section of pole.

Fig. 20-53 Crane starts lift of top section of steel pole for double circuit 161-kV transmission. Lineman directs crane operator.

Fig. 20-54 Top section of pole is guided into position by lineman working from a bucket truck. Note insulators were attached to the transmission-line structure before it was raised from the ground.

Fig. 20-55 Lineman in bucket guides top section of steel pole down over bottom section to provide proper joint overlap.

Fig. 20-56 Steel transmission pole assembly is completed. Top arms on pole without insulators will support static wires for lightning protection.

Fig. 20-57 Lineman has installed removable steps on steel-transmission-line pole. Lineman is descending pole after removing crane hitch from top of pole.

Fig. 20-58 Lifting completely assembled 115,000-volt tower onto its footings by means of a mobile power crane. The crane has a 60-ft boom. Note that the tower is being erected with the insulator strings attached to crossarms. They have to be securely fastened, however, to avoid swinging against the tower and breaking the porcelain on the insulator disks. (*Courtesy Virginia Electric and Power Co.*)

Fig. 20-59 Crane placing aluminum top section on galvanized-steel base having a height of 108 ft. Line is double circuit and operates at 115,000 volts. Insulators were installed on top section by linemen on the ground before starting lift. Use of aluminum superstructure for cage, crossarms, and goathead components of tower is to reduce maintenance costs. *(Courtesy Aluminum Company of America.)*

Fig. 20-60 Crane lifting steel transmission-line pole into position. Insulators are secured to steel arms in a manner to restrain them and prevent breakage. *(Courtesy L. E. Myers Co.)*

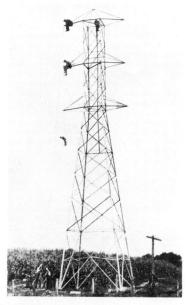

Fig. 20-61 Raising suspension insulator string to crossarm. This is done by the use of a hand line running over a single pulley fastened to the top crossarm. *(Courtesy Central Illinois Light Co.)*

concrete using a high-pressure pump. The poles can be erected very fast. The concrete reinforces the steel assembly when it has cured.

Insulator Installation   Insulators must be handled with care to prevent chipping or cracking the porcelain or glass or damaging epoxy assemblies. Excessive bending strain on the pin shanks or caps should be avoided. A cradle, or similar device, should be used to lift porcelain or glass insulators. Insulators that are damaged must be discarded. The insulators must be kept clean and free from grass, twigs, dirt, or other foreign matter. All cotter pins must be properly and adequately spread before the insulator strings are attached to the transmission-line structures. The suspension assemblies on all conductors of the line should be placed so that all nuts face the center of the structure and cotter pins are parallel to the line with the points of the pins facing inward or toward the space between the conductors. Installation of insulators is illustrated in Figs. 20-58 through 20-61.

Stringing Line Conductors

The installation of the line conductors on poles and towers must be accomplished in a manner to provide an electric circuit that will operate reliably and not endanger the public. Conductors with surface scratches or defects energized at high voltages will have large corona losses and generate radio interference voltages that will be transmitted through the atmosphere. The conductors may be strung using the slack or the tension method.

Slack Conductor Stringing   Slack conductor stringing is usually limited to short lengths of line operating at low voltages utilizing conductors with a weatherproof covering or installations where scratches on the surface of the conductor are not important.

The reels of wire are mounted on a vehicle in such a manner that the reels are free to rotate. The ends of the conductors are fastened to a pole, tower footing, or other fixed object. The vehicle is then slowly propelled along the route of the line, allowing the conductors to unwind as the reels are moved forward. In this method the conductors are not dragged over the ground, causing them to become scratched or damaged. They are simply payed out on the ground without being pulled over it. The method, however, cannot be used on one-circuit tower lines or on X-braced H-frame lines because the center or middle conductor must be placed over the internal framework of the tower or over the X braces of the H frames. This obviously is impossible if the reels are on the lead end of the conductor.

If the reels on which the conductors are wound remain in a fixed location and are raised off the ground or supported in their carriages in such a way that they are free to rotate, the conductors can then be pulled out, thereby rotating the reels and unwinding the conductors, as illustrated in Fig. 21-1.

The conductor should never be payed out from a nonrotating reel or coil, as each turn removed gives the conductor a complete twist which may cause kinks or other damage.

Unreeling conductors from stationary reels provides two possibilities, one simply to draw the conductors forward, sliding or dragging them over the ground, and the other to keep the conductors suspended in the air in tension so that they will not touch the ground. When the conductors are pulled forward sliding over the ground, they are apt to become scratched and nicked, especially if the ground is rough and covered with surface stone. Aluminum conductors are more easily injured in this manner than copper.

Tension Conductor Stringing   The process of installing overhead line conductors in a manner which keeps the conductors off the ground, clear of vehicular traffic and other structures that might damage the conductors, and clear of energized circuits is called tension stringing (Fig. 21-2). The linemen install conductor stringing blocks on the bottom of the transmission-line insu-

Fig. 21-1 Reel of triplexed low-voltage secondary cable. Axle in reel is supported by winch line on boom of line truck. Axle has a swivel joint permitting reel to rotate easily. Lineman is restraining rotation as cable is payed out. Cable and other line material was transported to job site in material trailer behind line truck.

lator strings to permit pulling the conductors in under tension (Fig. 21-3). The conductor stringing blocks are built for the installation of one to four conductors for each point on the tower (Fig. 21-4). The sheaves on the stringing blocks are usually lined with a conductive-type neoprene, or urethane, to protect the phase conductors which are usually aluminum or aluminum and steel (ACSR). The blocks used for static, or ground wires, are not lined since the static wires are usually galvanized steel, copperweld, or alumoweld, which are not easily damaged.

The conductive neoprene lining of the stringing blocks permits the conductors to be effectively grounded by a jumper from the stringing block to a ground wire on the metal tower structure. Effectively grounded stringing blocks will eliminate induced, static, or impulse voltages that

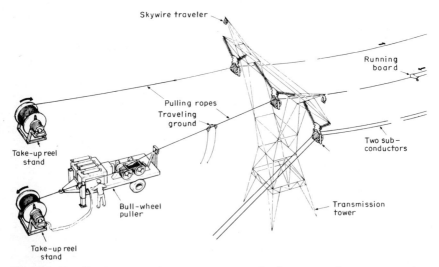

Fig. 21-2 Typical equipment set up for tension stringing a two-bundle transmission line. *(Courtesy Timberland Equipment Limited.)*

might be present during construction. The stringing blocks, often called "travelers," can be equipped with outriggers to permit the installation of the pilot line with a helicopter.

When the linemen install the stringing blocks on the insulators, they normally place a light-weight rope called a "finger line" over the traveler (Fig. 21-5). The finger line must be long enough to reach the ground on both sides of the traveler. Finger lines are used by the linemen to pull the pilot line through the traveler from the ground. If the pilot line is installed with a helicopter, finger lines are not used. Pilot lines are light-weight rope used to pull pulling ropes through the travelers as shown in Fig. 21-2. A pilot-line winder (Fig. 21-6), is used to provide the power necessary to pull the conductor pulling rope from the reel on the take-up reel stand through the travelers. Tension is kept on the pulling rope, often referred to as the "bull line," to prevent interference with objects on the ground.

The conductor reels, tensioners, and pulling machines must be in line before the conductor pulling is started. The tensioner often called a "bull-wheel" tensioner (Fig. 21-7), and the "bull-wheel" puller (Fig. 21-8), should be set up as near to midspan as possible (Fig. 21-9). The slope of the conductors between the equipment and the stringing blocks at the first tower must not be steeper then five horizontal to one vertical.

The transmission-line conductor retarding bull wheels should have a minimum of five turns of the conductor over the bull wheel and have multiple grooves lined with neoprene or other approved nonmetallic resilient material so that the conductor will cushion into the lining to prevent flattening, or otherwise damaging the conductors as they are payed out (Fig. 21-10). The diameter of the bull-wheel grooves must not be less than conductor diameter, plus 25 percent. The bull wheels should have a minimum bottom groove ("root") diameter of 25 times the conductor diameter.

The conductors should be pulled directly from the cable reel onto the tensioners and into the stringing blocks without touching the ground. The conductor pulling tension should not exceed 70 percent of the conductor sagging tension. Care must be taken at all times to ensure that the conductors do not become kinked, twisted, abraded, or damaged and that foreign matter does not become deposited on them. Dependable communication must be maintained between the linemen operating the pulling equipment and the tensioning equipment and observing at intermediate points at all times during the wire-stringing operations (Fig. 21-11).

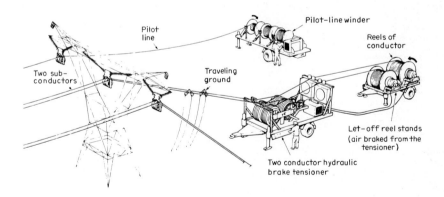

Fig. 21-3 A 765-kV transmission-line steel H-frame tower. Linemen are installing stringing blocks on top of tower for installation of static wires. Stringing blocks for phase conductors are in place on bottom of insulator strings. (Courtesy A. B. Chance Co.)

Fig. 21-4 Three-conductor-bundle stringing block. (Courtesy Sherman & Reilly, Inc.)

Fig. 21-5 Finger line looped through traveler extends to the ground. (Courtesy American Electric Power Co.)

Fig. 21-6 Linemen using a pilot-line winder to pull conductor pulling rope through travelers. Note frame of winder is grounded. *(Courtesy Sherman & Reilly, Inc.)*

Fig. 21-7 Bull-wheel tensioner and conductor reels set up to pull three conductors per phase. The three conductors are called subconductors and form a three-conductor bundle per phase. *(Courtesy Sherman & Reilly, Inc.)*

Fig. 21-8 Linemen setting up bull-wheel pullers prior to starting the conductor pulling operation. *(Courtesy L. E. Myers Co.)*

Fig. 21-9 Tensioners and reels of conductor set up for installing conductors on transmission line in background. *(Courtesy L. E. Myers Co.)*

Bundled conductors are pulled simultaneously by one pulling line with the use of a unidirectional articulated running board (Fig. 21-12). Special bundle-conductor-type stringing blocks are required. For stringing two conductors (see Fig. 21-13) the block consists of two sheaves and one drum, all of which turn independently of one another. The two outer sheaves support the conductors, and the center portion of the drum carries the pulling line. The drum is designed to accept the running board. For stringing a bundle of three conductors, a block with three sheaves and a two-part drum is used, as shown in Fig. 21-4. The pulling line rides in the center groove as does the center conductor, shown in Fig. 21-14. The sheave spacing corresponds to the final spacing of the bundle conductor.

Fig. 21-10 Lineman operating bull-wheel tensioner used for two subconductors per phase. *(Courtesy L. E. Myers Co.)*

Fig. 21-11 Lineman with two-way radio directing conductor-pulling operation. *(Courtesy L. E. Myers Co.)*

Fig. 21-12 Articulated running board designed for use in stringing bundled conductors. Vertical links hold running board in horizontal position for correct entrance into stringing block. *(Courtesy Sherman & Reilly, Inc.)*

The tensioner for a two-conductor bundle consists of two pairs of bull wheels, one pair for each conductor. The bull wheels are geared together in such a way that a small auxiliary engine can be used to even up the conductors if one of them should sag more than the other. The engine is also used to air-cool the brakes and to reverse the bull wheels slowly if this should become necessary.

When stringing a bundle of four conductors it is well to string the bottom two conductors first and leave them in soft sag. The top conductors are then strung over them.

The conductors should be grounded while they are being installed. Traveling grounds can be used at the tensioner and puller sites. The use of travelers with a conductive-type neoprene lining permits grounding the conductors at each structure. The grounding cable must be large enough to adequately conduct fault current to ground without fusing.

Fig. 21-13 Pulling board about to be drawn over two-conductor stringing block. Central drum carries pulling line, and two outer sheaves carry the two-bundled conductors. Wedge-shaped running board ensures block centering and placing conductors in grooves. *(Courtesy Sherman & Reilly, Inc.)*

Fig. 21-14 Three-conductor-bundle block supported on suspension insulator string. Block has three sheaves. Center sheave carries pulling line. After running board passes block, third conductor follows pulling line on center sheave. *(Courtesy Sherman & Reilly, Inc.)*

When the conductors being installed reach the pulling equipment, they must be secured to prevent them from developing slack which might let them touch the ground (Fig. 21-15).

New distribution circuit conductors are usually installed using tension-stringing methods, as described, except the equipment for pulling and tensioning the conductors is smaller and lighter (Figs. 21-16 through 21-18). Distribution circuits are shorter in distance than most transmission lines and the conductors are usually smaller with a limit of one conductor per phase. Finger lines can be used to pull the pilot line through the stringing blocks (Fig. 21-19). However, care must be taken to tie the finger lines off high enough on the pole to keep them out of reach of the public.

Distribution lines can be reconductored using tension-stringing procedures. The energized conductors must be set out in the clear by linemen using hot extension arms (Figs. 21-20 through 21-22).

Pilot lines can be installed for distribution-line tension stringing using the "Spider System." Pilot-line controllers are installed on the lead pole by a lineman approximately 15 to 20 ft above the ground. A pilot-line storage reel is hung on each controller (Fig. 21-23). The pilot line comes off the storage reel over an arm that releases the brake on the pilot-line controller when the rope

Fig. 21-15 Conductors secured temporarily with conductor clamps and steel cables anchored to frame of bull-wheel pullers. *(Courtesy L. E. Myers. Co.)*

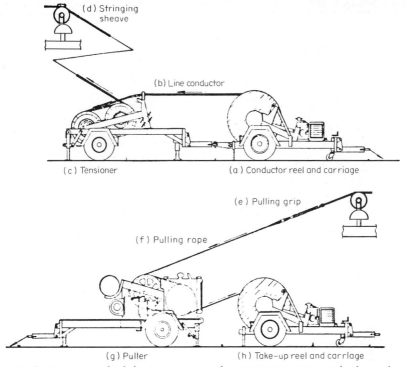

Fig. 21-16 Diagrammatic sketch showing arrangement of tension stringing equipment when line conductor is strung in tension. (*a*) Conductor reel and carriage. (*b*) Line conductor. (*c*) Tensioner. (*d*) Stringing sheave. (*e*) Pulling grip. (*f*) Pulling rope. (*g*) Puller. (*h*) Take-up reel and carriage. (*Courtesy Petersen Engineering Co., Inc.*)

Fig. 21-17 Tension stringing of line conductors with four-reel payout tensioner operated in conjunction with four-reel take-up puller. Each reel is provided with brake to produce tension in conductor. Lineman kneeling using a walkie-talkie two-way radio to communicate with lineman at puller in order to coordinate pulling and tensionsing. (*Courtesy Union Electric Company.*)

Fig. 21-18 Tension stringing of line conductors with four-reel rope take-up puller equipped with automatic level winder. Separate gasoline engine drives take-up reels. Each reel can hold 3000 ft of ¾-in rope. Tension stringing suspends conductors above ground in each span. *(Courtesy Union Electric Company.)*

Fig. 21-19 Finger lines have been installed in stringing blocks for new distribution circuit. Pilot line is being pulled through block for static wire. Lineman is checking equipment on first pole. Normally pilot line will be pulled through stringing blocks with finger lines from the ground. *(Courtesy A. B. Chance Co.)*

Fig. 21-20   Groundman has taken hot-extension arm from material and equipment trailer. The hot-extension arm will be sent up pole with hand line to lineman.

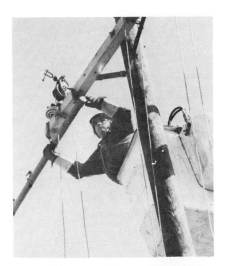

Fig. 21-21   Lineman has removed energized conductor covered with insulated line hose from insulator on crossarm. Hot extension arm is being installed on crossarm. Energized conductor will be secured in the clear by clamp on hot extension arm. Note stringing block clamped to crossarm with finger line in place.

Fig. 21-22   All three energized phase conductors on pole in foreground have been set out in the clear on hot extension arms. Stringing blocks for phase conductors are installed on crossarm. Stringing block for neutral ground wire is clamped to pole below crossarm.

Fig. 21-23 Spider System of installing pilot line for tension stringing of distribution circuit conductors. Reels are mounted high enough on pole to prevent pedestrian contact and overnight vandalism. *(Courtesy Sherman & Reilly, Inc.)*

Fig. 21-24 Attaching the kellum grip on a 795,-000-cir mil ACSR cable. This grip is used for pulling the cable over the stringing sheaves on the structures. When the grip is in place, it is taped down to the conductor to keep it from pulling back and releasing. *(Courtesy Detroit Edison Co.)*

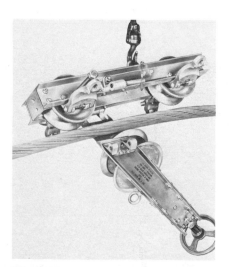

Fig. 21-25 Running ground device designed to maintain positive contact with the conductor being strung regardless of tension. *(Courtesy Sherman & Reilly, Inc.)*

Fig. 21-26 In case conductors have to be strung over a live line, guard poles which hold the wires above the live line are used. Shown here are two guard poles carrying three line conductors and two ground wires of a 132,000-volt line over a 4800-volt single-phase distribution feeder. *(Courtesy Wisconsin Electric Power Co.)*

is pulled and engages the brake when the rope is slack. This eliminates "backwinding" and holds tension during the installation of the pilot line.

All pilot lines can be carried up each pole and placed in the stringing blocks when the blocks are placed on the pole. As this is accomplished, the pilot lines can be tightened and left clear of any obstructions below, such as tree limbs, transformers, and street lights.

The pulling line is positioned at the opposite end of the Spider controller and storage reels and is tied to the end of the pilot line. The partially empty storage reel is then mounted on a line truck winch or electric winch by use of one of the winch adapters.

The winch turns the reel and rewinds the pilot line, thus hauling in the pulling line under tension and eliminating the need for linemen to walk through carrying the pulling line over each structure.

Fast, efficient, safe, and field-tested, the "Spider System" is used by many utilities. The pilot line is used to pull the pulling line, or bull rope, through the stringing blocks. The pulling rope is connected to the conductor with a conductor grip (Fig. 21-24).

The conductors should be effectively grounded while they are being installed (Fig. 21-25). The grounding cable connecting to the running ground must be large enough to safely conduct fault current to the point of ground attachment.

If distribution or transmission line conductors are to be strung over highways or energized circuits, guard poles should be used to protect the public and the workmen (Fig. 21-26).

Sagging Line Conductors

Sagging operations are completed as soon as the conductor stringing is completed to establish the proper conductor tension for the conditions that exist at the time the work is performed. The correct conductor tension and sag for various sag or control spans will be detailed in the specifications for construction of the transmission line or the standards for the distribution circuit to provide proper clearances. The weather conditions, including temperature and wind velocity, must be taken into consideration to complete the process. The proper conductor tension can be obtained by establishing the proper sag in the control spans by sighting methods using a surveyor's transit, or targets. The tension in the conductor can be measured with a dynamometer, or determined with timing waves.

All conductor grips, come-alongs, stringing blocks, and pulley wheels used in the sagging operations, must be of a design which will not damage the conductors by kinking, scouring, or unduly bending them (Fig. 22-1). Sag sections must be selected before the operations are started. The length of sag sections are limited by dead ends, angle structures, and the terrain. The sag sections should not normally be longer than 4½ miles.

Sag sections must have control spans. The control spans will be used to measure the sag for each sag section. If the control spans are properly sagged, the conductors in the sag section should all be at the proper tension. The control spans should be at approximately the length of the ruling span. Ruling spans are a calculated dead-end span length assumed to have the same conductor tension as all other spans in the sag section even though they vary in length. Control spans should be located away from the dead-end sections and large line angles in a level portion of the line if possible. If only one span is used for checking the sag, it should be approximately in the middle of the section being pulled to tension. Where two or more spans are used for checking the sag, they should be located approximately equidistant from each other and from each end of the sag section. The sag of spans on both sides of points in the line where the grade varies by more than 10 degrees should be checked.

The use of more than one sag control span in each sag section will eliminate problems from cumulative sheave effect and human error. One control span is permissible in short sag sections of five spans or less. A minimum of two control spans should be used in sag sections of five to eight spans in length. A minimum of three control spans should be used in sag sections of nine or more spans. Sag sections are normally limited to approximately twenty spans. The control spans selected should be checked to be sure there are no major discrepancies between the actual span lengths and those on the plan and profile sheets included with the specifications.

All wires on new lines being sagged under deenergized conditions should be grounded (Fig. 22-2). When a bundle of three or four conductors is sagged, the top conductors are sagged first and bottom conductors are sagged last. The tension in the bottom conductors should be about 3 percent

Fig. 22-1 Close-up view of lineman attaching come-along to line conductor. Note that cable used for sagging, runs through a separate sheave. Note method of supporting lineman from ladder hooked over crossarm. Also note length of suspension insulator string consisting of 18 suspension units. This line is to operate at 345,000 volts. *(Courtesy American Electric Power System.)*

Fig. 22-2 Ground device installed on moving conductor while sagging and stringing operations are completed. Grounding cable is attached with hot-line tool. Rope secured to pole or tower prevents ground device from moving with conductor. *(Courtesy Everly Ever-Ground Co.)*

less than the top conductors. All wires must be pulled up to the proper sag without exceeding the necessary tension. The wires must not be pulled up to a tension greater than the sagging tension. The conductors must be sagged to the initial sag specifications.

Temperature of the conductor must be determined with a thermometer that is normally inserted into a piece of conductor approximately 30 in long and suspended at the approximate elevation of the wire to be sagged. Two thermometers should be used at different locations near sag spans in the sag section. The thermometers should be in position for a minimum of 15 min before the temperature is recorded to determine the proper wire sag. After the conductors have been properly sagged, intermediate spans should be inspected to be sure the sags are uniform and correct. The wires are sagged from the tensioner end of the sag section to the puller end (Figs. 22-3 through 22-7).

The foreman initiates and supervises the sag operations maintaining communication with the

Fig. 22-3 Tensioner end of sag section of 765-kV transmission line. Conductors have been pulled in and are hanging in travelers. Sagging operations are ready to be started. *(Courtesy American Electric Power System.)*

Fig. 22-4 Conductor secured at tensioner end of sag section with wire clamp for sagging operations. *(Courtesy L. E. Myers Co.)*

Fig. 22-5 Installing the pulling grip or come-along on a 1,275,000-cir mil ACSR expanded cable having an outside diameter of 1.6 in on pulling end of sag section. Cable will be drawn up by the caterpillar tractor shown. Note men holding cable to keep it from kinking or becoming damaged. Voltage of line is 345,000 volts. *(Courtesy American Electric Power System.)*

linemen at each sag span and other points under observation in the sag section between the tensioner end and the pulling end (Figs. 22-8 through 22-11). The sagging operations cannot be completed on days with strong winds because of the conductor uplift from wind pressure. The aerodynamic effect of the wind on the conductors can cause sagging errors.

Special precautions must be taken to sag distribution conductors that are near energized circuits or are energized (Figs. 22-12 through 22-19).

Fig. 22-6 Close-up view of come-along attachment to sagging cable. *(Courtesy American Electric Power System.)*

Fig. 22-7 Cables are threaded through sagging sheaves on crossarms. Caterpillar tractor is being used as motive power for sagging the wires. *(Courtesy American Electric Power System.)*

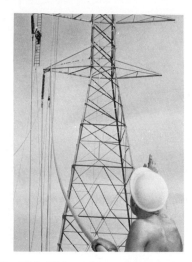

Fig. 22-8   Foreman supervises sagging operations by portable two-way radio with pulling tractor, and tower men, whose positions can be alerted simultaneously. *(Courtesy American Electric Power System.)*

Fig. 22-9   While sagging continues, tower men keep supervisor informed of conditions at cable sheave while conductor is being pulled up. *(Courtesy American Electric Power System.)*

Fig. 22-10   Close-up view of tower man as he observes the wire sagging and keeps supervisor informed of conditions at cable sheaves. *(Courtesy American Electric Power System.)*

Fig. 22-11   Line conductors pulled up close to correct sag and snubbed off to earth anchors. The snub must be made far enough out from the tower to keep the down pull on the crossarms to a safe value. Lineman shown is using a coffing hoist with wire grips to pull conductors up to exact sag. The conductors are 795,000-cir mil ACSR cables and have been strung over 14-in stringing sheaves. The desired sag is equivalent to a tension of 7000 lb. *(Courtesy Detroit Edison Co.)*

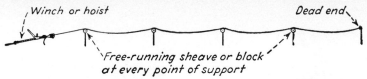

Fig. 22-12 Schematic diagram of distribution line being pulled up to proper sag. Note use of free-running sheave or snatch block at every point of support.

Fig. 22-13 Distribution phase conductor in stringing block mounted on insulated epoxy stand-off bracket in position for sagging operation. *(Courtesy Sherman & Reilly, Inc.)*

Fig. 22-14 Distribution circuit conductors in stringing blocks mounted on line post insulators. Circuit can be energized in this position while sagging operations are completed if proper precautions are taken. *(Courtesy Sherman & Reilly, Inc.)*

Fig. 22-15 Rope basket employed to guard against unstrung line conductors dropping low or falling onto street. *(Courtesy Union Electric Co.)*

Fig. 22-16 "Cumalong" grip depends on wedge action to grip conductor. The greater the pull on the wedge, the harder the conductor is forced against the steel grip insert in the body. "Cumalongs" can be installed in live-line work with a tie stick using the lifting ring provided on the top. *(Courtesy James R. Kearney Corp.)*

Fig. 22-17 Pulling grip or come-along being installed on an energized circuit using a hot-line tool. Note manner in which wire will be held by grip when tension is applied to linkage by attaching a hoist to the eye of the linkage *(Courtesy A. B. Chance Co.)*

Fig. 22-18 Lineman pulling up on energized conductor with hoist and grip or come-along to properly sag energized conductors. Personal and line protective equipment are used while working energized circuit from insulated bucket truck.

Fig. 22-19 Linemen sagging energized distribution conductors. Note linemen wear rubber gloves and sleeves and perform work on an insulated platform. All conductors except the one being sagged, with the use of rope blocks, are covered with protective equipment. *(Courtesy A. B. Chance Co.)*

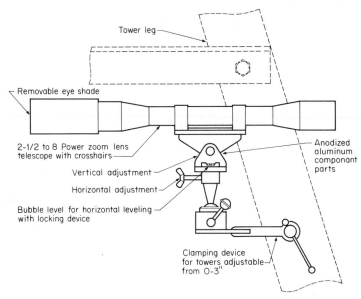

Tower leg

Removable eye shade

2-1/2 to 8 Power zoom lens
telescope with crosshairs

Vertical adjustment

Horizontal adjustment

Bubble level for horizontal leveling
with locking device

Anodized
aluminum
componant
parts

Clamping device
for towers adjustable
from 0-3"

Fig. 22-20   Sag scope with zoom capabilities for wire sagging. *(Courtesy Hi-line Industries, Inc.)*

Sagging Line with Transit   The method of determining conductor sag used commonly by linemen is to climb one structure and place targets at the specified sag level for the conditions that prevail. The lineman then climbs the structure on the opposite end of the control span and mounts a surveyor's transit, or an especially designed sag scope (Fig. 22-20) at a point equal to the sag distance below the conductor support. The lineman then sights through the transit and observes conductor sag. When the conductor is pulled up to the proper level, he tells the foreman supervising the operations to prevent overstressing the conductor (Fig. 22-21). If the structures on each end of the control span are not the same height or not located on level ground, special adjustments must be made in the target and transit locations to compensate for the differences in elevations.

In the case of an H-frame line, the line conductors are on one or the other side of the poles. A lineman stationed on one pole and looking to the corresponding pole a span length away cannot include the lowest point of the conductor in his line of sight. To get around this difficulty, a transit is securely fastened to the pole at a distance equal to the desired sag below the conductor support.

Fig. 22-21   Lineman sighting through a transit telescope at a target on the next structure while a line conductor is being drawn up during preliminary sagging operation. The thin wire hooked over the line wire near the snatch block is one lead of a portable telephone which is being used to communicate with the crew that is doing the pulling. The coil of rope on the hand line is used in the final sagging of the conductor by timing. *(Courtesy Wisconsin Electric Power Co.)*

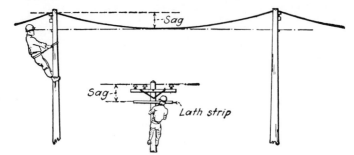

Fig. 22-22 Lineman sighting for sag by means of two lath strips nailed to poles at proper distance below the conductor when resting on the insulator. *(Courtesy Wisconsin Power & Light Co.)*

Then the transit is leveled. To observe the sag, the transit is sighted at the conductor at mid-span and then is swung around until in line with the pole a span length away. The sag is then observed, and the line conductor is drawn up to the specified distance.

A simple and accurate method of measuring the sag is by the use of targets placed on the poles below the crossarm, as shown in Fig. 22-22. The targets may be a light strip of wood, like a lath, nailed to the pole at a distance below the conductor when resting on the insulator equal to the desired sag. The lineman sights from one lath to the next. The tension on the conductor is then increased until the lowest part of the conductor in the span coincides with the lineman's line of sight. Careful attention must be paid to the temperature at the time of sagging in of the conductor as the sag varies considerably with temperature. This is illustrated in Fig. 22-23, where typical values of sag are shown for the three values of temperature. The reason, of course, that the sag is greater at the higher temperature is because of the expansion of the metal in the conductor, causing it to be elongated.

The method of sighting for sag gives very satisfactory results when the sag is not less than 6 in and when the visibility is good. If the visibility is not good, it is often helpful to provide a suitable target at the distant pole for visual contrast with the conductor. This may be a piece of wood or metal approximately 2 ft square, painted white with a horizontal black line about 1 in wide across the center. For the sighting end, a similar device may be used, having a sighting slot in place of the black line (Fig. 22-24). The distant target and the slot are set below the support according to the desired sag, and the conductor sag is varied until the low point of the conductor is in line with the slot and the black line.

Sag Measurement by Timing Accurate sagging by sighting is difficult if the spans are long or if the sags are quite small. Furthermore, if the poles used in sighting are not at the same level, it is even more difficult. A mechanical wave, instigated near one support, will travel to the next support, be reflected in reverse phase, and pass back and forth repeatedly between the supports. Although the wave is attenuated continuously, the time required for the wave to return is independent of the span length or the size or type of conductor. The sag of the conductor can be

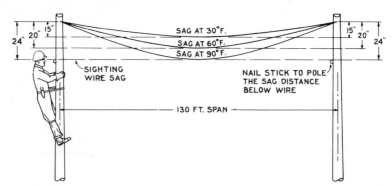

Fig. 22-23 Sketch showing how sag increases with rise in temperature also illustrates correct manner of sighting for sag.

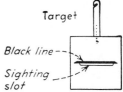

Fig. 22-24 Sighting target about 2 ft square and painted white. Black line is about 1 in wide. Sighting slot is also about 1 in wide.

determined by timing the wave returns with a standard stop watch, and converting the time measured to sag from the equation

$$\text{Sag} \ (D) \ = \ 48.3 \left(\frac{t}{2n} \right)^2$$

where sag (D) is in inches, t is time in seconds, and n is the number of return waves counted.

The number of returns depends primarily upon the conductor weight or size and the length of the span being sagged. For long spans and/or larger conductor sizes, the wave energy imparted to the conductor will dissipate more rapidly and the number of return waves which can be counted accurately is less than for light-weight conductors and/or short spans.

Attach a rope to the conductor about 3 ft from the support. Then give the rope a sharp jerk as in Figs. 22-25 and 22-26, and at the same time start a stop watch or note the reading of the second hand on your pocket watch. Striking the conductor causes a wave or ripple to travel along the

Fig. 22-25 Sag measurement by timing. The rope over the wire is used for applying an impulse to the conductor which creates a wave that travels along the wire. The lineman on the ground is holding a stop watch in one hand and the rope with the other. The lineman on the pole communicates directions to the pulling crew. If sag is too large, more pull must be applied; if sag is too small, pull must be relaxed. *(Courtesy Wisconsin Electric Power Co.)*

Fig. 22-26 Close-up view of lineman on ground measuring the conductor sag by timing. Lineman holds stop watch in one hand and rope in the other. The taut rope which passes over the line conductor signals each return of the wave. The time required for either 3, 5, or 10 returns of the wave is measured with the stop watch. *(Courtesy Wisconsin Electric Power Co.)*

conductor until it reaches the next pole. When it reaches the next pole, it is reflected and returns to the pole where you are stationed. Here it is also reflected, and thus starts its second round trip. This will continue until the energy of the blow has been expended. The length of time in seconds required for the wave to return to the near support corresponds to a definite sag which can be calculated or read from prepared tables or graphs. The time is independent of the span length or the size or type of the conductor.

Observe the travel of this wave, and count the number of returns until the wave has returned three, five, ten, or fifteen times. Upon the arrival of the third, fifth, tenth, or fifteenth return wave read the stop watch or your pocket watch and note the time elapsed in seconds. A sufficient number of tests should be made until at least three identical readings are obtained. Then refer to Fig. 22-27 and read the corresponding sag opposite the time in seconds. If this is not the desired sag, change the pull on the conductor and repeat the timing operation.

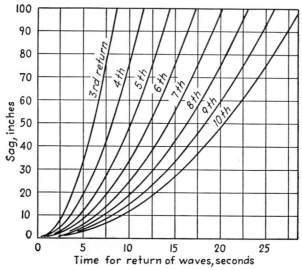

Fig. 22-27   Graphs giving relation between time in seconds and sag in inches for third to tenth return of wave. *(Courtesy Copper Wire Engineering Association.)*

The choice of the number of return waves depends principally upon the span length and size of conductor. For long spans and large conductors the number of return waves which can be accurately counted is less than for short spans and small conductors. The distance of the "jerk" line from the support of a conductor will largely determine the force of the impulse given the conductor as well as the ease with which the lineman with the stop watch can feel the returning waves. A distance of 3 ft or more is commonly used. The largest number of return waves which can be observed should be used as this minimizes errors in recording the time.

The return wave may be felt by a man on the pole by placing a finger lightly on the conductor if the line is deenergized and effectively grounded. Or readings may be made from the ground by throwing a light dry insulated rope over the conductor about 3 ft from the support, as shown in Figs. 22-25 and 22-26. This light rope may also be used to give the impulse initiating the wave. The use of the dry insulated rope is essential when the line is "hot," that is, energized.

Care must be taken not to count "one" when the impulse is given to the line, but to count "one" on the first return of the line. In other words count, "hit," one, two, three, etc.

This sagging method is most satisfactory if the line is not in motion. Any vibration of the line as might be caused by working on it or by a strong wind makes it difficult to determine the exact time of the return wave. The conductor should rest on the crossarm or in the snatch block to reflect the wave, although it need not be tied in.

Unisagwatch   The watch (Fig. 22-28) developed by Philip C. Evans takes the place of the stopwatch and sag-time tables or curves. The Unisagwatch carries logarithmic scales calibrated in feet and decimals for both third and fifth returns of a mechanical wave, providing direct readings of sags up to 100 ft. Sagging with this instrument eliminates the need for conversion tables and setting and sighting targets. The Unisagwatch is suitable for use in poor visibility conditions due to weather or in a selectively cleared right-of-way. The sag of an energized conductor may be readily checked by an observer on the ground.

Sagging Line with Dynamometer   When conductors are to be strung with unusually small sags, the measuring of the sag may not be as accurate or convenient as measuring the conductor tension. This is done by means of a dynamometer which is inserted in the pulling equipment. Such a dynamometer is shown in Fig. 22-29, and the manner of using it is shown in Fig. 22-30. One of the pointers indicates the pull at all times, and the other will remain at the point of maximum load after the tension is released. The pull is increased until the tension is obtained for the desired conductor type and size, span length, loading district, and prevailing temperature.

Any convenient span in the section of line may be used for measuring the tension, but care should be taken that the tension is uniform in all spans. After stringing, a common practice is to

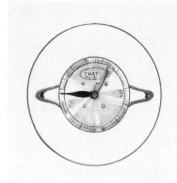

Fig. 22-28 The Unisagwatch dial reading gives sag in feet and decimals corresponding to the time required for three or five returns of the mechanical wave. *(Courtesy Sagline, Inc.)*

Fig. 22-29 Dynamometer used in pulling up line conductors to desired tension. Tension is indicated on scale on dial. *(Courtesy John Chatillon & Sons.)*

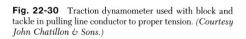

Fig. 22-30 Traction dynamometer used with block and tackle in pulling line conductor to proper tension. *(Courtesy John Chatillon & Sons.)*

Fig. 22-31 Front view of shunt dynamometer attached to line conductor for measurement of tension. *(Courtesy John Chatillon & Sons.)*

allow the conductor to remain untied for several hours before final measurement and tying. On a long section of line it is advisable to check the sag or tension at several spans to ensure the uniformity of the tension in the line.

The shunt dynamometer illustrated in Figs. 22-31 and 22-32 is so designed that it can be applied to the conductor without breaking in on the line to be measured for tension. It may also be left on the line while the dial shows the desired reading. The operating principle of the shunt dynamometer is based on the relation of the tension in the conductor to the force necessary to displace it in a direction perpendicular to the axis of tension.

Clipping-In Conductors   Conductors and shield wires should be clipped in as soon as practical after the sagging is completed. In all cases the conductors and static wires, or ground wires, should be clipped in within two days after the sagging is completed (Figs. 22-33 through 22-36). Conductors should be lifted from the stringing sheaves using a standard suspension clamp or a large plate hook. Rope slings should not be used. The inside diameter of the plate-hook seat should be at least ⅟₁₆ in greater than the diameter of the conductor to be lifted.

The center longitudinal portion of the plate-hook seat must be bent down at substantially the same curvature as found on the suspension clamp. The inside surfaces of the plate hook must be free of all burrs and rough edges. The conductors must be attached to the insulator assemblies at all suspension and deadend points in accordance with the specifications or the standards for the line under construction. Armor rods and vibration dampers are usually specified for all high-voltage transmission lines (Figs. 22-37 through 22-41). The dampers should be placed on the conduc-

Fig. 22-32 Rear view of shunt dynamometer attached to line conductor. Tension in line conductor compresses spring. To release dynamometer, handle is turned which lowers spring away from conductor. *(Courtesy John Chatillon & Sons.)*

Fig. 22-33 Ascending tower on steps provided on corner leg to apply armor rods, attach corona shield, and clamp conductor to insulator string. Tower is 148 ft high. *(Courtesy American Electric Power System.)*

Fig. 22-34 Hoisting work platform to linemen who are to apply armor rods, attach corona shield, and clamp conductor to insulator string. *(Courtesy American Electric Power System.)*

tors as soon as the conductors are clipped in to avoid the risk of conductor damage as a result of critical wind conditions.

Any conductor supported above ground in long spans under relatively high tension is subject to vibration. Such vibration is caused by the passage of air over the conductor at right angles to the line. As the air blows over the conductor, eddy currents are set up on the leeward side. These eddy currents oscillate, first in one direction and then in the other. When the frequency of these eddies corresponds to some natural frequency of the line conductor, a tendency to vibrate exists.

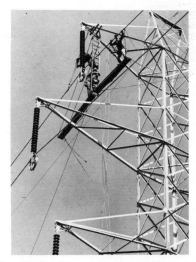

Fig. 22-35 Lineman on portable work platform removing sagging sheaves, attaching corona shield, conductor clamp, armor rods, and conical vibration damper. *(Courtesy American Electric Power System.)*

Fig. 22-36 Close-up view of lineman on work platform removing sagging sheaves, attaching corona shield, conductor clamp, armor rods, and conical vibration dampers. *(Courtesy American Electric Power System.)*

Fig. 22-37 Installing preformed armor rods on a 115,000-volt two-circuit transmission line. Note special yoke used to support conductor while armor rods are being applied. *(Courtesy Virginia Electric and Power Co.)*

Changes in velocity or direction of wind dampen the original vibration and set up new ones. It is for this reason that a steady gentle breeze may set up far more destructive vibrations than high winds of unsteady force and direction.

Under certain conditions these vibrations may build up an amplitude which produces alternating stresses large enough to cause fatigue failure. The failure usually takes place at the point of support, very seldom at a dead end or at a joint. Failure may not be experienced over a period of years, or it may develop in a few months. When a vibration passes through a wire, the conductor tends to conform to the wave form of the vibration. When this vibration encounters some obstacle such as a supporting clamp, it may be reflected back on itself with consequent increase in destruc-

Fig. 22-38 Installing preformed aluminum armor rods on 795,000-cir mil ACSR conductor. The armor rods will reduce conductor damage due to vibration of conductor in wind. Men are working from a hook ladder supported from the crossarm. *(Courtesy Detroit Edison Co.)*

Fig. 22-39 Applying armor rods. When all rods are partially twisted on, the assembly is lowered into the suspension clamp and clipped in. The rope and hoist are then removed. The installation of the rods is then completed. Oiling the rods helps installation. *(Courtesy Detroit Edison Co.)*

Fig. 22-40 Installing armor rods on 345,000-volt line. Conductor has been placed in insulator clamp. Note length of 18-unit insulator string. Also note chain hoist which was used in transferring conductor from the stringing sheave to the conductor clamp. Linemen are supported from ladder hooked over crossarm. *(Courtesy American Electric Power System.)*

Fig. 22-41 Close-up view of linemen wrapping armor rods on cable each side of insulator clamp. Note large corona shield which is also supported by the insulator clamp. Voltage of line 345,000 volts. *(Courtesy American Electric Power System.)*

tive stresses. More destructive still is the effect of two vibrations of similarly directed bending tendencies meeting at opposite ends of a common supporting clamp.

Vibration fatigue of conductors may be prevented either by reinforcing the conductor at points of stress to resist the effects of repeated bending or by reducing the vibration to negligible and harmless amplitudes with damping devices. One form of damper known as the "Stockbridge" damper is illustrated in Fig. 22-42. It consists of a resiliently supported weight with a suitable clamp for attaching it to the conductor. The action of the damper is to absorb continuously the energy of the vibrations started by winds of all kinds and to prevent these vibrations from building up to damaging proportions. The damper is attached a suitable distance ahead and back on the line from various types of conductor supports, dead ends, etc.

Distribution-circuit conductors are fastened to insulators by the use of either (1) wire ties or (2) clamps. Wire ties are used in distribution systems where the pin insulator is employed. Insulator clamps are used on some high-voltage pin-insulator lines and entirely on suspension insulators.

Conductors should occupy such a position on the insulator as will produce minimum strain on

Fig. 22-42 The Stockbridge damper clamped to transmission-line conductor to reduce conductor vibration by changing natural frequency of line conductor. *(Courtesy Aluminum Company of America.)*

the tie wire. The function of the tie wire is only to hold the conductor in place on the insulator, leaving the insulator and pin to take the strain of the conductor.

In straight-line work the best practice is to use a top-groove or *saddle-back* insulator. These insulators all carry grooves on the side as well. When the conductor is placed in the top groove (Fig. 22-43a), the tie wire serves only to keep the conductor from slipping out. The groove practically relieves the tie wire of any strain and allows the entire strain to be placed on the insulator and pin. When side-groove insulators are used in straight-line work, the wires should be placed

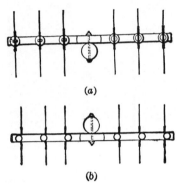

(a)

(b)

Fig. 22-43 Position of wires on insulator in straight-line work for (a) top-groove insulators, and (b) side-groove insulators. Top-groove insulators are generally used with conductors size 0 and larger and side-groove insulators with size 2 and smaller.

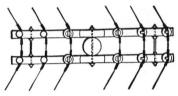

Fig. 22-44 Position of wires on both top- and side-groove insulators at angles. All conductors are placed on side of insulator so that pull is against insulator.

Fig. 22-45 Prefabricated conductor tie kit for top-tie insulator. *(Courtesy Preformed Line Products Co.)*

Fig. 22-46 Installation of prefabricated conductor tie on deenergized circuit. The neoprene pad is centered in the insulator groove between the conductor and the insulator. The center of the tie is positioned over the insulator so that both legs are parallel to the conductor. *(Courtesy Preformed Line Products Co.)*

as shown in Fig. 22-43b. The conductors on the pins nearest to the pole are placed on the side away from the pole. This is to increase the climbing space for the lineman. All other wires are placed on the pole side of the insulator. This prevents the conductors on the end pins from falling in case their tie wires break.

On corners and angles where the wires are not dead-ended, the conductors should be placed on the outside of the insulators (Fig. 22-44) on the far side of the pole and on the inside of the

Fig. 22-47 The prefabricated tie is rotated around the insulator so that one leg is around the neck of the insulator wrapped onto the conductor and the end snapped in place. *(Courtesy Preformed Line Products Co.)*

Fig. 22-48 Completed application of prefabricated tie with both legs wrapped around the conductor and the ends of the tie snapped in place. *(Courtesy Preformed Line Products Co.)*

insulators on the near side of the pole, irrespective of the type of insulator employed. This puts the conductor pull against the insulator instead of away from the insulator.

Distribution-circuit conductors are usually secured to the insulators with prefabricated ties. (See Figs. 22-45 through 22-52.) The ties are packaged and identified by the manufacturer to facilitate their correct application. The prefabricated ties can be installed on energized circuits with proper protective equipment in place to insulate the tie from grounded equipment or other energized

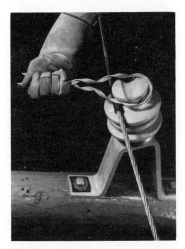

Fig. 22-49 Prefabricated conductor tie for application as a side tie. Neoprene pad is placed around the conductor to prevent conductor contact with insulator. The legs of the tie are squeezed together to enlarge the loop, enabling it to be pushed over the head of the insulator. *(Courtesy Preformed Line Products Co.)*

Fig. 22-50 The side tie is positioned around the neck of the insulator so that the conductor is between the legs of the tie. *(Courtesy Preformed Line Products Co.)*

Fig. 22-51 The legs of the side tie are pulled firmly and wrapped around the conductor. *(Courtesy Preformed Line Products Co.)*

Fig. 22-52 Prefabricated conductor side-tie installation completed. When both legs are completely wrapped around the conductor, the ends of the legs must be snapped into position against the conductor. *(Courtesy Preformed Line Products Co.)*

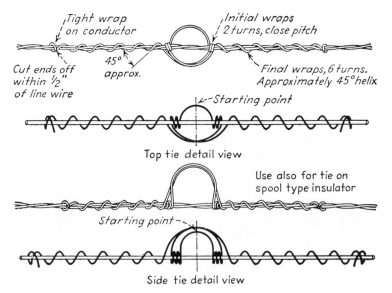

Fig. 22-53 Single-pin-type insulator tie for copper conductors. *(Courtesy Copper Wire Engineering Association.)*

phase conductors. Hot-line tools or rubber gloves and rubber sleeves would be used, depending on the voltage applied to the conductor.

Ties used for tying conductors to pin-type insulators should be relatively simple and easy to apply. They should bind the line conductor securely to the insulator and, in addition, should reinforce the conductor on both sides of the insulator. Any looseness between the conductor, tie wire, and insulator will result in chafing and injury to the conductor.

The hand-wrapped ties illustrated in Figs. 22-53 through 22-59 have been widely used and when properly applied give excellent service. These ties should be applied by hand without the

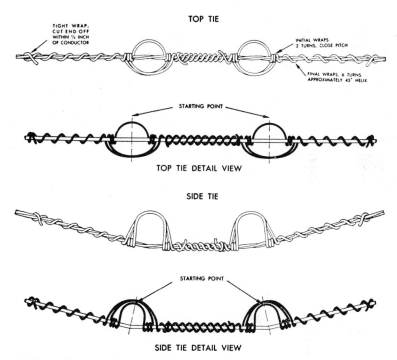

Fig. 22-54   Double-pin-type insulator tie for copper conductors. *(Courtesy Copper Wire Engineering Association.)*

use of pliers, and care should be taken to use the proper length and size of fully annealed tie wire specified for each conductor.

In general, the tie wire should be the same kind of wire as the line wire. If the line wire is a bare conductor, the tie wire should be bare also; if the line conductor is covered, the tie wire should also be covered. Copper tie wires should be used with copper line conductors and aluminum tie wires with aluminum line conductors. The tie wires, however, should always be made of soft-annealed wire as the hard-drawn tie wire would be too brittle and cannot be wrapped snugly. A hard tie wire might also injure the line conductor.

A tie wire should never be used the second time.

Good practice is to use No. 6 tie wire for line conductors of sizes No. 4 and smaller, No. 4 tie wire for line conductors No. 1 to No. 4, and No. 2 tie wire for line conductors No. 0 and larger, as shown in Table 22-1.

The length of the tie wire varies from 3 ft for a simple tie on a small insulator to 25 ft for a stirrup tie on a 50,000-volt insulator. The approximate lengths for various ranges of conductor sizes for some of the common ties are given in Table 22-2.

Table 22-1   Size of Tie Wire to Use with Various Sizes of Line Conductors

Size of line conductor, AWG	Size of tie wire, AWG
No. 4 and smaller	No. 6
No. 1 to No. 4	No. 4
No. 0 and larger	No. 2

Table 22-2   Approximate Length of Tie Wires Required for Different Types of Ties

Type of tie	Length of tie wire, in
Western Union	{ No. 000 bare cable, 54 { 2,000,000 cir-mil cable, 87
Bridle	{ No. 0000 bare cable, 54 { 2,000,000-cir mil cable, 87
Horseshoe	All sizes, 39
Armored Western Union	All sizes, 102
Armored top	All sizes, 114
Stirrup	All sizes, two pieces—each 54

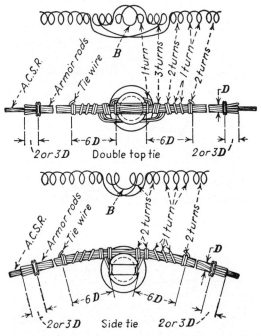

Fig. 22-55   Single-pin-type insulator tie for ACSR and aluminum cables equipped with armor rods. *(Courtesy Aluminum Company of America.)*

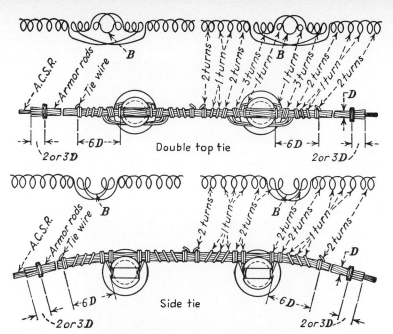

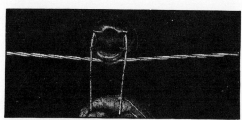

Fig. 22-56 Double-pin-type insulator tie for ACSR aluminum cables equipped with armor rods. *(Courtesy Aluminum Company of America.)*

Bend tie wire around insulator *above* conductor to form a U. Both legs of the tie wire should be of equal length after binding.

Holding tie wire tightly against insulator, throw two tight close wraps around conductor on each side of the insulator—then cross legs of tie wire around insulator.

Finish like top tie with six tight 45-deg spiral wraps on each side of the insulator. Bend back ends and cut off.

Fig. 22-57 Step in making side tie. *(Courtesy Copperweld Steel Co.)*

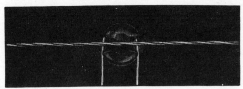

Bend annealed tie wire of proper size and length around insulator *under* the conductor, forming a U. Both legs of tie wire should be of equal length after bending.

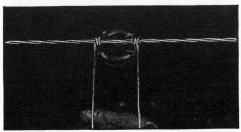

Holding the tie wire tightly against the insulator, throw two tight close wraps around the conductor on each side of the insulator, keeping these wraps snugly against the conductor.

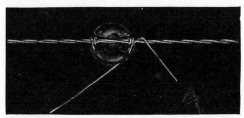

Cross the legs of the tie wire around the insulator, right to left and left to right.

With both legs of the tie wire crossed, tightly wrap each leg spirally around the conductor at an angle of 45 deg.

Fig. 22-58   Steps in making top tie.

Complete six spiral 45-deg wraps on each side of the insulator, bending back the ends and cutting them off short.

The finished top tie—made properly and tightly—will greatly reduce any possibility of conductor chafing at the insulator.

Fig. 22-59   Steps in making top tie *(continued)*. *(Courtesy Copperweld Steel Co.)*

Rules on good tying practice are as follows:

1. Use only fully annealed tie wire.

2. Use a size of tie wire which can be readily handled, yet one which will provide adequate strength.

3. Use a length of tie wire sufficient for making the complete tie, including an end allowance for gripping with the hands. The extra length should be cut from each end after the tie is completed.

4. A good tie should
   a. Provide a secure binding between line wire, insulator, and tie wire.
   b. Have positive contacts between the line wire and the tie wire so as to avoid any chafing contacts.
   c. Reinforce line wire in the vicinity of insulator.

5. Apply without the use of pliers.

6. Avoid nicking the line wire.

7. Do not use a tie wire which has been previously used.

8. Do not use hard-drawn wires for tying or fire-burned wire which is usually either only partially annealed or injured by overheating.

Steps in Making Ties   The simplest tie is the Western Union tie. It is universally used in distribution systems and is made as follows:

Bend tie wire around insulator *above* conductor to form a U. Both legs of the tie wire should be of equal length after binding.

Holding tie wire tightly against insulator, throw two tight close wraps around conductor on each side of the insulator—then cross legs of tie wire around insulator.

Finish like top tie with six tight 45-deg spiral wraps on each side of the insulator. Bend back ends and cut off.

Place the middle of the tie wire on one side of the insulator and the cable on the other side, both bearing in the side groove. Each end of the tie wire is then brought around the neck of the insulator, but underneath the cable, then up and around the cable, and served closely for about seven wraps.

The looped Western Union is quite similar except that the start is made on the cable side and the two ends are wound around the insulator once before being wrapped.

Bend annealed tie wire of proper size and length around insulator *under* the conductor, forming a U. Both legs of tie wire should be of equal length after bending.

Holding the tie wire tightly against the insulator, throw two tight close wraps around the conductor on each side of the insulator, keeping these wraps snugly against the conductor.

Cross the legs of the tie wire around the insulator, right to left and left to right.

With both legs of the tie wire crossed, tightly wrap each leg spirally around the conductor at an angle of 45 deg.

Fig. 22-60 An ACSR line conductor cable reinforced with armor rods. These armor rods reduce vibration and protect against chafing and flashover at point of support. Increased conductor size at support provides better grip for tie wire and also reduces danger of cable slippage. *(Courtesy Aluminum Company of America.)*

Complete six spiral 45-deg wraps on each side of the insulator, bending back the ends and cutting them off short.

The finished top tie—made properly and tightly—will greatly reduce any possibility of conductor chafing at the insulator.

Distribution-line conductors can be reinforced with armor rods, as illustrated in Figs. 22-60 and 22-61. This armor consists of a spiral layer of rods surrounding the conductor for a short distance (Fig. 22-62). The attachment of the conductor to its support is made in the middle of this armored length (Fig. 22-63). Since this provides a cable of much larger diameter than the conductor itself, the resistance to bending is greatly increased. This not only reduces the stress by distributing the bending but also strengthens the cable in the region of maximum bending stress.

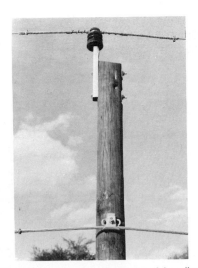

Fig. 22-61 Typical single-phase rural line illustrating the use of armor rods to minimize the effects of conductor vibration. *(Courtesy Ohio Brass Co.)*

Fig. 22-62 Lineman applying armor rods to ACSR conductor cable. *(Courtesy L. E. Myers Co.)*

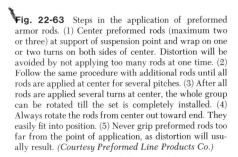

Fig. 22-63 Steps in the application of preformed armor rods. (1) Center preformed rods (maximum two or three) at support of suspension point and wrap on one or two turns on both sides of center. Distortion will be avoided by not applying too many rods at one time. (2) Follow the same procedure with additional rods until all rods are applied at center for several pitches. (3) After all rods are applied several turns at center, the whole group can be rotated till the set is completely installed. (4) Always rotate the rods from center out toward end. They easily fit into position. (5) Never grip preformed rods too far from the point of application, as distortion will usually result. (*Courtesy Preformed Line Products Co.*)

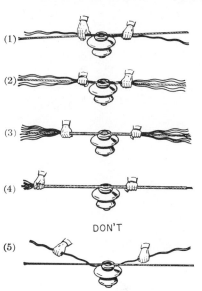

Clamp-top line-post insulators can be used on distribution circuits to eliminate tying the conductor to the insulator (Fig. 22-64). The conductor can be protected with armor rods if necessary before it is placed in conductor clamp. Dead-end clamps of the bolted, or compression, type are used to dead-end distribution conductors (Fig. 22-65).

Spacer Installation   Bundled conductor transmission lines use spacers to maintain uniform distance between the conductors for each phase (Figs. 22-66 and 22-67). The number of spacers to be installed at equal intervals in each span will be detailed in the specifications for the line.

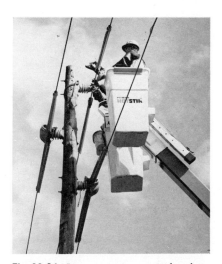

Fig. 22-64 Linemen securing energized conductor to clamp top-line-post insulator. Note protective equipment used on conductors. Linemen wear rubber gloves, rubber sleeves, hard hats, and safety glasses. Work is being performed from an insulated bucket truck. (*Courtesy A B. Chance Co.*)

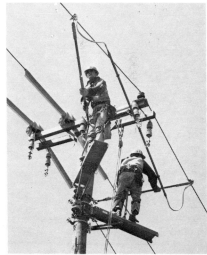

Fig. 22-65 Linemen cutting in dead end on an energized circuit. Note bolt-type dead-end clamps attached to bell insulators to be used to dead end the conductors. (*Courtesy A. B. Chance Co.*)

Fig. 22-66 Lineman tightening spacer on a two-conductor bundle. The spacer keeps the conductors of the bundle physically separated. Conductors shown are 1.6-in-diameter expanded ACSR. Voltage of line is 460,000 volts. *(Courtesy Aluminum Comapny of America.)*

Fig. 22-67 Lineman tightening spacer which maintains fixed separation among grouped conductors, four in this case. Two or more conductors acting as a single "wire" are called "bundled" conductors. Bundled conductors increase the power capacity of a line without the need for building a second parallel line of towers. *(Courtesy Aluminum Company of America.)*

Fig. 22-68 Linemen installing spacers on a 765-kV transmission line with four bundled conductors per phase using self-propelled aerial cart. *(Courtesy American Electric Power System.)*

Fig. 22-69 Lineman operating self-propelled aerial cart used to install conductor spacers. The cart travels on grooved wheels which can be engaged and disengaged in sequence to negotiate past insulator strings at the suspension points. *(Courtesy Sherman & Reilly, Inc.)*

Fig. 22-70 Lineman installing corona rings after completing installation of tapered armor rod assembly on a conductor bundle of 1.75-in-diameter ACSR. Lineman is working from a ladder hooked to the tower crossarm. *(Courtesy Aluminum Company of America.)*

Spacers should be installed immediately after clipping-in operations are completed in each pull to minimize conductor damage due to contacts between the subconductors (Figs. 22-68 and 22-69). The bolted spacers should be installed with the conductor clamp free to rotate on one conductor without binding, and then the second clamp should be bolted loosely on the second conductor. After the spacer is in place properly, the lineman should hold the clamp firmly in place and tighten all bolts to the proper tension.

Corona Shield Installation Extra- and ultra-high-voltage transmission lines require corona shields at each point of conductor support. The corona shields are normally installed after the conductors are clipped in and spacers installed (Fig. 22-70).

Joining Line Conductors

Transmission- and distribution-line conductors must be joined together with full tension splices. Compression, internally fired, and automatic-type splices are used to obtain the strength and electrical conductivity needed. The number of locations where the conductors are joined should be kept to a minimum. All splices in transmission-line conductors should be made at least 50 ft, or more, from the suspension, or dead-end points. Splices should not be made in spans crossing over railroads, rivers, canals, or inter-state highways. It is advisable to avoid splices in spans crossing over communication circuits or electric transmission and distribution lines, if possible.

The splices must be applied in accordance with the transmission-line specifications, or distribution standards, and in accordance with splice manufacturer's instructions. The conductors of a transmission line requiring a full tension splice should be laid out straight for a distance of 40 ft each way from the joining point and straightened at the ends to prepare the conductors for splicing (Fig. 23-1). The conductors should be supported, and if necessary laid on polyethylene sheets to keep them clean if laid along the ground in preparation for splicing.

Compression Joint for ACSR Conductor   A compression joint for an aluminum conductor with steel reinforcing actually consists of two separate joints, one for the steel core and the other for the overall cable (Fig. 23-2). The complete joint, therefore, consists of a small steel sleeve compression joint for the central steel core and a much larger aluminum compression sleeve joint for the overall aluminum conductor. The plugs are for sealing the holes in the aluminum sleeve after the paste filler has been injected.

The steps in the making of the compression joint on an ACSR conductor are as follows (see Fig. 23-3):

Caution: Before proceeding, make sure that the bores in the two sleeves are perfectly clean.

1. A compression connector shall be used only for the size of conductor for which it is specified. Neither the connector nor the conductor shall be altered to attempt installation of a fitting on a conductor for which it is not designed.

2. Select the correct press and dies for the conductor and compression connector.

3. Slip the aluminum compression sleeve over one cable end, and back it out of the way along the cable.

4. Cut off the aluminum strands from each cable end, exposing the steel core for a distance of a little more than half the length of the steel compression sleeve. Use care not to nick the steel core. Before cutting, secure the cable just back of the cut.

5. Insert the steel core ends into the steel compression sleeve, making sure that the ends are jammed against the stop in the middle of the sleeve.

6. Compress the steel sleeve over its entire length, making the first compression at the center and working out toward the ends, allowing dies always to overlap their previous position (Fig. 23-4).

Fig. 23-1 Lineman cutting conductor with cable cutter in preparation for completing splice. The conductor to be inserted in the splice is held with a clamp to prevent the wire strands from unraveling. *(Courtesy L. E. Myers Co.)*

7. Remove restraint from the cable and measure from the center of the steel joint the distance of one-half the length of the aluminum sleeve and mark the cable with tape. This will center the aluminum sleeve when slipped over the steel joint. Brush the aluminum conductor with a wire brush to a bright finish. Coat the brushed conductor with an inhibitor-type compound.

8a. Slip the aluminum sleeve up over the steel joint to the tape mark. This will center the aluminum sleeve over the steel joint.

    b. Using the calking gun equipped with the tapered nozzle provided with the compressor, inject filler paste through both holes provided in the aluminum joint until the paste is visible at the ends of the aluminum joint. The preferred filler paste contains zinc chromate (Fig. 23-5).

    c. Insert the plugs in the filler holes and hammer them firmly in place. They will be securely locked in compressing the aluminum joint.

9. Compress the aluminum sleeve. Make the first two compressions with the inner edges of the dies matching the positions stenciled on the aluminum sleeve. Make additional compressions advancing to the ends, allowing the dies always to overlap the previous position (Figs. 23-6 and 23-7).

Compression Joints for Distribution Conductors Single sleeve full tension splices are usually used for aluminum, aluminum alloy, ACSR, and copper distribution-line conductors (Fig. 23-8). Compression joints for smaller distribution conductors can be made with light-weight hand-powered compression tools like those shown in Figs. 23-9, 23-10, and 23-11. Power-driven and hand-operated pump-type compression tools are used for the larger distribution conductor joints (Figs. 23-12 and 23-13).

Single sleeve compression joints should be completed in the following manner:

1. Select the proper connector for the type and size of distribution-line conductor to be spliced.

2. Use the proper compression tools and dies fo the connector and the wire.

3. Clean the conductor thoroughly to remove corrosion and film. Aluminum conductors should be brushed with a wire brush until the wire has a bright finish.

4. Cover the conductor with an inhibitor compound.

5. Cut off the ends of the conductors to be joined with a tube cutter or a hack saw to obtain flush ends on the conductors. The splice should be made at least 12 in from the edge of the nearest tie wire, armor rod, or dead-end clamp on the distribution conductor.

Fig. 23-2 Parts of compression sleeve for ACSR conductor. *(Top)* Steel sleeve for steel core, flanked by plugs for sealing holes in the aluminum sleeve after paste filler has been injected. *(Bottom)* Aluminum sleeve for complete cable. *(Courtesy James R. Kearney Corp.)*

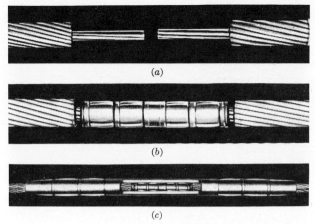

(a)

(b)

(c)

Fig. 23-3 Showing several steps in the making of a compression joint on an ACSR conductor: (a) *Step 4:* Aluminum strands cut back a distance equal to one-half of the length of the steel sleeve. (b) *Step 6:* Completed compression joint on the steel core. (c) *Step 9:* Completed compression joint on ACSR conductor. Cutaway view shows inner steel core compression joint. *(Courtesy Burndy Corporation.)*

6. Support the conductors to be joined in a straight line on each side of the joint to keep the splice straight. The wires should be straight for a distance of 10 ft or more if possible.

7. Wrap two turns of tape around one conductor exactly one-half the length of the sleeve from the end of the conductor if the sleeve has no center stop.

8. If the sleeve is not completely filled with inhibitor compound, apply additional compound to the conductor so that the sleeve will be completely filled when installed.

9. Slip the ends of the conductors into the sleeve with one end of the sleeve touching the tape and the conductor ends touching inside the sleeve or against the stop if one is provided in the sleeve.

10. Make the required number of compressions on the sleeve beginning at the center and working toward the ends of the sleeve.

Fig. 23-4 Joining conductor by means of a power-driven hydraulic compressor. At the instant shown, the sleeve is being compressed onto the steel core. The larger sleeve shown at right will be compressed onto the aluminum strands after the sleeve on the steel core is completed. The conductor shown is a 795,000-cir mil ACSR cable. *(Courtesy Detroit Edison Co.)*

Fig. 23-5 Impregnating compression-sleeve cable splice with zinc chromate to prevent oxidation of joint. Note compressor has been opened to expose central portion of joint. *(Courtesy American Electric Power System.)*

Fig. 23-6 Compressing aluminum sleeve onto an ACSR cable by means of a power-driven hydraulic press. The cable being spliced is a 1,275,000-cir mil expanded ACSR cable having an outside diameter of 1.6 in. Note wrapping on cable to keep strands from unraveling. *(Courtesy American Electric Power System.)*

Fig. 23-7 Completed compression-sleeve joint on a 1,275,000-cir mil expanded ACSR cable having an outside diameter of 1.6 in. This cable is being installed on a 345,000-volt transmission grid. The average span over level country is 1200 ft and in mountainous terrain is 1700 ft. The large diameter is necessitated to hold the corona loss to a reasonable value. *(Courtesy American Electric Power System.)*

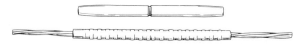

Fig. 23-8 Compression-type splicing sleeve. Upper view shows sleeve before compression, and lower view shows sleeve fully compressed. Markings on sleeve are produced by die used in compression. *(Courtesy Copperweld Steel Co.)*

Fig. 23-9 Hand-operated compression tool modeled after bolt cutter uses principle of lever to develop force needed to compress sleeve onto conductor. Wood handle provides insulation for use on live low-voltage lines. *(Courtesy James R. Kearney Corp.)*

Fig. 23-10 Hand-operated compression tool using principle of the hydraulic press. Less than 20 strokes of the handle move the dies from wide open to full compression. Tool is capable of producing 21,000-lb thrust. A pressure-relief valve prevents overcompression. Long wood handles can be used for use in liveline work.

Fig. 23-11 Lineman using hand-operated compression tool to connect service drop to secondary main. Compression tool operates on hydraulic principle. *(Courtesy Aluminum Company of America.)*

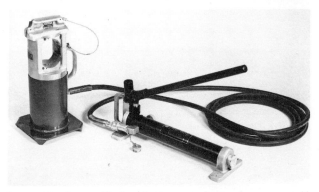

Fig. 23-12 Compression tool operated by foot, electric, or pneudraulic power pump, capable of developing 60-tons compression force. It can be used to install compression sleeves on conductors as large as 1,780,000 cir mils and weighs 38 lb. *(Courtesy Burndy Corporation.)*

Fig. 23-13 Line conductors joined wth hand-powered hydraulic compression tool. Compressor is capable of providing a compression load of 40,000 lb. Dies are interchangeable to fit various sizes of splicing sleeves. *(Courtesy Copperweld Steel Co.)*

Compression tools can be used with the proper accessories in a manner similar to the procedures described to dead-end conductors, complete taps, and install terminals or lugs (Figs. 23-14 through 23-22).

Internally Fired Connector System Internally fired connectors can be used for both transmission and distribution lines. The connectors are made for joining stranded aluminum, aluminum alloy, and aluminum-steel composite (ACSR) conductors. The housing of the connectors is made of high-tempered aluminum alloy, tapered at the ends where the conductors enter (Fig. 23-23). The housing contains a high-strength steel powder chamber that is loaded with a fast-burning propellant charge. Igniting the charge creates instantaneous high pressure in the chamber. This pressure drives cylindrical sets of wedge-shaped serrated aluminum jaws (into which the conductor ends have been inserted) at high velocity into the tapered ends of the housing (Fig. 23-24).

The jaws clamp and lock the conductor ends in position, providing the required holding strength and establishing a low resistant current path across the housing. When joining ACSR

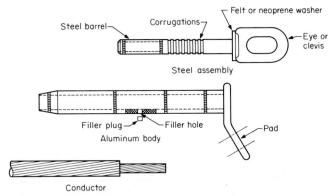

PROCEDURE:

A. Prepare the conductor and dead end as stated in procedure for splicing ACSR conductors.

B. Slide aluminum body over conductor until sufficient working length protrudes from pad end.

C. Cut back aluminum strands a distance equal to the depth of wire bore in steel barrel plus 1/2 inch. Do not nick steel strand. File off all burns or sharp edges from ends of steel and aluminum strands.

D. Insert steel strands in steel barrel to full depth.

E. Select die size for compressing steel barrel. The die size number on die and die size number for steel barrel marked on steel assembly must be the same.

F. Make the required number of compressions on steel barrel beginning at knurl adjacent to corrugations and working toward mouth of barrel.

G. Coat steel assembly with inhibitor compound up to felt or neopreme washer.

H. Slide aluminum body over conductor and steel assembly until pad end butts solidly against felt or neoprene washer. Washer should butt against shoulder of clevis or eye.

I. Align eye or clevis of the steel assembly so that the pad of the aluminum body will be in the proper position and the eye or clevis will line up properly with the attaching hardware.

J. Select die size for compressing aluminum body. The die size number or die and die size number for aluminum marked on aluminum body must be the same.

K. Hold aluminum body in position against felt or neopreme washer and compress onto the steel assembly by making the required number of compressions beginning at the knurl mark nearest pad and working to the second knurl mark from pad end. No compressions shall be made between the second and third knurl mark from the pad end.

L. Inject Imibitor compound into filler hole until compound is visible at the end of aluminum body. Insert and drive filler plug into hole and peen edge of hold over top surface of plug.

M. Compress aluminum body onto the conductor by marking the required number of compressions beginning at third knurl from pad end and working toward outer end.

N. Finish the installation to provide the compressed portion of the connector with a smooth uniform appearance. Remove any flashing with file or emery cloth and crocus cloth. Wash the connector and adjacent wire with solvent to remove all inhibitor compound or other possible causes of corona.

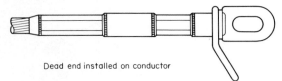

Dead end installed on conductor

Fig. 23-14 Compression dead end for ACSR conductor.

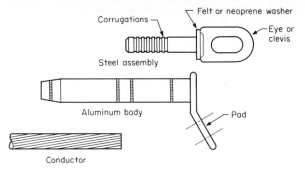

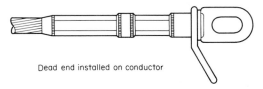

PROCEDURE:

A. Prepare the conductor and dead end as stated in procedures for splicing distribution conductors.

B. Coat steel assembly with inhibitor compound up to felt or neoprene washer.

C. Slide aluminum body over steel assembly until pad end butts solidly against felt or neoprene washer. Washer should butt against shoulder of clevis or eye.

D. Align eye or clevis of the steel assembly so that the pad of the aluminum body will be in the proper position and the eye or clevis will line up properly with the attaching hardware.

E. Select die size for compressing aluminum body. The die size number on die and die number for aluminum marked on aluminum body must be the same.

F. Hold aluminum body in position against felt washer and compress onto the steel assembly by making the required number of compressions beginning at the knurl mark nearest pad and working to the second knurl mark from pad end. No compressions shall be made between the second and third knurl mark from the pad end.

G. Insert conductor to the full depth of the bore and mark conductor at end of body. Check by measuring.

H. Inject sufficient inhibitor compound into the aluminum body to insure that excess compound will be visible when body is completely compressed.

I. Insert clean, brushed end of the conductor into the body to the mark on the conductor.

J. Compress aluminum body onto the conductor. Using same die as used in step E, by making the required number of compressions beginning at third knurl mark from pad and working toward outer end.

K. Finish the installation to provide the compressed portion of the connector with a smooth uniform appearance. Remove any flashing with file or emery cloth and crocus cloth. Wash the connector and adjacent wire with solvent to remove all inhibitor compound or other possible causes of corona.

Dead end installed on conductor

Fig. 23-15 Compression dead end for aluminum and aluminum alloy conductors.

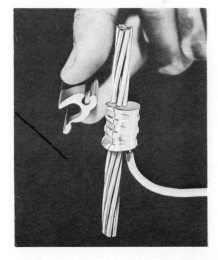

Fig. 23-16 Compression tap and completed tap joint used for service drop, secondary or primary tap, and grounding installations on all combinations of aluminum to aluminum, or aluminum to copper conductors. *(Courtesy Somerset Division of Homac Manufacturing Co.)*

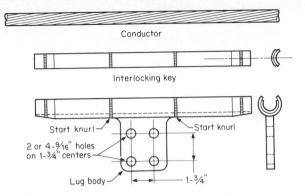

Conductor

Interlocking key

Start knurl · · Start knurl

2 or 4-⁹⁄₁₆" holes on 1-¾" centers

Lug body · ← 1-¾" →

PROCEDURE:

A. Prepare the conductor and tee lug as stated in procedure for splicing distribution conductors.

B. Coat the clean, brushed conductor and inner surface of key and body with sufficient compound so that the groove will be completely filled.

C. Wire sizes smaller than 477 MCM do not have interlocking key.

D. Insert clean, brushed conductor and interlocking key into groove.

E. Select correct die size for tee lugs. The die size number on die and die size number marked on tee lug must be the same.

F. Make the required number of compressions for tee lugs. The center portion which is over the pad is not compressed.

Make initial compression beyond "start" knurl on one end of the barrel of tee lug. Make second compression beyond "start" knurl on opposite end and complete compressions to the end of the barrel of tee lug. Then complete compressions on starting end.

G. Clean contact areas of pad and connecting fitting. Coat with inhibitor compound. Wire brush through compound and assemble joint without removing compound.

H. Finish the installation to provide the compressed portion of the connector with a smooth uniform appearance. Remove any flashing with file or emery cloth and crocus cloth. Wash the connector and adjacent wire with solvent to remove all inhibitor compound or other possible causes of corona.

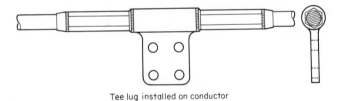

Tee lug installed on conductor

Fig. 23-17 Compression tee-lug connector for installation on aluminum conductor.

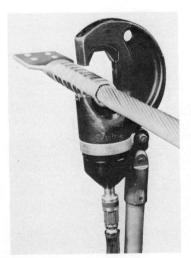

Fig. 23-18 Compression straight lug terminal installed on large aluminum distribution conductor with electric pump operated compression tool. Hydraulic head and dies are supported with hot stick. Hydraulic hose and fluid have a high dielectric for use on energized conductors. (*Courtesy Somerset Division of Homac Manufacturing Co.*)

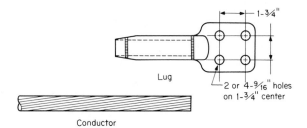

Lug

Conductor

2 or 4-9/16" holes
on 1-3/4" center

PROCEDURE:

A. Prepare the conductor and lug as stated in procedures for splicing distribution conductors.

B. If lug is prefilled with inhibitor compound, mark conductor end with pencil or crayon at a distance equal to length from barrel end of connector to the knurl at pad end plus 3/8 inch.

C. If lug is not prefilled with innibitor compound, insert conductor to the full depth of the bore and mark conductor with pencil or crayon at end of barrel. Remove conductor after marking. Inject sufficient inhibitor compound into the barrel to insure that excess compound will be visible when barrel is completely compressed.

D. Insert clean, brushed end of the conductor into the barrel to the mark on the conductor.

E. Select correct die size for straight lugs. The die size number on die and die size number marked on lug must be the same.

F. Make the required number of compressions. for straight lugs. Beginning at pad end and working to outer end.

G. Clean contact areas of pad and connecting fittings. Coat with inhibitor compound. Wire brush through compound and assemble joint without removing compound.

H. Finish the installation to provide the compressed portion of the connector with a smooth uniform appearance. Remove any flashing with file or emery cloth and crocus cloth. Wash the connector and adjacent wire with solvent to remove all inhibitor compound or other possible causes of corona.

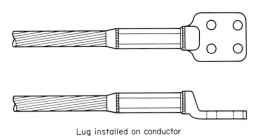

Lug installed on conductor

Fig. 23-19 Compression straight lug terminal for aluminum conductor.

Fig. 23-20 Compression-pin terminals and completed pin-terminal joint used to connect aluminum conductors to copper alloy equipment clamps, designed for use with copper. Cold flow problems of aluminum conductor in copper clamps are eliminated. (*Courtesy Somerset Division of Homac Manufacturing Co.*)

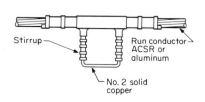

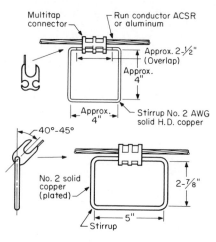

Notes:

1. Clean all conductors thoroughly with soft wire brush or with No. 70 emery cloth.

2. After the cleaning operation, coat the conductors with corrosion-inhibiting grease before installing connector. Apply grease not later than one hour after the cleaning operation. If connectors are factory-packed with grease, be sure any supplementary grease used on run or tap conductors is similar or compatible.

3. Be sure connector contact surfaces are clean. (Consider connectors factory-packed with grease to have clean contact surface when received.)

4. Always place copper conductor below aluminum conductor where possible.

5. Install connectors directly on line conductor where possible instead of over armor rods.

6. Use compression tool with proper dies to install connector.

Fig. 23-21 Compression stirrup connector. Stirrup permits use of hot-line taps on distribution line without damaging line conductor.

Fig. 23-22 Compression stirrup installed on aluminum distribution line. Hot-line tap installed on stirrup permits readily removable connection for distribution transformer. *(Courtesy Sherman & Reilly, Inc.)*

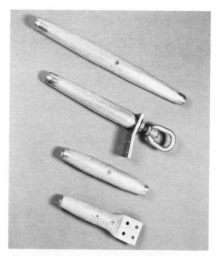

Fig. 23-23 Internally fired splice, dead-end, jumper sleeve, and terminal shown in sequence top to bottom. Splice and dead end are full tension connectors. *(Courtesy Amp., Incorporated.)*

conductors, the aluminum jaw assemblies of full tension splices and dead ends contain a cylindrical set of steel jaws to grip the conductor steel core (Fig. 23-25).

The conductors must be cut and cleaned in the same manner as described for compression fittings. The aluminum surface of the conductor should be brushed with a wire brush to a bright finish, and coated with an inhibitor compound containing abrasive particles before inserting the conductor in the connector housing. A remote-firing collar consisting of a plug at the end of 10 ft of firing leads fastened to a belt is used to ignite the powder charge. The plug is inserted into the ignitor hole, by the lineman, and secured in position by wrapping the belt around the connector. The lineman connects the firing leads to the terminals of a 6-volt lantern battery to fire the charge. Completed dead-end assemblies for a two-conductor bundled transmission line is shown in Fig. 23-26.

Internally fired jumper terminals and lubs are manufactured for ignition with a battery (see Fig. 23-23) or on impact (Fig. 23-27). The impact internally fired lug is actuated by the lineman by removing the safety pin and then hitting the lug end with a hammer. The lugs and jumper terminals are used to complete connections between the dead-end terminals and equipment (Fig. 23-28).

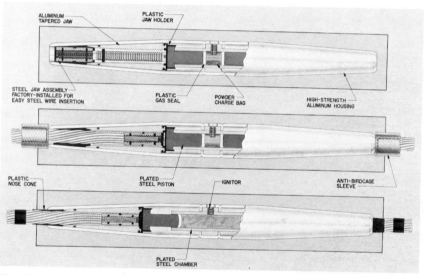

Fig. 23-24 Sectional view of internally fired splice shown before, during, and after installation. Splice is used for full tension applications. *(Courtesy Amp., Incorporated.)*

The Ampact tool is convenient for the lineman to use to install propelled tap connectors (Fig. 23-29). The tool is loaded with a specially designed color-coded shell that contains a small propellant charge. The tap connector consists of a tapered C-shaped member and a wedge (Fig. 23-30). The lineman hooks the "C" member over the main and tap conductors and places the wedge in position. The joint can be completed on energized conductors using hot sticks, or with rubber glove and sleeve procedures (Fig. 23-31).

The lineman clamps the tool over the tap and detonates the propellant in the tool by striking the end of the tool with a hammer. The tool drives the wedge at high velocity between the main conductor and the tap wire within the support of the "C" member. This action spreads the "C" member. As it attempts to revert to its original shape, the "C" member exercises a permanent retentive force on the electrical contact areas. The locking tab ensures the wedge remains in position even under severe service conditions (Fig. 23-32). The Ampact tool can be used to remove the taps. The propellant reduces the effort needed to remove the tap. The main conductor is free of connector bodies after the tap has been removed.

Automatic Tension Splice   This type of joint is made with a single-bore sleeve having internal gripping jaws at each end (Fig. 23-33).

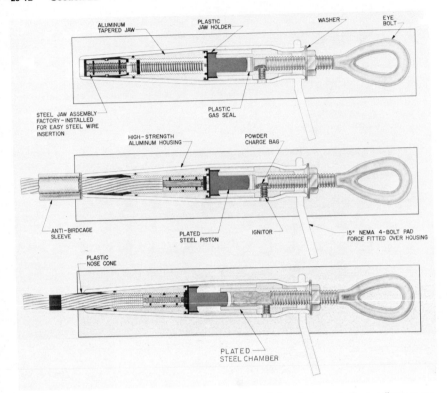

Fig. 23-25 Sectional view of internally fired dead end shown before, during, and after installation to terminate on ACSR conductor. Dead end is used for full tension applications. *(Courtesy Amp., Incorporated.)*

Fig. 23-26 Two-conductor bundled transmission circuit terminated on steel pole with internally fired deadend connectors. *(Courtesy Amp., Incorporated.)*

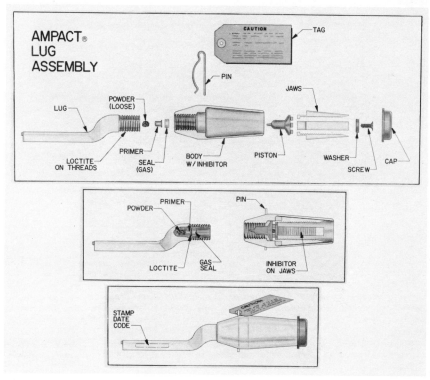

AMPACT® LUG ASSEMBLY

Fig. 23-27 Sectional view of internally fired impact-type jumper lug. Conductor is prepared in the normal manner as previously described before it is inserted in the lug housing. *(Courtesy Amp., Incorporated.)*

Fig. 23-28 Internally fired jumper terminals connect the line conductors to disconnect switch pad at dead-end structure of substation. *(Courtesy Amp., Incorporated.)*

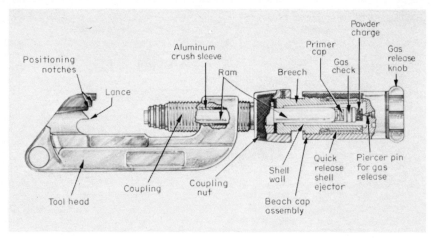

Fig. 23-29   Detail and sectional view of Ampact tool. *(Courtesy Amp., Incorporated.)*

Automatic line splices may be used in any span where the tension during installation exceeds 15 percent of the rated breaking strength of the conductor. It is necessary to force the teeth of the jaws into the conductor to obtain reliable gripping. This should be accomplished by imposing several severe jerks on the conductor or by exerting enough pulling tension on the conductor to obtain 15 percent of the rated breaking strength. Automatic line splices *shall not be used* in taps or jumpers.

Splices should not be made in spans crossing railroads and preferably not in adjacent spans which are depended upon for withstanding the longitudinal load of the crossing conductors. Splices shall not be installed within 12 in of a tie wire or armor rod.

Automatic line splices must remain free of dirt which would impair the proper operation of the internal jaws. Splices which have lost their original protective equipment, such as plastic end

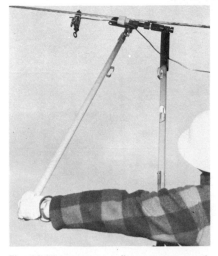

Fig. 23-30 Ampact tap component parts, assembly, and completed tap joint. *(Courtesy Amp., Incorporated.)*

Fig. 23-31 Lineman installing Ampact top with hot-line tools. Ampact tool is in position to complete the connection. A piggyback hot-line clamp is used to hold the tap conductor in place while regular tap is completed. *(Courtesy Amp., Incorporated.)*

inserts, must be carefully inspected for the presence of dirt and such other contaminants which would impair the operation of the jaws.

During installation, automatic line splices must not be dragged through dirt, mud, ice, or water and must remain free of fouling until the conductor reaches its final installed tension. Any splice with a deformed or dented barrel shall not be used. The above conditions interfere with the proper seating of the internal jaws.

Fig. 23-32 Dead-end pole with conductors terminated with internally fired dead-end connectors. Jumpers are connected between conductors with Ampact taps. *(Courtesy Amp., Incorporated.)*

Fig. 23-33 Automatic tension splice for use on aluminum, aluminum-alloy, and copper conductors. Ends of conductors are surrounded by free-floating self-aligning chucks. The spring serves to position the chucks against the conductors. *(Courtesy Reliable Electric Co.)*

The making of good splice with an automatic line splice requires following these important procedures:

1. Choose the proper size and type of splice for the conductor size and material.
2. Clean the conductor with a wire brush and remove all burrs but *do not "grease"* (automatic splices are pregreased with a material which is grit free. The presence of grit prevents proper jaw operation).
3. Be sure the conductor strands are in normal lay and not deformed.
4. Insert the conductor with one sharp continuous thrust, forcing it in as far as possible.
    a. Aluminum splices are provided with a pilot cup for retaining and guiding the strands into the jaws of the chuck. This pilot cup must remain in place and will be driven into the interior of the splice until it strikes the center barrier.
    b. Copper splices are provided with slotted openings which allow visible examination to determine if the conductors are properly inserted into the jaws.
5. It is extremely important to set the jaws into the conductor by applying one or more momentary jerks to the conductor forcing the jaws to bite into the conductor.

When an attempt is made to withdraw the conductors, the jaws clamp down on the conductors because of a taper in the bore of the sleeve and on the jaws. Tension causes a wedging action which increases with the pull applied on the conductor. A loss of tension could cause the jaws to release their grip and allow the conductor to drop out. This type of splice is therefore used where

Fig. 23-34 Lineman working in insulated bucket using rubber gloves, rubber sleeves, safety glasses, and insulated hard hat to install bolted parallel groove clamp to complete jumper between energized dead-ended conductors. Temporary insulated jumper with hot-line taps is used to by-pass regular connection being completed. Conductors are terminated with bolted-type dead-end clamps.

the wires are kept in tension. It is also used where service must be restored quickly. It is especially suited for live-line splicing.

Bolted Joints   Conductors can be joined together with bolted connectors if the wires are not under tension (Fig. 23-34). Bolted connectors are used successfully on both copper and aluminum conductors. The conductors must be cut to the proper length, cleaned, and coated with an inhibitor to obtain a good connection. Bolted parallel groove clamps, U-bolt clamps, tap lugs, and dead-end clamps are manufactured for use on a range of conductor sizes. The lineman must use care to obtain the proper tension when tightning the bolts on a bolted connector.

Section **24**

Live-Line Maintenance with Hot-Line Tools

The trend in electric-system management is to do more and more of the testing, repair, and maintenance work while the lines are live. The reason for this is to reduce the number of times the service is interrupted, or the number of "outages" as the interruptions in service are called. Live-line maintenance has been made possible by the development of special tools and procedures for such work.

Under live-line maintenance a great variety of work is included. The most common of these live-line operations are as follows:

1. Replacing insulators
   a. Pin
   b. Suspension
   c. Strain
2. Replacing crossarms
3. Replacing poles
4. Tapping a hot line
5. Cutting slack in or out
6. Splicing conductors
7. Installing vibration dampers
8. Installing armor rods
9. Phasing conductors

The most important single item is the replacement of insulators. Live-line work, therefore, resolves itself chiefly into the detection of the defective insulators and the replacement of these insulators while the system is live.

It will not be the object here to give detailed instructions covering the procedure of every hot-line tool live-line maintenance job. This would be impossible because of the great variety of jobs and types of line construction and special conditions. This section is, therefore, intended to give the reader only a general conception of the methods and procedures employed in the most fundamental operations of live-line maintenance work with hot sticks.

The maintenance of high-voltage lines while energized, or "hot," may appear on first thought to be hazardous, especially when compared with working on dead, or deenergized lines. Actually, however, the work is safe because the lineman is continually conscious of the fact that the lines are "hot" and of the need to be careful. Furthermore, there is no possibility of the line being "hot" when it was thought to be dead as when working on dead lines. Also there is no possibility of confusion with live duplicate circuits. For in live-line or "hot-stick" work every conductor is "hot" and every operation is planned and performed accordingly.

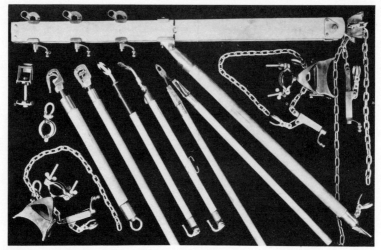

Fig. 24-1 Typical utility live-line tool set, designed for use on rural lines having voltages up to 13,200 volts. *(Courtesy James R. Kearney Corp.)*

TOOLS

"Hot-line" tools are of great variety, but the more generally used can be classified as follows (see Figs. 24-1 and 24-2):

1. Wire tongs
2. Wire-tong supports or saddles
3. Insulated tension links
4. Auxiliary arm
5. Strain carrier
6. Tie sticks
7. Platform
8. Insulated hoods
9. Insulated tools

Wire Tongs   A wire tong is a slender insulated pole provided with a hook clamp on one end and a swivel on the other end (see Fig. 24-3). Tongs are used primarily in holding live conductors. They are sometimes referred to as lifting sticks or holding sticks. The hook clamp is opened or closed by turning the pole. A live conductor is clamped by placing the hook over and turning the

Fig. 24-2 Special hot-line tool trailer. Note elevated cover. Tools are for live-line work on distribution and transmission lines. *(Courtesy A. B. Chance Co.)*

pole until the jaws of the hook are completely closed. The swivel clevis on the other end is for the application of lifting tackle.

The wire tong is used to hold the conductor away from the point where work is to be done, as when replacing an insulator or a crossarm. The simplest case is illustrated in Fig. 24-4, where the wire tongs are used to hold the live-line conductor on a rural line to one side while a transformer is being installed. A more complicated operation is illustrated in Fig. 24-5, where a number of wire tongs are used to hold the three conductors of a three-phase line away from the pole while the post insulators are replaced.

The long slender pole of a wire tong is made of either (1) wood or (2) fiberglass. In either case the pole serves as an insulator.

Fig. 24-3   Wire tong consisting of pole, conductor clamp, and swivel. Pole is made of either wood or fiberglass. *(Courtesy A. B. Chance Co.)*

Fig. 24-4   Lineman using two wire tongs to hold live conductor on rural line to one side so as to clear working space needed in installing transformer. *(Courtesy A. B. Chance Co.)*

Fig. 24-5   Lineman moving third line conductor into the clear, preparatory to changing post insulators. One pair of wire tongs is used to hold each line conductor away from the pole. *(Courtesy A. B. Chance Co.)*

If the pole is of wood, it must be kept dry and clean. Scratches, scars, or other abrasions on the stock will damage the wood surface and permit moisture and dirt to enter the fibers of the wood. Moisture and dirt will form conducting paths along the hot stick which will render it hazardous for use.

If the pole is of fiberglass, it will not absorb moisture and therefore does not have to be dried periodically. Fiberglass is not affected by creosote, gasoline, oil, grease, or most solvents. The smooth pole does not have any grooves to trap dirt or moisture. Fiberglass poles are stronger and a better insulator by one-third than wood. They can also be molded in bright colors for improved visibility.

Wire-Tong Supports   A wire-tong support serves the special purpose of guiding and holding a wire tong in a fixed position after the conductor has been moved. A typical tong support is

Fig. 24-6   Typical wire-tong base for anchoring tong to pole. *(Courtesy James R. Kearney Corp.)*

shown in Fig. 24-6. The support consists essentially of a base or saddle fastened by a chain to the pole or tower. A swivel clamp is mounted on the saddle. The wire tong slides between the jaws of this clamp, while the swivel action permits movement of the tong in all directions. Wing nuts are used to tighten the jaw of the clamp onto the tong.

A multiple form of tong support is illustrated in Fig. 24-7. When a number of saddles are needed on the same side of a pole or tower, a slide type of wire-tong support may be used. This support consists of a round rod secured to the pole at each end by means of a chain. On the rod are placed the tong holders. These holders slide up and down the rod freely but can be clamped in any position. One to six tongs can be supported from one fastening, as shown in Fig. 24-5. The rod is available in lengths up to 6 ft.

Insulated Tension Link   An insulated tension link is a wood or fiberglass strain insulator with a conductor clamp on one end and a swivel on the other, as shown in Fig. 24-8. It is commonly used between the line conductor and the rope tackle when pulling conductors in the clear

Fig. 24-7 Multiple wire-tong support. Each support can be clamped in any position along the rod. *(Courtesy A. B. Chance Co.)*

of structures, when changing insulators, crossarms, or poles. Figure 24-9 shows two tension links in use when changing post insulators. The wire tongs hold the conductor at the desired distance, and the links keep the tongs from swaying toward the pole. Figure 24-10 shows the use of six tension members to hold the six line conductors clear of the pole while changing crossarms. Figures 24-11 and 24-12 show the use of two hot sticks in tension to support the line conductor on a high-voltage transmission line.

Fig. 24-8   Insulated tension link used to spread line conductors for pole installations and removals, when pulling conductors in the clear of structures for maintenance work, and when handling extremely heavy conductors on H frame and tower lines. *(Courtesy James R. Kearney Corp.)*

Fig. 24-9   Two tension links holding line conductors at proper distance from pole. *(Courtesy James R. Kearney Corp.)*

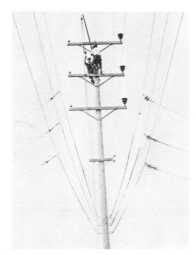

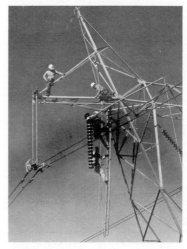

Fig. 24-10   Six tension links holding line conductors clear of pole while changing crossarms. *(Courtesy Safety Live Line Tool Co.)*

Fig. 24-11   Hot-line tool strain carrier assembly utilized to change suspension insulator string for 345,000-volt circuit. *(Courtesy A. B. Chance Co.)*

Fig. 24-12 Using hot sticks to change defective insulators on a live 132,000-volt line. Note two tension hot sticks used to support the line conductor, and one hot stick to keep the line conductor from swaying. A fourth hot stick is being used to pull the insulator string away from the line so that the defective insulator units may be replaced. Note use of insulator shields on ends of insulator strings and use of Stockbridge dampers on line conductors. *(Courtesy American Electric Power System.)*

Fig. 24-13 Auxiliary arm used as lifting crossarm to hold line conductors well above normal position to clear working space while a new crossarm is being installed. *(Courtesy A. B. Chance Co.)*

Fig. 24-14 Auxiliary arm used as side arm to hold two line conductors to one side. The third conductor is held in the clear on the opposite side with wire tongs. *(Courtesy A. B. Chance Co.)*

Fig. 24-15 Lineman transferring second line conductor from main crossarm to the auxiliary crossarm by the use of wire tongs. The third line conductor will be held off on the opposite side of the pole with the wire tongs. *(Courtesy American Electric Power System.)*

Auxiliary Arm   An auxiliary arm is an arm that is used to support the line conductors temporarily while the insulators or crossarms are being replaced. It may be used to raise and support the conductors vertically, or it may be used to support one or more conductors to one side horizontally. Figure 24-13 shows an auxiliary arm used as a lifting crossarm, and Fig. 24-14 shows it used as a side arm. Figure 24-15 shows a lineman transferring a second line conductor from the main crossarm to the auxiliary arm by means of wire tongs. An auxiliary arm is especially useful where it is desired to install a pole of greater height or where low-voltage lines must be cleared preparatory to working on higher voltage lines above them. Clamps are provided on the top of the arm to hold the conductors securely.

Strain Carrier   The strain carrier is used to take the load of strain or suspension insulators while the insulators are changed. Two types of strain carriers are in use, one for light loads up to about 2000 lb and one for heavy loads up to about 10,000 lb.

The light-duty strain carrier consists of two rocker arms hinged on a bar (Fig. 24-16). By bringing together the two rocker arms, the insulator string is compressed, and the insulator units may be removed and replaced.

The heavy-duty strain carrier consists of two parallel insulating tension rods which are set astride the insulators. The strain is relieved by taking up screws on the ends of the rods, thereby

Fig. 24-16   Light-duty strain carrier of the rocker-arm type. Illustration shows strain carrier mounted on corner pole and used to relieve the insulator string of conductor pull. *(Courtesy James R. Kearney Co.)*

compressing the insulator string. The insulator string is supported in a cradle consisting of two wooden rods below the insulator string shown in Fig. 24-17.

The strain carrier may be used on either pole or tower construction, as shown in Figs. 24-18 and 24-19.

A so-called "trolley-pole" method of changing suspension insulators is illustrated in Fig. 24-20.

Tie Sticks   Tie sticks are used to fasten or unfasten the tie wire which holds the conductor to the pin-type insulator (see Fig. 24-21). Tie sticks are usually also equipped with a small double hook called the tie-assistant. It is fastened to the tie stick a short distance from the end. It has several uses, such as prying loose ends of old tie wires that may be wrapped close to the conductor, holding the conductor down in the insulator groove while being tied in, or as a means of hanging the tie stick on the conductor when not in use.

Platform   The platform is small and collapsible and may be secured to the pole or tower. Figure 24-22 shows a small utility platform, and Fig. 24-23 shows a more elaborate type designed for hot-line work. This platform is equipped with a hand railing. Its use enables the lineman to reach farther out from a pole or tower and thus perform the work with greater ease.

Insulated Hoods   Insulated hoods are available to place over conductors and insulators to add to the safety of linemen doing hot-stick work (see Figs. 24-24 and 24-25).

Insulated Tools   Many other tools can be adapted to hot-line work by mounting them on the ends of insulated poles. Among these are pliers, wire cutters, tree trimmers, cotter-key pullers,

Fig. 24-17   Insulator string supported by cradle while being lowered into vertical position. Note two wooden tension rods supporting conductor. *(Courtesy A. B. Chance Co.)*

Fig. 24-18   Heavy-duty strain carrier mounted on dead-end pole. Note two-pole cradle provided to support insulator string after being uncoupled from line-conductor clamp. *(Courtesy James R. Kearney Co.)*

Fig. 24-19 Illustration shows suspension string moved from vertical to horizontal position away from live conductor for replacement of defective insulators. *(Courtesy A. B. Chance Co.)*

Fig. 24-20 Trolley-pole method of changing suspension insulators on high-voltage tower line. Insulator string is moved toward tower on trolley pole so that the whole string can be lowered to the ground for fast insulator replacement. *(Courtesy A. B. Chance Co.)*

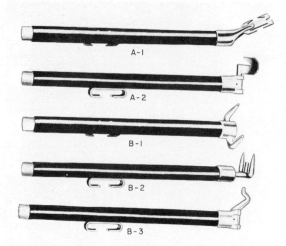

Fig. 24-21   Tie sticks for use on live lines. (*A*) Blade type: (1) fixed, (2) rotary. (*B*) Prong type: (1) two prong, (2) three prong, (3) rotary prong.

Fig. 24-22   Small utility platform. (*Courtesy A. B. Chance Co.*)

Fig. 24-23   Insulated platform designed for live-line maintenance work. (*Courtesy James R. Kearney Co.*)

Fig. 24-24   Insulated insulator cover. (*Courtesy James R. Kearney Co.*)

Fig. 24-25   Polyethylene insulator hood, conductor covers, and crossarm guard in use on live-line maintenance work on 34.5-kV line. *(Courtesy A. B. Chance Co.)*

screwdrivers, fuse pullers, saws, wrenches, brushes, and clamps. Some of these are illustrated in Fig. 24-26.

Required Clearance   Table 24-1 gives the minimum safe working distances from the conductors or from the hot end of sticks to the lineman.

CHANGING PIN INSULATORS

Order of Steps   The principal parts of a pole line are the conductor, insulator, crossarm, and pole. Since the conductor hardly ever fails, live-line maintenance resolves itself into replacing insulators, crossarms, and poles. To replace one of these requires that the conductors must be

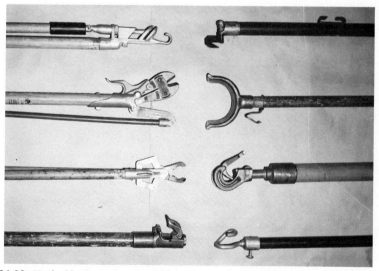

Fig. 24-26   Heads of hot-line tools used in holding off and repairing lines. Hot-line tools must be inspected and tested periodically for safe use. *(Courtesy American Electric Power System.)*

Table 24-1

Line voltage, volts	Minimum clearance		Line voltage, volts	Minimum clearance	
	Feet	Inches		Feet	Inches
2,100 to 15,000	2	0	161,000 to 169,000	3	8
15,100 to 35,000	2	4	230,000 to 242,000	5	0
35,000 to 46,000	2	6	345,000 to 362,000	7	0
46,000 to 72,500	3	0	500,000 to 552,000	11	0
72,600 to 121,000	3	4	700,000 to 765,000	15	0
138,000 to 145,000	3	6			

untied from the insulators, removed, and held in the clear. When so held the insulators or cross-arms, or both, can be removed and replaced. If the pole is to be replaced also, the best practice is to set the new pole first and then to mount all wire tongs and saddles and other tools on the new pole. In this way the old pole, which usually is weak, will not have to be specially guyed for this operation.

The steps in the process are performed in the following order:

1. Fasten wire tongs to conductor.
2. Untie conductor from insulator.
3. Move conductor into clear.
4. Remove old insulator, or crossarm, or both.
5. Mount new insulator, or crossarm, or both.
6. Return conductor to insulator.
7. Tie in conductor.
8. Remove wire tongs or auxiliary arm.

Moving Line Conductor   Three methods for moving line conductors into the clear are available:

1. Wire-tong method
2. Auxiliary-arm method
3. Combination of wire-tong and auxiliary-arm method

Wire-Tong Method   In the wire-tong method the conductor is moved and supported by the use of two wire tongs, one being used for lifting, called the lifting tong or stick, and the other for holding, called the holding tong or stick (see Fig. 24-27). The lifting stick is clamped onto the conductor to be moved, and its saddle is fastened to the pole in such a position that several feet of swivel end of the lifting stick extend through the saddle. The saddle clamp is then tightened onto the stick by turning the wing nut. Next a set of blocks is attached between the saddle and the swivel on the end of the lifting stick. This block and tackle will later be used to lift and move the conductor out of reach.

Another saddle for the holding stick is fastened to the pole just below the crossarm braces. The holding stick is now clamped onto the conductor and placed in the top saddle.

The conductor may now be untied from the insulator by means of the insulated tie sticks. After the conductor is untied, it is ready to be pushed out of reach. To move the conductor, pull up on the set of blocks, thereby raising the conductor off the insulator. If the outer conductor is being moved, the distance raised should be about 10 in. If the center conductor is moved, it should be raised enough to clear the end of the crossarm. When the conductor has been moved outward to a safe working distance, the saddle clamps on both sticks are tightened firmly. The blocks are then removed.

The same procedure is repeated for the other two conductors. The other outside conductor is moved next, and the inside or center conductor is moved last, as it must be moved above one of the outer conductors. The saddles for the sticks supporting the outer conductors may be placed on opposite sides of the poles, while a saddle extension may be used for the saddles supporting the middle conductor. Figure 24-28 shows three live conductors of a three-phase four-wire circuit moved in the clear by means of wire tongs preparatory to changing the pole and crossarm.

After the insulators, or crossarms, or both have been replaced, the above procedure is reversed and the conductors are brought back in. As the conductors are returned, they are tied in, using the special tie sticks.

For heavier construction such as wishbone, H frame, or at angles in the line, the same procedure may be followed except that an insulated link stick with blocks may be used to help move the conductor away from the insulator. The top holding stick should still be used because it helps to stabilize the conductor on the line stick.

Auxiliary-Crossarm Method   Where construction permits, the work of moving conductors may be simplified by the use of the auxiliary crossarm. The auxiliary crossarm may be used to lift all conductors from a crossarm vertically at one time, or it may be used to hold two or more conductors out on one side of the pole.

Mast Arm   When the auxiliary arm is used as a mast arm, that is, in lifting the conductors up from a crossarm, two saddles are mounted on the pole. The first saddle is mounted just beneath the crossarm braces. The assembled mast arm is then placed in position in this saddle and pushed

Fig. 24-27   Two wire tongs being used to hold live conductor on a rural line while lineman is preparing to replace the pole-top pin. The lower longer tong is the lifting tong and supports the conductor; the shorter upper tong is the holding tong and is used to hold the conductor firmly, at the proper distance from the pole. (*Courtesy American Electric Power System.*)

Fig. 24-28   Three sets of wire tongs holding three live conductors firmly in the clear while a new pole and crossarm are being installed. The tongs are all supported on the old pole. (*Courtesy American Electric Power System.*)

up against the conductors and clamped into position. The lower saddle is then mounted on the pole below the point where the bracing is fastened to the mast arm, and the main lifting pole is clamped in position. Saddle extensions should be used if necessary.

The conductors are then untied from the insulators with the tie sticks and placed in the holders on the mast arm provided for that purpose. When all the conductors have been transferred, the whole mast crossarm assembly is lifted up until the conductors are in the clear.

If the line makes an angle greater than 5 deg, it is advisable to "side guy" the mast arm, using an insulated link stick and a rope.

Side Arm   When the auxiliary arm is used as a side arm, it should be placed on the two-conductor side of the pole. One end of the side arm is mounted on a saddle at a point just below the crossarm braces. The conductors are then transferred to the holders on the side arm provided for that purpose with the use of lifting and holding tongs. The outside conductor is moved first, as shown in Fig. 24-29. The two tongs have just been attached to the live conductor. In Fig. 24-30 the conductor is being placed in the outermost holder on the auxiliary arm. A block and tackle fastened to the upper end of the lifting tongs is also used to assist in moving the conductor. In Fig. 24-31 the tongs are shown fastened to the second conductor preparatory to transferring it to the auxiliary arm.

The third conductor on the opposite side of the pole is moved and supported with wire tongs.

Removing the Tie Wire   The tie wire is untied by starting to unwind the end of the tie wire with the tie-wire blade on the tie stick, as shown in Fig. 24-32. If the tie wire has no end which

Fig. 24-29 First step in transferring outermost conductor to auxiliary arm consists in attaching lifting and holding tongs to conductor. Lineman is shown tightening saddle clamp on lifting tongs. *(Courtesy A. B. Chance Co.)*

Fig. 24-30 Moving first conductor out onto auxiliary arm by means of wire tongs. Note use of block and tackle on lifting tongs to assist in supporting weight of conductor. *(Courtesy A. B. Chance Co.)*

can be easily reached with the blade of the tie-wire stick, an end is made by cutting the tie wire at some convenient point with a hack saw. The tie wire is then unwound by pulling the end of the tie wire, if it points away from the pole. If the end of the tie wire points toward the pole, it is started by pushing up with the tie-wire stick. After this has been done, the old tie wire is easily unwrapped with the head of the tie stick. If it has a long wrapping, it should be cut with the hack saw as soon as the untied portion becomes difficult to handle. If it gets too long, there also is danger of its touching the other conductor and shorting the line.

During the process of removing old tie wires and tying in, the lineman should not be allowed to sit on the crossarm or put his legs or arms through the braces, but should be made to stand with his full weight in the climbing hooks, his feet as close together as possible, and should be instructed to lean back in his safety belt and keep his hands off the pole, crossarms, and crossarm braces. If he will do this, he will benefit from the insulation properties of the pole and the crossarm.

Fig. 24-31 Beginning the transfer of the second conductor to the holder on the auxiliary arm. Note use of block and tackle fastened to upper end of lifting tongs. This provides steadier control of conductor movement, both outward and return. *(Courtesy A. B. Chance Co.)*

Fig. 24-32   Removing tie wire with tie stick. *(Courtesy A. B. Chance Co.)*

Tying in Conductor   Particular mention should be made of the procedure employed in tying the conductor onto the new insulator. Special provision is made for this by providing the new insulator with special tie wires before the insulator is screwed in place. Figure 24-33*d* shows the insulator in place with the tie wires fastened to the insulator. Figure 24-33*a*, *b*, and *c* shows the details of the tie wires and the method of fastening them to the insulator. Each piece of the tie wire is placed in the neck of the insulator and bent until the ends meet (both ends being the same length). They are then given a couple of twists close to the neck of the insulator and allowed to extend horizontally in the same direction as the top groove of the insulator, one pair of tie-wire ends extending out in opposite direction for the other side of the insulator, as shown in Fig. 24-33*c*. These ends are now bent over the top of the insulator. as shown in Fig. 24-33*d*, so that in screwing the insulator on the pin the tie-wire ends will not come in contact with any part of the live line. The insulator is then sent up on the pole and screwed onto the pin until it is tight and with the top groove of the insulator parallel to the line conductor. The line conductor is then placed in the top groove of the insulator, care being taken that both ends of the tie wire in the same pair of ends come on the same side of the line conductor and that the two pairs of tie-wire ends come on opposite sides of the line conductor. When in this position the insulator is ready to be tied in. The tie stick is used for this purpose. The man making the tie should be well below the crossarm. The tie stick is passed up on the far side of the line conductor from the pair of tie-wire ends on that side of the insulator. One horn of the tie stick is put through the loop in the end of the tie wire nearest the insulator. This end is then pulled downward. The same is then done with the other end of the tie wire. When both ends are then pointing downward, the end of the tie wire which was first pulled down is pushed upward with a stiff punch on the other side of the line

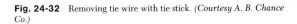

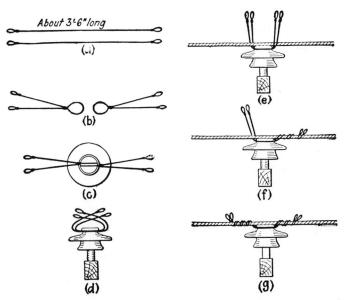

Fig. 24-33   Steps in tying in conductor by use of live-line maintenance tools. Tie wires are attached to insulator before insulator is screwed in place on pin. *(Courtesy Johnson Mfg. Co.)*

conductor; the other tie wire is likewise pushed up. By repeating this operation, both ends of the tie wire are thus served up to the loops in their ends. These loops can then be removed if desired, although there is no reason for removing them, by cutting them with a hack saw and breaking them off or by twisting them with a tie stick until they break off close to the line conductor. Obviously, the same method can be used in tying the line conductor in the neck of the insulator as will be necessary on angles. In the case of double arms, half of the above tie is made on one insulator and half on the other insulator.

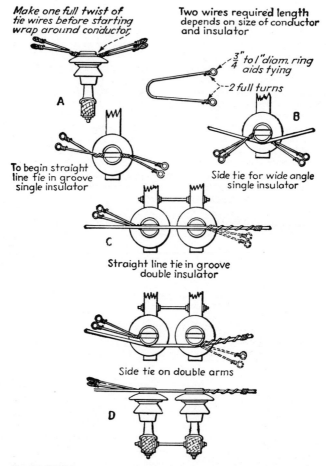

Fig. 24-34   Most common types of hot-line ties. *(Courtesy A. B. Chance Co.)*

This is by no means the only tie which can be applied; any desired form of tie can be made. This particular tie is chosen here only because it lends itself to a clear and accurate description and is perhaps the easiest tie to learn.

The sketches in Fig. 24-34 illustrate the most common types of hot-line ties in use. Note that both wires on one side of an insulator are turned in the same direction and that both wires on the other side of the insulator are turned in the opposite direction. They are started from the top and bottom of the conductor as in B and D, (Fig. 24-34), or from opposite sides of the conductor as in A and C. This provides a self-snubbing action which prevents the conductor from rolling and assists in holding the conductor tightly against the cap of the insulator.

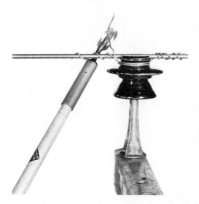

Fig. 24-35 Tying in a conductor using tie wires, provided with eyelets, and prong type of tie stick. *(Courtesy James R. Kearney Corp.)*

Fig. 24-36 Tying in line conductor with blade type of tie stick. *(Courtesy James R. Kearney Corp.)*

Note that eyes are shown on all ties. A lightweight stick with the prong-type head and this type of tie (Fig. 24-35) make a very satisfactory working combination. Some linemen prefer having the tie without eyes. In this case the blade-type head is more convenient (Fig. 24-36). The use of eyes is especially helpful in untying as it makes it unnecessary to pry up the ends of the tie wire.

A handy tie stick is one with the blade-type head on one end and the prong-type head on the other.

Figure 24-37 shows two linemen, each provided with a stick, tying in the middle line conductor. Both outside line conductors are already tied in. By working on opposite sides of the pole, all work is accessible to one or the other lineman. Figure 24-38 shows a lineman tying in the live conductor on a rural line.

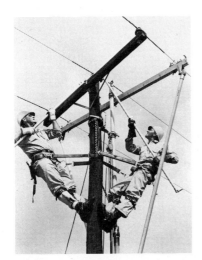

Fig. 24-37 Two linemen completing the insulator tie on the live middle conductor by the use of insulated tie sticks. *(Courtesy American Electric Power System.)*

Fig. 24-38 Lineman tying in conductor on rural line with tie stick. Conductor is held down in insulator groove by means of wire tongs. Note saddle which supports upper end of tongs. Also note that line conductor is provided with armor rods. *(Courtesy American Electric Power System.)*

Fig. 24-39 Lineman is installing Chance Superformed® Top-tie® to secure energized conductor to top of pin-type insulator. The pole is covered with insulated pole guard to provide safe working space. Lineman places unmarked leg of tie in a clamp stick one pitch in from the end and places tie on insulator and conductor. *(Courtesy A. B. Chance Co.)*

Fig. 24-40 Lineman uses hot stick with tie loop to wrap the marked tie leg to completion while holding the tie loop snug against the insulator with clamp stick. *(Courtesy A. B. Chance Co.)*

Prefabricated ties manufactured to fit specific insulators and conductors provide a fast, firm, and economical system for securing wires to pin- and post-type insulators. Hot-stick application of a prefabricated tie for a conductor installed on top of a pin-type insulator is illustrated in Figs. 24-39 through 24-46. Application of a prefabricated tie for a conductor installed on side groove of a horizontal line post insulator with hot sticks is shown in Figs. 24-47 through 24-51.

Removing Conductor from Clamp-Top Insulators In case the pin insulator is of the clamp-top type, the removal of the conductor from the insulator is greatly simplified. Likewise, the fastening of the conductor is also made easy.

Fig. 24-41 Lineman snaps marked leg of tie securely into place with tie loop on hot stick. Note lineman wears insulated hard hat and safety glasses. *(Courtesy A. B. Chance Co.)*

Fig. 24-42 Lineman slips the second tie of the set between the wrapped and unwrapped legs of the first tie using clamp stick on the marked tie leg. The marked leg must remain on top of the conductor parallel to the previously installed leg. The tie wires must not cross. Note clamp stick used to hold conductor in position on top of insulator. *(Courtesy A. B. Chance Co.)*

Fig. 24-43 Lineman removes clamp stick from marked leg and places it on the unmarked leg holding it while marked leg is wrapped to completion with tie loop. *(Courtesy A. B. Chance Co.)*

Fig. 24-44 Lineman applies unmarked leg of second prefabricated tie wire with tie loop. *(Courtesy A. B. Chance Co.)*

The clamp-top insulator, illustrated in Fig. 24-52, is similar to other pin insulators except that it has a bolted clamp cemented to the top. Two carriage bolts hold the keeper piece over the conductor.

To remove the conductor, therefore, it is only necessary to unscrew the two nuts. This is done by means of a special ratchet wrench and socket secured to the end of an insulated handle. Figure 24-53 shows this operation being performed, and Fig. 24-54 is a close-up view of the ratchet wrench and socket for loosening the nuts on the clamp-top bolts. Before this is done, the lifting and holding tools are attached to the conductor and secured to the pole.

Fig. 24-45 Lineman wraps remaining unapplied leg of first tie around energized conductor. *(Courtesy A. B. Chance Co.)*

Fig. 24-46 Lineman has completed installation of prefabricated tie securing conductor to insulator. Clamp stick holding conductor can now be removed. *(Courtesy A. B. Chance Co.)*

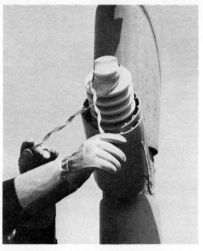

Fig. 24-47 Lineman places Superformed® Tygard® Line Tie over insulator side groove before energized conductor is lowered onto insulator side groove. Note plastic cover-up material on insulator support and pole. *(Courtesy A. B. Chance Co.)*

Fig. 24-48 Lineman rotates tie on insulator with tie loop on hot stick so legs of tie point skyward. The conductor supported by a clamp stick is lowered onto insulator side groove with conductor between tie legs. *(Courtesy A. B. Chance Co.)*

Fig. 24-49 Lineman applies left leg one pitch with tie loop on hot stick and holds tie while applying the right-hand leg. *(Courtesy A. B. Chance Co.)*

Fig. 24-50 Lineman completes application of left leg of tie after making sure right leg is snapped into place. *(Courtesy A. B. Chance Co.)*

Fig. 24-51 Lineman has completed application of prefabricated tie. Clamp stick supporting conductor and temporary protective equipment can now be removed. *(Courtesy A. B. Chance Co.)*

Fig. 24-52 Clamp-top type of pin insulator using a clamp instead of the usual tie wires to hold the conductor in position. *(Courtesy Ohio Brass Co.)*

Fig. 24-53 Unscrewing the two nuts that hold the keeper piece over the conductor. Note that the lifting and holding tools are already attached to the conductor and secured to the pole. *(Courtesy James R. Kearney Corp.)*

Fig. 24-54 Ratchet wrench and socket for loosening the nuts on clamp top bolts. *(Courtesy James R. Kearney Corp.)*

After the nuts are loosened and backed off about ⅝ in, the conductor is raised slightly, about ½ in, to take its weight off the clamp seat. The clamp is then lifted from the supporting pintles with a special wire cradle mounted on the opposite end of the lineman's insulated handle and slid along the conductor away from the insulator. A close-up view of this process is shown in Fig. 24-55, and a view of the complete operation is shown in Fig. 24-56. With the clamp slid away from the clamp-top seat, the hot conductor is free and can be moved to any desired position to clear the insulator. The completed untying is shown in Fig. 24-57.

For attaching the conductor to the insulator the reverse procedure is followed.

Fig. 24-55 Cradle for lifting the clamp, a suitably formed wire inserted in a standard hot-stick socket. *(Courtesy James R. Kearney Corp.)*

Fig. 24-56 Lifting the clamp from the supporting pintles with the wire cradle. *(Courtesy James R. Kearney Corp.)*

CHANGING STRING INSULATORS

The principal parts of a steel-tower line are the steel tower, steel crossarm, suspension insulator string, and conductor. At crossings, angles, and dead ends, strain insulators take the place of the suspension string. The most frequently performed live-line maintenance work on tower lines will be the replacement of suspension or strain insulators. The steel towers, steel crossarms, and conductors seldom fail.

In order to replace any defective units in the insulator string, the weight of the conductor must be taken off the string. This is accomplished by the use of the strain carrier. By drawing up on the strain carrier, the insulator string is relieved of the conductor load and can then be disconnected from the conductor clamp. The lineman is then free to remove and replace one or more of the insulator disks. When this is accomplished, the conductor load is transferred back to the insulator string and the strain carrier is removed. These steps can be listed as follows:

1. Attach strain carrier.
2. Draw up on strain carrier.
3. Disconnect insulator string.
4. Replace defective insulator units.
5. Reconnect insulator string.
6. Release tension from strain carrier.
7. Remove strain carrier.

Clearing Working Space   Before proceeding with the work the lineman should make sure that sufficient working space is available and that the required clearance from live conductors can be maintained. If protective gaps are in the way, they should be pushed out of the way or removed temporarily. Jumpers around dead-end strings should also be pushed out of the way and held away by use of a holding stick in the regular manner.

Use of Strain Carrier   The strain carrier is then placed in position. As the suspension string hangs vertically, the strain carrier also hangs vertically. No cradle for supporting the insulator string will be needed. The fixed end of the carrier is hooked onto the conductor on opposite sides of the clamp. The adjustable or take-up end is fastened to the crossarm, allowing enough slack in the tightening arrangement for taking the strain off the insulator string.

After the strain is taken off the insulator string, the insulators hang limply and may be disconnected from the strain clamp. This requires pulling out a cotter key or pin. This is performed with a cotter-key puller, which consists of a special cotter-key pulling hook fastened to the end of

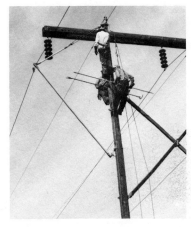

Fig. 24-57   With the clamp slid away from the clamp top seat, the hot conductor can be moved away from the insulator as shown. *(Courtesy James R. Kearney Corp.)*

Fig. 24-58   Wire tong also used in replacing defective suspension insulators. Note use of link stick and rope blocks. *(Courtesy A. B. Chance Co.)*

another hot stick. If the cotter key is hard to pull out, the conductor load on the string should be further reduced by additional tightening of the strain carrier.

With the insulator string now free from the clamp, the string can be removed and the defective insulator units replaced. It has been found preferable in most cases to remove the entire string of insulators rather than to replace the insulator disks singly. This is especially so when the insulator string is short, consisting of only two or three disks. The insulator string may be removed by using a shepherd hook, fork stick, lifting yoke, sling, or other suitable device.

After the insulators are replaced, the string is reconnected to the clamp. The cotter pin is installed by the use of a special cotter-key replacer. The cotter key is spread by means of a screwdriver fastened to the other end of the cotter-key replacer hot stick. The strain is now transferred to the insulator string by letting off on the strain carrier. The carrier and other tools should then be removed in the reverse order of their installation.

Wire-Tong Method   In many cases, especially on light pole lines, the suspension insulators may be replaced with the use of wire tongs or hot sticks instead of the strain carrier.

The procedure is similar to that used with pin-type insulators. The conductor and clamp are secured with a lifting tong. The assembly is then pushed away from the pole or tower a distance of about 6 in. The lifting tong is then secured in its saddle clamp. The holding tong is now fastened to the conductor, and the conductor is then pulled back to the pole or tower. This will cause the conductor to be raised slightly and thus transfer the conductor load onto the lifting tong. The insulator string will now hang limply. The insulator string may now be disconnected from its clamp and the line conductor moved farther out of the way (see Fig. 24-58). A link stick may be used in pulling the conductor away or in pulling it back, but the holding tong should always be used as it will help to stabilize the conductor in its position away from the pole or tower. After the defective insulator units are replaced, the conductor is pulled back in and the string is reconnected to the conductor clamp. All tongs and other tools are then removed in reverse order of their installation.

REPLACING STRAIN INSULATORS

The strain carrier is first placed in position as with suspension insulators. Since the strain-insulator strings are somewhat horizontal, the carrier can be slid over the insulators out to the line conductor. The fixed end of the carrier is attached behind the strain clamp on the span side of the clamp. The adjustable or take-up end of the carrier is fastened to the pole, crossarm, or tower, allowing enough slack in the tightening arrangement for taking the strain offf the insulator string. The

Fig. 24-59 Disconnecting strain insulator string from conductor clamp assembly for EHV bundled conductor line. (*Courtesy A. B. Chance Co.*)

cradle for supporting the slack string of insulators is then placed in position by hanging the adjustable hook over the conductor and fastening the other end to the crossarm. When the insulator string is short, that is, consists of only two or three disks, the supporting cradle need not be used, as the insulators can then be handled by means of a yoke arrangement on the end of a hot stick.

The next step is to take the conductor load off the insulators. This is accomplished by drawing up on the strain carrier. In the "buck-saw" type of strain carrier, this is done by pulling together the ends of the two hinged members, using a set of blocks. In the "cradle" type of strain carrier the conductor load is transferred to the carrier by taking up on the jackscrew at the rear of the two tension members.

After the strain is taken off the strain insulator string, the string may be disconnected from the clamp (see Fig. 24-59). The defective insulator disks may now be replaced and the complete string reconnected to the clamp. The conductor load is then returned to the strain-insulator string by letting off on the strain carrier. All hot-line tools should now be removed in reverse order of their placement.

PULLING SLACK

String Insulator   The same general procedure is followed when pulling or removing slack from a conductor when the line is energized as is used when the line is deenergized. When the lines are energized, properly insulated tools and insulated links must of course be used. The order of work is as follows: The pulling grip or come-along is attached to the conductor with a tie stick, holding stick, or other hot stick. An insulated link stick is next attached to the pulling eye of the come-along. A set of blocks is now placed between the end of the link stick and the pole, crossarm, or tower. The conductor is then secured by taking up on the set of blocks. After the conductor is thus secured, the clamp holding the conductor to the insulator string is loosened by unscrewing the nuts on the holding bolts by means of a socket wrench fastened on the end of a hot stick. The slack is then taken up by pulling further on the blocks. All pulling on the blocks should be done from the ground, if possible. When sufficient slack has been removed, the conductor clamp is tightened by use of the insulated socket wrench. If all the unwanted slack cannot be pulled in one operation, the foregoing proceudre is repeated until the desired conductor sag is obtained. When sufficient slack has been pulled, the blocks, insulated stick, and pulling grip are removed. The pulling grip is loosened by tripping the releasing lock with a tie stick or other hot stick.

Fig. 24-60 Removing come-alongs after completing a splice on a hot line. Note use of insulated link stick on come-alongs. *(Courtesy A. B. Chance Co.)*

Pin Insulator   The same procedure is followed when pulling slack in a line supported on pin insulators. Instead of unscrewing the nuts on the conductor clamp, the tie wires holding the conductor to the insulator are loosened but not removed. Then the slack is pulled up and held while the insulator ties are remade.

CUTTING OUT CONDUCTOR

If a short length of conductor needs to be cut out of the line, the slack can be pulled in both directions from the same pole or tower. The extra conductor may then be cut out, the conductor spliced and then refastened to the insulators.

In order not to interrupt the flow of current in the line when the conductor is cut, a jumper is placed around the section where the conductor is to be cut out. However, before the ends of the jumper are tapped onto the conductor, the jumper must be firmly held away from the lineman by means of a holding tong or stick securely fastened to the pole or tower. An insulated link stick with rope may also be used for this purpose. The ends of the jumper are then clamped onto the live conductor. The desired length of conductor is cut out of the line, and the ends are spliced. After splicing is completed, the jumper is removed.

If the line conductor is a stranded cable, extreme care must be taken to prevent the cable from unraveling and allowing the wires to get out of control.

SPLICING LIVE CONDUCTORS

The procedure for splicing a hot conductor with live-line tools is as follows: The conductors to be spliced are pulled tight with blocks, using link sticks on the come-alongs for insulation (see Fig. 24-60). The conductors are connected together with a temporary jumper to by-pass the current while the splice is completed. The ends of the conductors are held in position with holding sticks.

The splicing connector is next fastened to the end of one conductor. Then the end of the other conductor is brought over and placed in the connector. The connector is now firmly fastened, compressed, or fired, completing the splice. Insulated tools, of course, are used for all the operations.

TAPPING LIVE LINE

Taps are easily made to live connectors by the use of appropriate equipment. The conductor to be tapped to the live line is first fastened to the body of a special tapping clamp and is then held with a hot stick and hooked over the live-line conductor. The clamp is then tightened by turning the screw head of the clamp, using the hot stick for this purpose.

Fig.   24-61 Applying Stockbridge vibration damper to live line with special tool. *(Courtesy A. B. Chance Co.)*

Fig. 24-62 Vibration damper placed in special tool preparatory to fastening to live-line conductor. *(Courtesy A. B. Chance Co.)*

APPLYING VIBRATION DAMPERS

Vibration dampers of different types may be applied to lines while energized, using live-line tools for their installation.

A special tool simplifies the installation of the Stockbridge type of damper on live conductors, Fig. 24-61. The damper is placed in the head of the tool, as shown in Fig. 24-62. A spring holds the damper clamp in place whether it is open or closed. With the damper clamp in the open position, the damper is placed on the live conductor. The locking nut on the bottom of the clamp, which has previously been placed in a socket, is tightened by turning the hot stick. The tool is removed by pulling down on the hot stick, compressing the spring, and turning the stick a quarter turn to engage the socket in a locked position on a projecting lug on the head of the tool. The tool may then be easily lifted away. A ring is provided on the side of this tool for attaching the link stick and hand line or blocks for aid in handling heavy dampers of this type.

APPLYING ARMOR RODS

In order to install armor rods, it is first necessary to remove the conductor from the insulator in the usual manner and move it with wire tongs to a suitable working position, as shown in Fig. 24-63. In this case the conductors are supported on an auxiliary side arm. The armor rods are then applied with a special tool (Fig. 24-64). The tool consists essentially of two split wheels into which different size dies are set. The armor rods are placed into these dies, and the assembly is placed around the conductor by means of holding or clamping sticks. The wheels or holders are rotated in the proper direction, depending on the lay of the wire, one holder moving clockwise and the other counterclockwise. The rotation of the holders is accomplished by means of hooks fitted onto hot sticks engaging lugs on the holders. At the same time the holders are rotated, they are moved apart, starting near the center of the armor rods and moving toward the ends. The rotary movement of the holders is continued until the armor rods are completely and tightly twisted into place. The holders are then pushed off the rods and removed by means of clamp sticks or other hot sticks.

Armor clips are then inserted on each end of the armor rods and clamped into place by tightening the bolt nuts. Special spring sockets mounted on the ends of hot sticks are used to tighten the nuts, the spring being used to thrust the saddle into place as soon as the two halves of the clip are in alignment.

Fig. 24-63 Applying armor rods to live single-phase line. *(Courtesy A. B. Chance Co.)*

Fig. 24-64 Special armor rod wrenches and wrench holder used to apply armor rods while line is live. *(Courtesy A. B. Chance Co.)*

PHASING OUT

Phasing out is absolutely necessary when a new line is to be paralleled with another line, new or old, and after repairs or changes have been made on either of two lines which have previously operated in parallel. It is necessary in the latter case because of the possibility of interchanging conductors when making repairs or changes.

The process of "phasing out" consists of determining whether the phases of a given line or apparatus correspond with the phases of another line with which it is to operate in parallel. This problem arises most frequently at sectionalizing switches. The voltage across corresponding lines or phases should be zero. Therefore to determine if zero voltage exists across corresponding lines it is necessary to read the voltage. This is done with a "phasing-out" voltmeter.

Phasing-Out Voltmeter   A phasing-out voltmeter instrument consists of a high resistance connected in series with a voltage-indicating device. Using both a lamp and a voltmeter is useful because each serves as a check on the other. The indicating lamps used are the glow-lamp type which can withstand considerable over-voltage. A schematic diagram of the circuits and components used in this make of voltage detector is shown in Fig. 24-65.

Phasing-Out Procedure   Since the operation of high-voltage paralleling is usually done through a circuit breaker, the customary procedure in "phasing out" is to open the disconnects on one side of the circuit breaker and then close the circuit breaker so that the corresponding blades and contact jaws of the open disconnects are alive from the two sources that are to be paralleled. The instrument is checked for proper operation by connecting it between phases on each side of the open disconnect switch. This procedure verifies that both circuits to be phased out are energized if the instrument works correctly. The phasing-out voltmeter or voltage detector is then used to detect the presence of any difference of potential between a blade and the corresponding contact jaw of each disconnect.

If the lamp or voltmeter gives no indication as the two leads are successively applied to each pair of the three disconnect terminals, then no potential exists between each jaw and blade. This then shows that the correct phases are connected to each of the two parts of the respective disconnects. If the lamp should glow when the leads are applied to two of the three phases, then these two phases are interchanged; if the lamps glow on all three phases, then all three phases are interchanged. If no voltage is read across the open disconnects, the high-voltage voltmeter should

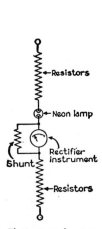

Fig. 24-65   Phasing-out voltmeter, with two series resistors, neon lamp, and voltmeter. A third resistor is shunted around the rectifier-type voltmeter to permit the lamp to glow in case of an open circuit in the rectifier instrument. *(Courtesy General Electric Co.)*

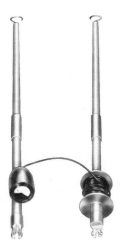

Fig. 24-66   High-voltage type of phase tester. Tester is supported on ends of insulated poles. Phase testing consists in measuring voltages between a phase in one line and the phases of the other line to determine the one that gives zero reading. It is the corresponding phase. *(Courtesy A. B. Chance Co.)*

be connected from the jaw of one switch to the blade of an adjacent switch verifying that the circuits are similar. Ungrounded neutral, or three-wire three-phase circuits without adequate capacitive coupling will require the use of two high-voltage voltmeters simultaneously on adjacent phases to obtain a proper reading.

Similar tests can be completed for proper phasing with a low-voltage meter if potential transformers or potential devices are connected to the lines on both sides of the separation between circuits. The voltmeter leads are connected to the low-voltage terminals of the potential devices in the same sequence as described for the high-voltage tester on energized primary lines. Transmission-line circuits are usually phased out in substations using the low-voltage terminals of potential devices installed in the substation for metering and relaying circuits.

High-Voltage Phase Tester   A phase tester for use on high voltages is illustrated in Fig. 24-66. Voltages up to 75,000 volts can be read with about 7 percent accuracy. This tester also consists of two high-resistance units connected in series with a voltmeter mounted on Epoxyglas-insulated housings. Insulated poles are bolted to the bottom of the housings. The housings are joined by some 20 ft of 15-kV insulated flexible cable coiled on a reel. Only the necessary length of cable should be unreeled to avoid contact with a ground that could distort the reading.

When two lines are phased together, the tester should first be connected across phases of the same line to be sure that the line is energized, and the instrument is working properly, and across phases of one line to the other line to be sure that both circuits are energized and properly coupled. This should then be repeated for the other line to make sure that it is also energized. If both lines are live, the tester is then connected from a phase of one line to each phase of the other line in turn to determine which one is the corresponding phase of that line. The phase which gives zero reading is the corresponding phase. This is repeated for each of the other two phases of the first line. The phases across which the readings are zero are the corresponding phases which may be safely connected together for parallel operation. Figure 24-67 shows the linemen reading the voltage on the tester when connected across a phase of one line and a phase of the other line. Note that the flexible cable is taut between poles. The tester should be removed from the lines as soon as the readings have been taken.

Insulator Testing   All live-line testing requires a knowledge of the voltage distribution over the units of a suspension string or across the shells of a multipart pin insulator. Unfortunately this distribution is not uniform, which would make the voltage across all units the same. Instead the

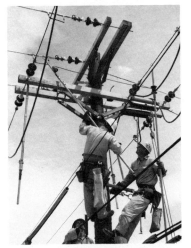

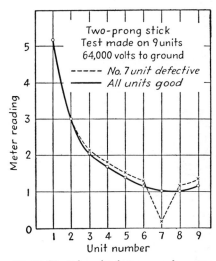

Fig. 24-67   Lineman using phase tester to measure voltage between a phase of one line and a phase of another line. Note that tester is hooked over conductors of the two lines. Also note taut cable between coil reels. (*Courtesy A. B. Chance Co.*)

Fig. 24-68   Voltage-distribution curves for a nine-unit suspension string. The solid line is for a string in good condition; the dashed line is for a string with No. 7 unit defective.

distribution is quite uneven, that is, the voltage across the unit nearest the line conductor is several times as great as the voltage across the unit nearest the crossarm or support. A typical distribution curve for a 64,000-volt line is shown in Fig. 24-68. The solid-line curve is the curve for a perfect insulator string. This shows almost five times as much voltage across the unit nearest the conductor as across the unit next to the support.

Now, if for any reason one or more units become defective, the distribution curve becomes distorted. This is illustrated in Fig. 24-68 by the dashed-line curve which is the curve when unit No. 7 is defective.

Therefore, if the voltage across each unit of a string can be conveniently measured, defective units can be easily located. Any unit whose voltage reading is lower than the reading for a like unit on a string in good condition would show that the former unit is not withstanding its share of the voltage and is therefore defective. Actually, if a unit is very defective, the voltage drop across it will approach zero, as the broken-down insulation tends to act as a conductor. Ordinarily, if a reading is less than 60 percent of the average reading, the unit is probably defective. Similar distribution curves for a good and a defective pin insulator are shown in Fig. 24-69.

Insulator Tester   One make of tester employing a meter for reading the voltage distribution is illustrated in Figs. 24-70 and 24-71. The first is for pin insulators and is referred to as the "single-prong" instrument, and the latter is for suspension or strain insulators and is referred to as the "double-prong" instrument. The meter on which the voltage is read is mounted on the end of the insulated hot stick. Although this meter actually is an ammeter, being actuated by a condenser current, its readings are proportional to the voltage; therefore it may be looked upon as a voltmeter.

The procedure to be followed is obvious from the description of the tester and the illustrations

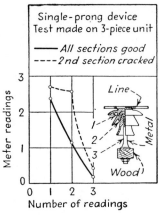

Fig. 24-69 Voltage-distribution curves for a good and a defective three-part pin insulator.

Fig. 24-70 Testing pin insulator with single-prong meter-type tester. *(Courtesy I-T-E Circuit Breaker Co.)*

shown. In the case of disk-type insulators in a string or insulators in a stack, the two prongs are touched to the meter portions on both sides of the insulator and readings are taken on the meter. Since the two prongs actually are the leads from the voltmeter, the readings give the voltage drop across the insulator. These readings are then compared with those for the corresponding units on a perfect insulator string. If any of the readings are much lower the units are defective.

In the case of multilayer pin-type insulators, the prong is inserted between the porcelain layers until it makes contact with the cement. This prong is one lead from the voltmeter. The other lead from the meter connects to a condenser made of several metal tubes located in the hollow of the stick. The metal tube is one plate of the condenser, and the earth is the other plate. The condenser is made variable by providing two metal tubes of different length. The longer one is used at all times unless the reading goes off the scale. Then the shorter one is used. On the other hand, if the reading is too small and needs to be increased, both tubes may be joined and used.

On a two-piece insulator a reading of the upper part will give the condition without testing the bottom part. If the upper part should be bad, the meter reading will drop. If the lower part is bad, the meter reading on the upper part will be above normal.

Fig. 24-71 Testing suspension insulator with double-prong meter-type tester. *(Courtesy I-T-E Circuit Breaker Co.)*

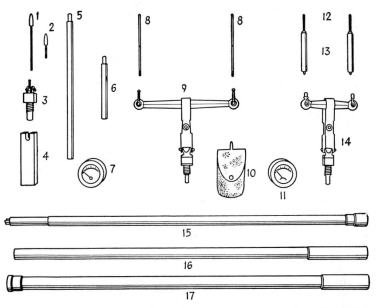

Fig. 24-72   Parts of meter-type live-line insulator tester. *(Courtesy I-T-E Circuit Breaker Co.)*

This type of insulator tester is made with different type heads and a variety of contact tips to fit the many types of insulators in use (see Fig. 24-72). The holding stick is furnished in three sections for convenience in reaching the insulator.

The phasing-out high-voltage voltmeter can be used to complete the tests, following the same procedure as described for the insulator tester, if the voltages are within the range of the voltmeter capacity.

Live-Line
Maintenance from
Insulated Aerial Platforms

General   Mobile extension ladders have long been useful for tree trimming, as shown in Fig. 25-1, and for the replacement of burned-out lamps in street-lighting fixtures. In the insulated aerial platform the extension ladder is replaced with a lifting boom the upper member of which is made of insulating material, thereby insulating the lineman from ground. This makes the insulated aerial platform an improved elevated platform for street-lighting maintenance (Fig. 25-2) and a safer platform for tree trimming near energized lines (Fig. 25-3), but its greatest field of usefulness is for live-line maintenance. In this field it can serve as a platform from which to perform live-line operations in three ways:

1. Using rubber gloves on distribution lines.
2. Using conventional hot-line tools on distribution and transmission lines.
3. Using bare hands on medium- and high-voltage transmission lines.

For tree trimming or replacement of street lamps, the platform usually consists of a single bucket or basket with room for only one lineman, while for live-line maintenance, single and double buckets are employed.

The use of the aerial platform is limited only by accessibility to the work area in the overhead lines. As access is sometimes impossible, the aerial platform can only supplement regular line crews using conventional maintenance procedures.

Description of Insulated Aerial Rig   The lifting boom pedestal is placed behind the truck cab. Outriggers from the pedestal base are used to stabilize the boom when it is extended. The boom and baskets are positioned over the cab while traveling. The upper member of the boom and the baskets are made of fiberglass to provide the required insulation from ground. A fiberglass section may be installed in lower boom to provide protection for groundmen if the noninsulated section of the boom should contact an energized line. All movements of the lift are controlled by the lineman in the basket. For bare-handed work the baskets must be lined with a wire mesh. The illustrations to follow illustrate the various features mentioned.

Maintenance with Rubber Gloves   Live-line maintenance from an insulated aerial platform with rubber gloves is limited to the distribution voltages. Some utilities work on lines with voltages through 20 kV to ground or 34.5 kV phase-to-phase using rubber gloves and sleeves. Figure 25-4 shows two linemen in an insulated aerial double basket making conductor repairs at midspan on a 4-kV line using rubber gloves, and Fig. 25-5 shows two linemen working on insulated platforms installing dead-end insulators on a 13,200Y/7620-volt line utilizing rubber gloves and sleeves. Other operations that are usually performed with rubber gloves and rubber sleeves are installing transformers, switches, and cutouts, refusing cutouts, and inspecting equipment.

Fig. 25-1 Mobile extension ladder used as lift for tree trimming before the advent of the insulated aerial platform. *(Courtesy Davey Tree Surgery Co., Ltd.)*

Fig. 25-2 Replacing street lamp from an insulated aerial platform. *(Courtesy Pitman Mfg. Co.)*

Work on distribution circuits operating at 13,200Y/7620 volts, or higher, requires the use of insulated aerial devices or insulated platforms to supplement the rubber gloves and sleeves. Rubber gloves and sleeves must be manufactured and rated for a voltage greater than the distribution voltage to be worked. Distribution voltages lower than 13,200Y/7620 may be worked using rubber gloves while the lineman's gaffs are engaged in wood poles.

Maintenance with Hot-Line Tools   Live-line maintenance from an insulated aerial platform using hot-line tools is more conveniently and more rapidly done than when working from the pole or tower. The ease of reaching elevated positions and the reduction in the required protective covering, as well as the reduced number of linemen required, favor this method. Some-

Fig. 25-3 Trimming trees with a power pruner from an insulated aerial basket. The mechanized unit shown also includes a branch chopper and a hopper for hauling the chopped branches. *(Courtesy Davey Tree Surgery Co., Ltd.)*

Fig. 25-4 Rubber-glove work at midspan on 4-kV line making conductor repairs. *(Courtesy A. B. Chance Co.)*

Fig. 25-5 Rubber-glove and rubber-sleeve work on 13,200/7620-volt line using insulated platforms. *(Courtesy A. B. Chance Co.)*

Fig. 25-6 Linemen changing a twist-on sleeve to a compression sleeve in midspan. Note cable used to bypass current. Also note covered line phase. *(Courtesy A. B. Chance Co.)*

times a combination of linemen working both from the platform and from the pole or tower is preferred. The following figures illustrate the manner and variety of operations that can be performed: Figure 25-6 shows two linemen installing a compression sleeve at midspan. Figure 25-7 shows two linemen changing a string of insulators in a 44-kV line. Figure 25-8 shows two linemen adding a fourth insulator to the string prior to raising the voltage of the line. Figure 25-9 shows one lineman on the platform and one on the pole covering up the middle and outside phases with polyethylene cover-up equipment preparatory to cutting in dead ends on the opposite phase, Fig. 25-10 shows two linemen replacing a damaged insulator on urban distribution, and Fig. 25-11 shows two linemen cutting in a set of dead ends and air-break switches into an energized line.

Fig. 25-7 Changing string of insulators on 44-kV line. Conductor is supported with hot-line tools attached to the pole. *(Courtesy A. B. Chance Co.)*

Fig. 25-8 Adding a fourth insulator to the insulator string so that line can be stepped up in voltage. *(Courtesy A. B. Chance Co.)*

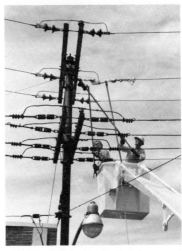

Fig. 25-9 Covering up middle and outside phase conductors with linear polyethylene cover-up equipment preparatory to cutting in dead ends on the opposite phase. *(Courtesy A. B. Chance Co.)*

Fig. 25-10 Replacing damaged insulator on urban distribution. The use of the platform permitted the linemen to move around the underbuild without having to cover up the line. *(Courtesy A. B. Chance Co.)*

Maintenance Using "Bare-Hand" Method   When live-line work is to be done "bare-handed" from an insulated aerial platform, the linemen in the basket must be at the same potential as the live conductor on which the repair is to be made. This is accomplished by bonding the metal mesh lining of the basket to the live-line conductor. The mesh lining, also referred to as basket shielding, is brought to line potential by means of a bonding lead one end of which is

Fig. 25-11 Cutting in a set of dead ends and air-break switches on an energized line. Note covered middle phase and pole guy. *(Courtesy A. B. Chance Co.)*

Fig. 25-12 Linemen in two-man bucket preparing to work bare-handed on energized 345,000-volt line. Lineman on left is using a hot stick to make initial contact with live conductor to bond the insulated aerial lift shield. Note arc between conductor and end of bonding wire. Also note bullet hole in stranded conductor. *(Courtesy Indiana and Michigan Electric Company of American Electric Power System.)*

Fig. 25-13 View showing linemen in an elevated position working bare-handed on a 34,500-volt pole line. Linemen have removed a conventional pin insulator to be replaced with a post insulator. Note outriggers on truck to stabilize boom. Also note part of upper arm of boom made of fiberglass. *(Courtesy Ohio Power Company of American Electric Power System.)*

Fig. 25-14 Linemen working bare-handed from twin one-man insulated buckets on a 34,500-volt pole line. Linemen are about to install post insulator on crossarm. Note outriggers on truck to stabilize boom and leads clamped onto live conductor to bond basket shield. *(Courtesy Ohio Power Company of American Electric Power System.)*

permanently attached to the shielding while the other end is clamped onto the energized conductor as the basket approaches it. A short hot stick is used to fasten the metal clamp (see Fig. 25-12). One lead is employed for each lineman. Since the two leads are electrically interconnected, the second lead serves as a safety measure in case one should accidentally be knocked off. If both were removed accidentally, the lineman would receive a charging current shock each time he touched or removed his hand from the energized conductor.

Fig. 25-15 Linemen working bare-handed on live conductor of energized 34,500-volt line. Linemen are cleaning conductor damaged by rifle fire preparatory to installing a compression sleeve. Note leads which bond bucket shield to live conductor. Linemen are supported in twin one-man buckets. *(Courtesy Ohio Power Company of American Electric Power System.)*

The linemen in turn are also "bonded" to the shielding by means of a spring-type ankle clamp or by the wearing of conductive soled shoes.

While the linemen and their shielding are thus bonded to the energized conductor, the linemen are maintained at conductor potential. They can move around in the baskets and touch the con-

Fig. 25-16   Linemen applying compression sleeve to damaged line conductor at midspan on energized 34,-500-volt line. Note leads clamped onto line conductor which maintain conductor and buckets at same potential. Buckets are insulated from ground by means of fiberglass boom. *(Courtesy Ohio Power Company of American Electric Power System.)*

Fig. 25-17   Linemen working bare-handed from insulated twin one-man buckets on live 34,500-volt line. Linemen are clamping live-line conductor to top of post insulator. Note bonding leads clamped on line conductor to maintain bucket shield at line potential. *(Courtesy Ohio Power Company of American Electric Power System.)*

ductor and any metal accessories which are at the same potential without having any physical awareness of the fact that they are charged at line potential. As long as the lineman is at the same potential as the conductor he is working on, and as long as he is insulated from ground, no current can flow from conductor to man. The lineman can work barehanded on the conductor or on any metal fittings at the same potential as the conductor without experiencing any discomfort. The rule is "Touch only that which you are clipped to."

In many locations it is impossible to position an insulated bucket truck at the work site. Bare-hand live-line work can be performed by linemen in special conducting suits working from special insulated ladders. The conducting suit is connected to the energized conductors, as shown in Fig. 25-23.

Fig. 25-18 Lineman replacing dead-end insulators on energized 115-kV line using bare-hand methods. Note both buckets are connected to energized conductor. Hot-line dead-end tool assembly is used to remove strain on insulators and hot-line insulator cradle is used to support insulators. *(Courtesy A. B. Chance Co.)* ˜

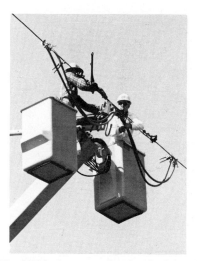

Fig. 25-19 View showing linemen in twin baskets after having removed two insulators from each phase of 69,000-volt line which was originally insulated for 115,000 volts. Weight of conductor is supported by the insulated boom mounted between buckets. Note bonding leads clamped on conductor. *(Courtesy Ohio Brass Co.)*

Fig. 25-20 Linemen working bare-handed making a splice at midspan on an energized 115,000-volt conductor. Note heavy cable for bypassing current. *(Courtesy Ohio Brass Co.)*

Fig. 25-21 View showing linemen in two-man bucket repairing line conductor at midspan on energized 345,000-volt line. Linemen are working barehanded. Bucket shield and line conductor are electrically connected to keep them at same potential. Fiberglass boom insulates bucket from ground. *(Courtesy Indiana and Michigan Electric Company of American Electric Power System.)*

Fig. 25-22 Linemen in two-man insulated bucket completing repair of bullet hole in live conductor of 345,000-volt line by installing "peanut" clamps on ends of compression sleeve. Note wire leads which bond bucket shield to live-line conductor, thereby maintaining bucket and line conductor at same potential. *(Courtesy Indiana and Michigan Electric Company of American Electric Power System.)*

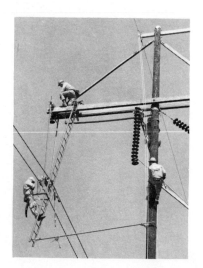

Fig. 25-23 Lineman, working from an insulated ladder, performing bare-handed live-line maintenance on a 345,000-volt circuit in a conductive suit. *(Courtesy A. B. Chance Co.)*

Touching live conductors with bare hands when insulated from ground is similar to the contact birds make when resting on live conductors, except that the birds would call it the "barefeet" method. The birds apparently experience no uncomfortable effects, or they would leave the conductor. Of course, the birds are thoroughly insulated from ground, since they require no auxiliary support.

Figures 25-13 through 25-23 illustrate linemen performing live-line maintenance operations on 34.5-, 69-, 115-, and 345-kv lines. Various details are noted in the legends.

Section **26**

Grounding

The grounded static wires for transmission lines and the grounded common neutral wire for primary distribution circuits originate at substations. The static wires for transmission lines will normally originate and terminate at a substation. The substation may be located at a generating station.

The substation grounding system governs the proper functioning of the whole grounding installation and provides the means by which grounding currents are conducted to remote earth. It is very important that the substation ground have a low-ground resistance, adequate current carrying capacity, and safety features for personnel. The substation grounding system normally consists of buried conductors and driven ground rods interconnected to form a continuous grid network. The surface of the substation is usually covered with crushed rock or concrete to control the potential gradient when large currents are discharged to ground and to increase the contact resistance to the feet of personnel in the substation. Substation equipment should be connected to the ground grid with large conductors to minimize the grounding resistance and limit the potential between the equipment and the ground surface to a safe value under all conditions. All substation fences should be constructed inside the ground grid and connected to the grid frequently to protect the public and workmen.

Ground Resistance   The electrical resistance of the earth is largely determined by the chemical ingredients of the soil and the amount of moisture present. Measurements of ground resistances completed by the Bureau of Standards are summarized in Table 26-1.

The resistance of the soils tested varied from 2 ohms to 3000 ohms. The type of soil, the chemical ingredients, and the moisture level surrounding an electrode determines the resistance, as

Table 26-1   Resistance of Different Types of Soil

Grounds tested	Soil	Resistance, ohms		
		Average	Minimum	Maximum
24	Fills and ground containing more or less refuse such as ashes, cinders, and brine waste	14	3.5	41
205	Clay, shale, and adobe, gumbo, loam, and slightly sandly loam with no stones or gravel	24	2.0	98
237	Clay, adobe, gumbo, and loam mixed with varying proportions of sand, gravel, and stones	93	6.0	800
72	Sand, stones, or gravel with little or no clay or loam	554	35	2700

SOURCE: Bureau of Standards Technologic Paper 108.

illustrated in Fig. 26-1. If the soil has uniform resistivity, the greatest resistance is in the area immediately surrounding the electrode which has the smallest cross section of soil at right angles to the flow of current through the soil. The area of the current path 8 to 10 ft from the electrode is so large that the resistance is negligible compared to the area immediately surrounding the ground rod. Measurements show that 90 percent of the total resistance surrounding an electrode is generally within a radius of 6 to 10 ft. A variation of a few percent in moisture will make a great difference in the effectiveness of a ground connection made with electrodes of a given size for moisture contents less than approximately 20 percent. Experimental tests made with red clay soil indicated that with only 10 percent moisture content, the resistivity was over 30 times that of the same soil having a moisture content of about 20 percent (Fig. 26-2).

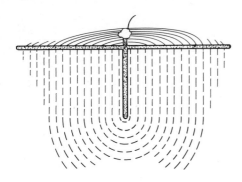

Fig. 26-1 Resistance of earth surrounding an electrode. This may be pictured as successive shells of earth of equal thickness. With increased distance from the electrode, the earth shells have greater area and therefore lower resistance. (*Courtesy Copperweld Steel Co.*)

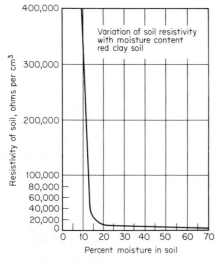

Fig. 26-2 Variation of soil resistivity with moisture content. (*Courtesy Copperweld Steel Co.*)

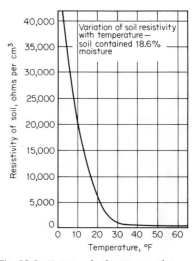

Fig. 26-3 Variation of soil resistivity with temperature. (*Courtesy Copperweld Steel Co.*)

The soil resistivity also varies greatly with the temperature (Fig. 26-3). The water in the soil freezes when the temperature is below 32°F. causing a large increase in the resistance of a ground connection. Grounding electrodes, which do not extend below the frost line where the soil freezes, will have a large variation in resistance as the seasons change. The depth of the ground electrode is important to the electrical performance. Driven ground rods should be long enough to reach the permanent moisture level of the soil (Fig. 26-4). Soil is seldom of uniform resistivity at different depths. The first few feet of soil near the surface normally has a relatively high resistance since it is subject to changes in moisture content as a result of rainfall.

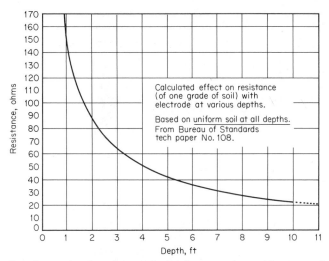

Fig. 26-4 Chart showing the relation between depth and resistance for a soil having a uniform moisture content at all depths. In the usual field condition, deeper soils have a higher moisture content and the advantage of depth is more pronounced. *(Courtesy Copperweld Steel Co.)*

The ground rod should be driven deep enough to be in perpetually moist earth. In most localities a depth of 8 to 10 ft is usually necessary to meet this requirement. A ground rod that penetrates perpetually moist earth will have a fairly low resistance. The codes require that the resistance of a driven ground shall not exceed 25 ohms. Still lower values of resistance are desirable and essential for the proper functioning of lightning arresters. Therefore, the ground resistance should be measured as the ground rod is driven as shown in Figs. 26-5 and 26-6. The ohmmeter shown resting on cable reel gives a continuous indication of the ground resistance as the rod is driven into the earth.

Fig. 26-5 Driving ground rod with gasoline-powered trailer rig. Note ohmmeter on cable reel indicating ground resistance in ohms as driving progresses. *(Courtesy Copperweld Steel Co.)*

Fig. 26-6 Ohmmeter used to indicate ground-rod resistance as ground rod is driven into earth. *(Courtesy Copperweld Steel Co.)*

If when the rod is driven to its full length it is found that the resistance is not well below 25 ohms, resort can be had to one of the two following procedures:

1. Extend the length of the ground rod.
2. Drive additional multiple rods.

A sectional ground rod is one that makes possible adding section upon section in order that the rod can be driven deeper into more favorable moist soil, thereby lowering its ground resistance. Figure 26-7 shows the parts of a typical sectional ground rod and gives details of its construction. The sections are joined by a heavy tapped bronze coupling. A special impact-resisting steel driving stud is screwed into the top section to take the hammer blows while driving. It protects the threading for subsequent attachment of additional sections of rod if needed. The sections are threaded at each end, one of which is pointed. The usual section length is 8, 10, or 12 ft.

Fig. 26-7 Typical sectional ground rod showing driving stud, couplings, rod sections, and driving point. *(Courtesy Copperweld Steel Co.)*

(*a*) Removal driving stud of special steel takes the blows, protects the thread.

(*b*) Cutaway view showing coupling and driving stud installed.

(*c*) The steel core gives strength and rigidity for driving. The molten-welded copper covering provides high conductivity, permanent protection against corrosion.

(*d*) A strong union and tight contact—easily made as successive sections are joined by the heavy bronze coupling.

(*e*) Bottom end of sectional rods are pointed for easier installation. Rods are threaded on both ends.

The installation procedure of sectional ground rods is as follows. It is the same whether hand-driven or power-driven.

1. The coupling and driving stud are screwed on the top end of the sectional rod.
2. This assembly is then driven until the coupling on the rod reaches the ground level.
3. The driving stud is then removed.
4. The lower end of the second sectional rod is then screwed into the coupling.
5. Another coupling is screwed onto the top end of the second sectional rod.
6. The driving stud is screwed into the second coupling.
7. Driving is now continued until the second section reaches ground level.
8. This sequence is repeated until resistance measurements indicate that the desired value of resistance has been obtained.

The other means of securing lowered ground resistance is by use of additional multiple rods. When two or more driven rods are well spaced from each other and connected together, they provide parallel paths to earth. They become, in effect, resistances in parallel or multiple and tend to follow the law of metallic parallel resistances. Figure 26-8 shows installations of one, two, three, and four ground rods installed at each pole, respectively. The rods are connected together with wire no smaller than that used to connect to the top of the pole.

If the rods were widely separated, two rods would have half or 50 percent of the resistance of one rod, but because they are usually driven only 6 ft or more from each other, the resistance of two rods is about 60 percent of that of one rod. Three rods, instead of having one-third or 33 percent of one rod, have on an average about 40 percent. Four rods, instead of having 25 percent of one rod, have about 33 percent. These values are shown graphically in Fig. 26-9.

Multiple rods are commonly used for arrester grounds and for station and substation grounds. In addition to lowering the resistance they provide higher current-carrying capacity and can thus handle larger fault currents.

Transmission Line Grounds   High-voltage electric transmission lines are designed and constructed to withstand the effects of lightning with a minimum amount of damage and interruption of operation. When the lightning strikes an overhead ground or static wire on a transmission line, the lightning current is conducted to ground through the metal tower or the ground wire installed along the pole. The top of the structure is raised in potential to a value determined by the magnitude of the lightning current and the surge impedance of the ground connection. If the impulse resistance of the ground connection is high, this potential may be many thousands of

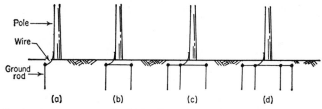

Fig. 26-8   Sketches showing one ground rod (*a*) and two, three, and four ground rods at poles (*b*), (*c*), and (*d*), respectively. Rods at each pole are connected together electrically.

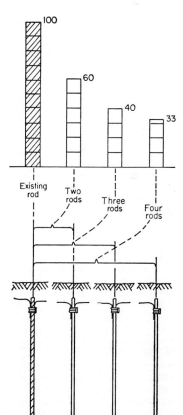

Fig. 26-9   Installations of multiple ground rods. Upper graphs show extent to which ground resistance is reduced by the use of two, three, or four rods connected in multiple.

volts. If the potential exceeds the insulation level of the equipment, flashover will result causing a power arc which will initiate the operation of protective relays and removal of the line from service. If the transmission structure is well grounded, and proper coordination exists between the ground resistance and the conductor insulation, flashover can usually be avoided. The transmission line grounds are installed in accordance with the specifications provided. Ground rods are usually used to obtain a low-ground resistance (Figs. 26-10 through 26-13).

A pole butt grounding plate (Fig. 26-14) or a butt coil can be used on wood pole structures. A butt coil is a spiral coil of bare copper wire placed at the bottom of a pole as shown in Fig. 26-15. The wire of the coil continues up the pole as the ground wire lead. Sometimes it is wound around the lower end of the pole butt a few times as a helix (see Fig. 26-16) to increase the amount of wire surface in contact with the earth. Butt coils should have enough turns to make good contact with the earth. A coil having seven turns and 13 or more ft of wire would provide satisfactory ground contact. Number 6 or larger AWG soft-drawn or annealed copper wire should be used.

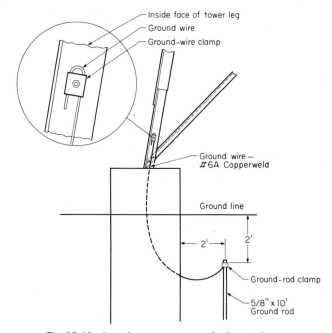

Fig. 26-10   Ground wire connects ground rod to metal tower.

To prevent the coil from acting as a choke coil, the ground lead from the innermost turn is stapled to all the other turns.

If the soil has a high resistance, a grounding system called a "counterpoise" may be necessary. The counterpoise for an overhead transmission line consists of a special grounding terminal which reduces the surge impedance of the ground connection and increases the coupling between the ground wire and the conductors. Counterpoises are normally installed for transmission line structures located in areas with sandy soil or rock close to the surface. The types of counterpoises used are the continuous, or "parallel" type, and the radial, or "crowfoot," type (Fig. 26-17). The continuous, or "parallel" counterpoise, consists of one or more conductors buried under the transmission line for its entire length, or under sections with high-resistance soils. The counterpoise wires are connected to the overhead ground, or static wire, at all supporting structures. The radial-type counterpoise consists of a number of wires extending radially from the tower legs. The number and length of the wires will depend on the tower location and the soil conditions. The continuous counterpoise wires are usually installed with a cable plow at a depth of 18 in or more (Fig. 26-18). The wires should be deep enough that they will not be disturbed by cultivation of the land.

Distribution Circuit Grounds   A multigrounded, common neutral conductor for a primary distribution line is always connected to the substation grounding system where the circuit originates and to all grounds along the length of the circuit. The primary common neutral conductor must be continuous over the entire length of the circuit and should be grounded at intervals not to exceed ¼ mile. The primary neutral conductor can serve as a secondary neutral conductor. If

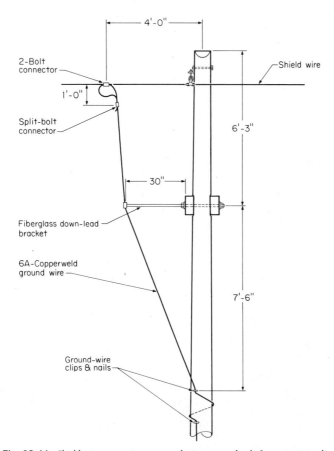

Fig. 26-11   Shield wire connection to ground wire on wood pole for transmission line.

separate primary and secondary neutral conductors are installed, the conductors should be connected together if the primary neutral conductor is effectively grounded. All equipment cases and tanks should be grounded (Figs. 26-19 and 26-20). Lightning arrester ground terminals should be connected to the common neutral if available and separate ground installed as close as is practical. Switch handles, street light fixtures, down guys, and transformer secondaries must be properly grounded (Fig. 26-21). The common conductor on the transformer secondary for a single-phase 120/240-volt service must be grounded. A three-phase delta or open-delta connected transformer secondary for 240 or 480-volt service would have the center tap of one of the transformer secondary terminals grounded. If the center tap of one of the transformer secondaries is not available,

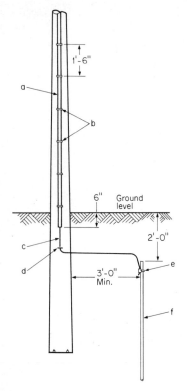

Fig. 26-12 Ground wire connection to ground rod for wood pole transmission line structure. (*a*) ground-wire molding, (*b*) staples molding, (*c*) 6A Copperweld, (*d*) staples fence, (*e*) ground-rod clamp, (*f*) ⅝″ x 10′ ground rod.

one-phase conductor is normally grounded. The neutral wire for three-phase 4-wire wye-connected secondaries for 208Y/120 or 480Y/277-volt service is grounded (Fig. 26-22). If the primary circuit has an effectively grounded neutral conductor, the primary and secondary neutrals are connected together as well as to ground. The neutral conductor of the service to the customer should be connected to ground on the customer's premises. A metal water pipe provides a good ground. Care must be taken to avoid plastic, cement-lined, or earthenware piping. Driven grounds may be installed near the customer's service entrance.

Installing Driven Grounds Since the most common ground in distribution systems is the driven ground, its installation will be briefly described. A galvanized-steel rod, stainless steel rod, or copperweld rod is driven into the ground beside the pole at a distance of 1 to 2 ft as shown in Fig. 26-23. The rod should be driven until the top end is 4 to 6 in below the ground level (see Fig. 26-24). If driving is difficult, a bar may be used to form the hole before the rod is driven. Another form of ground-rod driver is shown in Fig. 26-25. This driver can accommodate a rod of any length, as the driving hammer slides over the rod. If many ground rods have to be driven, a power hammer can be used, as shown in Fig. 26-26. This greatly reduces the effort required and speeds up the work.

After the rod is driven in place, connection is made with the ground wire from the pole. The actual connection to the ground rod is usually made with a heavy bronze clamp as shown in Fig. 26-27. The clamp is placed over the ground-rod end and the ground-wire end. The size of the ground wire should not be less than No. 6 AWG copper wire. Number 4 wire is generally used. When the setscrew in the clamp is tightened, the ground wire is squeezed against the ground rod. Several precautions are taken in making the actual connection to minimize the possibility of the ground wire being pulled out for frost action or packing of the ground above it. The ground wire may be brought up through the clamp from the bottom. It is then bent over the clamp so that it cannot pull out. The wire is trained loosely so that there is plenty of slack between the ground-wire molding and the rod.

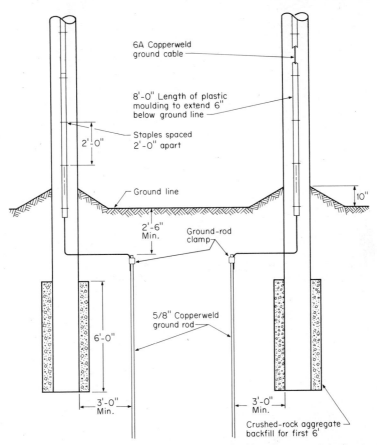

6A Copperweld
ground cable

8'-0" Length of plastic
moulding to extend 6"
below ground line

Staples spaced
2'-0" apart

2'-0"

Ground line

10"

2'-6"
Min.

Ground-rod
clamp

5/8" Copperweld
ground rod

6'-0"

3'-0"
Min.

3'-0"
Min.

Crushed-rock aggregate
backfill for first 6'

Fig. 26-13   Ground wire connections to ground rods for wood pole H-Frame transmission line structure.

Fig. 26-14   Pole butt grounding plate attached to pole of a transmission or distribution line before the pole is set. The grounding plate is made from copper plate with moisture retaining cups to maintain a good ground. *(Courtesy Homac Manufacturing Co.)*

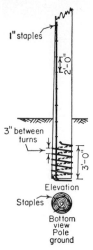

Fig. 26-15 Butt coil in position on butt of pole. Entire weight of pole helps to maintain contact with earth below pole. The spiral coil is stapled to the pole butt. Wire from inner spiral is stapled to all turns as it crosses them to shunt out the turns.

Fig. 26-16 Views of butt-coil ground. Elevation shows six twin helix with 3 in between turns. Bottom view shows spiral pancake coil of seven turns.

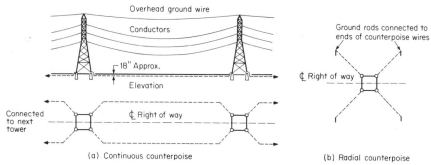

Fig. 26-17 Two general arrangements of counterpoise installations. *(Courtesy Copperweld Steel Co.)*

Fig. 26-18 Lineman using vibrating plow to install continuous counterpoise wire. *(Courtesy The Charles Machine Works, Inc.)*

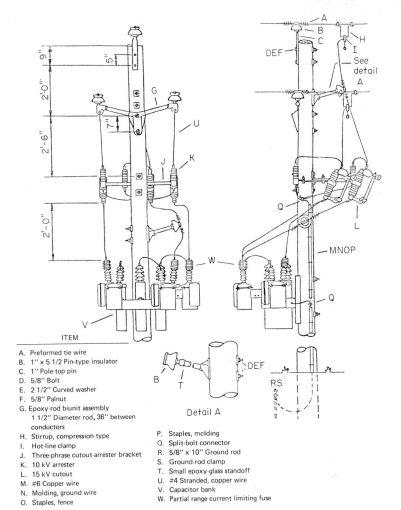

ITEM

A. Preformed tie wire
B. 1" x 5 1/2 Pin-type insulator
C. 1" Pole-top pin
D. 5/8" Bolt
E. 2 1/2" Curved washer
F. 5/8" Palnut
G. Epoxy-rod biunit assembly
   1 1/2" Diameter rod, 36" between
   conductors
H. Stirrup, compression type
I. Hot-line clamp
J. Three-phrase cutout-arrester bracket
K. 10 kV arrester
L. 15 kV cutout
M. #6 Copper wire
N. Molding, ground wire
O. Staples, fence

P. Staples, molding
Q. Split-bolt connector
R. 5/8" x 10" Ground rod
S. Ground-rod clamp
T. Small epoxy-glass standoff
U. #4 Stranded, copper wire
V. Capacitor bank
W. Partial range current limiting fuse

Detail A

Fig. 26-19 Capacitor bank installation for 13.8 kV circuit with grounding details illustrated.

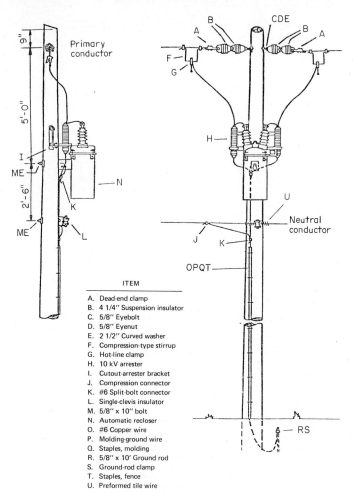

ITEM

A. Dead-end clamp
B. 4 1/4" Suspension insulator
C. 5/8" Eyebolt
D. 5/8" Eyenut
E. 2 1/2" Curved washer
F. Compression-type stirrup
G. Hot-line clamp
H. 10 kV arrester
I. Cutout-arrester bracket
J. Compression connector
K. #6 Split-bolt connector
L. Single-clevis insulator
M. 5/8" x 10" bolt
N. Automatic recloser
O. #6 Copper wire
P. Molding-ground wire
Q. Staples, molding
R. 5/8" x 10' Ground rod
S. Ground-rod clamp
T. Staples, fence
U. Preformed tile wire

Fig. 26-20   Single-phase rural distribution automatic recloser installation with grounding details illustrated.

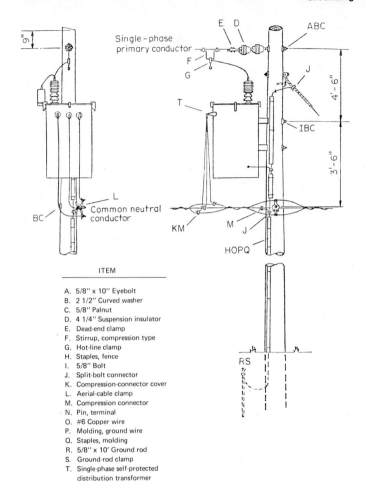

Single-phase primary conductor

Common neutral conductor

ITEM
A. 5/8" x 10" Eyebolt
B. 2 1/2" Curved washer
C. 5/8" Palnut
D. 4 1/4" Suspension insulator
E. Dead-end clamp
F. Stirrup, compression type
G. Hot-line clamp
H. Staples, fence
I. 5/8" Bolt
J. Split-bolt connector
K. Compression-connector cover
L. Aerial-cable clamp
M. Compression connector
N. Pin, terminal
O. #6 Copper wire
P. Molding, ground wire
Q. Staples, molding
R. 5/8" x 10' Ground rod
S. Ground-rod clamp
T. Single-phase self-protected distribution transformer

Fig. 26-21   Distribution transformer installed on dead-end pole with grounding details illustrated.

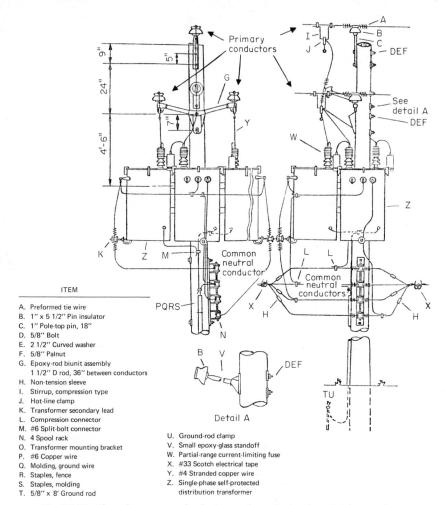

ITEM

A. Preformed tie wire
B. 1" x 5 1/2" Pin insulator
C. 1" Pole-top pin, 18"
D. 5/8" Bolt
E. 2 1/2" Curved washer
F. 5/8" Palnut
G. Epoxy-rod biunit assembly
   1 1/2" D rod, 36" between conductors
H. Non-tension sleeve
I. Stirrup, compression type
J. Hot-line clamp
K. Transformer secondary lead
L. Compression connector
M. #6 Split-bolt connector
N. 4 Spool rack
O. Transformer mounting bracket
P. #6 Copper wire
Q. Molding, ground wire
R. Staples, fence
S. Staples, molding
T. 5/8" x 8' Ground rod

U. Ground-rod clamp
V. Small epoxy-glass standoff
W. Partial-range current-limiting fuse
X. #33 Scotch electrical tape
Y. #4 Stranded copper wire
Z. Single-phase self-protected
   distribution transformer

Fig. 26-22   Three-phase wye-wye distribution transformer bank with quadruplex secondaries.

Fig. 26-23 Lineman helper driving the ground connection. The rod is copperweld, has a diameter of ⅝ in, and is 8 ft long. Lineman helper is using a ground-rod driver which slips over the end of the rod. The driver must be operated with short sharp strokes. Care must be taken not to lift the driver above the head of the rod and have it slide off, as this might cause injury. *(Courtesy Wisconsin Electric Power Co.)*

Fig. 26-24 Lineman helper driving the ground rod with a sledge after he has taken it as far down as he can go with the ground-rod driver. A driving head is placed over the ground-rod end, as shown in the figure, to prevent burring or mushrooming of the ground-rod end. Before the use of the driving head, mushrooming occurred whenever the rod was driven into hard rocky ground. In very hard or rocky ground the ground rod has a tendency to vibrate excessively when struck with a sledge because of its small diameter (⅜, ½, ⅝, or 1 in). In such a case a second man must steady the rod. This is done with a pair of long tongs or a stiff piece of wire. An ordinary piece of No. 6 W.P. copper wire is good for this purpose. One end is wrapped around the rod once or twice while the man holds the other end. The rod is never steadied by hand because of the possibility of an accident. *(Courtesy Wisconsin Electric Power Co.)*

Fig. 26-25 Another form of ground-rod driver. A heavy chuck slid over the ground-rod grips the rod. A long hollow handle is moved up and down by the operator. Each downward blow strikes the chuck housing and forces the rod farther into the earth. The chuck is automatically moved up on the rod as it is driven. Any length rod can be driven since it is fed through the hollow handle from the top.

Fig. 26-26   Lineman driving sectional ground rod by means of gasoline-driven hammer. *(Courtesy Hubbard & Co.)*

Fig. 26-27   Lineman helper clamping ground wire to ground rod, which has been driven 6 in below the surface of the ground. This connection must be made before the lightning arresters are completely connected up on the pole. If this were the last connection—the one that completes the lightning-arrester circuit to ground—the helper could get a high-voltage electric shock if one of the lightning arresters should be defective. The last connection is always made at the lightning arrester by the lineman, either to its top or its bottom terminal.

Ground-Wire Molding   The molding protects the wire as well as the lineman. The molding should extend well down below the surface of the ground, so that in case the ground rod bakes out, that is, fails to make contact with moist earth, there will not be any chance of a child or animal coming in contact with the ground wire and the ground at the same time. When the installation is finally covered up, the ground wire should be completely out of sight.

Another reason for covering the ground wire is to protect persons in case of lightning-arrester trouble. If a lightning arrester should break down and allow power current to flow, the ground wire will become "hot" and remain so for an indefinite period of time if the resistance of the ground connection is high. Anyone touching the wire while it is energized could get an electric shock.

Protective Grounds

Protection of the lineman is most important when a transmission or distribution line or a portion of a line is removed from service to be worked on using deenergized procedures. Precautions must be taken to be sure the line is deenergized before the work is started and remains deenergized until the work is completed. The same precautions apply to new lines when construction has progressed to the point where they can be energized from any source.

The installation of protective grounds and short-circuiting leads protect against the hazards of static charges on the line, induced voltages, and accidental energizing of the line.

When a deenergized line and an energized line parallel each other, the deenergized line may pick up a static charge from the energized line because of the proximity of the lines. The amount of this static voltage "picked up" on the deenergized line depends on the length of the parallel, weather conditions, and many other variable factors. However, it could be hazardous, and precautions must be taken to protect against it by grounding the line at the location where the work is to be completed. This will drain any static voltage to ground and protect the workman from this potential hazard.

When a deenergized line parallels an energized line carrying load, the deenergized line may have a voltage induced on it in the same manner as the secondary of a transformer. If the deenergized line is grounded at a location remote from where the work is being done, this induced voltage will be present at the work location. Grounding the line at the work location will eliminate any induced voltage.

Grounding and short-circuiting protects against the hazard of the line becoming energized from either accidental closing in of the line or accidental contact with an energized line which crosses or is adjacent to the deenergized line.

The procedures established to control the operation of equipment in an electric system practically prevent the accidental energizing of a transmission or distribution line. Hold off tagging procedures have proven to be very effective. If a circuit should be inadvertently energized, the grounds and short-circuits on the line will cause the protective relays to initiate tripping of the circuit breaker at the source end of the energized line in a fraction of a second and deenergize the "hot" line. During this short interval of time, the grounds and short-circuits on the line being worked on will protect the workmen (Fig. 27-1).

The possibility of another energized line making contact with a deenergized line being worked on is remote. However, if the line being worked on is grounded and short-circuited, fault current in the line making the contact will initiate operation of equipment to deenergize the circuit. The protective grounds on the line being worked on will protect the workmen during this short time interval.

Fuz Testing The lineman must test each line conductor to be grounded to be sure it is deenergized (dead) before the protective grounds are installed. The circuit can be tested with high-voltage voltmeters previously described or by fuzzing.

Fuzzing procedures are reliable to detect voltages of 13.2Y/7.62 kV or higher. The metal end of a hot-line switch stick or a fuzzing ring held in a Grip-All hot stick can be used for the fuzzing operation (Figs. 27-2 and 27-3). Each conductor of the line must be tested to be sure it is deenergized immediately prior to the installation of the grounding conductors. When bringing the metal device on the end of the hot-line tool close to the conductor, the lineman must watch and listen to determine if the conductor is energized.

Fuzzing indications between an energized and a deenergized line are distinctly different, and there should be no doubt as to the results. If, however, there should be any question, make a comparison check on a conductor known to be energized.

If the line is alive a very audible crackling, buzzing, or frying sound will be heard as the fuzzing attachment is brought slowly toward or away from the conductor. The sound will be definite and unmistakable, especially within inches of the conductor, but may vary in intensity depending upon weather conditions such as temperature and humidity.

An arc will usually accompany the sound. The arc will continue as long as the fuzzing device is held close to the conductor.

If the line is deenergized, there will normally be no indication when the fuzzing attachment is brought into the conductor.

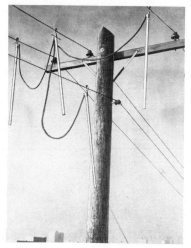

Fig. 27-1   Grounding clamps installed on four-wire three-phase line. Line conductor mounted on side of pole is grounded neutral. *(Courtesy James R. Kearney Corp.)*

Fig. 27-2   Fuzzing ring made from ⅜-in round copper bar stock in the form illustrated for use with a Grip-All clamp stick.

If an induced voltage or static charge on the line is great enough, it will produce a slight buzzing noise and perhaps a visible arc when the fuzzing device is slowly moved away from the conductor being tested. A high-voltage voltmeter can be used to check for static or induced voltage if the lineman is unable to positively determine that the line is safe for grounding. The high-voltage voltmeter connected between ground and conductor energized with a static charge will give an indication of voltage, and then indicate no voltage, as the static charge is drained off to ground through the high-voltage voltmeter.

Protective Ground Installation   After the testing is completed, the protective grounds should be installed in the following sequence:

1. Connect one end of the grounding conductor to an established ground.

2. Connect the other end of the grounding conductor, using a hot-line tool, to the bottom conductor on vertical construction or the closest conductor on horizontal construction (Fig. 27-4).

3. Install grounds or jumpers from a grounded conductor to the ungrounded conductors in sequence until all conductors are grounded and short-circuited together.

When the work is completed and the protective grounds are to be removed always remove the grounds from the line conductors in reverse sequence removing the connection to ground last.

When a connection is made between a phase conductor and ground, the grounding lead, connection to the earth, and the earth itself become a part of the electrical circuit. All parts of the grounding circuit must be adequate to provide protection to the workmen under this severe condition (Figs. 27-5 through 27-7).

Protective Ground Requirements An inadequate or poorly installed ground can be a safety hazard. It can give a lineman a false sense of security without actually protecting him under severe conditions. The protection that is provided by a ground is as good (or as bad) as the care that is taken to make sure it is installed properly. The requirements of a good ground are

1. A low resistance path to earth.
2. Clean connections.
3. Tight connections.
4. Connections made to proper points.
5. Adequate capacity of grounding equipment.

A good available ground (earth connection) is the system neutral; however, it usually is not available on a transmission line. However, if there is a common neutral at the location where the

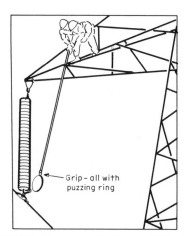

Grip–all with puzzing ring

Fig. 27-3 Lineman using hot-line tool with fuzzing ring to detect energized conductor.

Fig. 27-4 Lineman completing the installation of protective grounds using a hot-line tool. Grounding operation was completed by connecting the first jumper connection to the common neutral which is effectively grounded and located at the lowest point on the pole. Jumpering was completed in sequence from bottom to top providing the lineman with a zone of protection as the protective grounds are installed.

line is to be grounded it must be connected to the grounding system on the structure so that all grounds in the immediate vicinity are connected together. If the structure is metal, the common neutral should be bonded to the steel, either with a permanent connection or a temporary jumper. If the structure is a wood pole that has a ground wire connection to the static wire, the common neutral should be permanently connected to the ground wire. If there is no permanent connection, a temporary jumper should be installed. If there is no ground wire on the pole, the common neutral should be temporarily bonded to whatever temporary grounding connection is used to ground the line.

Static wires are connected to earth at many points along the line and provide a low resistance path for grounding circuits. These are normally available for use as the ground-end connection of grounding leads. If the line has a static wire, it is usually grounded at each pole. A ground wire runs down the pole and connects to a ground rod. All metal poles and towers are adequately grounded when they are installed and serve as a good ground connection point, provided the connection between the metal structure and the ground rod has not been cut or disconnected. If there is an exposed lead from the structure to the ground, it should be checked for continuity.

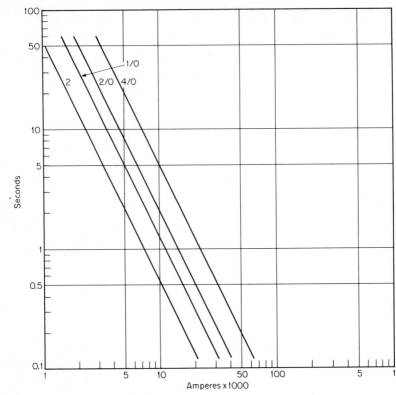

Fig. 27-5 Suggested maximum allowable fault currents for copper grounding cables *total* current-on time. For minimal off time between consecutive current-on periods—reclosures. These values exceed IPCEA recommendations for cable installations by 1.91. (Based on tests of 10 to 60 cycle duration, 30°C ambient) *(Courtesy A. B. Chance Co.)*

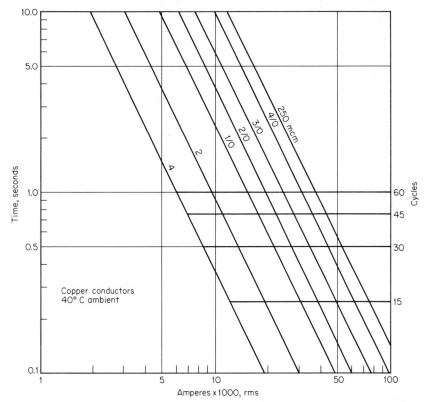

Fig. 27-6   Fusing current time for copper conductors, 40°C ambient temperature. *(Courtesy A. B. Chance Co.)*

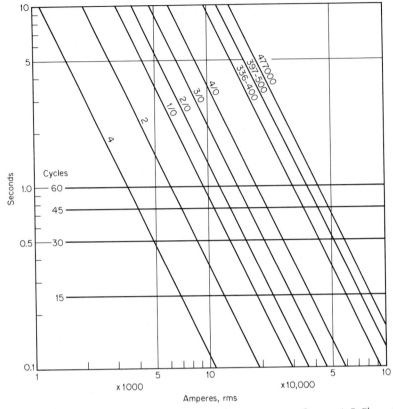

Fig. 27-7   Fusing currents, aluminum conductor, 40°C ambient temperature. *(Courtesy A. B. Chance Co.)*

Fig. 27-8   Floating ground connector designed to keep the conductor grounded as stringing operations are in process. A rope is used to secure the connector in place. *(Courtesy Everly Ever-Ground Co.)*

If a common neutral, static wire grounded at the pole, or grounded structure is not available, use any good metallic object, such as an anchor rod or a ground rod, which obviously extends several feet into the ground. If no such ground is available, the lineman must provide a ground by driving a ground rod, or screwing down a temporary anchor until it is firm and in contact with moist soil.

Protective Ground Locations   A line is grounded and short-circuited whenever it is to be worked on deenergized. It is grounded before the work is started and must remain so until all work is completed.

Special devices are used to permit the grounding of conductors during stringing operations (Fig. 27-8). Deenergized lines are grounded for the protection of all the men working on the line. The protective grounds are installed from ground in a manner to short-circuit the conductors so that the lineman and everything in the working area will be at "equal potential." This is the "man-shorted-out" concept, since the grounding, short-circuiting, and bonding leads will carry any current which may appear because of potential differences in the work area.

When a line is to be worked on deenergized, the line must be grounded and short-circuited at the work location, even though the work location is within sight of the disconnecting means used to deenergize the line.

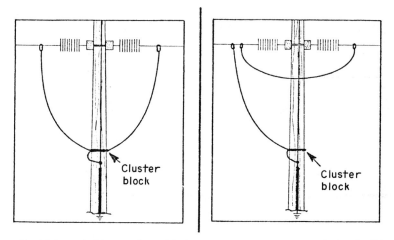

Fig. 27-9   Method for grounding and bridging conductors across an open point is illustrated.

When work is to be done at more than one location in a line section, the line section being worked on must be grounded and short-circuited at one location, and only the conductor(s) being worked on must be grounded at the work location.

When a conductor is to be cut or opened, the conductor must be grounded on both sides of the open or grounded on one side of the open and bridged with a jumper cable across the point to be opened (Fig. 27-9).

A cluster block can be installed on a wood-pole structure to facilitate the ground connections (Figs. 27-10 through 27-12).

Protective grounds installed on a steel transmission-line structure to connect conductors to ground and together are illustrated in Fig. 27-13.

Underground System Protective Grounds   Underground transmission and distribution cable conductors are not readily accessible for grounding. Transmission cables originating and terminating in substations can normally be grounded with portable protective grounding equipment at the cable termination points in the same manner as described for overhead transmission circuits. If the cable termination points are not accessible, special equipment manufactured and provided for grounding the circuit must be used. Underground distribution cables that originate in substation switchgear to supply complete underground circuits must be tested and grounded

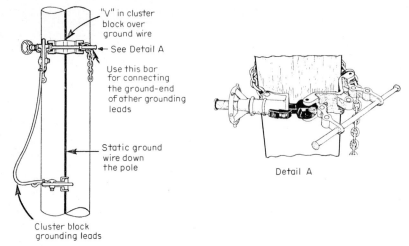

Fig. 27-10   Metal cluster block equipped with 4 ft chain binder. The connection bar will accommodate four ground lead clamps.

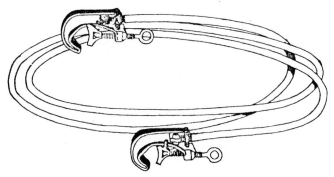

Fig. 27-11   Grounding and short-circuiting jumper equipped with C-type clamps designed for operation with hot-line tool. Jumper cable is 1/0 copper conductor, or larger, with a large number of strands to provide flexibility.

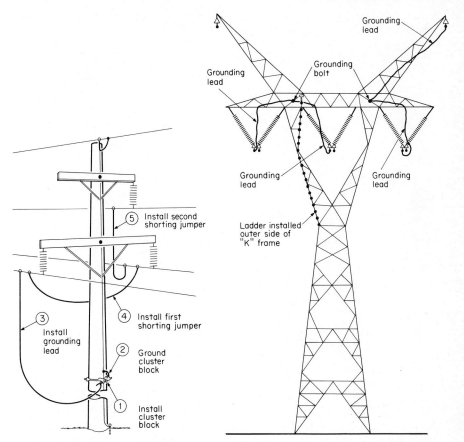

Fig. 27-12 Protective grounds installed on conductors of a single-circuit wood pole in a subtransmission line.

Fig. 27-13 Protective grounds installed on conductors of a single-circuit steel transmission-line tower.

Fig. 27-14   Lineman testing cable circuit on riser pole with high-voltage-testing voltmeter prior to installation of protective grounds. *(Courtesy Iowa-Illinois Gas and Electrc Co.)*

Fig. 27-15   Lineman installing grounding jumper to common neutral conductor. *(Courtesy Iowa-Illinois Gas and Electric Co.)*

with special equipment provided at the substation before work is performed on the cable conductors. Underground equipment installed in vaults may be equipped with oil switches which can be used to isolate and ground the cable conductors. Underground cable circuits originating at a riser pole in an overhead line can be grounded at the riser pole with portable protective grounding equipment (Figs. 27-14 through 27-16). Underground cable circuits originating in pad-mounted switchgear illustrated in Fig. 27-17 can be grounded at the switchgear near the cable termination. The switches in the switchgear must be open to isolate the cable circuit. Opening the switch provides a visible break in the circuit. A high-voltage-test voltmeter can be used by the lineman to be sure the circuit does not have a feedback from the remote end proving that the cables are deenergized. A special conductor is available above the cable terminations for the application of portable grounding devices in the switchgear.

Fig. 27-16   Lineman completing grounding operation of cable riser circuit. Lineman uses hot stick to connect grounding lead to conductor attached to cable pothead. All three cutouts between the distribution line and the cable potheads are open. The grounding jumpers connect each pothead to the common neutral conductor grounding and short circuiting all three underground cables. *(Courtesy Iowa-Illinois Gas and Electric Co.)*

Fig. 27-17 Pad-mounted switchgear with compartment doors open showing cable terminations, conductor for testing, and grounding above cable terminations and switches. Switches are in the closed position in the picture. *(Courtesy S. & C. Electric Co.)*

Fig. 27-18 Using hot-line tool to mount feed-through device on bracket in high-voltage compartment of pad-mounted transformer. *(Courtesy Iowa-Illinois Gas and Electric Co.)*

Fig. 27-19 Removing cable to be grounded from bushing with hot-line tool hooked to elbow connector terminating cable and installing elbow connector on feed-through device. *(Courtesy Iowa-Illinois Gas and Electric Co.)*

Fig. 27-20 High-voltage-test voltmeter with special probe for testing cable circuit through feed-through device. *(Courtesy Iowa-Illinois Gas and Electric Co.)*

Fig. 27-21 Lineman testing cable circuit with high-voltage-test voltmeter through feed-through device. Note one lead of the high-voltage-test voltmeter is connected to grounding conductor in pad-mounted transformer compartment. *(Courtesy Iowa-Illinois Gas and Electric Co.)*

Fig. 27-22 Connecting special grounding jumper with elbow connector on one end and "C" clamp on the other end to grounded neutral cable in pad-mounted transformer compartment. "C" clamp is connected to ground lead. *(Courtesy Iowa-Illinois Gas and Electric Co.)*

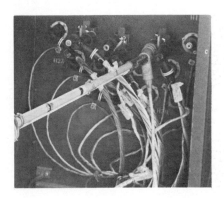

Fig. 27-23 Lineman installs elbow connector on grounding jumper on feed-through-device bushing, grounding cable. *(Courtesy Iowa-Illinois Gas and Electric Co.)*

Underground cable conductors terminating in pad-mounted transformers can be isolated, tested, and grounded at the transformer location. The lineman opens the pad-mounted transformer compartment door and installs a feed-through device with a hot-line Grip-All clamp stick to prepare for the isolating and grounding operation (Fig. 27-18). The cable to be worked on must be isolated to separate it from a source of energy at the remote end. The lineman or cableman identifies the cable to be worked on at the pad-mount transformer with the markings on the cable and maps of the circuit. When the proper cable has been identified, the elbow cable terminator can be separated from the bushing on the transformer with the Grip-All clamp stick. The lineman then installs the elbow on the bushing on the feed-through device with the hot stick (Fig. 27-19). A high-voltage tester equipped with a special connection can be used to test the isolated cable through the second bushing on the feed-through device (Fig. 27-20). If the cable is found to be deenergized, the lineman can connect a special grounding conductor to the ground wires in the pad-mounted transformer compartment and place it on the bushing of the feed-through device with a Grip-All clamp stick (Figs. 27-21 through 27-23). A cable is now properly grounded so that the necessary work can be performed safely.

Street Lighting

Street-light systems have developed gradually from candles in the seventeenth century to gas lighting in the eighteenth century and electric lighting in the twentieth century. The first electric lights were electric arc lamps installed in Paris, France. The development of the incandescent lamp initiated by Thomas Edison greatly improved street lighting. The modern street light system reduces traffic hazards, helps to prevent crime, and enhances the beauty of our surroundings. The incandescent street lighting fixture that was commonly used in the early part of the twentieth century is gradually being replaced by electric discharge lamps of the mercury-vapor, fluorescent, high-pressure sodium, and metal-halide types. The mercury-vapor lamp has wide use because of its good efficiency, long life, and good light maintenance. High-pressure sodium and metal-halide lamps have a greater light output per watt of electric power input than the mercury-vapor lamps.

Street-Lighting Terms

Light   A form of radiant energy. Measurements are based upon a unit of luminous intensity equal to the light emitted by a "standard candle" in a horizontal direction.

Lamp   A source of light.

Luminaire   The device which directs, controls, or modifies the light produced by the lamp. It consists of a light source and all necessary mechanical, electrical, and decorative parts.

Candlepower   The amount of light that will illuminate a surface 1 ft distant from the light source to an intensity of 1 footcandle.

Lumen   The unit quantity of light output defined as the amount of light which falls upon an area of 1 sq ft, every point of which is 1 ft distant from a source of 1 candlepower. A uniform 1-candlepower source of light emits a total of 12.57 lumens.

Footcandle   The unit of illumination produced on a surface all points of which are 1 ft distant from a uniform point source of 1 candle, or the illumination of a surface 1 sq ft in area on which a flux of 1 lumen is uniformly distributed.

Ballast   An auxiliary device used with vapor lamps, on multiple circuits, to provide proper operating characteristics. It limits the current through the lamp and may also transform voltage.

Mast Arm   An attachment for a pole used to support a luminaire.

Parts of Street-Lighting System   A street-lighting system consists of the following:

1. Circuit
    a. Multiple or parallel
    b. Series
2. Light-fixture support
    a. Mast arm
    b. Post

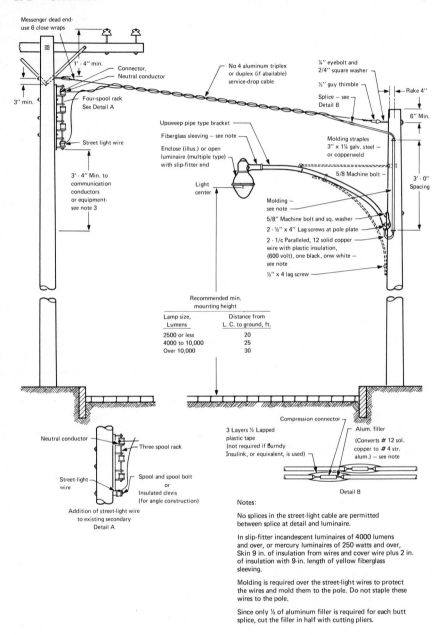

Messenger dead end-
use 6 close wraps

1' - 4" min.

Connector.
Neutral conductor

3" min.

Four-spool rack
See Detail A

Street light wire

3' - 4" Min. to
communication
conductors
or equipment-
see note 3

No 4 aluminum triplex
or duplex (if abailable)
service-drop cable

¼" eyebolt and
2/4" square washer

½" guy thimble

Splice — see
Detail B

Rake 4"

6" Min.

Molding straples
3" x 1¼ galv. steel —
or copperweld

5/8 Machine bolt —

3' - 0"
Spacing

Upsweep pipe type bracket

Fiberglass sleeving — see note

Enclose (illus.) or open
luminaire (multiple type)
with slip-fitter end

Light
center

Molding —
see note

5/8" Machine bolt and sq. washer

2 - ½" x 4" Lag screws at pole plate

2 - 1/c Paralleled, 12 solid copper
wire with plastic insulation,
(600 volt), one black, onw white —
see note

½" x 4 lag screw

Recommended min.
mounting height

Lamp size, Lumens	Distance from L. C. to ground, ft.
2500 or less	20
4000 to 10,000	25
Over 10,000	30

Neutral conductor

Three spool rack

Street-light
wire

Spool and spool bolt
or
Insulated clevis
(for angle construction)

Addition of street-light wire
to existing secondary
Detail A

3 Layers ½ Lapped
plastic tape
(not required if Burndy
Insulink, or equivalent, is used)

Compression connector

Alum. filler

(Converts # 12 sol.
copper to # 4 str.
alum.) — see note

Detail B

Notes:

No splices in the street-light cable are permitted
between splice at detail and luminaire.

In slip-fitter incandescent luminaires of 4000 lumens
and over, or mercury luminaires of 250 watts and over,
Skin 9 in. of insulation from wires and cover wire plus 2 in.
of insulation with 9-in. length of yellow fiberglass
sleeving.

Molding is required over the street-light wires to protect
the wires and mold them to the pole. Do not staple these
wires to the pole.

Since only ½ of aluminum filler is required for each butt
splice, cut the filler in half with cutting pliers.

Fig. 28-1 Multiple street-light installation on wood pole and method of extending a multiple circuit across a street or road. *(Courtesy American Electric Power System.)*

3. Luminaire
    a. Incandescent
    b. Electric discharge
        (1) Mercury vapor
        (2) Metal halide
        (3) Low-pressure sodium vapor
        (4) High-pressure sodium vapor
        (5) Fluorescent
4. Light control
    a. Reflector
    b. Refractor
    c. Diffuser
5. Time control
    a. Manual
    b. Automatic time switch
    c. Light-sensitive relay
6. Circuit switching
    a. Pilot wire
    b. Cascading

Multiple-Circuit Street Lights   The lamp for a multiple street-light fixture is manufactured to operate on a constant line voltage of 120, 208, 240, 277, or 480 volts. The incandescent-type lamps operate directly from the line-voltage supply. Electric-discharge-type lamps require a ballast designed for the particular type lamp to provide starting voltage, sustaining voltage during lamp warm-up period, and operating voltage. The ballast for an electric-discharge lamp must provide a relatively high voltage to start the lamp and control the current in the lamp during the warm-up period. The impedance of the lamp decreases as the arc in the lamp gets hot. Electric discharge lamps take longer to reach full brilliance in cold weather as a result of the warm-up requirement. The lamps or ballasts for use on a multiple-type circuit are connected across a constant voltage source usually obtained from the secondary of a distribution transformer (Fig. 28-1). In a multiple circuit the lamps are connected in parallel across the circuit as shown in Fig. 28-2. The distribution transformer delivers a constant low voltage, usually 120 volts, to the circuit. All lamps, therefore, have the same voltage impressed across them.

A multiple circuit requires that both wires be brought to each lamp. This makes each lamp independent of the other; that is, any lamp can be switched on or off or burn out without affecting the other lamps. In this respect the multiple circuit is more satisfactory than the series circuit. The lamp socket does not have to be equipped with a film cutout as is required for series operation. Modern luminaires built today for use on multiple circuits do not utilize the plug-in type socket

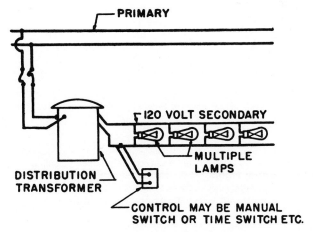

Fig. 28-2   Multiple street-lighting circuit.

but use a permanently mounted socket. However, if luminaires are converted from series to multiple, then an insulator is used to separate the prongs of the series socket permanently as shown in Figs. 28-3 and 28-4.

Generally, filament lamps operate from 120-volt two-wire circuits or 120/240-volt three-wire circuits, whereas electric discharge lamps using ballasts operate at the higher voltages. These voltages are safer for both the lineman and the public than the higher voltage of a series circuit.

In the early days of street lighting the series system was used almost wholly. Today the multiple systems are most common.

Fig. 28-3   Series lamp socket adapted for use with multiple lamp. Film disk terminals are permanently separated with insulator. *(Courtesy Joslyn Mfg. and Supply Co.)*

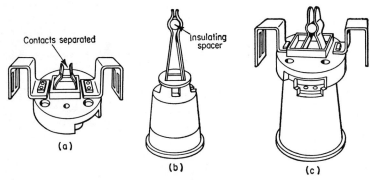

Fig. 28-4   Series lamp socket and receptacle. (*a*) Series receptacle; (*b*) series socket; (*c*) assembly. Contacts in receptacle and in cutout are permanently separated.

Ornamental street-lighting fixtures installed in commercial areas or residential subdivisions with underground distribution lines are usually connected in multiple (Fig. 28-5).

Series-Circuit Street Lights   In a series circuit all the lamps are connected in series as shown in Fig. 28-6. The same current, therefore, flows through all the lamps. The most common value of current used is 6.6 amp. Sometimes the lamps and circuits are designed for 20 amp and occasionally for 15 amp. The lamps may be either operated directly in series in the 6.6-amp circuit or supplied through insulating transformers connected onto the 6.6-amp circuit as shown in Fig. 28-7.

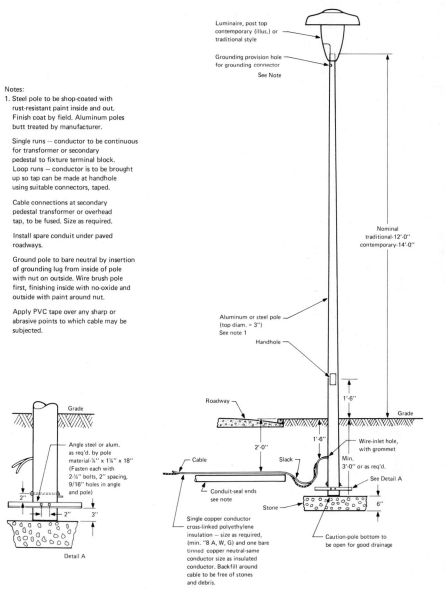

Notes:

1. Steel pole to be shop-coated with rust-resistant paint inside and out. Finish coat by field. Aluminum poles butt treated by manufacturer.

   Single runs — conductor to be continuous for transformer or secondary pedestal to fixture terminal block. Loop runs — conductor is to be brought up so tap can be made at handhole using suitable connectors, taped.

   Cable connections at secondary pedestal transformer or overhead tap, to be fused. Size as required.

   Install spare conduit under paved roadways.

   Ground pole to bare neutral by insertion of grounding lug from inside of pole with nut on outside. Wire brush pole first, finishing inside with no-oxide and outside with paint around nut.

   Apply PVC tape over any sharp or abrasive points to which cable may be subjected.

Luminaire, post top contemporary (illus.) or traditional style

Grounding provision hole for grounding connector
See Note

Nominal
traditional-12'-0"
contemporary-14'-0"

Aluminum or steel pole (top diam. = 3")
See note 1

Handhole

Roadway

Grade

1'-6"

Grade

Angle steel or alum. as req'd. by pole material-¼" x 1¼" x 18" (Fasten each with 2-½" bolts, 2" spacing, 9/16" holes in angle and pole)

Cable

Slack

1'-6"

Wire-inlet hole, with grommet

2'-0"

Min. 3'-0" or as req'd.

See Detail A

2"

2"   3"

Conduit-seal ends
see note

Stone

6"

Detail A

Single copper conductor cross-linked polyethylene insulation — size as required, (min. "8 A, W, G) and one bare tinned copper neutral-same conductor size as insulated conductor. Backfill around cable to be free of stones and debris.

Caution-pole bottom to be open for good drainage

Fig. 28-5   Multiple street light installation with post-top luminaire and underground supply. *(Courtesy American Electric Power System.)*

PRIMARY

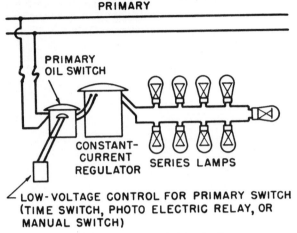

LOW-VOLTAGE CONTROL FOR PRIMARY SWITCH (TIME SWITCH, PHOTO ELECTRIC RELAY, OR MANUAL SWITCH)

Fig. 28-6 Series street-lighting circuit.

Since the lamps are in series, only one wire is needed, as shown in Fig. 28-8a. Some of the lamps are connected in the outgoing wire, and the rest are connected in the return wire. This is called an "open-loop" series circuit.

In order to make it possible to locate a fault like an open circuit or a ground, it is desirable to bring the outgoing and return conductors close together in numerous places so that the circuit can be conveniently short-circuited. Such a circuit is called a "closed-loop" circuit and is illustrated in Fig. 28-8b. Sometimes the circuit is arranged to combine the open- and closed-circuit features as in Fig. 28-8c. The open-loop circuit requires the least wire, but troubleshooting is difficult. The combination of the open and closed loop helps overcome this shortcoming.

Enough lamps are connected in series, sometimes a hundred or more, to make the circuit a high-voltage circuit. The voltage of the circuit is equal to the sum of the voltages of all the lamps plus the voltage drop in the wire. Voltages between 2400 and 7200 are commonly employed. The circuit and lamp fixtures must therefore be well insulated, and the circuit must always be treated as a high-voltage circuit even though the lamps themselves are low-voltage lamps.

Any break in the circuit, like a burned-out filament in a lamp, interrupts the entire circuit. To keep the remaining lamps burning, a film-disk cutout is provided in the socket of each lamp, as shown in Figs. 28-9 and 28-10. The cutout consists of two metal disks separated by a thin film of insulating material. The insulating film is held in place by the spring pressure of the contact disks. Under normal lamp voltage of, say, 50 volts, depending on lamp size, the film acts as an insulator between the two contact disks. In case of a lamp burnout or a broken filament, all the lamps go out and the entire circuit voltage of several thousand volts appears across the insulating film. This

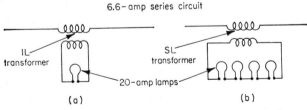

Fig. 28-7 Showing manner of supplying 20-amp lamp from 6.6-amp series circuit by use of an individual lamp series transformer referred to as an IL transformer. The use of such a transformer also insulates the 20-amp lamp from the high-voltage series circuit. A type SL transformer can be used to supply several lamps as shown in (b). Film disk cutouts are not necessary when current is supplied through an IL or SL transformer, as the main 6.6-amp circuit is not interrupted.

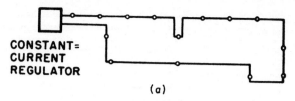

(a)

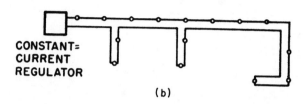

(b)

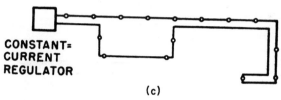

(c)

Fig. 28-8   Diagrams of (a) open loop, (b) closed loop, (c) combined open and closed loop, series street-lighting circuits.

Fig. 28-9   Series lamp socket showing film cutout. Insulating film is held in position between two metal disks supported by springs connected across lamp terminals. *(Courtesy General Electric Co.)*

is more than sufficient to puncture the film. The puncture closes the circuit between the two metallic disks and restores the continuity of the series circuit by bypassing the defective lamp.

The series circuit should not be opened without taking the constant-current transformer out of service. If the constant-current transformer is not first switched off, a serious arc is drawn which may result in injury to the lineman.

Constant-Current Transformer   The constant current of 6.6 amp required by the series system is supplied by a so-called constant-current transformer or regulator. The regulator is a special type of transformer having a movable secondary which is adjustable in position (see Fig. 28-11). Alternating voltage impressed on the primary winding induces a voltage in the secondary winding. The value of the induced voltage, however, depends upon the separation between the primary and secondary windings. The voltage is greatest when the windings are close together and least when they are farthest apart. When the windings are close together, most of the mag-

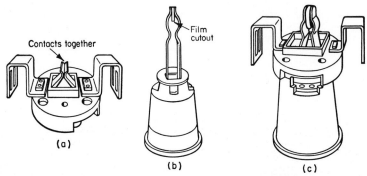

Fig. 28-10   Series lamp socket and receptacle. (*a*) Series receptacle; (*b*) series socket; (*c*) assembly. Cutout consisting of two metal disks and film shown on base of socket.

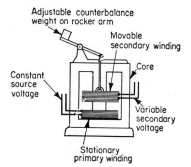

Fig. 28-11   Schematic diagram of constant-current transformer. Note movable secondary winding and weight on rocker arm which counterbalances part of weight of secondary winding.

netic flux produced by the primary current links or threads through the secondary winding. When the windings are separated, some of the flux produced by the primary leaks across the air gap without linking or threading through the secondary. Hence less voltage is induced in the secondary winding.

An adjustable weight counterbalances most of the weight of the secondary winding. The remainder of the lifting force is provided by the magnetic repulsion between the primary and secondary currents.

The constant current is maintained by the changing position of the secondary. If lamps burn out, lowering the resistance of the series lighting circuit, the secondary current increases momentarily. This increases the repelling force between the two windings, and the secondary winding is moved farther away from the primary winding. The wider separation of the primary and secondary windings increases the leakage flux and hence causes the induced voltage to decrease, allowing the secondary current to fall to its former 6.6-amp value.

The constant-current transformer, therefore, supplies a constant current at variable voltage from a source of constant voltage and variable current. A typical constant-current transformer is shown in Fig. 28-12.

Light-Fixture Support   The light fixture can be supported above the street in two different ways:

1. Supported on top of a post
2. Supported from a bracket attached to a post (mast arm)

If the lighting circuit is placed underground along the street, the most convenient support for the lighting fixture is a pole or post. The circuit enters the pole or post underground and passes up in the inside of the hollow post. The light fixture can be mounted directly on the top of the post as illustrated in Fig. 28-13 or supported by a bracket or mast arm out over the street as shown in Fig. 28-14. The latter is properly called a side "slip-fitter" mounting and is being preferred for most new installations. The overhang places the light fixture partially out over the street, and the higher mounting height makes for better light distribution.

Many presently installed street lights are still supported on brackets attached to utility poles as shown in Fig. 28-15.

Fig. 28-12 Typical constant-current transformer showing movable secondary winding and counterweight. *(Courtesy General Electric Co.)*

Fig. 28-13 Street lamp and fixture supported on top of post. Power is supplied from buried underground cable. *(Courtesy Line Material Industries.)*

Street-Light Lamps   Lamps used for street lighting are of two principal types:

1. Incandescent
2. Electric discharge:
    a. Mercury vapor
    b. Metal halide
    c. Low-pressure sodium vapor
    d. High-pressure sodium vapor
    e. Fluorescent

Incandescent Lamp   In an incandescent lamp a wire filament is sealed in a glass bulb from which the air is extracted. Electric current flowing through the filament heats it to incandescence and produces light. It is necessary to heat the filament to about 5000°F to produce full brilliance. The metal tungsten has a melting point of 6120°F, well above the operating temperature, and is therefore well suited for use in incandescent lamps. The absence of oxygen prevents the filament from burning up. A mixture of inert gases is often put into the bulb. These gases carry any tungsten that evaporates from the filament to the top of the bulb. A typical incandescent series street lamp, the PS 25, rated 2500 lumens, is shown in Fig. 28-16.

Fig. 28-14 Fluorescent lamp and fixture supported from pole. Power is supplied from underground circuit. *(Courtesy Line Material Industries.)*

Fig. 28-15 Street light supported from bracket attached to utility pole. *(Courtesy Joslyn Mfg. and Supply Co.)*

Lamps are rated as to light output in lumens. A lumen is defined as one unit of light output. Table 28-1 lists the incandescent lamps in common use in series circuits and gives their lumen output rating.

Incandescent lamps used on multiple circuits are operated at 120 volts and are rated in watts input as well as lumens output. Table 28-2 lists the multiple incandescent lamps most commonly used.

Electric Discharge Lamps In electric discharge lamps electrons flow between electrodes through an ionized gas or vapor. All gaseous conduction lamps have a negative resistance characteristic. That means that the lamp resistance decreases as the lamp heats up. As the resistance decreases, the current increases. In fact, the current will increase indefinitely unless a current-limiting device is provided. All gaseous conduction lamps, therefore, have current limiters called "ballasts."

The main advantage of a gaseous discharge lamp is that it gives out more light than an incandescent lamp of comparable rating. Mercury and sodium are the most economical chemicals used in electric discharge lamps.

Vapor Lamps The most common of these is the mercury-vapor lamp. This type of lamp contains a small quantity of argon gas, which permits a starting discharge between the special starting electrode and one of the main electrodes. During the starting interval the liquid mercury gradually vaporizes, and eventually the full lamp current flows between the main electrodes, causing the mercury vapor to give off a considerable amount of light—extremely rich in the blue, green, and yellow colors.

Where more normal color appearance of people or objects is important, improvement is obtained by the use of mercury lamps having fluorescent materials coated on the inside of the

Table 28-1   Ratings of Series Incandescent Lamps in Common Use

Bulb shape	Volts	Amperes	Initial lumens	Rated life, hr
PS 25	22.0	6.6	2,500	3000
PS 35	34.2	6.6	4,000	3000
PS 40	50.5	6.6	6,000	3000
PS 40	84.0	6.6	10,000	3000

Table 28-2   Ratings of Multiple Incandescent Lamps in Common Use

Bulb shape	Volts	Watts	Initial lumens	Nominal lumens	Rated life, hr
PS 25	120	189	2,920	2,500	3000
PS 35	120	295	4,950	4,000	3000
PS 40	120	405	6,900	6,000	3000
PS 40	120	620	10,800	10,000	3000

outer bulbs. The phosphors convert some of the invisible ultraviolet energy into visible light, in colors that supplement those from the mercury-vapor discharge.

A typical mercury lamp is shown in Fig. 28-17, and the essential parts of its construction are indicated in Fig. 28-18. Note the main electrodes and the starting electrode, the starting resistor, the inner arc tube, and the "mogul" screw base. A typical 400-watt mercury lamp is rated at 135 volts and 3.2 amp and in its clear design gives out an average of 20,500 lumens initially when operated in a vertical position. When it is operated horizontally, its light output is slightly reduced, but this burning position is commonly used for street lighting because it permits a more effective distribution of light from the fixture. The mercury-vapor lamp has an output of approximately 50 lumens per watt and a life expectancy of more than 24,000 burning hours.

The metal-halide lamp is similar in construction and operation to the mercury-vapor lamp. The electrodes in the lamp are modified to permit use of the metallic halide compounds. Inert gases are placed within the arc tube and halogen compounds are introduced into the mercury, increasing the efficiency of the lamp. Ballasts must be constructed to produce a starting voltage higher than that normally utilized for mercury-vapor lamps. The lamp has an output of approximately 75 lumens per watt and a life of approximately 12,000 burning hours.

The low-pressure sodium-vapor lamp operates in a similar manner electrically. However, its light is limited to an intense yellow-orange color. This lamp type has little present-day use in the United States.

The high-pressure sodium-vapor lamp operates at high pressure and temperature. The lamp is elliptical and smaller than a mercury-vapor lamp of equal lumen output. The lamp is constructed with a ceramic arc tube utilizing sodium and other metallic traces and is contained by a borosilicate glass envelope. High voltage for starting the lamp is generated by high-impulse electronic

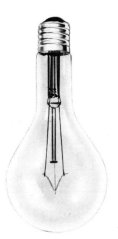

Fig. 28-16   Typical incandescent series street lamp rated 2500 lumens. *(Courtesy General Electric Co.)*

Fig. 28-17   Typical mercury-vapor lamp rated 400 watts. *(Courtesy General Electric Co.)*

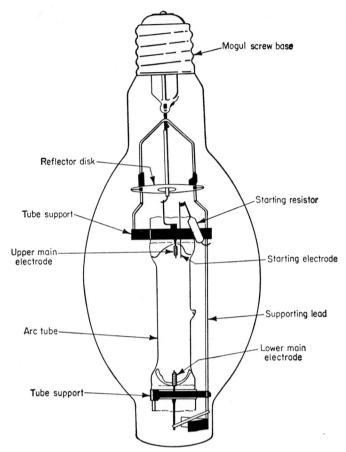

Mogul screw base

Reflector disk

Starting resistor

Tube support

Upper main electrode

Starting electrode

Arc tube

Supporting lead

Lower main electrode

Tube support

Fig. 28-18   Construction details of the mercury-vapor lamp shown in Fig. 28-17. *(Courtesy General Electric Co.)*

equipment. The ballast is designed specifically for the high-pressure sodium lamp and is not interchanged with ballasts for other electric discharge lamps. The high-pressure sodium-vapor lamp has an output of approximately 100 lumens per watt and a life expectancy of approximately 20,000 burning hours.

Fluorescent Lamp   The fluorescent lamp is built in the shape of a long, slender glass tube with an electrode at each end. See Fig. 28-19 for details of construction.

In the fluorescent lamp the flow of current takes place between electrodes through a mercury vapor as in a mercury-vapor lamp. The inside of the glass tube, however, is coated with a thin

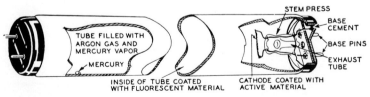

STEM PRESS

BASE CEMENT

TUBE FILLED WITH ARGON GAS AND MERCURY VAPOR

BASE PINS

MERCURY

EXHAUST TUBE

INSIDE OF TUBE COATED WITH FLUORESCENT MATERIAL

CATHODE COATED WITH ACTIVE MATERIAL

Fig. 28-19   Details of construction of.fluorescent lamp. *(Courtesy General Electric Co.)*

layer of fluorescent material called phosphor. This material will glow and give off visible light when struck by the electrons flowing through the mercury vapor between electrodes. The light produced is thus a secondary effect of the current flow. A "cool white" color is usually preferred, although other colors can be obtained by the use of other phosphors.

A small measured amount (about a drop) of mercury placed in the arc tube permits the lamp to operate, and a ballast in the form of a choke coil in the circuit limits the current. Lamps requiring starters provide a temporary heating circuit. This circuit opens after a few seconds' operation.

The fluorescent lamp has an output of approximately 70 lumens per watt and a life expectancy of 9000 or more burning hours.

Street-Light Control   Bare lamps emit light in many directions. Light in the skyward direction is largely wasted, and crosswise light on streets is also partly wasted.

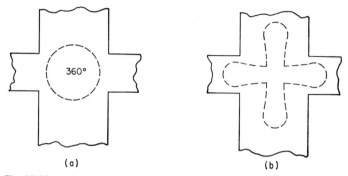

Fig. 28-20   Light-distribution patterns at street intersection: (*a*) uniform; (*b*) four-way.

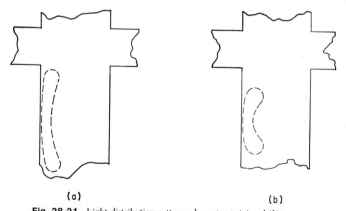

Fig. 28-21   Light-distribution patterns along street: (*a*) and (*b*) two-way.

Street-lighting fixtures (luminaires) at street intersections can be designed to direct their light equally in four directions, and luminaires along the sides of streets should emit light in two principal directions. Such light-distribution patterns are illustrated in Figs. 28-20 and 28-21. Figure 28-20*a* shows uniform distribution in all directions, and Fig. 28-20*b* a light pattern in four main directions. Figure 28-21*a* and *b* shows light distribution in two principal directions along the street.

The required control of the light emitted from the lamp is obtained by the use of reflectors, refractors, and diffusing globes or combinations thereof.

A reflector is a surface which stops the light ray and redirects it in the desired direction. Figure 28-22 shows a typical light fixture equipped with a reflector. Such a reflector will stop the light

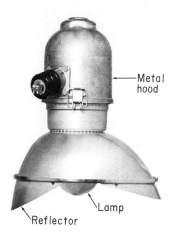

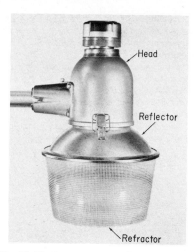

Fig. 28-22 Lamp and reflector-type fixture. *(Courtesy General Electric Co.)*

Fig. 28-23 Luminaire using reflector and refractor for light control. Note photoelectric control mounted on top of luminaire. *(Courtesy General Electric Co.)*

rays shining upward and redirect them downward to the ground. Reflectors may be made of porcelain, metal, or silvered glass. They are made in various shapes to give different types of light distribution.

A refractor is made of a transparent material, usually glass or acrylic plastic, which passes light rays but bends or deflects them in the desired direction. The enclosing globe is actually a series of optical prisms which bend the light rays. A bowl refractor is illustrated in Fig. 28-23. It is used with pendant luminaires, while dome refractors are used on upright post luminaires.

Fig. 28-24 Luminaire mounted on ornamental street-light pole in a residential area with underground distribution utilizing reflector and diffusing bowl for light control.

Fig. 28-25 Motor-driven time switch with astronomical dial. *(Courtesy General Electric Co.)*

Refractors must be positioned correctly to get the desired light distribution in the horizontal plane. They are adjusted vertically with respect to the lamp filament.

Sometimes the enclosing globe is made to diffuse the light, that is, scatter the light in many directions. Diffusing the light reduces the glare by spreading the apparent light over a larger area at lower brightness. The emitted light from the lamp so lights up the globe that the globe itself appears to be the light source.

Special types of light fixtures often combine the several methods of light control in the same fixture. Such a combination fixture is shown in Fig. 28-24.

Methods of Time Control   As the purpose of street lighting is to illuminate streets and intersections when natural lighting from the sun is not sufficient, the lights should be switched on as darkness approaches and off when daylight returns. This can be accomplished either manually or automatically by the following methods:

1. Manual
2. Clock-operated time switch
3. Photoelectric control

Manual   This method requires that some member of the operating personnel be given the specific task of closing the lighting circuits at dusk and opening them at dawn, as well as at other daylight hours when unusual weather conditions indicate the need for artificial light. On series circuits the switch on the primary side of the constant-current transformer, shown in Fig. 28-6, is operated. On multiple circuits fed by transformers whose loads are street lighting only, the primary side of the transformer is frequently switched. However, if the transformer supplies power to other loads in addition to street lighting, then the street-lighting secondary is switched as shown in Fig. 28-2.

Automatic Clock Control   Clock-operated time switches can be:

1. Motor driven
2. Spring driven—hand rewound
3. Spring driven—motor rewound

In case of a power interruption the motor-driven time switch stops and has to be reset to the correct time manually. The spring-driven hand-rewound clock has to be rewound regularly. The spring-driven motor-rewound clock rewinds itself automatically. The last arrangement is the most satisfactory when clock-driven time switches are employed.

Time switches are usually equipped with astronomical dials which automatically change the "turn-on" and "turn-off" times to conform to the seasonal changes of sunrise and sunset. This gradual change is reflected in the shape of a cam which is driven by a small electric motor. The cam allows the contacts to be made or broken in accordance with the shape of the cam. The dial is adjusted at the factory for the locality in which the time switch is to be used. Figure 28-25 shows a typical time switch equipped with an astronomical dial.

Photoelectric Cell Control   A photoelectric-cell-controlled relay may be employed to control street-lighting circuits by turning on the lights whenever the natural illumination falls below a given level. The photocell activates a relay which switches the lights on or off.

A photoelectric cell is a vacuum tube containing two electrodes, a cathode and an anode. The cathode (see Fig. 28-26) is in the shape of a circular surface covered with light-sensitive material. When light shines upon this material, electrons are ejected which are drawn over to the anode, since the positive anode attracts the negative electron. The anode is maintained at a positive potential by the battery shown. This flow of electrons from cathode to anode constitutes a current which flows through the relay and holds its contacts open. When darkness approaches, the light

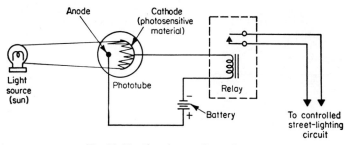

Fig. 28-26   Photoelectric cell control circuit.

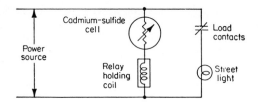

Fig. 28-27 Cadmium-sulfide cell control circuit.

shining upon the light-sensitive surface falls off, the number of electrons ejected becomes less, the current flow is reduced, and the relay contacts are permitted to close. This energizes the light-control circuit, and the street lights come on. In the morning, when daylight returns, the amount of light shining on the photocell increases, more electrons are emitted, the current flow increases, and the relay contacts are pulled open. This opens the control circuit, and the lights are turned off. A time delay of 5 to 7 sec is usually provided to prevent lightning or passing cars from actuating the relay.

Cadmium-Sulfide Cell Control   Originally the photoelectric cell described above was utilized for automatic street-lighting circuit control. More recently the cadmium-sulfide cell has come into general use. This device can be mounted on the luminaire itself as illustrated in Fig. 28-23, thereby eliminating the need of control circuits wherever secondary voltage is available.

A diagram of the basic circuit for this type of control is shown in Fig. 28-27. Its operation is based upon the variable resistance of the cadmium-sulfide cell rather than upon the small current-generating characteristic of the photoelectric tube. The cadmium-sulfide cell has a very high resistance (about 15,000,000 ohms) when dark, but when exposed to sunlight it has a relatively low resistance, less than 1000 ohms. The cell is placed in series with the holding coil of the relay. Thus, in the daytime the resistance of the cell is low, and sufficient current flows through the relay coil to hold the contacts open and keep the lights off. As evening approaches and the resistance of the cell begins to increase, owing to the decreasing light level, the current in the holding coil decreases. When the natural light level reaches the control setting, the current of the holding coil lets the relay contacts close and the street light is energized.

The next morning, as the light level begins to increase, the resistance of the cell again is reduced. When the light level reaches the control setting, the current in the relay holding coil has become large enough to cause the relay to open the contacts and deenergize the light. The relay holds these contacts open again throughout the day until the following evening.

Fig. 28-28 Assembled view of cadmium-sulfide cell street-light control unit. Cadmium-sulfide cell is in lower control location directly behind window in cover. *(Courtesy General Electric Co.)*

A typical cadmium-sulfide cell control unit is shown in Fig. 28-28. The cell is located directly behind the window in the cover.

Switching Several Circuits   Two methods are in use to switch multiple lighting circuits fed from different secondaries or transformers on and off. These are

1. Pilot wire
2. Cascading

In the first method a pilot wire operates a relay in each lighting circuit. The pilot wire can be either energized or deenergized at night. If the pilot wire is energized at night when lighting is needed, the relay must be of the circuit-closing type as shown in Fig. 28-29. Energizing the pilot wire energizes the relays, and the relays close the lighting circuits. If the pilot wire is deenergized at night, the relays must be of the circuit-opening type. Normally the relays are closed. Deenergizing the pilot wire reduces the pull on the relays; the armatures fall back and in so doing close the lighting circuits. This method has the advantage that, should a defect occur in the pilot wire or relays during the daytime, the lights will come on. The burning lamps indicate immediately that a fault has developed and repairs can be made during the daylight hours.

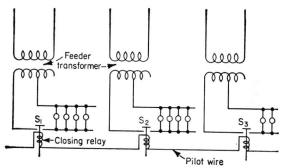

Fig. 28-29   Pilot wire control of multiple lighting circuits supplied from several feeder transformers.

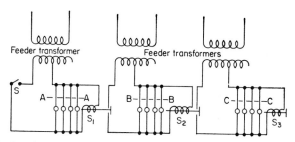

Fig. 28-30   Cascade control of multiple lighting circuits supplied from several feeder transformers.

Switching lighting circuits on and off by the cascade method means that the first circuit upon being energized activates a relay which in turn energizes the second circuit. Another relay activated by the second circuit energizes a third circuit, and so on, as illustrated in Fig. 28-30. The only extra wiring required is from the end of one circuit to the relay of the next circuit. Figure 28-30 clearly shows that when circuit A is energized, the relay S_1 for circuit B is activated, energizing circuit B. This in turn causes the relay S_2 for circuit C to operate, which energizes circuit C. The word "cascade" means to put one after the other.

Design of Street-Lighting Installations   For information on the design of street-lighting systems the reader is encouraged to refer to the "Practice for Roadway Lighting" ANSI/IES RP8-1977 published by the American National Standards Institute (ANSI) under the sponsorship of the Illuminating Engineering Society. This provides information on the design of street-lighting systems and the measurement of their performance. Copies may be obtained from the Illuminating Engineering Society, 345 East 47th Street, New York 10017.

Section **29**

Underground System

Underground transmission and distribution are resorted to when:

1. Space is not available for overhead lines, as in the congested downtown areas of large cities.

2. The hazard of high-voltage overhead lines is too great, as in the heavily built-up areas of large cities.

3. The appearance of numerous heavy overhead lines would be unsightly, as in dense downtown areas or in the new or redeveloped residential districts of cities.

In general, underground lines cost more than the equivalent overhead lines. Underground lines are therefore used only when necessity demands them, as stated above, or when municipal, state, or federal authorities request or require them.

Parts of Underground System   An underground system consists of five essential parts:

1. Conduits or ducts
2. Manholes
3. Cables
4. Transformer vaults
5. Risers

The first three parts listed above are illustrated in Fig. 29-1. The conduits or ducts are the hollow tubes connecting the manholes, one with another. The cable is the electrical power circuit placed in the conduit through which power flows, and the manhole is the chamber where cable ends are spliced together to make a continuous circuit. Transformer vaults (Fig. 29-2) are other underground rooms in which the power transformers, network protectors, voltage regulators, circuit breakers, meters, etc., are housed. Cables terminate in transformer vaults or customers' substations or connect with overhead lines in potheads, as illustrated in Fig. 29-3. The last termination is called a riser.

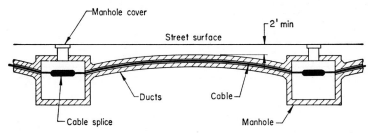

Fig. 29-1   Principal parts of an underground conduit system for underground electrical power transmission and distribution.

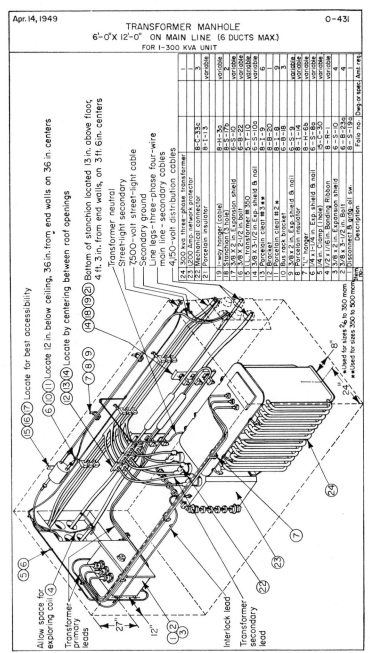

Apr. 14, 1949

TRANSFORMER MANHOLE
6'-0" X 12'-0" ON MAIN LINE (6 DUCTS MAX.)
FOR 1-300 KVA UNIT

O-431

Item no.	Description	Folio no.	Dwg. or spec. no.	Amt. req.
24	300 kva three-phase transformer			1
23	1,200 Amp. network protector			—
22	Mechanical connector	8-C-33c		3
21	Porcelain insulator	8-I-13		variable
19	1-way hanger (cable)	8-H-3a		variable
18	Stanchion (3 hole)	8-S-17b		2
17	3/8 x 2 in. Expansion shield	6-S-10		variable
16	3/8 X 2-1/2 in. bolt	6-B-22		variable
15	L. transformer #350	5-T-1		variable
14	3/8 X 3-1/2 in. Exp. shield & nail	6-S-10a		6
13	Porcelain cleat #3 **	8-I-9		—
12	Bracket	8-B-20		1
11	Porcelain cleat #2 *	8-I-8		9
10	Bus rack bracket	6-B-18		3
9	3/8 x 2 in. Exp. shield & nail	6-S-9		variable
8	Porcelain insulator	8-I-14		variable
7	"L" hanger	8-H-6b		variable
6	1/4 x 1-1/4 in. Exp. shield & nail	6-S-8a		variable
5	1/4 in. Clamp (1 hole)	15-C-30		variable
4	1/2 x 1/16 in. Bonding Ribbon	8-R-1		variable
3	3/8 x 2 in. Expansion shield	6-S-10		4
2	3/8 x 3-1/2 in. Bolt	6-B-23a		4
1	Disconnect & grdg. oil sw.	8-S-19a		1

* Used for sizes 2/6 to 350 mcm
** Used for sizes 350 to 500 mcm

Allow space for exploring coil (4)

Transformer primary leads

(5)(6)(7) Locate for best accessibility

(6)(10)(11) Locate 12 in. below ceiling, 36 in. from end walls on 36 in. centers

(12)(13)(14) Locate by centering between roof openings

(7)(8)(9)

(14)(18)(19)(21) Bottom of stanchion located 13 in. above floor, 4 ft. 3 in. from end walls, on 3 ft. 6in. centers

Transformer neutral
Street-light secondary
7,500-volt street-light cable
Secondary ground
Line legs – three-phase four-wire
main line – secondary cables
4,150-volt distribution cables

Interlock lead
Transformer secondary lead

Fig. 29-2   Typical transformer vault. (*Courtesy Public Service Electric and Gas Company of New Jersey.*)

Fig. 29-3   Typical cable riser with potheads, fused cutouts, and lightning arresters. *(Courtesy S. & C. Electric Co.)*

Conduits

Types   The hollow tubes or ducts running from manhole to manhole can be made of fiber, plastic (polyvinyl-chloride, PVC), tile, concrete, asbestos cement, or a composition such as transite. Of these, fiber, plastic, and transite are in the most common use because of their lesser cost.

Sizes   Ducts are made in varying sizes with inside diameters from 2 to 6 in. The size of duct selected depends on the size of the cable to be drawn into it or to be installed in the future. In general, when a single cable is installed in a duct, the diameter of the duct should be at least ½ to ¾ in greater than the diameter of the cable. When more than one cable is drawn into one duct, the diameter of the duct should be at least ½ to ¾ in greater than the diameter of the circle enclosing all the cables. The duct opening should always be large enough to make the pulling in of the cable as easy as possible.

Duct runs should preferably be short, straight runs between manholes. This facilitates pulling in of the cable. When this is impossible, as in going around obstacles, the bends should be laid out with the greatest possible radius of curvature, thereby avoiding sharp curves.

Duct runs should be slightly graded; that is, they should slope gently toward one or both manholes, thus allowing any water that may seep in to drain to one of the manholes. Figure 29-4 illustrates a duct run on level ground. It has a double slope; that is, half of the run drains to each manhole. Good practice requires that the slope be no less than 3 in for every 100 ft of length. Figures 29-5 and 29-6 illustrate the manner of providing proper grade on moderate and steep slopes. In these cases the drainage is all in one direction.

In general, the top of a duct run should not be less than 2 ft below the street surface, as indicated in the diagrams.

Manholes

A manhole is the opening in the underground duct system which houses cable splices and which cablemen enter to pull in cable and to make splices and tests. The manhole is therefore often called a splicing chamber or cable vault.

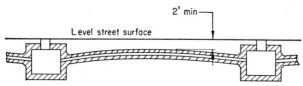

Fig. 29-4   Double-slope duct run on level ground. Water drains toward manholes from point midway between manholes.

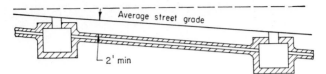

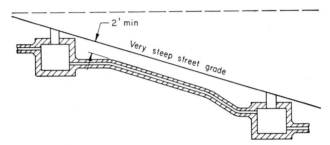

Fig. 29-5 Single-slope duct run between manholes on sloping street.

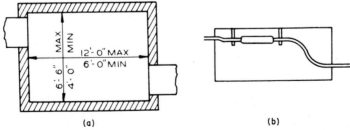

Fig. 29-6 Single-slope duct run on street with steep grade. Note that duct run leaves at bottom of one man-hole and enters near top of next manhole.

Manholes are generally built of reinforced concrete or brick, and the covers are made of steel. The opening leading from the street to the manhole chamber is called the "chimney" or "throat." An opening having a diameter of 32 in is recommended. The size opening is large enough for a man to enter on a ladder and also to pass the equipment needed for splicing and testing.

Manholes are usually located at street intersections so that the ducts can run in all four directions. The distance between manholes seldom exceeds 500 ft, as that is about as long a cable as can be pulled into the duct. In long blocks additional manholes are placed in the middle of the block.

Two-Way Manholes   The shape and size of manholes depend on the number and size of cables to be accommodated and the number of directions in which ducts leave the manhole. The simplest shape of manhole is illustrated in Fig. 29-7, where ducts and cables enter and leave in only two directions. Such a manhole is called a "two-way" manhole. Note in Fig. 29-7b that the cables do not pass through the manhole in a straight line but instead are offset; that is, they enter and leave at opposite corners. This is to provide opportunity to rack the cables around the sides of the manhole. Cables expand and contract in length with changes in temperature caused by changes in electrical power flow through the cable. The most practical way to take care of this expansion and contraction is to provide a large reverse curve in the cable as it passes through the manhole. This reverse curve, consisting of two large-radius 90-deg bends, enables the cable to take up the expansion movements, thereby reducing to a minimum the danger of cracking or buckling the lead sheath on the cable.

Fig. 29-7 Plan views of two-way manhole. Note how cable is racked around sides of manhole with double reverse curve.

A dimensioned side view of such a two-way manhole is given in Fig. 29-8. Note that the inside vertical height should not be less than 7 or more than 8 ft. These dimensions allow for the racking of the cables with their splices around the side of the manhole and sufficient space and headroom for the splicer to work.

In case ducts and cables do not enter and leave the manhole at the same level, the manhole design would appear as in Fig. 29-9. Note the change in level of the cables as they pass through the manhole in the side view and the reverse curve of the cables in the plan view.

A "two-way" 90-deg turning manhole with cables at the same level would appear as in Fig. 29-10. Only one curve is needed here to allow for cable expansion. Such a manhole is usually called a corner manhole.

Fig. 29-8 Side view of two-way manhole. Ducts enter and leave at same level.

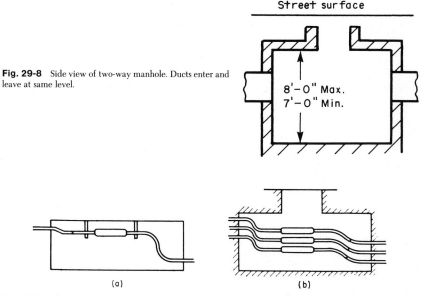

Street surface

8'-0" Max.
7'-0" Min.

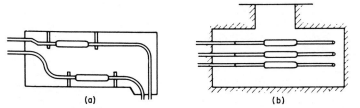

(a)   (b)

Fig. 29-9 Plan view (a) and side view (b) of manhole when ducts and cables enter and leave manhole at different levels.

(a)   (b)

Fig. 29-10 Plan view (a) and side view (b) of two-way manhole when ducts and cables make a 90-deg turn. Such a manhole is usually called a corner manhole.

Three- and Four-Way Manholes Manholes designed for "three-way" duct and cable lines are illustrated in Fig. 29-11, and manholes designed for "four-way" lines are illustrated in Fig. 29-12. The headroom in each case need not exceed 8 ft but should not be less than 7 ft.

Manhole Drainage Drainage should be provided in manholes to keep them free of water. Oftentimes a "dry well" is resorted to. This requires that a part of the manhole floor be left open, the dirt excavated to a depth of a foot or two, and the hole filled with stones. This permits the water to seep away gradually. In case a connection to the city storm sewer line can be made conveniently, that is often done. Figure 29-13 shows a cross section or side view of a typical sewer connection. In areas as along waterfronts where the soil is so moist or the water table so high that water will not drain out of a manhole, the only alternative remaining is to build the manhole walls

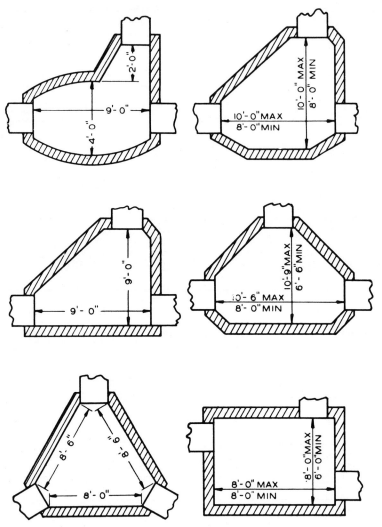

Fig. 29-11   Possible variations of three-way manhole.

and floor of waterproofed concrete. In addition, the concrete is usually painted with a waterproof paint. In extreme cases, an automatic pump is installed which keeps the water in a sump below floor level.

Distribution Manholes   When only a few trunk ducts are required for primary feeders to serve only the local distribution system, smaller manholes are satisfactory. The cables are usually smaller in diameter and therefore more easily racked. The dimensions for a two-way distribution manhole are given in Fig. 29-14. This manhole could accommodate a maximum of six trunk ducts. Note that the headroom is reduced to 7 ft.

A similar distribution manhole providing for primary and secondary circuits radiating in four directions at a street intersection is shown in Fig. 29-15.

The same type of cable manhole may be used exclusively for secondary distribution at locations where it is only necessary to accommodate several circuits. This occurs frequently in congested areas adjacent to buildings requiring large and multiple service from either or both the primary and secondary circuits.

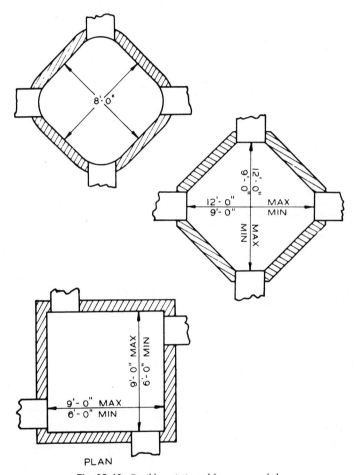

Fig. 29-12   Possible variations of four-way manhole.

Handholes   When small splicing chambers are necessary on lateral two-way duct lines, so-called handholes are used. They should be restricted to one-conductor cables only because the cable makes a complete loop once around in the hole, as shown in Fig. 29-16. The handhole cover should be large enough to permit working from above the handhole.

Cable

On systems operating above 2 kV to ground, the design of the conductors or cables installed in nonmetallic conduit should consider the need for an effectively grounded shield and/or sheath.

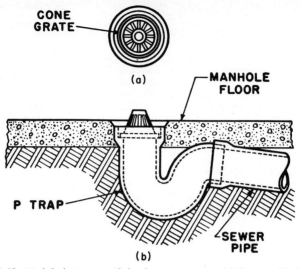

Fig. 29-13 Manhole drainage provided with sewer connection. (*a*) Plan view; (*b*) elevation.

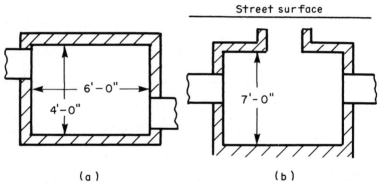

Fig. 29-14 Two-way distribution manhole. (*a*) Plan view; (*b*) elevation.

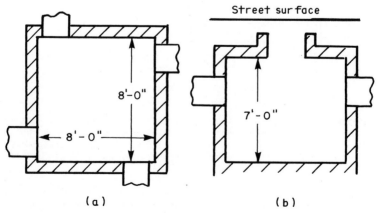

Fig. 29-15 Four-way distribution manhole. (*a*) Plan view; (*b*) elevation.

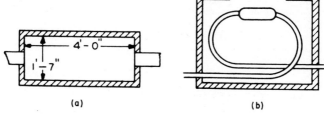

Fig. 29-16   Distribution handhole. (*a*) Plan view; (*b*) elevation.

Bending during handling, installation, and operation shall be controlled to avoid damage to any of the components of the supply cable. Pulling tensions and side-wall pressures of cable should be limited to avoid damage to the supply cable. Manufacturer's recommendations may be used as a guide. Ducts should be cleaned of foreign material which could damage the supply cable during pulling operations. Cable lubricants shall not be detrimental to cable or conduit systems. On slopes or vertical runs, consideration should be given to restraining cables to prevent downhill movement. Supply, control, and communication cables shall not be installed in the same duct unless the cables are maintained or operated by the same utility.

Cable supports shall be designed to withstand both live and static loading and should be compatible with the environment. Supports shall be provided to maintain specified separation between cables. Horizontal runs of supply cables shall be supported at least 3 in above the floor or be suitably protected. This rule does not apply to grounding or bonding conductors. The installation should allow cable movement without destructive concentration of stresses. The cable should remain on supports during operation. Adequate working space shall be provided. Where cable and/or equipment is to be installed in a joint-use manhole or vault, it shall be done only with the concurrence of all parties concerned. Supply and communication cables should be racked from separate walls. Crossings should be avoided. Where supply and communication cables must be racked from the same wall, the supply cables should be racked below the communication cables. Supply and communication facilities shall be installed to permit access to either without moving the other. Clearances shall be maintained as specified in Table 29-1.

These separations do not apply to grounding conductors, and they may be reduced by mutual agreement between the parties concerned when suitable barriers or guards are installed. Cables shall be permanently identified by tags or otherwise at each manhole or other access opening of the conduit system. This requirement does not apply where the position of a cable, in conjunction with diagrams or maps supplied to workmen, gives sufficient identification. All identification shall be of a corrosion-resistant material suitable for the environment. All identification shall be of such quality and located so as to be readable with auxiliary lighting. Where cables in a manhole are maintained or operated by different utilities or are of supply and communication usage, they shall be permanently marked as to company and/or type of use.

Insulation shielding of cable and joints shall be effectively grounded. Cable sheaths or shields which are connected to ground at a manhole shall be bonded or connected to a common ground. Bonding and grounding leads shall be of a corrosion-resistant material suitable for the environment or suitably protected.

Although fireproofing is not a requirement, it may be provided in accordance with each utility's normal service reliability practice to provide protection from external fire.

Table 29-1   Minimum Separation between Supply and Communications Facilities in Joint-Use Manholes and Vaults

Phase-to-phase supply voltage	Inches surface to surface
0 to 15,000	6
15,001 to 50,000	9
50,001 to 120,000	12
120,001 and above	24

Special circuits operating at voltages in excess of 400 volts to ground and used for supplying power solely to communications equipment may be included in communications cables under the following conditions:

1. Such cables shall have a conductive sheath or shields which shall be effectively grounded, and each such circuit shall be carried on conductors which are individually enclosed with an effectively grounded shield.

2. All circuits in such cables shall be owned or operated by one party and shall be maintained only by qualified personnel.

3. Supply circuits included in such cables shall be terminated at points accessible only to qualified employees.

4. Communications circuits brought out of such cables, if they do not terminate in a repeater station or terminal office, shall be protected or arranged so that, in event of a failure within the cable, the voltage on the communications circuit will not exceed 400 volts to the ground.

5. Terminal apparatus for the power supply shall be so arranged that live parts are inaccessible when such supply circuits are energized.

6. Such cables shall be identified and the identification shall meet the pertinent requirements. The requirements do not apply to supply circuits of 550 volts or less which carry power not in excess of 3200 watts.

Cable Construction   Power cables are constructed in many different ways. Cables are generally classified as nonshielded or shielded cables. Nonshielded cables normally consist of a conductor, conductor strand shielding, insulation, and a jacket. The use of nonshielded power cables is limited to the lower primary voltages which normally do not exceed 7200 volts. Shielded power cables are generally manufactured with a conductor, conductor strand shielding, insulation, semiconducting insulation shielding, metallic insulation shielding, and a sheath. If the sheath is metallic, it may serve as the metallic insulation shielding and be covered with a nonmetallic jacket to protect the sheath (Fig. 29-17). Shielded cables are commonly used on circuits operating at 4,160Y/2400 volts or higher (Fig. 29-18).

Multiconductor cables used for three-phase power circuits consist of three single-conductor cables with a common sheath and/or jacket (Fig. 29-19).

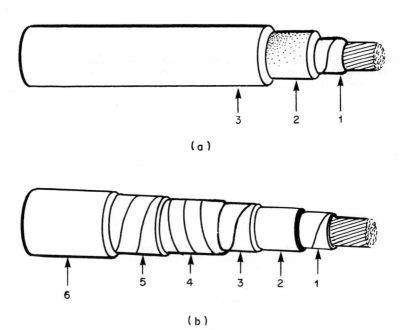

Fig. 29-17   Cable construction. (*a*) Nonshielded cable. (1) Strand shielding, (2) insulation, (3) cable jacket. (*b*) Shielded power cable. (1) Strand shielding, (2) insulation, (3) semiconducting material, (4) metallic shielding, (5) bedding tape, (6) cable jacket.

layers a viscous, moisture repelling, nondrying slipper compound is used. Varnished-cambric insulation has moderately high dielectric and impulse strength; is ozone, corona, and oil resistant; and is used extensively in transformers and oil switches. Varnished-cambric insulation tends to have a high power factor and low ionization level, and it must be covered by a lead sheath for moisture protection if used on a cable in a nonprotected atmosphere.

Rubber insulation is generally used in cables from 600 to 15,000 volts. Rubber insulations are made from either synthetic rubber, natural rubber, or butyl synthetic rubber. The compounds may contain as little as 20 percent or as high as 90 percent rubber. The balance of the compound consists of suitable organic or inorganic materials which facilitate manufacture and develop desirable properties not possessed by rubber alone. The rubber insulations are rated anywhere from 60°C to 130°C maximum.

Oil-base rubber insulation is the oldest type of the rubber compounds. It consists of a vulcanized vegetable oil and bitumen base which is compounded with other ingredients to form an ozone-resistant insulation. This type of insulation carries 75°C conductor temperature rating and is furnished on power cables for operation up to 35 kV.

Butyl synthetic rubber insulation has molecules that have been tailor-made to meet the requirements for the highest grade vulcanized insulation for high-voltage applications. It is basically a heat- and ozone-resisting rubber. Butyl rubber has excellent resistance to moisture, excellent dielectric strength and insulation resistance, and low dielectric loss factor. Butyl dielectrics are rated at 90°C up to 2000 volts, 85°C up to 15,000 volts, and 80°C up to 28,000 volts by the Insulated Power Cable Engineers Association (IPCEA).

Silicone rubber is a synthetic compound that has excellent heat, moisture, and ozone resistance. Cables insulated with silicone rubber can be operated at temperatures up to 125°C for power applications and are flexible down to −54°C. The tensile strength, elongation, and abrasion resistance of silicone rubber at ordinary temperatures are not as great as those of organic rubber compounds. Silicone rubber is extremely resistant to flame in the sense that when exposed to direct flame it will gradually burn and leave an ash which is electrically nonconducting, and if contained by a glass braid will serve as an insulator until the emergency is over and the wiring can be replaced. It is more expensive than most other types of cable insulation.

Ethylene-propylene rubber-base (EPR) insulations are vulcanizable compounds which can be made with one of the EPR elastomer gums developed by several chemical companies. Cables made with EPR insulations show great promise of combining in one dielectric a moisture resistance approaching that of polyethylene, the dielectric strength retention of oil base, and the heat and form stability of butyl insulations. These insulations are rated at 125°C for emergency or hot spot operation and a normal operating temperature of at least 90°C.

Polyvinyl chloride (PVC) is the thermoplastic most commonly used for low-voltage power and control cables. Its power factor and dielectric loss limit its use to low voltage, 600 volts or less, for general-purpose wiring. From the standpoint of performance, PVC differs from the vulcanized rubberlike compounds in its sensitivity to temperature and melting when in high heat. Thermoplastic-building-wire insulation is well known as type THW or TW.

Polyethylene cable insulation contains relatively small amounts of material other than the base polymer. It may contain coloring materials to protect it from exposure to sunlight and weather, and it should contain a small amount of carbon black. Polyethylene has a low power factor and dielectric loss, high dielectric and impulse strength, excellent moisture resistance, and high impact strength and abrasion resistance.

Cross-linked polyethylene cable insulation has compounds consisting of the polyethylene resin, fillers, antioxidants, and vulcanizing agents. It has most of the electrical advantages of polyethylene plus much better mechanical properties such as superior resistance to heat deformation and environmental stress cracking. It should be noted that the insulation thickness used on these cables is as thin as that of polyethylene power cables and in many cases much thinner as a result of its exceptional electric strength. Most cross-linked polyethylenes are rated at 90°C at 15 kV with a 100 hour per year overload temperature of 130°C for 5 years of its life. For the same voltage rating, the insulation thickness of the rubber type is generally thickest, cross-linked polyethylene the thinnest, and straight polyethylene somewhere between these two. The specific thicknesses for rubber and polyethylene are listed in the NEMA-IPCEA manual.

Cable Insulation Shielding   Insulation shielding confines the dielectric field within the cable to obtain symmetrical radial distribution of voltage stress within the dielectric. The shielding protects cables connected to overhead lines or otherwise subject to induced potentials and limits radio interference. If the shield is grounded, it reduces the hazard to shock and provides a ground return in case of a cable failure.

Conducting nonmetallic tape is used directly over the insulation of shielded power conductors to provide more intimate contact between the insulation and metallic shield. It is at least 0.0025 in thick and consists of carbon-black-filled cloth, rubber, or neoprene. Extruded semiconducting polyethylene is used over the insulation of concentric neutral underground cables. The extruded jackets are superior to tapes in both durability and moisture-penetration resistance.

Insulation shieldings are either metal, tape, braid, or wire. In general, rubber-covered cable with a synthetic jacket should be shielded when applied to systems over 3000 volts. Requirements for shielding are given in the following industrial specifications: "Solid Type Impregnated-Paper-Insulated Lead-Covered Cables," issued by AEIC, and "Wire and Cable with Rubber, Rubber-like and Thermoplastic Insulations," issued by IPCEA and NEMA. The metallic tape is generally 5 mils thick and is applied over the conducting nonmetallic tape.

Cable Bedding Tapes   Bedding tapes are generally used in a cable to add mechanical strength. They are located in various places by different manufacturers—under the sheath, under the shielding, and between the conductor and insulation. If one is not sure whether a tape is bedding or conductive, it should always be considered to be simiconductive and handled accordingly.

Cable Filler Tapes   Filler tapes of suitable materials generally are used in the interstices of cables, where necessary, to give the completed assembly a substantially circular cross section. Fillers are normally either treated hemp rope, synthetic rubber compound, or thermoplastic material.

Cable Jackets, Sheaths, or Coverings   The jacket or sheath over the cable insulation is a protective covering which provides mechanical protection; in some degree it is used for electrical reasons (Fig. 29-22). Sheaths are made from basically four different materials—fibrous, rubber and rubberlike, thermoplastic, and metallic.

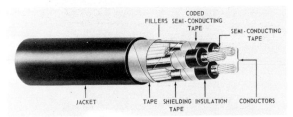

Fig. 29-22   Three-conductor rubber-insulated shielded cable with polyvinyl chloride jacket. *(Courtesy Reynolds Metals Co.)*

Fibrous jackets made with cotton braid provides a relatively low cost flexible covering with good mechanical strength. They are the most extensively used type of protective covering for electrical wires and cables installed in a dry location. Various materials, such as compounds of petroleum asphalt and stearin pitch, are used for saturating and finishing cotton braids to provide moisture resistance and nonflammability. A final coating of paraffin wax yields a nontacky finish that facilitates pulling the cables into conduits.

Jute wrapped on the outside of cable acts as a cushion and binder, protects galvanizing from abrasion during installation, and acts as a structure to hold intact corrosion-resisting asphalt coatings. The most common use of jute in the cable industry is on underground and submarine cables. Jute is usually spiral wrapped rather than woven. It has been used to provide a bedding under metallic coverings.

Basket-weave glass-fiber braids are often used in preference to cotton braids, particularly where better resistance to deterioration at elevated temperatures is desirable. They are sometimes used for underground cable because of the good stability of the glass fiber under wet service conditions.

Asbestos basket-weave braids are often used in preference to cotton braids where flameproofing is desired. It is sometimes used for underground and aerial cables because of the long life of treated asbestos braids under wet conditions or when exposed to the weather. Vulcanized natural rubber compounds are adversely affected by oils, ozone, and sunlight. Neoprene jackets have been used on cables instead of natural rubber compounds.

Synthetic rubber is a good heavy-duty jacket for aerial, duct, and buried installations. Two types of jackets commonly used are neoprene and Buna-N. The use of Buna-N has been limited to special service conditions, such as extremely low temperature locations or where maximum oil resistance is desired. Its strength and flexibility at low temperatures and the oil resistance are excellent. Neoprene has excellent weather resistance and good aging and mechanical properties. Black neoprene properly compounded will not deteriorate on long exposure to sunlight and weather or in wet locations, such as underground. Heavy-duty portable cords and cables are frequently made with neoprene-compounded sheaths. Neoprene is frequently used in place of cotton braids and other fibrous coverings.

Polyvinyl chloride (PVC) has excellent resistance to acids, alkalies, cutting, flame, moisture, oils, ozone, sunlight, and weathering. It is a good heavy-duty jacket when installed under conditions that eliminate harmful plastic-flow. It is not as flexible as neoprene and cannot be used at extremely low temperatures. PVC compound jackets are often used over lead and interlocked armor sheaths for protection against oils, gasoline, acids, and alkalis. TW wire is insulated with PVC.

Polyethylene is quite stable physically throughout its temperature range and resistant to many destructive chemicals and oils up to at least 70°C. Excellent moisture resistance makes polyethylene suitable as a nonmetallic sheath in damp locations and for aerial installation. In thin sections polyethylene is weak in abrasion resistance and often deforms at high temperatures. It will also cold flow under pressure.

Fig. 29-23   Three-conductor varnished-cambric-shielded lead-sheathed cable. *(Courtesy The Okonite Company.)*

Fig. 29-24   Three-conductor lead-sheathed paper-insulated power cable showing various layers of shielding and insulation. *(Courtesy The Okonite Company.)*

Nylon is a thermoplastic of the polyamide type. A thin layer film of nylon is sometimes extruded over a wall of polyethylene insulation to supplement the physical properties of the latter. The nylon is not subject to cold flow and, therefore, can be used as a protective covering. Nylon is mechanically strong and is not affected by oils.

Metallic sheaths provide mechanical protection to the cable and may also function as a current return. Lead sheaths are used with rubber, varnished-cambric, and paper-insulated cable as mechanical protection and to prevent entrance of moisture or loss of insulation impregnant (Fig. 29-23).

Lead is particularly suited to the purpose since it may be applied by extrusion presses in long continuous lengths. It forms a flexible covering, allowing the cable to be bent easily during handling and installation. Experience has shown that an intact lead sheath is the most effective protection for insulation against moisture, weather, and oxidation (Fig. 29-24).

The physical properties of aluminum are quite good. It is lighweight and capable of withstanding high internal pressures. It is not as ductile a metal as lead and, therefore, requires a larger bending radius. It is also necessary to protect it against chemical corrosion, particularly if it is drawn into ducts or laid directly on the ground. A thermoplastic or neoprene jacket is sometimes applied over the aluminum to eliminate this corrosion. Aluminum has a higher ac loss in the sheath as compared to lead; this will influence the effective ac resistance of the conductors when calculating the current-carrying capacity of the cable.

Steel armors for cables consist of flat-metal-tape armor, interlocked-metal-tape armor, steel-wire armor, or basket-weave armor. The metallic armor is usually applied over a jute bedding which is applied over a fibrous covering, jacket, or lead sheath for mechanical strength and mechanical outer protection. Single-conductor cables, when armored with steel, are subject to the magnetic properties of the steel. This condition affects the reactance of the circuit and causes heat losses in the armor which, in turn, affect the current-carrying capacity of the cable.

Rubber Cable Specification   This specification covers a typical 15-kV single conductor, rubber insulated, shielded, jacketed power cable with semicompressed concentric-stranded aluminum conductor. Application of this cable is for 60-Hz Y-connected system with grounded neutral operating at 13,200Y/7620 volts.

The cable shall be suitable for installing in ducts or direct-burial in the earth in wet or dry locations.

Cable shall be furnished with an aluminum, 750 kCMIL, stranded conductor.

Maximum conductor temperature shall be 90°C normal, 130°C emergency, 250°C short circuit.

Conductor: Alloy 1350 aluminum. Stranding shall be class B stranded per ASTM B231 and semicompressed.

Insulation System: The insulation shall be heat, moisture, ozone, corona, and impact resistant thermosetting rubber based elastomeric compound. The insulation shall be non black in color. When examined under 15 power magnification there shall be no inclusions, contaminants, or voids. The insulation shall consist of a conductor energy suppression layer, a primary wall of insulation and an outer energy suppression layer.

Conductor Energy Suppression Layer: The conductor energy suppression layer shall be extruded over the stranded conductor with good concentricity and shall be a black insulating material which is compatible with the conductor and the primary insulation. The extruded energy suppression layer shall have a specific inductive capacitance of ten or greater. In addition to its energy suppression function, this layer shall provide stress relief at the interface with the primary insulation. The energy suppression layer shall be easily removed from the conductor with the use of commonly available tools and shall have a minimum average thickness of 0.020″ and a minimum thickness at any part of the cable 0.016″ when measured over the top of the strands. The outer surface of the extruded stress suppression layer shall be cylindrical, smooth, and free of significant irregularities.

Primary Insulation: The primary insulation shall be a thermosetting rubber based elastomeric compound and shall be non black in color. The primary insulation shall be extruded directly over the conductor energy suppression layer with good concentricity and shall be compatible with the conductor energy suppression layer. The outer surface of the extruded primary insulation shall be cylindrical, smooth, and free of significant irregularities.

Insulation Energy Suppression Layer: The insulation energy suppression layer shall be extruded directly over the primary insulation with good concentricity and shall be a black insulating material compatible with the primary insulation and adjacent compounds of the insulation shielding system. This extruded stress suppression layer shall have a specific inductive capacitance of ten or greater. The extruded energy suppression layer shall have a minimum average thickness of 0.035″ and a minimum thickness at any part of the cable of 0.030″ when measured over the outer surface of the insulation. The outer surface of the extruded layer shall be cylindrical, smooth, and free of significant irregularities.

Insulation Thickness: The minimum average thickness of the insulation system shall be 0.225″. The inner and outer energy suppression layers are to be included with the primary insulation when measuring the insulation system thickness since they are insulating in nature.

Insulation Shielding System: The cable insulation shielding system shall consist of a nonmagnetic metallic conducting shielding tape, and an underlying nonmetallic semiconducting cloth tape.

The underlying semiconducting tape shall consist of cloth, coated, impregnated or calendered with a carbon black semiconducting compound and shall be compatible with the insulation system and the metallic shielding tape. The semiconducting tape shall be applied smoothly, tightly, and completely cover the insulation system with a minimum 10% lap on itself. The tape shall have an average thickness of 0.006 inches and be legibly identified as being semiconducting by surface printing throughout the length of the cable.

The metallic shielding tape shall be 0.005 inch thick tinned copper. The shielding tape shall be helically applied, smoothly and tightly, and completely cover the underlying semiconducting tape with a 20% lap upon itself. The shielding tape shall be compatible with the semiconducting underlying tape and the jacket which will be extruded over the shielding tape. The metallic tape shall be free from burrs, wrinkles, or contaminates and shall be made electrically continuous by welding, soldering, or brazing.

Jacket: The jacket shall be polyvinyl chloride extruded over the metallic shield. The minimum average thickness shall be 0.095 inches and the minimum thickness at any point on the cable shall not be less than 80% of the minimum average.

Identification: The outer surface of each jacketed cable shall be durably marked by sharply defined surface printing. The surface marking shall be as follows:

Manufacturer's name
Conductor size
Conductor material
Rated voltage
Year of manufacture

Cable Diameter Dimensional Control: The overall diameter of the cable shall not exceed 1.716 inches.

Production Tests: The following production tests on each length of cable shall be performed.

Conductor Tests: The conductor shall be tested in accordance with ASTM B-231 and IPCEA Standard S-19-81 (NEMA WC-3, section 6.3).

Conductor Energy Suppression Layer: The integrity of the conductor energy suppression layer shall be continuously monitored prior to or during the extrusion of the primary insulation using an electrode system completely covering the full circumference of the conductor energy suppression layer at a test voltage of 2 kV dc.

Insulation Tests

Preliminary Tank Test: Each individual length of insulated conductor, prior to the application of any further coverings, shall be immersed in a water tank for a minimum period of 24 hours. At the end of 16 hours (minimum), each length of insulated conductor shall pass a dc voltage withstand test, applied for five minutes. The applied dc voltage to be 70 kV.

At the end of 24 hours (minimum), each length of insulated conductor shall pass an ac voltage withstand test applied for five minutes. The applied ac voltage to be 35 kV. Each length of insulated conductor following the five minute ac voltage test shall have an insulation resistance, corrected to 15.6°C (60°F), of not less than the value of R, as calculated from the following formula (after electrification for one minute):

$$R = K \log_{10} D/d$$
R = insulation resistance in megohms/1000 ft
K = 21,000 megohms/1000 ft
D = diameter over insulation
d = conductor diameter

Final Test on Shipping Reel: Each length of completed cable shall pass an ac voltage withstand test applied for one minute. The applied ac voltage to be 35 kV. Each length of completed cable, following the one minute ac voltage test, shall have an insulation resistance not less than that specified.

Dimensional Verification

Each length of insulated conductor shall be examined to verify that the conductor is the correct size.

Each length of insulated conductor shall be measured to verify that the insulation wall thickness is of the correct dimension.

Each length of completed cable shall be measured to verify that the jacket wall thickness is of the correct dimension.

Conductor and Shield Continuity: Each length of completed cable shall be tested for conductor and shield continuity.

Test Reports: Certified test reports of all cable will be required. One copy of test reports to be sent to ordering company purchasing department.

Shipping Lengths: The shipping lengths will be specified on the inquiry and purchase order. Unless otherwise specified on the purchase order, the cable shall be furnished in one continuous length per reel.

Reels: Reels shall be heavy duty wooden reels (nonreturnable, if possible) properly inspected and cleaned to assure that no nails, splinters, etc., may be present and possibly damage the cable. Each end of the cable shall be firmly secured to the reel with nonmetallic rope. Wire strand is not acceptable for securing cable to the reel. Deckboard or plywood covering secured with steel banding is required over the cable.

Cable Sealing: Each end of the cable shall be sealed hermetically before shipment, using either heat-shrinkable polyethylene caps or hand-taping, whichever the manufacturer recommends. Metal clamps which may damage pulling equipment are not acceptable. The end seals must be adequate to withstand the rigors of pulling-in.

Preparation for Transit: Reels must be securely blocked in position so they will not shift during transit. The manufacturer shall assume responsibility for any damage incurred during transit which is the result of improper preparation for shipment.

Reel Marking: Reel markings shall be permanently stenciled on one flange of the wooden reel. Such markings shall consist of the purchaser's requirements as indicated elsewhere in this specification, but shall consist, as a minimum, of a description of the cable, purchaser's order number, manufacturer's order number, date of shipment, footage, and reel number. An arrow indicating direction in which reel should be rolled shall also be shown.

Cable Terminations As pointed out in the beginning of this section, underground transmission is resorted to only when necessary because of congestion in downtown areas or where overhead lines detract from the appearance of the community. However, it is possible that once the cables have passed underground through some crowded area, they may change to overhead lines. This is arranged either by continuing the cable as an aerial cable supported from a messenger wire or by connecting the cable to an open overhead line through a riser and pothead.

Aerial Cable An aerial cable is illustrated in Fig. 29-25. The cable is supported at close intervals by rings attached to a tightly strung steel cable called a "messenger." Instead of rings being used, the cable can be supported by means of a continuous wire wrapped around both the cable and the messenger in spiral fashion with a spinning machine.

Cable Riser When it is desired to connect an underground cable to an overhead line, the cable is brought out of a manhole to a pole through a run of conduit which is usually extended 10 ft or more up the pole in metallic pipe for mechanical protection. If the cable is low voltage, it terminates in a so-called "weather head," and if it is a high-voltage cable, it terminates in a so-called "pothead." The whole assembly is called a "riser." Figure 29-3 shows such an assembly consisting of the pipe conduit attached to the pole and the cable coming through it and terminating in the pothead. When the cables are single-conductor cables, they are fanned out just below the crossarm and terminate in single-conductor potheads like that shown in Fig. 29-26. When the cable is a multiconductor cable, it terminates in a multiconductor pothead like that shown in Fig. 29-27. From the pothead the conductors spread out into the overhead line. The pothead is supported by means of bolts fastened to the pole or crossarm.

Fig. 29-25   Three-conductor rubber-insulated shielded self-supporting aerial cable. *(Courtesy The Okonite Company.)*

Fig. 29-26   Single-conductor pothead. *(Courtesy G. & W. Electric Specialty Co.)*

Fig. 29-27   Typical three-conductor pothead. *(Courtesy G. & W. Electric Specialty Co.)*

Fig. 29-28   Pipe-type cable on reels before being pulled into the pipe. *(Courtesy Anaconda Wire and Cable Co.)*

Fig. 29-29 Cablemen installing pipe directly in the ground for pipe-type cable. Joint in pipe is being completed prior to backfilling the ditch. Pipe and joint are covered to prevent corrosion. *(Courtesy L. E. Myers Co.)*

Fig. 29-30 Cableman completing splice of pipe-type cable in manhole. *(Courtesy Anaconda Wire and Cable Co.)*

Fig. 29-31 Cableman and lineman terminating pipe-type cable with potheads. The porcelain insulator is in the process of being placed over the cable termination for one phase. *(Courtesy Anaconda Wire and Cable Co.)*

Weather Head   A "weather head" is a simple underground-overhead fitting which provides a cap or roof for the vertical conduit to prevent rain from entering it. Its use is restricted to connecting underground secondary cables to overhead secondary lines or service wires.

Pothead   A "pothead" (Fig. 29-27) provides for the connection between an underground cable and an overhead wire. It serves to separate the conductors from their close position to one another in the cable to the much wider separation in the overhead line without permitting arcing between them. It also seals off the end of the cable so that moisture cannot enter the insulation of the cable.

The pothead consists essentially of a pot, from which it gets its name, and a cover. The pot has a hole in the bottom through which the cable enters. The pot is attached to the sheath of the cable by a wiped joint or a clamp. The conductors of the cable are made to flare out to give the necessary separation. The pot is then filled with molten insulating compound which hardens on cooling. The conductors of the cable are connected to terminals encased in porcelain attached to the cover. These projections of porcelain have "petticoats" on them the same as pin insulators and naturally serve the same purpose. The overhead line conductors are connected to these terminals.

Pipe-Type Cable Underground transmission circuits with voltages in the range of 69,000 to 345,000 volts are usually installed utilizing pipe-type cable. The cable consists of a copper or aluminum conductor, conductor shielding, paper-tape insulation, and insulation shielding (see Fig. 29-28). Three conductors are pulled into the pipe to form a three-phase circuit. The pipe is installed direct buried or in a tunnel (see Fig. 29-29). Manholes are installed for splices. Figure 29-30 shows the three insulated conductors in the pipe with a cableman completing a splice in a manhole. The pipe is coated to protect it from electrolysis damage. Cathodic protection is normally installed to protect the cable pipe. The pipe is filled with insulating oil and the oil is continually pumped through the pipe to provide a cooling medium and maintain a high dielectric strength between the energized cables and ground. The cables are terminated in potheads, as illustrated in Fig. 29-31.

Laying Conduit

Conduit systems should be located so as to be subject to the least disturbances practical. Conduit systems extending parallel to other subsurface structures should not be located directly over or under other subsurface structures. If this is not practical, the rules on clearances should be followed. Conduit alignment should be such that there are no protrusions which would be harmful to the cable. When bends are required, the minimum radius shall be sufficiently large to prevent damage to cable being installed in the conduit. The maximum change of direction in any plane between lengths of straight rigid conduit without the use of bends should be limited to 5 deg.

Routes through unstable soils such as mud or shifting soil or through highly corrosive soils should be avoided. If construction is required in these soils, the conduit should be constructed in such a manner as to minimize movement and/or corrosion. When conduit must be installed longitudinally under the roadway, it should be installed in the shoulder or, to the extent practical, within the limits of one lane of traffic. The conduit system shall be located so as to minimize the possibility of damage by traffic. It should be located to provide safe access for inspection or maintenance of both the structure and the conduit system. The top of a conduit system should be located not less than 36 in below the top of the rails of a street railway or 60 in below the top of the rails of a railroad.

Where unusual conditions exist or where proposed construction would interfere with existing installations, a greater depth than specified above may be required. Where this is impractical or for other reasons, this clearance may be reduced by agreement between the parties concerned. In no case, however, shall the top of the conduit or any conduit protection extend higher than the bottom of the ballast section which is subject to working or cleaning. At crossings under railroads, manholes, handholes, and vaults should not, where practical, be located in the roadbed. Submarine crossings should be routed and/or installed so they will be protected from erosion by tidal action or currents. They should not be located where ships normally anchor.

The clearance between a conduit system and other underground structures paralleling it should be as large as necessary to permit maintenance of the system without damage to the paralleling structures. A conduit which crosses over another subsurface structure shall have a minimum clearance sufficient to prevent damage to either structure. These clearances should be determined by the parties involved. When conduit crosses a manhole, vault, or subway tunnel roof, it may be supported directly on the roof with the concurrence of all parties involved. Conduit systems to be occupied by communications conductors shall be separated from conduit systems to be used for supply systems by 3 in of concrete, 4 in of masonry, or 12 in of well-tamped earth. Lesser separations may be used where the parties concur. If conditions require a conduit to be installed parallel to and directly over a sanitary or storm sewer, it may be done provided both parties are in agreement as to the method.

Where a conduit run crosses a sewer, it shall be designed to have suitable support on each side of the sewer to avoid transferring any direct load onto the sewer. Conduit should be installed as far as is practical from a water main in order to protect it from being undermined if the main breaks. Conduit which crosses over a water main shall be designed to have suitable support on each side as required to avoid transferring any direct loads onto the main. Conduit should have sufficient clearance from fuel lines to permit the use of pipe maintenance equipment. Conduit and fuel lines shall not enter the same manhole. Conduit should be so installed as to prevent detrimental heat transfer between the steam and conduit systems.

The bottom of the trench should be undisturbed, tamped, or relatively smooth earth. Where the excavation is in rock, the conduit should be laid on a protective layer of clean, tamped backfill. All backfill should be free of materials that may damage the conduit system. Backfill within 6 in of the conduit should be free of solid material greater in maximum dimension than 4 in or with sharp edges likely to damage it. The balance of backfill should be free of solid material greater in maximum dimension than 8 in. Backfill material should be adequately compacted.

Duct material shall be corrosion-resistant and suitable for the intended environment. Duct material and/or the construction of the conduit shall be designed so that a cable fault in one duct would not damage the conduit to such an extent that it would cause damage to cables in adjacent ducts. The conduit system shall be designed to withstand external forces to which it may be subjected by the surface loadings except that impact loading may be reduced one-third for each foot of cover, so no impact loading need be considered when cover is 3 ft or more. The internal finish of the duct shall be free of sharp edges or burrs which could damage supply cable.

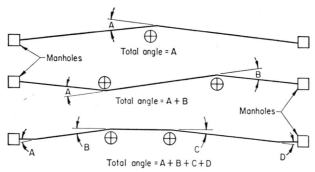

Fig. 30-1 Typical outlines of duct runs showing manner of avoiding obstructions ⊕. Angles *A*, *B*, *C*, and *D* are degree deflections from straight line.

Conduit, including terminations and bends, should be suitably restrained by backfill, concrete envelope, anchors, or other means to maintain its design position under stress of installation procedures, cable pulling operations, and other conditions such as settling and hydraulic or frost uplift. Ducts shall be joined in such a way as to prevent solid matter from entering the conduit line. Joints shall form a sufficiently continuous smooth interior surface between joining duct sections that supply cable will not be damaged when pulled past the joint. When conditions are such that externally coated pipe is required, the coating shall be corrosion-resistant and should be inspected and/or tested to see that it is continuous and intact before backfilling is begun. Precautions shall be taken to prevent damage to the coating when backfilling.

Conduit installed through a building wall shall have internal and external seals intended to prevent the entrance of gas into the building insofar as practical. The use of seals may be supplemented by gas venting devices in order to minimize building up of positive gas pressures in the conduit. Conduit installed in bridges shall be such as to allow for expansion and contraction of the bridge. Conduits passing through a bridge abutment should be installed so as to avoid or resist any shear due to soil settlement. Conduit of conductive material installed on bridges shall be effectively grounded. Conduit should be installed on compacted soil or otherwise supported when entering a manhole to prevent shear stress on the conduit at the point of manhole entrance.

Laying Out Line   In general, a conduit run should be as straight as possible. When this is impossible due to obstructions, the run is made up of combinations of straight sections and curved sections. The combinations of straight sections and curves can follow the typical outlines shown in Fig. 30-1 depending on the number and position of the obstructions to be avoided. After the

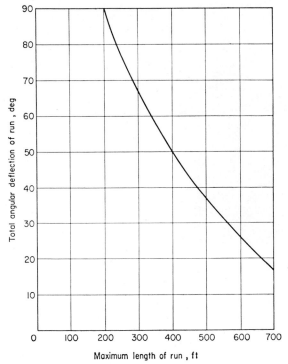

Fig. 30-2 Chart showing relation between total angular deflection of duct run and permissible maximum length of run.

outline for a particular set of conditions has been made the total angular displacement is determined by adding the degrees at the angles A, B, C, and D as illustrated in Fig. 30-1, thus,

$$\text{Total angle} = A° + B° + C° + D°$$

If there is a displacement in a vertical plane also, this is included in the summation of the total angle. This total angle limits the length of the run between manholes as shown in the chart of Fig. 30-2. The greater the total angle the shorter the permissible duct run, and the smaller the angle the longer the permissible run. In general, the maximum length of run even when entirely straight is not much over 500 ft.

The radius of the curve connecting the straight sections should be as large as possible to facilitate the pulling in of the cable (see Fig. 30-3). In general, a radius of 500 ft is preferred. The smallest permissible radius is four-tenths of the length of the run and never less than 10 ft.

Since the curved section is made up of short straight sections, these straight sections should be short enough to approximate a smooth curve. Table 30-1 shows the preferred lengths of sections for various lengths of radii of curvature.

Table 30-1

Length of radius of curvature, ft	Length of short straight sections of conduit, in
10 to 20	6
20 to 30	9
30 to 150	18
150 and over	36

Fig. 30-3   View showing a conduit run making a turn with a large radius of curvature. Note use of spacers to keep ducts in correct alignment. *(Courtesy Line Material Industries.)*

Grading the Line   Conduit runs are graded so that any water seeping into the conduit will drain into the manhole. Good practice requires that the minimum grade be no less than 3 in for every 100 ft of run.

Grades are established by an engineer with a level and usually marked with wood grade stakes driven 5 ft apart on the center line of the bottom of the duct trench. The top of the grade stakes indicates the top surface of the bottom conduit sheathing.

The methods of grading are double slope and single slope. Figure 29-4 illustrates the double-slope grade. This is the preferred method on level streets as it reduces the distance any seep water needs to drain. The duct run is highest midway between manholes, coming to within 2 ft of the street surface, and grades off in each direction. Figure 29-5 illustrates single-slope grading. The entire conduit run is graded in only one direction. This is the logical method when the street slopes from one manhole to the other. Sometimes obstructions also prevent the use of double-slope grading, as is illustrated in Fig. 30-4. Here the single-slope grade is increased in order to have the conduit pass under the obstruction.

Arrangement of Ducts   Ducts can be arranged in the trench in various ways, as shown in Fig. 30-5, depending on the number of ducts in the run. The number of ducts built into a duct run depends on the number of ducts actually needed and the number of spares to be provided for future load growth. A rule often followed is to double the number initially filled with cable. This would therefore provide for a 100 percent increase in load.

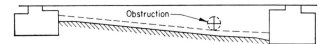

Fig. 30-4   Manner of grading conduit run in order to pass under obstruction ⊕.

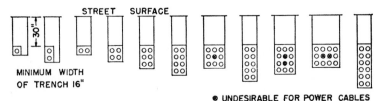

● UNDESIRABLE FOR POWER CABLES

Fig. 30-5   Various arrangements of ducts in a duct run. Interior ducts are sometimes called "dead" ducts because they do not radiate heat so well as outside ducts. They are therefore generally used for street-lighting or control cables.

Fig. 30-6 Backhoe being used to dig trench for conduit run. *(Courtesy Iowa-Illinois Gas and Electric Co.)*

Since cables carrying electric power become heated from the electric losses in the conductors they should be so installed that cooling can take place by radiation. When a large number of cables are close together the ability to radiate heat is greatly reduced. Hence, the number of cables in one duct run should be kept low. A recommended maximum of ducts in a single run is nine. When a greater number is used the heat radiation per cable becomes greatly reduced so that the total power carried in the duct run is only slightly increased.

Excavating the Trench   The trench for the duct run is first marked off with a crayon if the street is paved, and with batter boards and mason twine if the street is unpaved. Both sides of the trench must be marked off if the trench is to be dug by hand. If a digging machine is to be used (Fig. 30-6), only one side of the trench need be marked. Cuts in the pavement itself must conform exactly to the size and shape of the required excavation. The width of the cut is made equal to the overall width of the conduit run. The sides of the excavation are shored where necessary (see Fig. 30-7). The trench is dug true to the marked line with perpendicular sides. The sides are trimmed smooth to provide for a uniform sheath of concrete around the conduits. The entire trench is opened between manholes before any conduit is laid in order to determine if any obstructions exist that must be bypassed.

When the required depth of the trench is known for all points along its length the grade may be established. After the grade stakes are put in place, the bottom of the trench is trimmed 3 in below the top of the stakes to provide space for the concrete envelope. Trenches which have been dug too deep at any point are partially refilled and tamped solid before pouring the bottom concrete sheath.

Laying the Ducts   When the conduits are to be encased in a concrete envelope two methods of laying the ducts are in general use. These are known as:

1. The built-up method
2. The tier-by-tier method

In both methods the laying of the ducts is begun midway between manholes where duct spacing is standard, and then working toward the manholes where the ducts may flare out. In each method the pieces of conduit are laid with joints staggered at least 6 in, both vertically and horizontally (see Fig. 30-8). Spacing between adjacent ducts is usually 1, 1½, 2, or 3 in. The concrete envelope

Fig. 30-7   Shoring the sides of the trench to prevent cave-in.

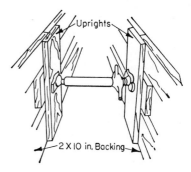

surrounding the duct structure is usually 3 in thick on top, bottom, and sides (see Fig. 30-9). The concrete mixture used is generally one part cement, three parts clean sand, and five parts broken stone or gravel. The stone or gravel used must pass through a ¾-in screen.

Procedure in "Built-Up" Method   In this method all the ducts in the run are placed in a self-supporting structure before the concrete envelope is poured. The conduits rest either on a concrete base or on spacers. If a concrete base is used, the concrete is poured into the bottom of the trench and tamped into place to form an evenly graded layer 3 in thick to the top of the grade stakes previously established. If spacers (Figs. 30-10 to 30-12) are used they are placed crosswise in the bottom of the trench at approximately 3- to 5-ft intervals and leveled even with the top of the grade stakes. These base spacers then provide a space of approximately 3 in between the bottom of the trench and the underside of the first duct, which space is later filled with concrete.

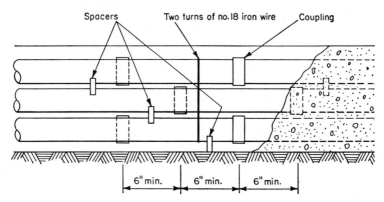

Fig. 30-8 Showing manner of staggering joints in conduits. *(Courtesy Johns-Manville Co.)*

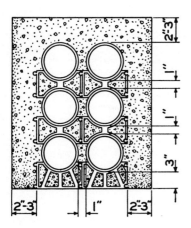

Fig. 30-9 Cross section of completed conduit bank constructed by the "built-up" method. Note spacers and ducts all encased in a 3-in concrete envelope. *(Courtesy Johns-Manville Co.)*

The laying of the conduit is then begun on either the concrete base or the base spacers. Intermediate spacers (Fig. 30-10b) are positioned horizontally near each end of the succeeding layers of duct. As the laying proceeds the individual ducts and layers are tied together with heavy cord or wire so as to securely hold the ducts in place during the pouring of the concrete. In this manner a rigid structure is built up to the height of the full conduit formation as shown in Fig. 30-12. Weights are placed on this formation to hold it in position while the concrete is poured (see Fig. 30-13). In the case of concrete or tile duct, their own weight is sufficient to hold them in place.

The concrete is poured around the ducts as soon as possible after they have been placed to protect them from mechanical damage. The concrete should be well mixed and not too wet. Very wet concrete has a tendency to float the conduits and change the alignment. Continuous spading is necessary to ensure a flow of concrete between and under the individual ducts. A long flat slicing tool or spatula worked carefully up and down between each vertical line of ducts will help to eliminate voids. Power-driven tampers shall not be used for this purpose. The pouring is continued to a height of 3 in above the top layer of ducts. If weights were used to hold the conduits down, they should be removed before the concrete sets and the voids left by them filled with concrete and spaded. Figure 30-9 shows a conduit bank after the pouring of the concrete is completed.

The open ends of the ducts should be closed with wooden, fiber, or plastic plugs to prevent animals or other objects from entering and becoming lodged (see Fig. 30-14). This should be done whenever the work is halted for any length of time, even for overnight.

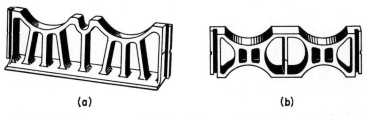

(a)   **(b)**

Fig. 30-10   Plastic (*a*) base spacer and (*b*) intermediate spacer. (*Courtesy Line Material Industries.*)

Fig. 30-11 Assembled bank of conduits being installed by "built-up" method. Note use of spacers to hold ducts in position prior to pouring of concrete. (*Courtesy Line Material Industries.*)

Fig. 30-12 Ducts resting on spacers before pouring of concrete envelope. (*Courtesy Line Material Industries.*)

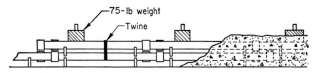

Fig. 30-13 Complete duct formation tied together with twine and held in place with weights before the pouring of the enclosing concrete sheath.

After the concrete has taken its initial set the trench is backfilled. The dirt shall be placed in the trench in 6- to 12-in layers, and each layer carefully tamped. In case a pavement was cut this is then replaced.

Procedure in Tier-by-Tier Method   In this method a concrete base or foundation is first poured to a thickness of 3 in and even with the top of the grade stakes. The lengths of duct are laid upon this base as in the "built-up" method to the proper line and separation. The correct separation between ducts is maintained by means of wooden or pipe comb separators, illustrated in Fig. 30-15. The width of the teeth used in these comb separators is the same as the required horizontal separation of the ducts and the crossbar of the comb has the same height as the vertical separation of the tiers. Separators are placed every 3 to 5 ft. Weights placed upon the separators may be used to hold the separators and ducts in place.

Fig. 30-14   Wooden plugs inserted in open end of conduits to prevent animals or objects from becoming lodged in them. *(Courtesy Edison Electric Institute.)*

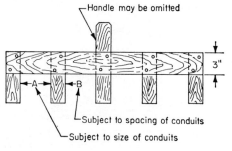

Fig. 30-15   Comb-type separator used to maintain correct spacing between ducts. *(Courtesy Orangeburg Mfg. Co.)*

As soon as the first tier of the ducts is laid and the temporary separators placed in position (see Fig. 30-16), the vertical spaces between ducts are filled with concrete (Fig. 30-17). The usual mixture consists of one part cement, three parts sand, and five parts stone or gravel aggregate as for the "built-up" method. This concrete is carefully spaded in and around the underside of the ducts with a wood spade.

The concrete for the sides is now carefully placed and worked down level with the top of the first tier of ducts. The horizontal layer of concrete between tiers is now poured level with the top of the crossbar of the comb separators. The separators are then withdrawn and the holes left by their removal are carefully filled ready for the next tier.

The next tier of ducts is now laid upon the concrete, and the temporary separators again put in place. Next the cement mortar is again filled in and spaded between and under the ducts. Then

Fig. 30-16 Showing use of wooden comb separators to maintain alignment while concrete is poured. *(Courtesy Line Material Industries.)*

Fig. 30-17 Pouring concrete on first tier of ducts being constructed by "tier-by-tier" method. Note use of separators to maintain duct alignment while workmen work cement under and between ducts. *(Courtesy Line Material Industries.)*

the concrete is built up on the sides to the top of this second tier. Next the horizontal layer of concrete is placed over the tier even with the top of the separators. The separators are then again withdrawn and the holes filled, ready for the third tier of ducts. This process (see Figs. 30-18 and 30-19) is continued until all the tiers are laid. A cross section of a conduit bank constructed by the tier-by-tier method is shown in Fig. 30-20. Note the dotted horizontal lines which separate the tiers. Backfilling the trench and tamping the dirt and replacing the pavement complete the conduit installation.

Joining Ducts Fiber and asbestos cement ducts are provided with factory-machined tapers on each end for insertion into a short sleeve coupling as shown in Fig. 30-21. The sleeve is provided with a mating taper on the inside of each end.

Fig. 30-18 Laying conduit by the "tier-by-tier" method. Each row of ducts is placed and concreted separately. *(Courtesy Orangeburg Mfg. Co.)*

In making the joint a coupling is placed on the first length of duct which is laid in position. The next length of duct with a coupling already on one end is inserted into the coupling of the first duct. With a block or board held against the last coupling the joint is driven together with a few blows using a hammer or light sledge (see Fig. 30-22). A coupling is always placed on the duct before driving it. Never hammer directly on the tapered duct end as so doing would damage it.

Plastic couplings are now being introduced. These have installation advantages because they are lighter, easier to handle, have no breakage, and do not require use of joint sealing compound. They also provide better joint flexibility without decrease in joint tightness.

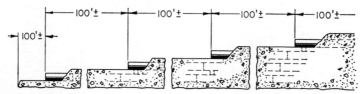

Fig. 30-19 Side view showing steps in laying conduit by the tier-by-tier method. If sufficient trench has been excavated, the laying of the ducts is usually done in 100-ft sections. The concreting process can thus be continuous. *(Courtesy Johns-Manville Co.)*

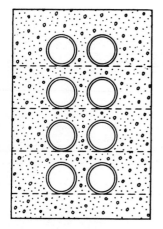

Fig. 30-20 Cross section of completed duct bank constructed by the tier-by-tier method. Ducts are completely surrounded by concrete, which helps to conduct heat produced in the cables to the surrounding earth. *(Courtesy Johns-Manville Company.)*

Fig. 30-21 Parts of a fiber duct joint. The tapered duct ends fit into ends of a similarly tapered sleeve coupling. *(Courtesy Orangeburg Mfg. Co.)*

Fig. 30-22 Tapping duct into coupling on opposite end. First coupling is placed on driven end, and then a block or board is held against coupling. Then tap lightly on board with hammer. *(Courtesy Johns-Manville Co.)*

Fig. 30-23 Cutting fiber conduit with a coarse-tooth saw. *(Courtesy Orangeburg Mfg. Co.)*

Fig. 30-24 Cutting taper on duct end with tapering tool consisting of an expanding chunk and cutting handle which holds a steel cutter blade. *(Courtesy Orangeburg Mfg. Co.)*

If other than standard lengths of duct are required the ducts can be cut with a coarse-tooth hand saw as shown in Fig. 30-23. New taper can be made on the job by means of a tapering tool as illustrated in Fig. 30-24. This tool consists of an expanding chuck and cutting handle on which is mounted a steel cutter blade.

Plastic utility ducts made from polyvinyl chloride plastic compound have many advantages. The ducts are available with diameters of 2, 3, 3½, 4, 5, and 6 in. The conduit comes with belled ends to permit easy joining of the ducts. Solvent cements can be used to seal the joints. The ducts can be direct-buried in soils or encased in concrete. The ducts are usually shipped in self-supporting framed units so that the ducts can be handled with a forklift truck.

Plastic ducts can be cut with a hand or power saw to obtain the proper length (Fig. 30-25). The burrs on the end of the conduit, as the result of the sawing, must be removed prior to completing the joint. Surfaces to be joined should be clean and free from dirt, foreign materials, and moisture. The joints should be sealed with the proper cement specified by the duct manufacturer. The

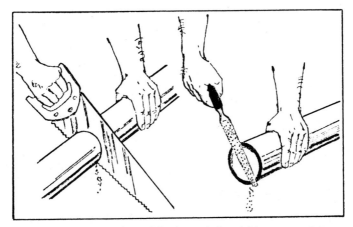

Fig. 30-25 Lineman cutting plastic duct with hand saw. The burrs left by sawing are being removed with a file.

plastic ducts can be installed in driven or bored steel casings (Figs. 30-26 through 30-28). Steel casings are frequently used when crossing under highways and railroads. Plastic utility duct is available in corrugated form and packaged in coils or on reels. Duct may be unreeled directly into the trench for installation. The flexibility of the corrugated duct permits its installation around corners without the use of prefabricated bends or the splicing of short sections of conduit (Fig. 30-29). The corrugated duct can be installed with plowing equipment if the soil is appropriate.

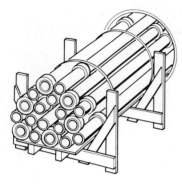

Fig. 30-26 Cradle holds plastic ducts in proper formation while circular-type spacers are placed over each duct prior to banding. Cradle also keeps duct formation lined up with casing entrance. *(Courtesy National Electrical Manufacturers Association.)*

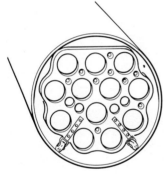

Fig. 30-27 Custom bore spacer with small openings for wire guides on each side. Wires prevent rotation of duct bank as it advances through the casing.

Fig. 30-28 Push plate, of either wood or metal, covers entire duct bundle and spreads load pressure evenly on the bundle. Winch line is threaded through center duct as well as the rear push plate so that formation can be pulled and pushed into the casing. *(Courtesy National Electrical Manufacturers Association.)*

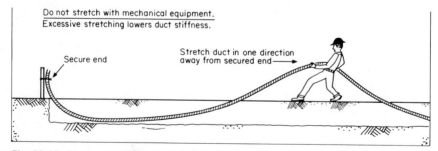

Fig. 30-29 Lineman installing corrugated plastic duct in trench. Duct is unreeled and placed in trench. *(Courtesy National Electrical Manufacturers Association.)*

Manhole Construction

Manholes, handholes, and vaults shall be designed to sustain all expected loads which may be imposed upon the structure. The horizontal and/or vertical design loads shall consist of dead load, live load, equipment load, impact, load due to water table, frost, and any other load expected to be imposed upon and/or occur adjacent to the structure. The structure shall sustain the combination of vertical and lateral loading that produces the maximum shear and bending moments in the structure.

In roadway areas, the live load shall consist of the weight of the moving tractor-semitrailer truck illustrated in Fig. 31-1. The vehicle wheel load shall be considered applied to an area as indicated in Fig. 31-2. In the case of multilane pavements, the structure shall sustain the combination of loadings which result in vertical and lateral structure loadings which produce the maximum shear and bending moments in the structure. Loads imposed by equipment used in road construction may exceed loads to which the completed road may be subjected.

In designing structures not subject to vehicular loading, the minimum live load shall be 300 lb/ ft^2. Live loads shall be increased by 30 percent for impact. When hydraulic, frost, or other uplift will be encountered, the structure shall either be of sufficient weight or so restrained as to withstand this force. The weight of equipment installed in the structure is not to be considered as part of the structure weight. Where pulling iron facilities are furnished, they should be installed with a factor of safety of 2 based on the expected load to be applied to the pulling iron.

A clear working space sufficient for performing the necessary work shall be maintained. The horizontal dimensions of the clear working space shall be not less than 3 ft. The vertical dimensions shall be not less than 6 ft except in manholes, where the opening is within 1 ft horizontally of the adjacent interior side wall of the manhole. Where one boundary of the working space is an unoccupied wall and the opposite boundary consists of cables only, the horizontal working space between these boundaries may be reduced to 30 in. In manholes containing only communications cables and/or equipment, one horizontal dimension of the working space may be reduced to not less than 2 ft provided the other horizontal dimension is increased, so that the sum of the two dimensions is at least 6 ft.

Round access openings in a manhole containing supply cables shall be not less than 26 in. in diameter. Round access openings in any manhole containing communication cables only, or manholes containing supply cables and having a fixed ladder which does not obstruct the opening, shall be not less than 24 in. in diameter. Rectangular access openings should have dimensions not less than 26 by 22 in. Openings shall be free of protrusions which will injure personnel or prevent quick egress.

When not being worked in, manholes and handholes shall be securely closed by covers of sufficient weight or proper design so they cannot be easily removed without tools. Covers should be suitably designed or restrained so that they cannot fall into manholes or protrude into manholes sufficiently far to contact cable or equipment. Strength of covers and their supporting structure shall be at least sufficient to sustain the applicable loads.

Vault or manhole openings shall be located so that safe access can be provided. When in the highway, they should be located outside the paved roadway when practical. They should be located outside the area of street intersections and crosswalks whenever practical to reduce the traffic hazards to personnel working at these locations. Personnel access openings in vaults or manholes should be located so that they are not directly over the cable or equipment. Where these openings interfere with curbs, etc., they can be located over the cable if one of the following is provided: a conspicuous warning sign, a protective barrier over the cable, or a fixed ladder. In vaults, other types of openings may be located over equipment to facilitate work on this equipment.

Where accessible to the public, access doors to utility tunnels and vaults shall be locked unless qualified persons are in attendance to prevent entry by unqualified persons. Such doors shall be designed so that a person on the inside may exit when the door is locked from the outside. This

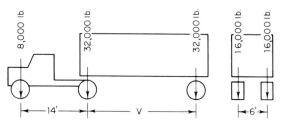

Variable spacing – 14 ft. to 30 ft. inclusive

Spacing to be used is that which results in vertical and lateral structure loading which produces the maximum shears and bending moments in the structure

Fig. 31-1   Vehicle loading requirements for manhole construction.

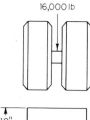

Fig. 31-2   Vehicle wheel loading requirements for manhole construction.

rule does not apply where the only means of locking is by padlock and the latching system is so arranged that the padlock can be closed on the latching system to prevent locking from the outside.

Fixed ladders shall be corrosion-resistant.

Where drainage is into sewers, suitable traps or other means should be provided to prevent entrance of sewer gas into manholes, vaults, or tunnels.

Adequate ventilation to open air shall be provided for manholes, vaults, and tunnels having an opening into enclosed areas used by the public. Where such enclosures house transformers, switches, regulators, etc., the ventilating system shall be cleaned at necessary intervals. This does not apply to enclosed areas under water or in other locations where it is impractical to comply.

Supply cables and equipment should be installed or guarded in such a manner as to avoid damage by objects falling or being pushed through the grating.

Manhole and handhole covers should have an identifying mark which will indicate ownership or type of utility.

Description Manholes are usually made of brick, concrete blocks, reinforced concrete, or a combination of these materials. The manhole cover is made of steel. As a rule, manholes are constructed in place. However, sometimes it is more economical and convenient to precast the manhole and transport it to the location in the street. This procedure eliminates the field parking of materials, pouring of concrete, etc., in crowded streets and avenues.

Precast manholes can be fabricated in various configurations such as a straight manhole, an "L" manhole and a "T" manhole (Figs. 31-3 through 31-5). Large manholes are usually precast in sections or in pieces (Fig. 31-6).

The area at the manhole installation site must be cleared of obstructions such as gas pipes and water mains. Mechanized equipment such as a backhoe can be used to excavate the area for the manhole. The bottom of the excavation should be firm undisturbed or compacted earth, leveled without any large rocks or obstructions that would prevent the manhole from setting properly

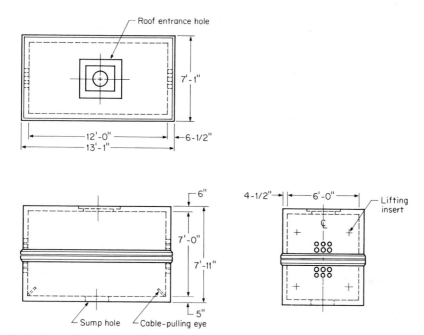

Fig. 31-3 Straight type precast cable manhole. Approximate weights top section 14,800 lb and bottom section 14,650 lb. Manhole is designed for use with 12-duct conduit run. *(Courtesy Commonwealth Edison Co.)*

(Fig. 31-7). The excavation should be shored to protect the worker preparing the excavated area for the manhole installation. The precast manhole can be moved to the site on a low-boy trailer. A mobile crane is normally used to lift the manhole from a trailer and place it in the excavated area (Figs. 31-8 through 31-10).

The hole around the precast manhole must be backfilled and thoroughly compacted. Sand, or the excavated material if it is fine and dry, can be used for backfilling.

Manhole frame and cover is installed after the precast manhole is in place. Precast manhole necks can be used to extend the manhole opening to the proper level for the surrounding area (Fig. 31-11).

As manholes differ in size and complexity owing to variations in the number of ducts entering and leaving; whether one-, two-, three-, or four-way; size of cables; length of splices; horizontal offsets of conduits; etc., only general principles of construction can be outlined.

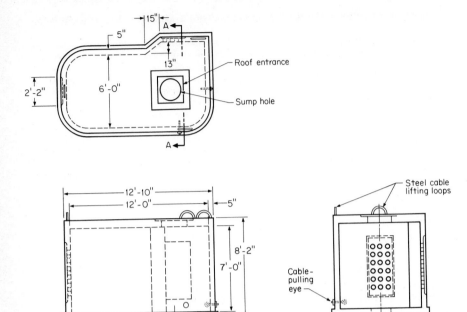

Fig. 31-4 An L-type precast cable manhole. Approximate weight 30,000 lb. Manhole is designed for use with 18-duct conduit run. *(Courtesy Commonwealth Edison Co.)*

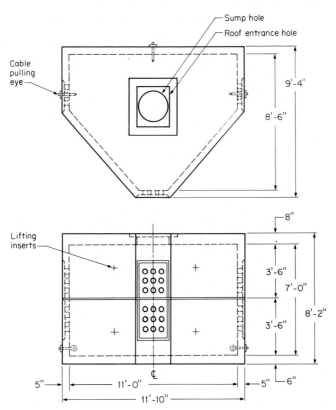

Fig. 31-5 A T-type precast cable manhole. Approximate weights top section 15,065 lb and bottom section 13,235 lb. Manhole is designed for use with 18-duct conduit run. *(Courtesy Commonwealth Edison Co.)*

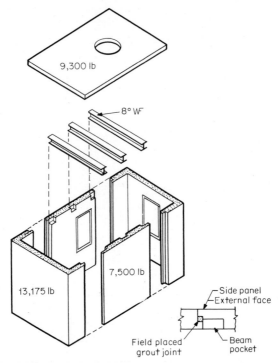

Fig. 31-6   Precast sections of concrete manhole. *(Courtesy Eugene Water and Electric Board.)*

Fig. 31-7   Workers checking bottom of excavation to be sure it is level before setting precast manhole.

Fig. 31-8 Placing precast sections in position in excavated manhole. *(Courtesy Eugene Water and Electric Board.)*

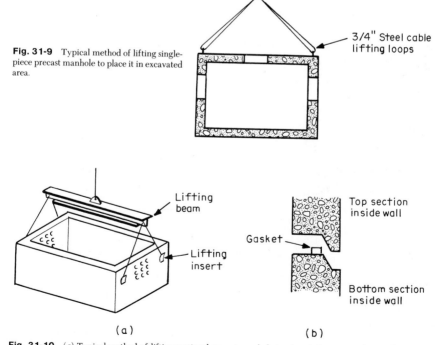

Fig. 31-9 Typical method of lifting single-piece precast manhole to place it in excavated area.

3/4" Steel cable lifting loops

Lifting beam

Lifting insert

Gasket

Top section inside wall

Bottom section inside wall

(a)   (b)

Fig. 31-10 (a) Typical method of lifting sectional precast manhole to place it in excavated area. (b) Detail of joint gasket installation.

Floors   Manhole floors are built of poured concrete approximately 6 in thick and then surfaced with cement mortar about ½ in thick. The mixture for the concrete is usually:

One part portland cement
Three parts clean sand
Five parts broken stone or gravel

The mixture for the cement mortar usually is:

One part portland cement
Three parts clean sand

The cement mortar may be placed after the walls have been completed or after the roof has been constructed.

The floor should slope toward the sewer drain or sump with a grade of about 1 in. in 6 ft.

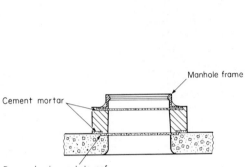

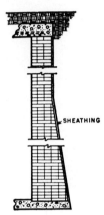

Fig. 31-11   Manhole frame and cover in place on precast manhole. A precast concrete manhole neck is used to extend opening proper distance so that cover will be flush with surface of terrain.

Fig. 31-12   Typical brick wall construction of manhole. Note poured concrete floor and roof.

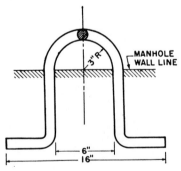

Fig. 31-13   Typical plastic duct bell used to terminate duct in manhole wall.

Fig. 31-14   Typical galvanized wrought-iron pulling eye built into manhole wall opposite duct runs.

Walls   Walls are commonly built of poured concrete, concrete blocks, or bricks. The thickness of the wall varies from 8 in to 2 ft depending on the height and length of the wall. It also depends on whether the soil is firm or fluid. Brick walls are usually half again as thick as concrete walls. Concrete walls may be reinforced if necessary. A typical brick wall is shown in Fig. 31-12.

The conduit runs should terminate at the interior side of the manhole wall in conduit bells. A conduit bell (Fig. 31-13) is a flared opening that makes it easier to pull in the cable and eliminates

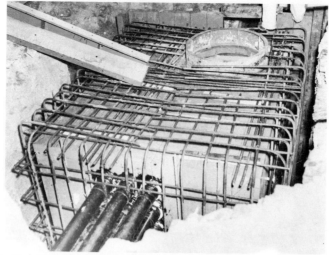

Fig. 31-15 Manhole constructed of reinforced concrete showing reinforcing rods before pouring of concrete. Note that conduit bank at manhole is being cast integrally with the manhole walls. *(Courtesy Edison Electric Institute.)*

the sharp corner that might injure the cable sheath as the cable is pulled in. It also protects the cable sheath as the cable moves in and out owing to load cycling. If the conduit run has not been installed at the time the manhole wall is being erected, an opening should be left in the wall for the conduit bells. The hole can be closed later, permitting correct alignment of bells and ducts.

Pulling eyes (Fig. 31-14) are usually embedded in the walls opposite the conduit runs to facilitate the pulling in of the cables.

Roofs Manhole roofs are usually constructed of reinforced concrete. If, however, it is necessary to restore traffic on the street as quickly as possible, a roof of steel beams and brick is used. This eliminates the delay in the time required for the concrete to set. A top-view sketch showing the reinforcing bars in place is shown in Fig. 31-15.

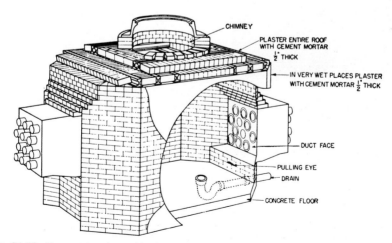

Fig. 31-16 Cutaway view of typical brick manhole showing concrete floor, brick walls and roof, chimney, two-way duct runs, 12 ducts, duct bells, pulling eye, and drain. *(Courtesy Edison Electric Institute.)*

Fig. 31-17 Installation of a 750-kVA oil-filled network transformer rated 14,400–480Y/277 volts in a vault. *(Courtesy General Electric Co.)*

Fig. 31-18 Linemen placing network transformer in vault constructed below sidewalk in a business area. *(Courtesy General Electric Co.)*

The concrete mixture for the roof is usually:
One part portland cement
Two parts clean sand
Four parts ¾-in crushed stone or gravel
The roof opening, or "throat" or "chimney" as it is sometimes called, must be located so all is clear directly below. The opening may be circular or rectangular in shape, depending on the type of cover frame used. A removable metal cover is used to close the opening.

A completed manhole is shown in the cutaway sketch of Fig. 31-16. The manhole is constructed with concrete floor and brick walls and roof. The sketch also shows the concrete-encased conduit run and the circular opening in the roof from which the manhole derives its name.

Vaults Large manholes are often referred to as vaults. Vaults are used for transformer and network protector installation. The construction of a vault is similar to that of the manhole and must meet the same construction specifications. Figures 31-17 and 31-18 show transformers being installed in vaults by linemen.

Pulling Cable

Installing Cable Racks   Before any cables are pulled into ducts, provision for supporting the cable ends on racks in the manholes should be made. Allowing the cable ends to lie on the manhole floor permits their being stepped on or damaged by falling objects. A typical cable rack is shown in Fig. 32-1. The channel which supports the rack proper is fastened to the manhole wall with expansion bolts. Holes for the expansion bolts are drilled 3 in deep. Safety goggles should be worn when concrete or brick is being drilled.

The spacing between racks adjacent to the proposed cable joint is usually 3 ft. In general, a cable should be supported at points 6 in from each end of the sleeve covering the joint. Additional racks are spaced from 3 to 4½ ft depending on the nature of the cable bends.

Selecting the Duct   Before the start of cable installation, the duct to be occupied should be selected throughout the entire length of the run. As far as possible, the same relative position in the duct bank should be maintained. In general, the longer cables should be installed in lower ducts and the shorter cables in the upper ducts. In this way the vacant spaces will be left in the upper ducts, where they can be utilized later on without disturbing the cables already installed.

Rodding the Duct   Rodding is the term used to describe the insertion of a number of short, jointed wooden rods or a flexible steel rod into a duct to make possible drawing a wire into it. The rods used are 3 to 4 ft in length and about 1 in. in diameter. They are fastened together with a screw connection or by means of a quick-coupling interlocking joint as illustrated in Fig. 32-2. Rodding consists of coupling rod to rod and pushing the assembled line of rods farther along in the duct (see Fig. 32-3). When the first rod appears at the other end of the duct in the next manhole, a No. 10 or 12 gauge steel wire is attached to the last rod and pulled into the duct as the rods are pulled forward and uncoupled in the next manhole. In case the duct in the next adjacent run is also to be rodded, it may be possible to direct the rods into that duct without uncoupling.

A long flexible steel rod can be used instead of short stiff wooden rods. Figure 32-4 shows such a steel rod on portable reels, and Fig. 32-5 shows the manner of rodding it into a duct. The size of the rod is such as to make it stiff enough for pushing. Coupling of rods is accomplished by use of a leader and pickup fittings on the ends, shown in Fig. 32-6.

Cleaning the Duct   Before the cables are installed, each duct must be checked to see that it is clean and free from all obstructions. For this purpose the wire drawn into each duct in the rodding operation is used to pull through a mandrel, a circular wire brush, and a swab. The mandrel is a cup cleaner (Fig. 32-7) about ¼ to ⅜ in smaller in diameter than the duct. A common mandrel is a steel or wooden tube rounded and fitted with a pulling eye at each end. It will remove serious obstructions such as concrete which has dripped through a joint by working it back and forth until the obstruction is cleared. The circular wire brush has the same diameter as the duct and is used to remove small pebbles or bits of concrete, etc. The swab is made by stuffing waste into a holding grip.

Fig. 32-1 Typical cable rack used to support cable and splices on walls of manhole. *(Courtesy Edison Electric Institute.)*

Fig. 32-2 Conduit rod equipped with interlocking quick-coupling joint. Hook cannot detach when rods are in use. *(Courtesy Line Material Industries.)*

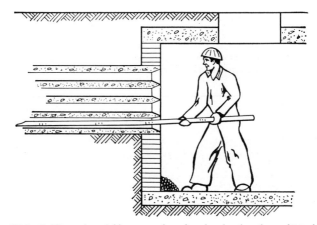

Fig. 32-3 Rodding a duct. Cableman couples rods and pushes them forward into duct.

Fig. 32-4 Flexible steel rod on reels. *(Courtesy Superior Switchboard and Devices Division.)*

Fig. 32-5   Feeding flexible steel rod to worker in the manhole, who pushes it into the duct. *(Courtesy Superior Switchboard and Devices Division.)*

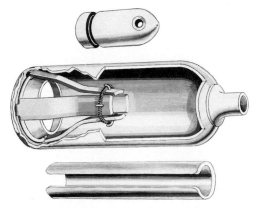

Fig. 32-6   Coupling fixtures for flexible steel rod. *(Courtesy Superior Switchboard and Devices Division.)*

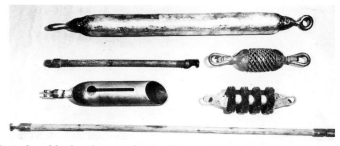

Fig. 32-7   Tools used for duct cleaning and testing. Note mandrel, wire brush, and swab. *(Courtesy Edison Electric Institute.)*

Wiring the Duct   If the conduits are to be used in the near future, a wire is attached behind the swab and drawn into the conduit as the cleaning tools are pulled forward. This wire is usually a galvanized steel wire of No. 10 or 12 gauge and is left in the duct for use by the cable-pulling crew. Besides serving this crew to pull in the heavier cable-pulling rope, it also provides a means of loosening or withdrawing the mandrel in case it should become lodged.

Pulling Lines   The cable-pulling rope is drawn into the duct by attaching it onto the wire that was left in the duct by the cleaning crew. As this wire is pulled out of the duct, the cable-pulling rope is drawn into the duct. For heavy pulls a ½-in steel wire rope is commonly used. For short pulls plain manila or hemp rope is preferred. The hemp rope varies in size from ½ to ¾ in depending on the length of the duct and the size of the cable.

Selecting the Pulling-In Manhole   The selection of the manhole from which the cable is to be pulled is important. Cables can be pulled on the first pull only as far as the pulling rig will permit. When the pulling eye in the cable end or on the basket grip reaches the lower sheave, the pulling must stop. The length of cable then in the manhole may not be sufficient to reach the cable rack on which the splice will be supported. Furthermore, additional cable length is needed to train the cable around the walls to the center of the splice on the rack. The shape of the manhole and the position of the splice on the wall will determine the amount of cable length needed. In

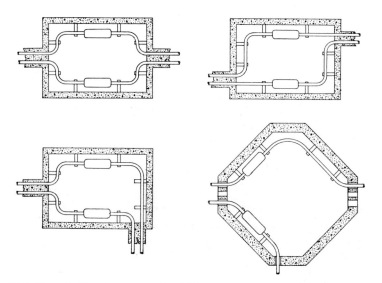

Fig. 32-8   Types of cable training in manholes of different shapes. *(Courtesy Edison Electric Institute.)*

a manhole where the ducts come in at one corner and leave at the diagonally opposite corner, the splice is always closer to one than the other (see Fig. 32-8). Also, when there is considerable difference in duct elevations, the length of cable from the joint to one entrance may be less at one end than the other. The rule, therefore, is that the manhole requiring the shorter cable length in the manhole to the center of the joint should be the pulling end, as any required length of cable can be left at the feeding end.

Manhole Rigging   The rigging used to pull in cable varies with the type of manhole, length of cable, and the conditions prevailing on the street.

The simplest rigging arrangement possible is a pulling eye that has been embedded in the manhole wall (Fig. 32-9). Here the pulling eye is directly opposite the duct, and the snatch block holds the pulling rope in line with the duct. In case the embedded pulling eye is located much below the duct level, a chain or rope of proper length is inserted between the snatch block and the pulling eye as shown in Fig. 32-10. This is the usual situation in a multiway manhole, where ducts enter the manhole on all sides and the eye must be placed below the duct level. A sheave in the manhole opening is often used in deep manholes to guide the pulling rope to the pulling winch on the street as shown.

When no embedded pulling eyes are available, the sheaves are supported on a so-called sheave stand. It consists of two timbers, channel irons, or I beams bolted together with a space between them for the sheaves. Holes through the two parallel members (see Fig. 32-11) permit placing the sheaves at the desired locations. Figure 32-12 shows such a sheave stand in use in an upright

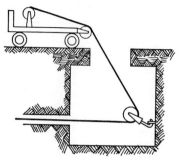

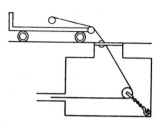

Fig. 32-9 Pulling cable by use of eye embedded in manhole wall opposite duct. Snatch block holds pulling rope in line with duct.

Fig. 32-10 Pulling eye located below duct level requires use of short rope or chain between snatch block and pulling eye.

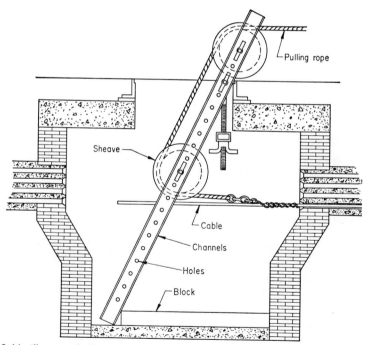

Fig. 32-11 Illustrating double-sheave stand made of two parallel channel irons. Holes in channel members permit placing sheaves at proper levels.

position, and Fig. 32-13 shows it used in an inclined position. In both cases the lower sheave is so placed that its bottom side is approximately even with the duct into which the cable is to be pulled. In Fig. 32-14 the sheave frame is anchored to the manhole roof with an adjustable bolt and hand nut for heavy pulling.

Other forms of rigging consist of a single sheave supported on wooden timbers cut to fit between the manhole floor and roof and held in place with wedges, blocks, and braces to suit given conditions. Figure 32-15 shows a single sheave on timbers cut to fit between floor and edge of manhole opening. Figure 32-16 shows a single sheave used with heavier bracing for heavier pulls, and Fig. 32-17 shows a single sheave used with overhead pulley arrangement. In each case the lower edge of the sheave is made to line up with the duct into which the cable is pulled.

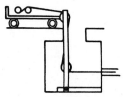

Fig. 32-12   Double-sheave stand in use in upright position. Lower sheave is placed directly opposite duct. Note blocking between foot of stand and manhole wall.

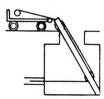

Fig. 32-13   Double-sheave stand in use in an inclined position.

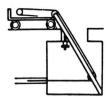

Fig. 32-14   Double-sheave stand anchored to manhole roof for heavy pulling.

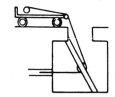

Fig. 32-15   Single-sheave stand.

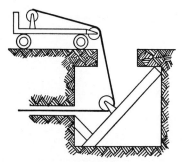

Fig. 32-16   Single-sheave stand held in place with timbers and wedges for heavy pulling.

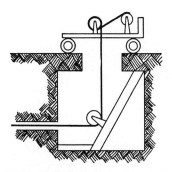

Fig. 32-17   Single-sheave used with overhead pulley arrangement for pulling cable.

Attaching Pulling Rope   The pulling rope is attached to the cable by means of a woven cable grip or by means of a clevis or eye.

The principle of a woven cable grip, or "basket" grip as it is sometimes called, shown in Fig. 32-18, is that the wires, which are in the form of an open weave, will contract laterally and thus hold fast to the cable sheath when a longitudinal pull is applied. Since the pull is thus applied to the sheath, this attachment is satisfactory only if the sheath is tight and if the pull is light.

When the sheath is loose or the pull heavy or there is not sufficient clearance in the duct for the wire mesh surrounding the cable, the pulling rope must be attached to a clevis or eye. The eye in turn is fastened to the conductors in the cable as well as to the sheath surrounding the cable (see Fig. 32-19). Such eyes are usually attached at the cable factory.

Pulling eyes can be attached in the field by first folding back about 4 in of sheath and removing the insulation. Then the conductor strands are wrapped around the lug of the eye for sweating. The sheath is then turned back and sealed with solder by wiping to the conductors and lug. The steps in this operation are illustrated in Fig. 32-20.

Since the pulling rope has a tendency to twist during the pull, a swivel (see Fig. 32-21) should be inserted between the pulling rope and the cable.

Feeding Tube   To prevent injury to the cable by scraping on the manhole frame or at the duct opening or in passing over other cables, a feeding tube is sometimes used. This guiding tube, or "baloney" as it is called, is usually made of flexible steel tubing about 4 in. in diameter with a flare at the reel end and a removable nozzle fitting at the duct end so it can be adapted to various size ducts. The tubes vary in length from 10 to 20 ft. Figure 32-22 shows typical feeding tubes.

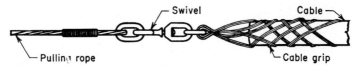

Fig. 32-18   Woven cable grip, sometimes called "basket" grip. Note swivel inserted between pulling rope and cable grip.

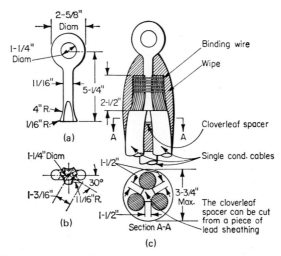

Fig. 32-19   Pulling eyes attached to cable end. Conductors are sweated onto lug of eye, and sheath is wiped onto conductors and lug. (*a*) Clevis type; (*b*) eyebolt type; (*c*) eyebolt type used with triplexed cables. (*Courtesy Edison Electric Institute.*)

Cable Lubrication   To protect the cable from excessive tension during pulling-in, the cable is lubricated. This reduces the friction between cable and duct walls, especially on curves in the duct run, by as much as 70 percent.

The lubricants for lead-sheath cables are greases, oils, soapstone, etc. The lubricant is applied with a paddle or brush just before the cable enters the feeding tube. A coating about 1/16 in thick is ample and will amount to 6 to 8 lb per 100 ft on 3-in cable.

No lubricant should be applied to the first and last 5 ft of cable for convenience and cleanliness in splicing.

Pulling the Cable   The cable is drawn into the duct by means of a winch, capstan, or truck located near the manhole at the pulling end.

The reel of cable must be properly placed at the feeding end to cause minimum flexing of the cable. It should always be located on the side of the manhole toward which the cable is to be pulled. This avoids putting a reverse bend in the cable and makes it unreel easily. Figure 32-23 is a diagrammatic sketch showing manhole, cable reel, feeding tube, and cable in correct relationship for pulling in cable.

The end of the cable is then threaded into the feeding tube extending from above the edge of the manhole down to the end of the duct. The pulling grip is now slipped over the end of the cable, the swivel is attached, and finally the pulling rope is attached. In case a pulling eye is provided on the cable, the swivel and pulling rope are attached to it.

The cable is inspected and greased as it enters the funnel of the feeding tube. Lead-covered cable must be completely covered with lubricant. On cables having a rubber belt on the outside,

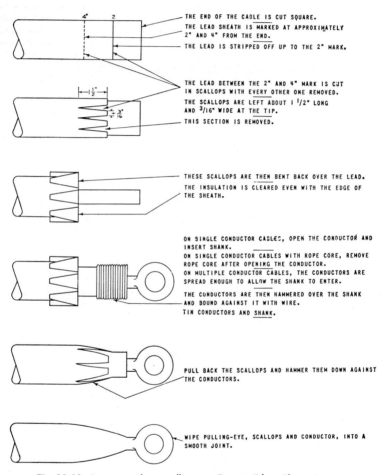

THE END OF THE CABLE IS CUT SQUARE.
THE LEAD SHEATH IS MARKED AT APPROXIMATELY 2" AND 4" FROM THE END.
THE LEAD IS STRIPPED OFF UP TO THE 2" MARK.

THE LEAD BETWEEN THE 2" AND 4" MARK IS CUT IN SCALLOPS WITH EVERY OTHER ONE REMOVED.
THE SCALLOPS ARE LEFT ABOUT 1 1/2" LONG AND 3/16" WIDE AT THE TIP.
THIS SECTION IS REMOVED.

THESE SCALLOPS ARE THEN BENT BACK OVER THE LEAD.
THE INSULATION IS CLEARED EVEN WITH THE EDGE OF THE SHEATH.

ON SINGLE CONDUCTOR CABLES, OPEN THE CONDUCTOR AND INSERT SHANK.
ON SINGLE CONDUCTOR CABLES WITH ROPE CORE, REMOVE ROPE CORE AFTER OPENING THE CONDUCTOR.
ON MULTIPLE CONDUCTOR CABLES, THE CONDUCTORS ARE SPREAD ENOUGH TO ALLOW THE SHANK TO ENTER.
THE CONDUCTORS ARE THEN HAMMERED OVER THE SHANK AND BOUND AGAINST IT WITH WIRE.
TIN CONDUCTORS AND SHANK.

PULL BACK THE SCALLOPS AND HAMMER THEM DOWN AGAINST THE CONDUCTORS.

WIPE PULLING-EYE, SCALLOPS AND CONDUCTOR, INTO A SMOOTH JOINT.

Fig. 32-20   Steps in attaching a pulling eye. *(Courtesy Edison Electric Institute.)*

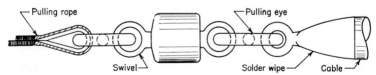

Pulling rope   Pulling eye
Swivel   Solder wipe   Cable

Fig. 32-21   Swivel inserted between pulling rope and cable to prevent twisting of cable. *(Courtesy Edison Electric Institute.)*

a mixture of soapstone and water is applied to the belt as the cable is drawn in. On all other types of nonmetallic sheath cables no pulling lubricant is used.

Workers stationed at the cable reel regulate the amount of slack in the cable so that it passes freely into the tube without being loose on the reel and without scraping on the manhole frame. A sketch showing the cable being drawn into the duct appears in Fig. 32-24.

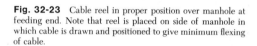

Fig. 32-22   Typical flexible-cable feeding tubes used to guide cable into duct and prevent abrasion. *(Courtesy Edison Electric Institute.)*

Fig. 32-23   Cable reel in proper position over manhole at feeding end. Note that reel is placed on side of manhole in which cable is drawn and positioned to give minimum flexing of cable.

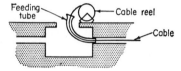

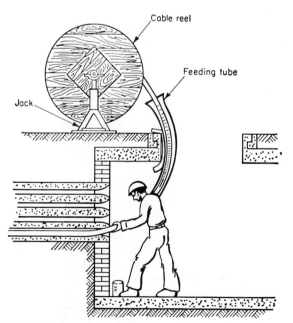

Fig. 32-24   Sketch showing cable being drawn into duct. Note reel, jacks, and feeding tube.

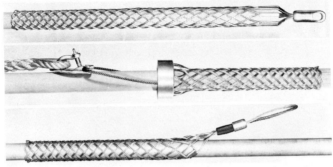

Fig. 32-25 Special basket grips applied to cable to pull up more slack at pulling end. *(Courtesy Edison Electric Institute.)*

A man should be stationed in the manhole at the pulling end to signal a stop to prevent pulling the cable over the sheaves. Another should be stationed at the top of the manhole to relay signals to the winch operator or pulling truck.

The speed at which the cable is drawn into the duct varies with conditions and cable size. It is good practice, however, to pull the cable slowly. In a straight duct a single cable can be pulled in at about 60 ft per min. If a higher speed is used, there is difficulty in greasing properly and in inspecting the cable. An average speed of about 40 to 40 ft per min, or ⅔ ft per sec, is preferable when a single cable is being pulled. When two or more cables are pulled into one duct, the speed should be further reduced to about 20 ft per min. A slow speed will prevent the possibility of crossing the conductors as they enter the duct.

The saving in time obtained with higher speeds is negligible, as the time required for drawing in a 400-ft length at 40 ft per min is only 10 min. This is a small part of the total time needed to prepare for the pulling-in operation.

Slack Pulling When the cable end has been drawn up to the manhole rigging, the first pulling operation must stop. If not enough cable length is then available in the manhole to reach to the center of the splice as the cable is trained around the wall, additional pulling is necessary. To obtain this additional slack in the cable, a new cable grip must be applied to the cable the required number of feet from the cable end. A special "basket" grip (Fig. 32-25) is used. This grip is open at both ends and can be slipped over the cable to the desired position. Care must be used to avoid injuring the cable.

When the cable has been pulled in as far as necessary, the forward end is placed on the cable supports. No attempt, however, should be made to train the cable into its final position. This will be done by the splicing crew. A check should be made to see that the seal at the end of the cable is still watertight. At the feeding end the cable is cut off with enough slack left to train it properly around the manhole walls. The end of the cable is then also placed on its supports, and the end of the cable sealed. The end of the cable remaining on the reel is also sealed.

Splicing Cable

Multiple lengths of cable are connected together to form a continuous length. These connections are called joints or splices. If underground ducts are used, the splices are normally completed in a manhole. When a cable is cut, in preparation for splicing or for any other reason, it must be protected from moisture and dirt. Cable ends exposed to the atmosphere will collect moisture and contamination. The cable ends should be properly sealed at all times except during the period when the splice is being completed or the cable termination is being installed. The cable ends should be thoroughly inspected before they are trained into final position for splicing.

If the splices are to be completed in a manhole, the cables and the splices should be supported on racks mounted on the manhole wall as illustrated in Section 29. The joint or splice is supported between two brackets located on the central portion of the manhole wall. Because of the expansion and contraction of the cables, caused by changes in the cable temperature, it is necessary to provide reverse curves or offset bends in the portion of the cable in the manhole. These reverse curves enable the cable to take up the expansion and contraction movements without cracking or buckling of the insulation or lead sheath. Therefore, the training of cables on the manhole wall is not only for making a neat arrangement in the manhole but more importantly for providing the expansion space needed to absorb the movements of the cable.

In general, the bending radius R (see Fig. 33-1) of the reverse curves in lead-covered cables must be no less than eight times the overall diameter of the cable. Thus, a 3-in cable should have a minimum bending radius of 24 in or more, and a 4-in cable should have a minimum bending radius of 32 in or more.

The range of values of R in terms of cable diameter varies from 5 to 12 and depends on size and number of conductors and whether it is a polyethylene, thermoplastic, rubber-, varnished-cambric-, or paper-insulated cable, as well as on the rated voltage of the cable.

At least 6 in of straight cable should extend beyond each end of the splice to provide space for resting on the saddles of the supporting racks. The clips holding the saddles on the bracket should be spread enough to permit the saddle to slide freely along the bracket.

All cables and joints should be so racked in the manhole that they are not directly under the manhole cover.

The exact makeup of a splice depends on the specific cable construction, that is, whether the cable is a single- or multiple-conductor cable; whether the insulation is rubber, polyethylene, thermoplastic, varnished cambric, or impregnated paper; or whether the cable insulation has a conducting shield; etc. The general installation procedure is similar to those described in this section. The cable manufacturer's specific instructions must be followed in all instances. The worker must maintain tools in good condition and keep them clean and dry. The working area and the area of the cable being worked on should be protected from moisture with a rubber blanket or other waterproof material. The cable splicer or lineman performing the work must maintain clean and dry hands.

Splices vary greatly with type and voltage of cable, so that drawings showing the essential dimensions for each type and voltage of cable splice are required. Figure 33-2 shows such a drawing. These drawings will give for each particular joint the length of lead sheath to be stripped back from the cable end, the amount of conductor insulation to be cut back so as to give the proper stepping or penciling, the diameter to which the individual conductor insulation should be built up, and the correct completed outside diameter of the insulation. The cable splicer must have the drawings and materials for the particular joint to be made on hand before starting.

The cable ends to be spliced are trained into position with ends overlapped. The bending radius is made as large as, or larger than, required. Paper- and varnished-cambric-insulated cables should not be bent when the temperature is below 14°F. In cold weather, the cables must be heated so that they are warm all the way through before they are subjected to bending.

Cables are then marked at the centerline of the splice. Cables are cut off squarely at this mark with a hacksaw or cable cutter. The new cable ends should now butt together squarely. Cables should not be retrained after the ends are cut off because the conductors and conductor strands will slide to unequal lengths.

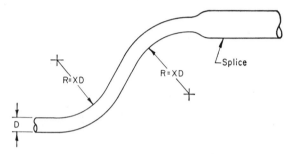

Fig. 33-1   Sketch showing minimum bending radius on reverse curves of cable as it is trained around wall in manhole. D is cable diameter; X is number of times D; R is bending radius.

Cable Jacket and Sheath Removal   Cables normally have a jacket or sheath or both. The most common cable jackets are metallic, rubber, or thermoplastic. It is necessary to remove the jacket the prescribed distance. The sheath of the cable should be cleaned to keep the shavings and contamination from coming in contact with the exposed cable insulation before the sheath is removed.

Lead cable sheaths can be cleaned by scraping with a shave hook or rasp. The entire surface must appear shiny and be free of all oil, cable-pulling compounds, and contamination.

The lead sleeve used to cover and protect the splice is cleaned inside and slid over one of the cable ends far enough to be out of the way while the splice is being made. Before this is done, each end of the sleeve should be thoroughly scraped with a rasp or steel brush for about 3 in. The scraped portion should then be coated with stearine flux to retard oxidation.

With a chipping knife mark the cable sheath where the ring cut is to be made according to the drawing. The length of sheath to remove from each cable end is usually 1½ in less than one-half the total length of the lead sleeve. Score the sheath at the ring cut by tapping the chipping knife lightly with a hammer around the sheath at the position marked (see Fig. 33-3). Cut about halfway through the lead. Slit the lead sheath lengthwise from the end of the cable to the score with the chipping knife held on a slant so that it will pass between the insulation or metallic braid, if shielded, and the sheath as the knife progresses (see Fig. 33-4). To remove the sheath, loosen the edge along the longitudinal cut from the insulation with the hammer (see Fig. 33-5). Then grip the edge of the sheath with pliers and tear off at the score with a twisting motion, thereby forming a slight bell in the end of the lead sheath. Great care should be used in scoring and removing the lead sheath to prevent cutting into or otherwise damaging the cable insulation. The new ends of the lead sheath are belled out approximately ¼ in with a special fiber or hardwood tool. The tool is driven under the sheath parallel to the cable to raise the end of the sheath. This will provide space for a binder tape to be carried into the bell for at least ¼ in.

Shielded cable ends are not belled, since the shielding tape is extended to the stress cone.

Rubber or thermoplastic jackets or sheaths must be cleaned of all contamination such as wax,

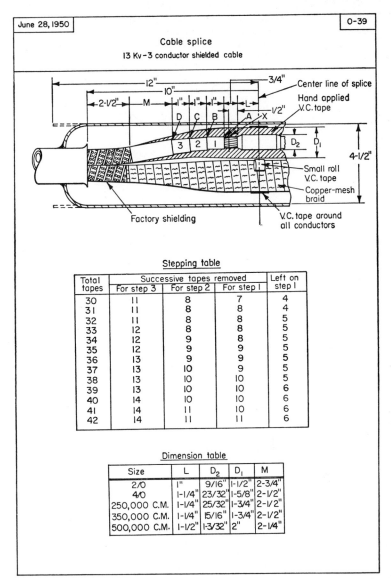

Cable splice

13 Kv – 3 conductor shielded cable

Stepping table

Total tapes	Successive tapes removed			Left on step 1
	For step 3	For step 2	For step 1	
30	11	8	7	4
31	11	8	8	4
32	11	8	8	5
33	12	8	8	5
34	12	9	8	5
35	12	9	9	5
36	13	9	9	5
37	13	10	9	5
38	13	10	10	5
39	13	10	10	6
40	14	10	10	6
41	14	11	10	6
42	14	11	11	6

Dimension table

Size	L	D_2	D_1	M
2/0	1"	9/16"	1-1/2"	2-3/4"
4/0	1-1/4"	23/32"	1-5/8"	2-1/2"
250,000 C.M.	1-1/4"	25/32"	1-3/4"	2-1/2"
350,000 C.M.	1-1/4"	15/16"	1-3/4"	2-1/2"
500,000 C.M.	1-1/2"	1-3/32"	2"	2-1/4"

Fig. 33-2 Drawing of 13-kV three-conductor shielded-cable splice. Accompanying table gives all essential dimensions for this particular splice. *(Courtesy Public Service Electric and Gas Co. of New Jersey.)*

cable pulling compound, and dirt in the area where the jacket is to be removed. This type of jacket can be cleaned by scraping with a knife, nonconductive abrasive cloth, or rasp. The jacket must be buffed and cleaned so that the tape or resin will bond and form a moisture tight seal. The cleaned area can be temporarily protected by covering it with a layer of vinyl tape. This tape picks up any loose particles from the surface when the inner portion of the splice is completed. The jacket can be removed by making a circular score approximately halfway through its thickness and then making a longitudinal cut from the end of the cable back to the score. The cable metallic shielding or insulation should never be cut during this operation. All bedding tapes should be removed back to the jacket.

Cable Metallic Shielding Removal   The metallic shielding should be terminated as evenly and smoothly as possible. A glass cloth tape can be wrapped around the shielding at the required extension beyond the cable sheath. This area can then be turned and held in place with solder. A ground strap or braid may be attached during the soldering process. It is important to be careful

Fig. 33-3   Making ring cut with chipping knife.

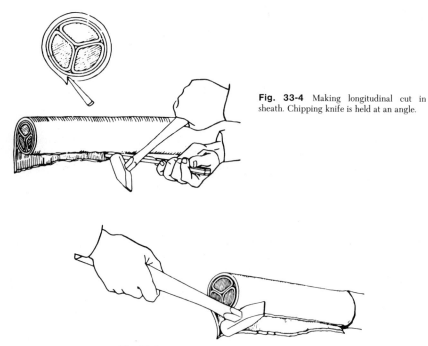

Fig. 33-4 Making longitudinal cut in sheath. Chipping knife is held at an angle.

Fig. 33-5   Loosening cable sheath from insulation.

not to overheat the cable insulation during the soldering process. Acid core solder or acid flux cannot be used. A special aluminum flux must be used when soldering aluminum shielding.

Metallic shielding can be removed by lightly holding a sharp knife against it at the point of removal and then pulling the shielding away from the cable against the blade of the knife. The shielding tapes should be rolled back at the terminating edge to eliminate sharp points.

Cable Semiconducting Material Removal All traces of semiconducting material must be removed from the exposed cable insulation back to within approximately ¼ in of the metallic shielding. The extension of the semiconducting material must not be overlapped when the insulation is built up in the cable splice area. The semiconducting material residue can be removed by scraping with a knife, a nonconductive abrasive cloth, or a rasp. The method of cleaning depends on the type of cable insulation.

Rubber insulations are best cleaned by first scraping with a knife or a rasp and then buffing with a nonconducting abrasive cloth. The semiconducting material normally strips rather cleanly from polyethylene or cross-linked polyethylene cables; therefore, they may only need to be buffed with an abrasive cloth.

The use of solvents for cleaning the cable insulation is a rather controversial subject. Improper solvents or poor workmanship may leave a conductive residue on the insulation, or the solvent may run under the shielding and semiconducting layer. Hidden solvents may cause the shielding tapes to crack or deteriorate. If solvents are to be used, the cable manufacturer's instructions must be followed carefully. Paper- or varnish-cambric-tape-insulated cables may have a black conducting or metallized paper tape under the shielding. These tapes must be removed to within approximately ¼ in of the edge of the insulation shielding.

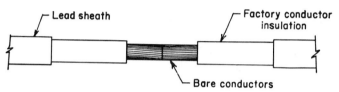

Fig. 33-6 Cable ends showing bare conductors after factory insulation has been removed. Note squarely fitting butt joint of conductors.

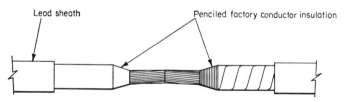

Fig. 33-7 Cable splice showing factory conductor insulation after being penciled.

Cable Insulation Removal The conductor must not be nicked or cut while removing the cable insulation. Cotton-type ties are made around the conductor insulation back of the trimming marks as shown on the manufacturer's drawing. The factory insulation is then removed on each conductor from the cable end for a length equal to one-half the connector length plus ½ in. Some drawings specify a square-end insulation trim, and others a sloping trim. The two cable ends will now appear as in Fig. 33-6.

To avoid a direct radial creepage path from the conductor outward, the ends of the factory insulation adjacent to the connector are "penciled" or "stepped."

On rubber, polyethylene, and varnished-cambric cables the factory insulation is usually penciled as shown in Fig. 33-7. Penciling insulation is similar to sharpening a pencil. Cuts are made at an angle, leaving the insulation tapered. A sharp knife should be used to avoid ragged edges. Keep the diagonal cut in a round and uniform cone shape (Fig. 33-8). After the pencil is formed, smooth with flint or nonconducting abrasive cloth. A mechanical penciling tool may be used to remove and taper the insulation of rubber, polyethylene, and cross-linked polyethylene insulation. It is very important to obtain a smooth and uniform pencil.

On paper-insulated cables "step" the insulation as shown in Fig. 33-9. Loop a steel piano wire with weights on the ends around the insulated conductor and tear off the layers of paper tape.

Start with the step farthest from connector. Tear tapes by count, progressing toward the connector as the required number of tapes are removed for each step. Temporarily bind the edges of paper tape in each step with saturated flax twine. The exposed insulation is covered with a muslin wrapping for protection.

Cable Paper-Insulation Tests   Moisture tests should be made on one or more fillers, several layers of paper insulation from each conductor, and several layers of belt insulation, from all cables rated above 5 kV. The test samples should be placed in a pail of paraffin which is maintained at a temperature of 300°F. If frothing or bubbling occurs at the surface of the paraffin within a few seconds, the insulation is wet and the condition should be reported before proceeding any further.

Cable Splice Connector   The proper choice of a connector is important to completing a reliable splice. A smooth tubular connector, such as a crimp or compression sleeve connector, or

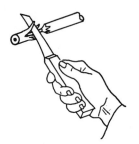

Fig. 33-8   Penciling conductor insulation.

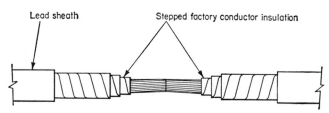

Fig. 33-9   Cable splice showing factory conductor insulation after being stepped.

Fig. 33-10   Typical split tinned copper cable connector. *(Courtesy Reliable Electric Co.)*

a split-soldered connector is used for most splices and are necessary for cables operating at voltages higher than 5 kV. Compression-type connectors must be used on thermoplastic-insulated cables to prevent damage to the insulation by heat. The conductor must be cleaned prior to applying the connector.

Apply stearine flux to the exposed surface of the strands if a solder-type connector is to be used. Ordinarily, solder flux in paste form is applied with a brush. If flux in stock form is used, melt it by holding it against a hot ladle and permit flux to drip over surface of conductor. Then open the slot of the connector (Fig. 33-10) sufficiently to pass over the conductors and slip the connector over the strands of one conductor (see Fig. 33-11). Then insert the end of the other conductor into the connector until it butts the end of the first conductor squarely. Work the connector until it is centered over the ends of both conductors with the slot on top. Tightly compress the connector around the conductor with pliers.

To make the soldered connection use two solder ladles. With one ladle containing hot solder held above the connector and the other dry ladle held underneath to catch the overflow, pour solder over the ends of connector and into the slot until the strands and connector are heated enough to make the solder flow freely through the cable strands and out of the ends of the connector (see Fig. 33-12). Return excess solder to the solder pot. Now hold the stock of soldering flux against the heated connector, and allow the flux to melt and run into the slot and down through the ends of the conductor. Continue to alternate pouring solder and applying flux until the connector is tinned to a bright, shiny cast. Allow the solder to cool in the ladle, and pour until the

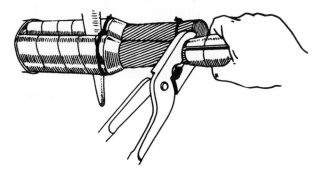

Fig. 33-11 Slipping split connector over conductor strands. Strands are held in the round with string and special eagle-claw pliers.

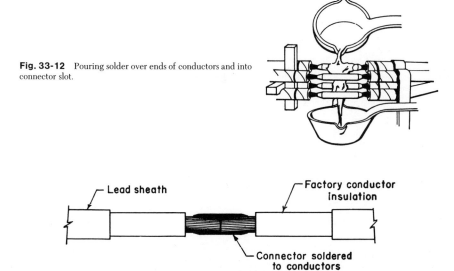

Fig. 33-12 Pouring solder over ends of conductors and into connector slot.

Fig. 33-13 Cable splice showing connector soldered to conductors.

solder becomes plastic and fills the slot. While the connector is still hot, smooth off burrs with dry cotton tape. While the solder is cooling, polish the connector with a loop of 1-in cotton tape drawn rapidly back and forth over it. The splice will now appear as in Fig. 33-13.

The compression connector is applied with the proper tool as in the manner specified by the manufacturer. The number of compressions or crimps varies with the type of tool used and the size of the conductor. Hydraulic-compression tools are used on the larger size conductors.

A semiconducting tape is normally applied over the compression connector and the exposed conductor. The semiconducting tape eliminates problems that might develop from the irregularities as a result of the indentations on the compression connector.

Installing Taped Cable Insulation The temporary protective cloth wrappings are removed. Then wrap rubber tape on rubber-insulated cables and varnished-cambric tape on varnished-cambric- or paper-insulated cables. Unless otherwise specified, the tape shall be applied with a half lap. One layer half-lapped is then equal to two thicknesses.

First fill in the space at both ends between the connector and the trimmed insulation. Then insulate over the connector and the pencils or steps to the diameter of the factory-applied insulation. The splice will now appear as shown in Fig. 33-14. Now continue taping over the entire exposed cable. The first layer of tape applied over the factory insulation should be wrapped in the same direction as the factory insulation and should extend back into the crotch as far as possible. In the case of high-voltage paper cable, one outer layer of factory-applied tape insulation is removed before the hand-applied taping is started. All varnished-cloth tape should be drawn up as tightly as possible without breaking the varnish over the fabric. All tape must be applied smoothly and carefully to expel all air and avoid wrinkles.

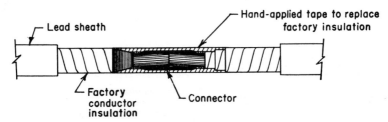

Fig. 33-14   Cable splice after factory conductor insulation has been replaced with hand-applied tape.

Fig. 33-15   Applying ½-in tape half-lapped over conductor strands. Note connector and penciled insulation on lower conductor. Completely wrapped middle conductor and wrapping in progress on upper conductor.

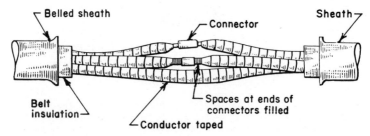

Fig. 33-16   Showing several stages in the splicing of a three-conductor cable. Upper conductor shows connector in place, middle conductor has spaces filled between connector and penciled conductor insulation, and lower conductor is completely taped.

When rubber tape or rubberlike tape is applied, the surface should be cleaned with the specified solvent and coated with rubber cement. Rubber tape should be applied with uniform and sufficient tension to reduce its original width approximately one-third. Rubber tape should be protected with friction tape and asphaltum paint or with a layer of weather-resistant tape as specified.

The joint specifications usually provide for approximately a 75 percent greater thickness of hand-applied insulation than is found in the factory insulation.

Figure 33-15 shows the manner of wrapping the tape. Figure 33-16 shows the three conductors of a three-conductor cable in various stages of the taping operation. The lower conductor is completely taped, the middle conductor has the space at the ends of the connector filled, and the upper conductor is without any insulation replaced.

At the completion of the taping operation, lock the last turn in place with a tape tie. This is made by passing the end of the tape under the preceding turn and pulling the tape tight (see Fig. 33-17). Cut off excess tape. A completely taped splice is shown in Fig. 33-18.

Prefabricated Cable Splice Installation Prefabricated cable splices are normally used on underground residential distribution cable installations with rubber, ethylene propylene rubber, polyethylene, or cross-linked polyethylene insulation (Fig. 33-19). The splices may be direct-buried or placed in enclosures to permit access. The prefabricated splice consists of a metallic connector, a connector shield, insulation, and an insulation shielding jacket.

Fig. 33-17 Wrapping ended in tape tie.

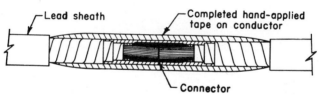

Fig. 33-18 Cable splice after completing the taping of the conductor. On rubber-insulated cables this should be equal to 1½ times factory insulation, and on varnished-cambric- or paper-insulated cables, it should be equal to 2 times factory insulation.

Fig. 33-19 Prefabricated splice installed on concentric neutral cross-linked polyethylene insulated URD cable. The cable and splice will be buried in the ground. *(Courtesy AMP, Inc.)*

The cables are prepared in the manner previously described for other splicing methods. Apply one half-lapped layer of vinyl tape over semiconductive jacket of the cables. Apply silicone grease over cable insulation and vinyl tape. Slide splice end caps onto each cable until against concentric wires. Slide splice body onto one of the cables. Place cable conductors into connector and crimp the connector with the proper tool in the manner specified by the manufacturer (Fig. 33-20). The

Fig. 33-20 Prefabricated splice on solid dielectric insulated, concentric neutral cable. Conductor connector has been crimped. *(Courtesy 3 M Co.)*

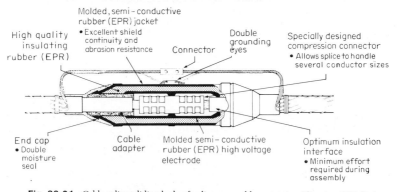

Fig. 33-21 Cable splicer sliding body of splice over cable connector. *(Courtesy 3 M Co.)*

splice body is slid into final position over cable connector leaving a small area of insulation exposed at both ends of the splice body (Fig. 33-21). The cable-splicer or lineman removes the vinyl tape applied to the semiconductive jacket prior to completing the splice assembly. The end caps are slid onto the splice body completing the semiconductive shielding jacket over the cable and splice insulation. The concentric neutral wires are extended over the splice and connected together with a compression-type connector. The concentric wires are connected to the grounding eyes on the splice body to ensure the splice jacket is grounded (Fig. 33-22). The insulation shield on a prefabricated splice normally consisting of a conductive jacket retains the electric field completely within the splice cable insulation. The electric field is properly distributed to eliminate any highly concentrated electric stress areas (Figs. 33-23 and 33-24).

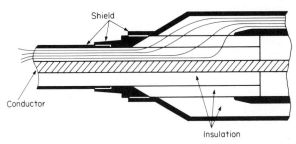

Fig. 33-22 Completed installation of prefabricated solid dielectric insulated concentric neutral cable splice. *(Courtesy 3 M Co.)*

Fig. 33-23 Electrical stress distribution for high-voltage solid-dielectric-insulated shielded cable with shield and insulation partially removed from cable conductor. Note high concentration of electrical stress at the edge of the cable shield.

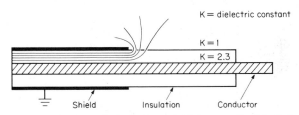

Fig. 33-24 Schematic of electrical stress distribution in a completed prefabricated splice installed on a solid-dielectric-insulated shield high-voltage cable. Note electric stress in transition area from cable to splice is evenly distributed.

Applying Cable Shielding Tape   Single-conductor and multiple-conductor cables rated at 13 kV and above have shielding tapes wrapped around the outside of the insulation of each conductor. The multiple-conductor cables have an additional shielding tape wrapped around the outside of the three conductors directly under the lead sheath, as shown in Fig. 33-25. Splices made on these cables must have the shielding continued over the applied insulation of each conductor.

This shielding is made with a butt lap wrapping of ¾-in copper tinsel braid applied from end to end where it is fastened to the factory shielding or to the lead sheath with a soldering iron. Adjacent turns are tacked together with solder, as necessary, to prevent the braid from sliding on the slopes of the applied insulation. Shielding braid tape may not be used in connection with splicing one-conductor 13-kV cables. These cables are shielded with a metallized paper tape,

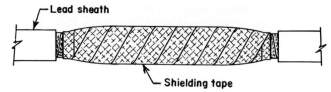

Fig. 33-25   Cable splice with shielding braid tape covering the insulated conductor or conductors.

which is torn off at the end of the lead sheath trim. In this case, the sheath end is belled out with a special belling tool. The ends of the lead sheath trims on other shielded cables are not belled at all, other than that resulting from the removing of the sheath.

Applying Cable Spacer and Binding Tapes   In three-conductor splices about five or six layers of dry varnished-cambric tape are wrapped around each conductor over the outside of the shielding tape and near the center of the joint or at two locations as shown on some drawings. This tape is installed to serve as a separator between conductors which permits the compound to flow freely between conductors. Then from three to six layers of dry varnished-cambric tape are wrapped around all three conductors and directly over the spacer tape in order to hold the conductors tightly together.

Installing Cable Lead Sleeve   After all the taping of the splice is completed, the lead sleeve which was slipped on the cable previously is now centered over the splice. As mentioned before, the sleeve serves to protect the splice from mechanical injury; seals the joint, thereby keeping out moisture; and furnishes a continuous path in the cable sheath for fault currents.

Wiping the Cable Sleeve   With the sleeve centered over the splice, mark the cable at the ends of the sleeve. Coat the scraped surface with stearine flux. Beat down the ends of the sleeve with a hardwood tool called a dresser until the sleeve fits snugly around the cable sheath with the sleeve centered over the tapped area.

Wipe the sleeve ends to the cable sheath using the stick and torch method. Melt small amounts of solder from the end of the stick, and apply over the scraped area (Fig. 33-26). Heat the solder with a torch, and scrub the wiping surface with a solder stick until the surfaces of the cable sheath and the sleeve are well tinned. Apply additional flux and solder, and work with a wiping cloth until a smooth solder joint is formed between the sleeve and sheath (Fig. 33-27).

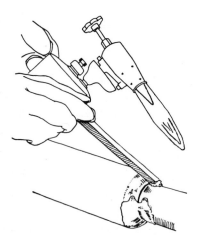

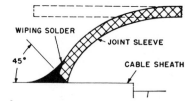

Fig. 33-26   Wiping lead sleeve onto lead sheath using stick and torch method.

Fig. 33-27   Completed wipe showing smoothly tapered shoulder of solder joining sleeve to cable sheath.

The wiped sleeve joint should be tested with gas or oil pressure to assure that the wipes are not porous and do not leak air. A leaky joint would admit moisture which would eventually cause the joint to fail. The pressure is built up to about 10 lb. Apply soapy water to the wipes. If bubbles form, the wipes leak and should be repaired by additional torching and rewiping until tight.

Filling Cable Sleeve with Compound   If the sleeve is not provided with fittings for filling as shown in Fig. 33-28, make two V-shaped cuts in the top of the sleeve, one near each end. Lift up the V flap formed. One opening is for pouring in the hot compound, and the other is for venting the air in the sleeve. Before the hot compound is poured in, the joint should be tilted slightly, that is, raised about 1 in at the vent end. Place the funnel into the filling hole at the lower end of the sleeve (Fig. 33-29). When the compound is poured in, the air in the sleeve is driven upward and out through the venting hole.

The compound usually consists of 70 percent Vinsol and 30 percent castor oil. It is first heated to about 250 to 300°F. It is then slowly and steadily poured into the funnel. Allow the compound to flow through the venting hole until it is free from bubbles. Then allow the joint to cool to about

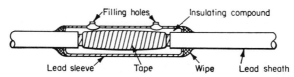

Fig. 33-28   Cross section of sleeve joint showing sheath, sleeve, tape, wipe, filling holes, and insulating compound.

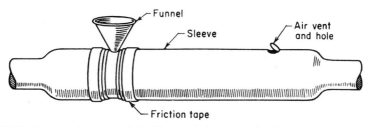

Fig. 33-29   Funnel in place in lower hole of sleeve ready for pouring of insulating compound. Air will escape through overflow hole. Funnel can be scaled with friction tape.

150°F. Now level the joint and refill the sleeve every 15 min. When the joint has cooled to 100°F, replace the fittings or seal the holes with solder. Wipes and seals are then painted with Victolac to prevent corrosion.

When the joint cools to the temperature in the manhole, voids will form inside the sleeve. However, when the cable carries electrical power, it will become heated again and the voids will become filled. If the voids were not present, the joint would burst when it became heated by the cable load.

Place Cable Splice on Rack   The cable with its completed joint (see Fig. 33-30) is now moved into its final position onto porcelain saddles on the rack. Lead sheaths are subject to electrolysis and therefore must be supported on porcelain or other insulating blocks. Some adjustment of the rack may be necessary.

Cable Bonding and Grounding   The purpose of bonding and grounding the cable sheaths is to maintain them at or near ground potential. A No. 2 AWG copper wire is generally used. It must be attached to the sheaths with a special bond clip which is soldered to the wire and sheath and connected to a low-resistance ground.

Bonding and grounding reduce the likelihood of arcing between the sheath of a faulted cable and other nearby sheaths. It thus reduces the danger to cablemen who may be in a manhole when a cable fault occurs. It also minimizes the harmful effects of corrosive action due to stray currents.

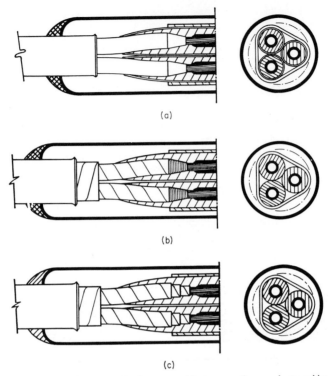

(a)

(b)

(c)

Fig. 33-30 Sectional views of completed splices on straight two-way three-conductor cables. (*a*) Rubber insulated; (*b*) varnished-cambric insulated; (*c*) impregnated-paper insulated. *(Courtesy G. & W. Electric Specialty Co.)*

Fig. 33-31 Cable joints protected with fireproofing tape to prevent spreading of damage from faulted cable to adjacent cables. *(Courtesy Public Service Electric and Gas Co. of New Jersey.)*

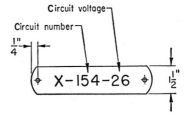

Fig. 33-32 Typical cable tag showing circuit number on left and voltage on right. *(Courtesy Public Service Electric and Gas Co. of New Jersey.)*

Fireproofing Cable Joint   Fireproofing tape is applied after bonding to limit the damage that could result from a failure on other cables in the manhole. Fireproofing is necessary only where there is the probability of damage to one or more cables due to failure of one of them. A group of fireproofed cables is shown in Fig. 33-31.

Tagging Cable   Every cable should bear an identifying mark or tag in each manhole through which it passes. This is usually in the form of a metal tag tied to the joint. The tag should indicate the size, voltage, origin, feeder designation, and any other pertinent information. A typical cable tag is shown in Fig. 33-32.

Section 34

Underground Residential Distribution

Electric distribution circuits have been installed underground in large cities for many years to serve the central business areas. The electric load density in the central business areas is high enough to justify the expenses associated with the conventional underground system employing the use of conduits encased in concrete; manholes; vaults with submersible transformers, network protectors, and paper-insulated, lead-covered, or polyethylene-insulated, shielded, and jacketed high-voltage cables.

The conventional underground system cannot be economically justified to serve the low electric load densities found in residential areas. Direct-buried cable systems with pad-mounted transformers, submersible transformers in fiber vaults, or direct-buried transformers can be installed in residential areas at costs that are economically feasible. Table 34-1 tabulates the average ratios of underground to overhead costs for installations in new subdivisions. This table contains information from the *1970 National Power Survey* prepared by the Federal Power Commission Technical Advisory Committee.

The converting of overhead electric distribution circuits to underground in residential areas is generally not economically feasible. Most of the conversion programs executed have been associated with renewal projects where it was necessary to remove the overhead electric facilities as a part of the demolition work. Table 34-2 provides the cost ratios for converting overhead electric circuits to underground operation as determined by the *1970 National Power Survey*. Underground lines are relatively immune to some of the major causes of failures in overhead circuits, such as power-pole accidents; damage from lightning, wind, ice, and snowstorms; or contacts to the wires by trees or other foreign objects.

Underground circuits have their own problems such as entrance of moisture, corrosion, cable dig-ins, insulation failures as a result of switching surges or corona, and damage during installation. Underground equipment must be designed for long life in below-ground enclosures that may be filled with water containing contaminates. Underground cables and equipment are both vulnerable to the entrance of moisture as a result of flaws in splices, terminations, gaskets, and connecting devices. Chemical corrosion or electrolysis can damage any of the exposed metal installed as a part of the underground distribution system.

Power Source In most cases the underground residential distribution circuits will be supplied from an overhead main feeder circuit installed adjacent to the subdivision. The loads in the residential area may be served by several underground cable laterals connected to the open wire overhead system at several cable riser poles. A typical cable riser pole showing cable, cable guard,

Table 34-1   Average Ratios of Underground to Overhead Costs for Extensions

Type of line	1967	Probable range, 1980–1990 Maximum	Minimum
URD-type lines°	1.8	1.5	1.3
Other types of lines†	5.0	4.0	3.3
All lines—			
weighted average	2.9	2.7	2.3

°URD (underground residential distribution) type lines are branch lines serving new residential subdivisions, predominantly single phase and with relatively small primary conductors.

†Other types of lines include distribution in new commercial and industrial developments, rural areas, and main primary circuits of large conductors through residential as well as other areas.

Table 34-2   Average Ratios of New Underground Cost to Original Overhead Cost for Distribution Line Conversions

Type of line	1967	Probable range, 1980–1990 Maximum	Minimum
URD-type lines	7.0	5.6	4.7
Other types of lines	10.0	8.0	6.7
Weighted averages			
Selective programs	9.7	7.8	6.5
General programs	9.7	6.8	5.7

pothead, lightning arrester, and fused disconnect is shown in Fig. 34-1. Figure 34-2 illustrates the essential components required to connect the underground cable to an overhead distribution circuit.

Total underground distribution circuits have been constructed in some locations. A total underground distribution circuit is shown schematically in Fig. 34-3. The three-phase feeder-main circuit extends from substation A to substation B under ground. The main feeder circuit has a capacity of 10 MVA (megavolt-amperes). The underground cables consist of 750 mCM (thousand

Fig. 34-1   Single-phase underground cable riser installed on pole with three-phase primary and street lights. The cable is connected to the overhead circuit through a fused cutout and protected by a lighting arrester. *(Courtesy S. & C. Electric Co.)*

circular mil) aluminum conductor with cross-linked polyethylene insulation rated 15 kV, shielding consisting of semiconducting tapes and copper tapes and a polyethylene jacket. Taps to the main circuit to serve residential areas or commercial customers are made in switching modules installed in pad-mounted switchgear or a submersible container.

Most of the underground residential distribution circuits tapping the main feeder are single-phase loop feeds with fuses to protect the main feeder circuit. Extending the three-phase main feeder circuit between two substations makes it possible to isolate a faulted cable section and

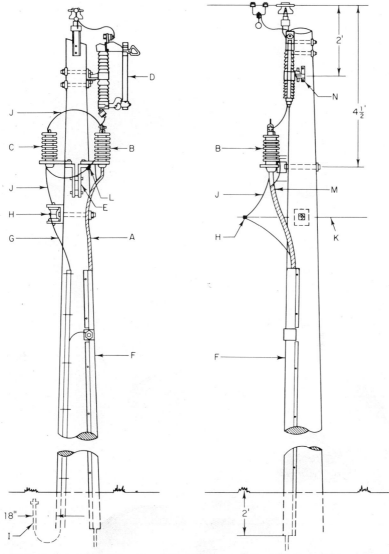

Fig. 34-2 Typical underground residential distribution riser pole for 7620-volt single-phase primary. (A) Cable, No. 2 AWG conductor, cross-linked polyethylene insulation, semiconducting polyethylene shield, 10 No. 14 AWG solid tinned, copper, concentric neutral; (B) pothead; (C) 10-kV lightning arrester, (D) 15-kV open type, fused cutout; (E) pothead and arrester bracket; (F) metal U-guard; (G) No. 6 AWG bare copper ground wire; (H) bimetal parallel grove clamp; (I) ground rod and clamp; (J) No. 2 AWG bare copper wire; (K) overhead circuit, common neutral; (L) compression connector; (M) concentric neutral; (N) bracket.

restore service to customers prior to completing the cable repairs. The loop-type feed serving the residential area makes it possible to isolate faulted cable circuits in order to restore service to the customers in the subdivision.

Types of Systems   Direct-buried underground distribution systems can be installed in several different ways. A common method of installation places the primary and secondary cables along rear lot lines and utilizes pad-mounted transformers. The pad-mounted transformers would normally serve from four to twelve customers depending upon the terrain, the customer loads, and the size of transformers installed. This type of installation is illustrated in Fig. 34-3.

Many times it is desirable to install the cables along the streets in front of the homes. This type of installation is commonly referred to as front lot line. The front-lot-line type of installation can

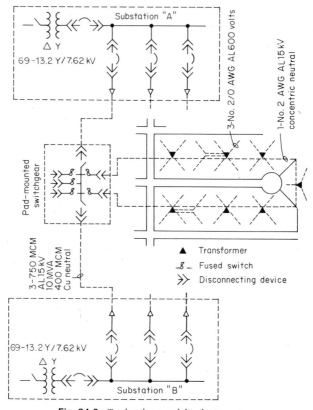

Fig. 34-3   Total underground distribution system.

use pad-mounted transformers, submersible transformers, direct-buried transformers, or unit residential transformers. (See Fig. 34-4.) The installation of pad-mounted transformers along the streets may be objectionable from an environmental point of view.

Submersible transformers are usually installed in fiber material vaults. The submersible transformer must be protected from corrosion. Maintenance of submersible transformers and cost of installation is usually greater than the costs associated with pad-mounted transformers. Direct-buried transformers have not been used extensively to date. It is necessary to provide corrosion protection for this type of installation.

The unit residential transformer is usually installed adjacent to a house and back from the street, where it is inconspicuous. The unit residential transformer usually consists of a dry-type core and coils installed in a metal housing for mounting on a concrete pad. Each house is served by an individual transformer. The cost of installing this type of equipment can be justified if the lots are large and the terrain is such that rear-lot-line construction is not practical. Different types of high-voltage switching arrangements have been installed to vary the flexibility and the cost associated with each different type of underground residential distribution installation.

Cables   Cable should be capable of withstanding tests applied in accordance with an applicable standard issued by a recognized organization such as the Association of Edison Illuminating Companies, the Insulated Power Cable Engineers Association, the National Electrical Manufacturers Association, or the American Society for Testing and Materials. The design and construction of conductors, insulation, sheath, jacket, and shielding shall include consideration of mechanical, thermal, environmental, and electrical stresses which are expected during installation and operation. Cable shall be designed and manufactured to retain specified dimensions and structural integrity during manufacture, reeling, storage, handling, and installation.

Cable shall be designed and constructed in such a manner that each component is protected from harmful effects of other components. The conductor, insulation, and shielding shall be

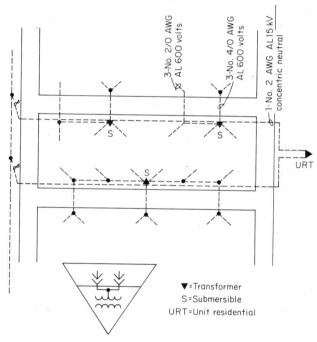

Fig. 34-4   Underground residential distribution front lot line with submersible and unit residential transformers.

designed to withstand the effects of the expected magnitude and duration of fault current except in the immediate vicinity of the fault. Sheaths and/or jackets shall be provided when necessary to protect the insulation or shielding from moisture or other adverse environmental conditions. Conductor shielding should be provided in accordance with an applicable standard issued by a recognized organization such as the Association of Edison Illuminating Companies, the Insulated Power Cable Engineers Association, and the National Electrical Manufacturers Association.

Insulation shielding shall be provided for cable operating at more than 5 kV to ground and is recommended for cables operating above 2 kV to ground. Shielding is not required for short jumpers which do not contact a grounded surface within enclosures or vaults provided they are guarded or isolated. Insulation shielding may be sectionalized provided that each section is effectively grounded. The shielding system may consist of semiconducting materials, nonmagnetic metal, or both. The shielding adjacent to the insulation shall be designed to remain in intimate contact with the insulation under all operating conditions. Shielding material shall either be designed to resist excessive corrosion under the expected operating conditions or shall be protected.

Voltages   The primary voltages used for underground residential circuits varies from 2400 to 34,500 volts. One of the most common primary voltages is 13,200Y/7620 volts. Other voltages used extensively are 4160Y/2400, 13,800Y/7960, 24,900Y/14,400, and 34,500Y/19,920 volts.

Utilization voltages for residential installations are normally 120/240 volts, single-phase, three-wire, grounded neutral system. Utilization voltages for commercial installations may be 120/240 volts single-phase three-wire grounded neutral system; 208Y/120 volts three-phase four-wire grounded neutral system; 240 volts three-phase three-wire ungrounded delta system; 480Y/277 volts three-phase four-wire grounded neutral system or 480 volts three-phase three-wire ungrounded delta system. In some instances a commercial or an industrial customer may receive service at the primary voltage which may be the same voltage used to supply power to a residential area. Large industrial customers may receive electric service at voltages commonly used for subtransmission or transmission circuits such as 69,000 to 345,000 volts three-phase.

Primary Cable   The high-voltage cables for a total underground system main feeder circuit usually consist of an aluminum or copper conductor with semiconducting tapes or an extruded conducting polyethylene conductor shielding; insulation consisting of cross-linked polyethylene, high-molecular-weight polyethylene, or ethylene propylene rubber; insulation shielding consisting of a semiconducting compound and copper shielding tapes or an extruded conducting polyethylene shield with copper shielding tapes and jacketed with an extruded polyethylene. (See Fig. 34-5.) A tinned, bare copper neutral conductor is usually installed with the high-voltage cables.

The main feeder cables are usually installed direct-buried at a depth greater than required by the National Electrical Safety Code to provide protection from dig-ins. The main feeder circuit

Fig. 34-5   Typical single-conductor high-voltage cable used for underground main feeder circuit. *(Courtesy Rome Cable Co.)*

Fig. 34-6   Typical two-conductor concentric cable used for underground primary. Conductor is insulated with polyethylene or synthetic rubber and surrounded by spiral-wound exposed tinned copper neutral ground wire. *(Courtesy Rome Cable Co.)*

cables may be installed in polyethylene plastic pipe. If fiber conduits encased in concrete are used for the primary feeder circuits, paper-insulated lead-covered cables with an extruded polyethylene jacket are often used. The primary cables supplying residential areas or commercial customers from the main feeder circuits, either overhead or underground, are usually the two-conductor concentric neutral type.

The high-voltage concentric neutral cable usually consists of aluminum or copper conductor with an extruded conducting polyethylene conductor shielding; insulation with cross-linked polyethylene, high-molecular-weight polyethylene or ethylene propylene rubber; insulation shielding consisting of a semiconducting compound and copper shielding tapes covered by conducting polyethylene jacket or an extruded conducting polyethylene shielding jacket surrounded by spiral-wound tinned-copper neutral conductors. The typical primary URD (underground residential distribution) cable is illustrated in Fig. 34-6.

Low-Voltage Cables   The secondary and service cables for underground electric distribution are generally larger than those applied on overhead systems because of the reduced capacity for a given conductor size as a result of the temperature limitations of the cable insulation, greater load density, and the need to provide for load growth. Aluminum conductors are generally used for underground residential distribution because of the lower cost compared to copper. The secondary cables often have a conductor of No. 4/0 AWG or 350 kcmil aluminum. Service cables often have a conductor of No. 2/0 AWG aluminum to provide 200-ampere capacity where National Electric Code ratings are not applicable. The underground residential distribution services will usually vary in size from No. 1/0 AWG to No. 4/0 AWG aluminum conductor. A nominal three-wire 120/240 volts single-phase grounded neutral system is normally installed to serve homes in areas supplied by underground residential distribution circuits.

A twin-concentric neutral cable is constructed with two 600-volt cross-linked insulated aluminum conductors in parallel configuration with an overall spirally applied concentric neutral of No. 14, 12, or 10 AWG tinned copper wires in sufficient number to provide the conductivity required. This type of cable construction is used where telephone and power conductors are installed random lay in a single trench.

Triplexed cables may be used for secondaries and services. This cable consturction consists of three single conductor aluminum cables with 600-volt cross-linked polyethylene insulation twisted together. Ths insulated aluminum neutral cable may be color coded, providing ease of identification and distinguishability from other cables (Fig. 34-7).

Secondary and service cables constructed in a flat parallel configuration are used in many installations. This cable assembly consists of two aluminum phase conductors and a neutral conductor,

Fig. 34-7 Reel of low-voltage insulated-triplexed cables mounted on back of truck to facilitate installation of underground secondaries and services. Cable is manufactured for direct-burial installation. Light-colored insulation identifies neutral conductor.

Fig. 34-8 Core and coils assembly for installation in transformer. *(Courtesy Westinghouse Electric Corp.)*

each individually insulated with cross-linked polyethylene and covered with a tight, contour-fitted polyethylene jacket. The jacket assembles the three cables in a flat parallel configuration, providing a web between the conductors for ease of separation when splicing or terminating the cable without danger of damaging the insulation.

Transformers Many component parts of transformers used for underground electric distribution are similar to those installed in transformers designed for pole mounting. The core and coils assemblies are interchangeable in many instances. The transformers can be equipped with no-load tap changers, high-voltage fuses, low-voltage circuit breakers, and pressure relief devices. The core and coils assembly is normally immersed in insulating oil to provide the major insulation and a cooling medium. A typical core and coils assembly is shown in Fig. 34-8.

Pad-mounted transformers are used more extensively than other types for underground residential distribution. The transformers are built in different sizes and configurations depending upon the kVA capacity of the transformer, the type of connections, and the accessory equipment

Fig. 34-9 Turf Hugger, single-phase pad-mounted transformer installed along rear lot line of a subdivision. *(Courtesy A. B. Chance Co.)*

installed. General Electric Co. classifies the transformers it manufactures as Micro-Mini-Pad, Mini-Pad, Lo-Profile, and Compad. The pad-mounted transformers vary in height from 18 in to 48 in or more. A low-profile pad-mounted transformer installed along the rear lot line in a new residential area is shown in Fig. 34-9.

Figure 34-10 illustrates a pad-mounted transformer with the hinged cover open. Many of the accessories can be seen in Figs. 34-10 and 34-11. The transformers are constructed with two high-voltage load-break bushings. In Fig. 34-11 one of the high-voltage load-break bushings is marked L_1. The high-voltage bushing is shown in Fig. 34-10 without the high-voltage cable elbow connector installed and in Fig. 34-11 after the connections have been made. The low-voltage bushings are marked X_1, X_2, and X_0. A parking stand for the high-voltage cable connections is located between the two high-voltage load-break bushings.

In Fig. 34-11 the lineman is installing a grounded parking stand that is used to ground a high-voltage cable for maintenance work. It is very important that the cables be tested with a high-voltage meter to be sure they are deenergized prior to placing the elbow connector on the grounded device installed on the parking stand. Other devices that may be accessible with the

Fig. 34-10 Pad-mounted transformer with cover open. Lineman is installing high-voltage cable load-break elbow connector with a hot-line tool on the high-voltage load-break bushing. *(Courtesy I. T. T. Blackburn Co.)*

Fig. 34-11 Closeup view of pad-mounted transformer with cover open, showing high-voltage and low-voltage terminal equipment and accessories. Lineman is installing a grounding device on high-voltage terminator parking stand with hot-line tool. *(Courtesy A. B. Chance Co.)*

transformer cover open include draw-out load-break fuses, operating handle for low-voltage circuit breaker, and pressure relief device.

Submersible distribution transformers are similar in appearance to pole-mounted transformers and contain similar equipment (Fig. 34-12). High-voltage cable connections are completed through load-break bushings using load-break elbow connectors installed on the high-voltage cables. The low-voltage connections are completely insulated. The transformers are designed to operate under water. Tanks of the transformers are made with stainless steel or mild steel and protected by special enamel finishes to prevent corrosion. Special care must be taken to protect the transformer finish from being scratched or damaged while the unit is being stored, transported, and installed. It is desirable to install cathodic protection equipment connected to the transformer tank to prevent electrolysis.

Direct-buried single-phase distribution transformers are available in sizes of 10 through 37.5 kVA. The transformers are usually installed as a part of a front-lot-line system with one transformer per customer. The direct-buried transformer reduces cost of installation by eliminating the vault and minimizes the maintenance compared to the requirements associated with submersible vault-installed transformers. The transformers are constructed with a metallic tank or a

Fig. 34-12 Cutaway view showing submersible single-phase distribution transformer installed in fiber vault. A street grate on top of the vault provides protection and ventilation openings. *(Courtesy Westinghouse Electric Corp.)*

Fig. 34-13 Direct-buried transformer with a metallic tank. *(Courtesy General Electric Co.)*

polyester tank with a metallic liner. Figure 34-13 illustrates a direct-buried transformer with a metallic tank and Fig. 34-14 illustrates a direct-buried transformer polymer tank with a metal lining.

Three-phase pad-mounted transformers are often used to serve shopping centers, schools, hospitals, churches, apartment complexes, and office buildings located in the vicinity of electric underground distribution circuits serving residential areas. The three-phase pad-mounted transformers are available in sizes of 75 through 2500 kVA, with primary voltages of 4160Y/2400 through 34,500Y/19,920 and secondary voltages of 208Y/120, 240, 480Y/277, and 480. Standard accessories such as high-voltage load-break bushings for use with elbow connectors, fuses, no-load tap changer, secondary breaker, pressure-relief device, liquid level gauge, and lightning arresters are available. A three-phase pad-mounted transformer installed at a shopping center is illustrated in Fig. 34-15.

Isolating Devices Equipment used to sectionalize three-phase underground main feeder circuits and provide a means for isolating single-phase and three-phase taps for underground residential and underground commercial circuits may be constructed for pad-mounting or submersible installation. The pad-mounted gear may be equipped with air, vacuum, or oil switches. The

Fig. 34-14 Direct-buried single-phase distribution transformer with a nonmetallic tank. The transformer tank is molded from reinforced polyester and lined with metal. *(Courtesy General Electric Co.)*

submersible isolating devices are constructed with vacuum or oil switches. The use of air or vacuum switches minimizes the maintenance of the equipment. The single and three-phase lateral taps originating at the switching cubicles are usually fused to protect the main feeder circuit from an outage as a result of failure of some of the equipment installed as part of one of the lateral circuits.

Installation An electric underground distribution system serving a residential subdivision normally consists of single-phase primary cable originating on riser poles (as shown in Figs. 34-1 and 34-2) connected to an overhead distribution circuit through fused cutouts and protected by lightning arresters. Mechanical protection for supply conductors or cables installed on a riser pole shall be provided. This protection should extend at least 1 ft below ground level. Supply conductors or cable should rise vertically from the cable trench with only such deviation as necessary to permit a reasonable cable bending radius. Exposed conductive pipes or guards containing supply conductors or cables shall be grounded.

Fig. 34-15 Three-phase pad-mounted transformer with underground connecting cables installed at shopping center. *(Courtesy General Electric Co.)*

Fig. 34-16   A pad-mounted three-phase gang-operated vacuum sectionalizing switch for an underground main feeder circuit installed adjacent to an office building. *(Courtesy Westinghouse Electric Corp.)*

The installation should be designed so that water does not stand in riser pipes above the frost line. Conductors or cables shall be supported in a manner designed to prevent damage to conductors, cables, or terminals. Where conductors or cables enter the riser pipe or elbow, they shall be installed in such a manner that shall minimize the possibility of damage due to relative movement of the cable and pipe. Risers should be located on the pole in the safest available position with respect to climbing space and possible exposure to traffic damage. The number, size, and location of riser ducts or guards shall be limited to allow adequate access for climbing.

The single-phase primary cables may be supplied by an underground distribution circuit (as shown schematically in Fig. 34-3) and originating at a sectionalizing switching device similar to those illustrated in Figs. 34-16 through 34-22. The underground residential distribution is usually installed along rear lot lines as shown schematically in Fig. 34-23. A typical schematic diagram

Fig. 34-17   A pad-mounted three-phase gang-operated vacuum sectionalizing switch for an underground main feeder circuit with compartment doors open. *(Courtesy Westinghouse Electric Corp.)*

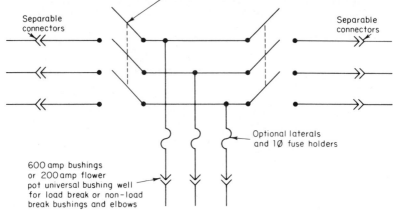

200-amp or 600-amp load break vacuum switch (two 3∅, open position)

Separable connectors

Separable connectors

Optional laterals and 1∅ fuse holders

600 amp bushings or 200 amp flower pot universal bushing well for load break or non-load break bushings and elbows

Fig. 34-18 Schematic diagram of three-phase gang-operated vacuum sectionalizing switch for an underground main feeder circuit pictured in Figs. 34-16 and 34-17. *(Courtesy Westinghouse Electric Corp.)*

Fig. 34-19 Submersible three-phase gang-operated vacuum sectionalizing switch for an underground main feeder circuit. *(Courtesy Westinghouse Electric Corp.)*

for the secondary distribution system is illustrated in Fig. 34-24. Figure 34-25 is an aerial view of a rear-lot-line underground residential distribution system installation. A front-lot-line underground residential distribution system is shown schematically in Fig. 34-4.

Most electric underground residential distribution systems employ direct-buried cables. The trenches for the direct-buried cables are usually dug with a small trencher, as shown in Figs. 34-26 and 34-27. In rural areas and urban locations free of obstacles under ground, cables are frequently installed with plowing equipment. If the cables are to be installed in conduits, cable preassembled in plastic pipe is usually installed by trenching or plowing. Both concentric neutral primary cable and secondary cables are available with this type of construction. The pipe or conduit may be either plain or corrugated for greater flexibility.

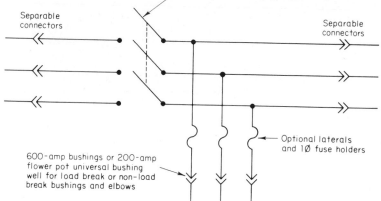

200-amp or 600-amp load break vacuum switch (3Ø, open position)

Separable connectors

Separable connectors

600-amp bushings or 200-amp flower pot universal bushing well for load break or non-load break bushings and elbows

Optional laterals and 1Ø fuse holders

Fig. 34-20 Schematic diagram of submersible three-phase gang-operated vacuum sectionalizing switch for an underground main feeder circuit pictured in Fig. 34-19.

Fig. 34-21 Pad-mounted three-phase gang-operated air sectionalizing switch for an underground main feeder circuit. Lineman is operating switch with crank handle. *(Courtesy S. & C. Electric Co.)*

The distribution facilities should be installed on utility easements or along the streets in public property where the utility has a franchise authorization to make the installation.

Cables should be located so as to be subject to the least disturbance practical. Cables to be installed parallel to other subsurface structures should not be located directly over or under other subsurface structures, but if this is not practical, the rules on clearances should be followed. Cables should be installed in as straight and direct a line as practical. Where bends are required, the minimum radius shall be sufficiently large to prevent damage to the cable being installed. Cable systems should be routed so as to allow safe access for construction, inspection, and maintenance.

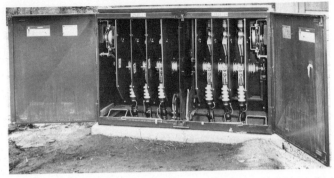

Fig. 34-22 Pad-mounted three-phase gang-operated air sectionalizing switch for an underground main feeder circuit with cubicle doors open to switching compartments. *(Courtesy S. & C. Electric Co.)*

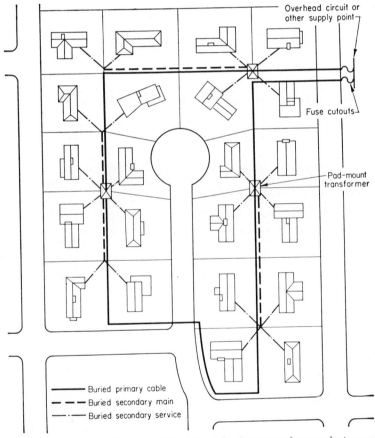

Fig. 34-23 Typical underground residential distribution system showing transformer and primary and secondary cables located on rear property lines. This makes possible serving four customers directly from each transformer and from most pedestals.

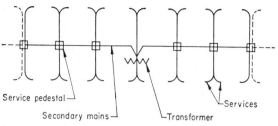

Service pedestal
Secondary mains
Transformer
Services

Fig. 34-24 Secondary distribution system showing transformer, secondary mains, pedestals, and service connections. Four customers are served directly from transformer secondary terminals and four from each pedestal except last one, where only two are served.

Fig. 34-25 Aerial view of residential area showing cables and transformers installed on rear lot lines, thus making service connections to homes short and convenient. ▼—transformers, ○—secondary service pedestals. *(Courtesy Line Material Industries.)*

Fig. 34-26 Trencher with digging arm raised. *(Courtesy Iowa-Illinois Gas and Electric Co.)*

The location of structures in the path of projected cable route shall, as far as practical, be determined prior to the trenching, plowing, or boring operation.

Routes through unstable soil such as mud, shifting soils, corrosive soils, or other natural hazards should be avoided. If burying is required through areas with natural hazards, the cable should be constructed and installed in such a manner as to protect it from damage. Such protective measures should be compatible with other installations in the area.

Supply cables should not be installed within 5 ft of a swimming pool or its auxiliary equipment. Cables should not be installed directly under building or storage tank foundations. Where a cable must be installed under such a structure, the structure shall be suitably supported to prevent the transfer of a harmful load onto the cable. The installation of cable longitudinally under the ballast section for railroad tracks should be avoided. Where cable must be installed longitudinally under the ballast section of a railroad, it should be located at a depth not less than 60 in below the top of the rail. Where a cable crosses under railroad tracks, the same clearances shall apply.

The installation of cable longitudinally under traveled surfaces of highways and streets should be avoided. When cable must be installed longitudinally under the roadway, it should be installed in the shoulder or, if this is not practical, within the limits of one lane of traffic to the extent practical. Submarine crossings should be routed and/or installed so that they will be protected from erosion by tidal action or currents. They should not be located where ships normally anchor.

Fig. 34-27 Trencher in operation conveying dirt to side of trench. (*Courtesy Iowa-Illinois Gas and Electric Co.*)

The horizontal clearance between direct-buried cable and other underground structures shall be controlled at a minimum of 12 in or larger as necessary to permit access to and maintenance of either facility without damage to the other.

Where a cable crosses under another underground structure, the structure shall be suitably supported to prevent transfer of a harmful load onto the cable system. Where a cable crosses over another underground structure, the cable shall be suitably supported to prevent transfer of a harmful load onto the structure. Adequate support may be provided by installing the facilities with sufficient vertical separation. Adequate vertical clearance shall be maintained to permit access to and maintenance of either facility without damage to the other. A vertical clearance of 12 in is, in general, considered adequate, but the parties involved may agree to a lesser separation.

If conditions require a cable system to be installed with less than 12 in horizontal separation or directly over and parallel to another underground structure, or if another underground structure is to be installed directly over and parallel to a cable, it may be done providing all parties are in agreement as to the method. Adequate vertical clearance shall be maintained to permit access to and maintenance of either facility without damage to the other.

Cable should be installed with sufficient clearance from other underground structures, such as steam or cryogenic lines, to avoid thermal damage to the cable. Where it is not practical to provide adequate clearance, a suitable thermal barrier shall be placed between the two facilities.

The bottom of a trench receiving direct buried cable should be relatively smooth, undisturbed earth, well-tamped earth, or sand. When excavation is in rock or rocky soils, the cable should be laid on a protective layer of well-tamped backfill. Backfill within 4 in of the cable should be free of materials that may damage the cable. Backfill should be adequately compacted. Machine compaction should not be used within 6 in of the cable.

Plowing in of cable in soil containing rock or other solid material should be done in such a manner that the solid material will not damage the cable, either during the plowing operation or afterward. The design of cable plowing equipment and the plowing-in operation should be such that the cable will not be damaged by bending, side-wall pressure, or excessive cable tension.

Where a cable system is to be installed by boring and the soil and surface loading conditions are such that solid material in the region may damage the cable, the cable shall be adequately protected.

The distance between the top of the cable and the surface under which it is installed shall be sufficient to protect the cable from injury or damage imposed by expected surface usage. Burial depths as indicated in Table 34-3 are generally considered adequate.

In areas where frost conditions could damage cables, greater burial depths than indicated in Table 34-3 may be desirable. Lesser depths than indicated may be used where supplemental protection is provided. Where the surface under which a cable is to be installed is not to final grade, the cable should be placed so as to meet or exceed the requirements indicated in Table 34-3 both at the time of installation and subsequent thereto.

Table 34-3   Burial Depths for Supply Conductors or Cables

Voltage	Depth of burial, in.
600 and below	24
601 to 22,000	30
22,001 to 40,000	36
40,001 and above	42

The conductors or cables of a supply circuit and those of another supply circuit may be buried together at the same depth with no deliberate separation between facilities provided all parties involved are in agreement (Fig. 34-28).

The conductors or cables of a communication circuit and those of another communication circuit may be buried together and at the same depth with no deliberate separation between facilities provided all parties involved are in agreement.

Supply cables or conductors and communication cables or conductors may be buried together at the same depth with no deliberate separation between facilities provided all parties involved are in agreement and the following requirements are met:

1. Grounded supply systems shall not be operated in excess of 22,000 volts to ground.
2. Ungrounded supply systems shall not be operated in excess of 5300 volts phase to phase.

A supply facility operating above 300 volts to ground must include a bare, grounded conductor in continuous contact with the earth. This conductor, adequate for the expected magnitude and duration of the fault current which may be imposed, shall be one of the following:

1. A sheath and/or shield.
2. Multiple concentric conductors closely spaced circumferentially.
3. A separate bare conductor in contact with the earth and in close proximity to the cable where such cable or cables also have a grounded sheath or shield not necessarily in contact with the earth. The sheath and/or shield as well as the bare conductor shall be adequate for the expected magnitude and duration of the fault currents which may be imposed.

Where a buried cable passes through a short section of conduit, as under a roadway, the contact with earth of the grounded conductor can be omitted provided the grounded conductor is continuous through the conduit. The bare conductors in contact with the earth shall be of suitable corrosion-resistant material. Cables of an ungrounded system operating above 300 volts shall be effectively grounded concentric shield construction in continuous contact with the earth. Such cables shall be maintained in close proximity to each other.

More than one cable system buried in random separation may be treated as one system when considering clearance from other underground structures or facilities.

The transformer pads and riser poles should be installed prior to placing the cable in the ground.

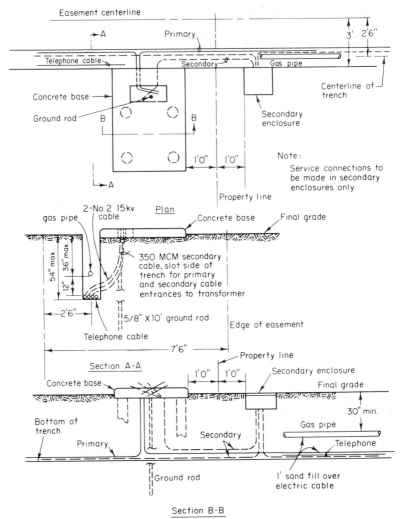

Fig. 34-28 Underground residential distribution transformer pedestal and cable layout for joint construction of telephone, gas, and 13,200Y/7620-volt electric distribution circuit.

Care exercised in handling the cable during installation, as illustrated in Figs. 34-29 and 34-30, will help to avoid trouble later, for damage sustained by the cable during installation has proved to be a major cause of subsequent cable failure.

The trench should be backfilled carefully. The cable can be best protected by using screened dirt or sand to provide several inches of cover over the cables. The backfill should be tamped to prevent settling (see Fig. 34-31). Tamping tools should not be allowed to hit the cables.

The cables installed on riser poles are usually terminated with a pothead, as shown in Fig. 34-32. The manufacturer's instructions must be followed carefully to prepare the cable for insertion in the pothead. It is necessary to remove the insulation shielding the correct distance and insert the cable so that the semiconducting insulation will contact the fabricated stress cone in the pothead properly.

The cables in switching cubicles and transformer enclosures are normally terminated with load-break elbow terminators as shown in Figs. 34-10 and 34-11. Figure 34-33 is a cross-sectional sketch of a typical load-break elbow terminator illustrating how cable is prepared and installed.

Fig. 34-29 Lineman installing cable in trench. *(Courtesy Iowa-Illinois Gas and Electric Co.)*

Figures 34-34 and 34-35 show a 7620-volt load-break elbow terminator being operated by a lineman with a hot-line tool to energize a pad-mount transformer.

Aluminum and copper cable is available in desired lengths, so that splices are seldom required. Prefabricated high-voltage splices are available. It is necessary to follow the manufacturer's recommendations to properly prepare the cable for installing a prefabricated splice. The semiconducting insulation shielding must be removed from a section of the insulation to permit the insulation buildup at the splice location. Stress cones in the prefabricated splice must contact the insulation shielding to maintain the shielding across the splice. Aluminum connectors and terminating devices should be used with aluminum conductors so as to avoid differential expansion and contraction that could result from the use of dissimilar metals.

Pad-mounted transformers are installed on concrete pads as illustrated in Fig. 34-36 or plastic pads as illustrated in Figs. 34-37 and 34-38.

Submersible transformers installed in vaults (see Fig. 34-12) and direct-buried transformers (see Figs. 34-13 and 34-14) must be protected for corrosion. Figure 34-39 is a suggested installation diagram for a cathodic protection system designed to prevent transformer tank corrosion. If the tank of the transformer is connected to the system ground, the copper neutral is part of the galvanic cell and the protective anode will attempt to protect it as well as the tank; thus the anode will be depleted rapidly. If the tank of the transformer is isolated from the system ground, the anode will protect the transformer tank only, giving satisfactory anode life.

Fig. 34-30 Primary and secondary mains buried directly in ground. Primary cable is two-conductor concentric cable having neutral ground conductor spiraled around outside of cable. *(Courtesy Commonwealth Edison Co.)*

Fig. 34-31 Helper is tamping backfill of trench with a hydraulic tamping tool.

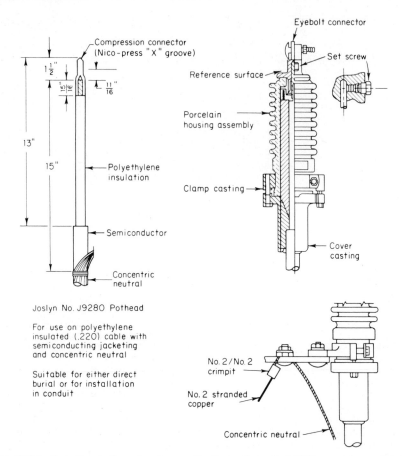

Joslyn No. J9280 Pothead

For use on polyethylene insulated (.220) cable with semiconducting jacketing and concentric neutral

Suitable for either direct burial or for installation in conduit

Fig. 34-32   Typical termination makeup for use with a two-conductor No. 2 AWG concentric neutral single-phase 7620-volt cable used for underground residential distribution. *(Courtesy Joslyn Mfg. and Supply Co.)*

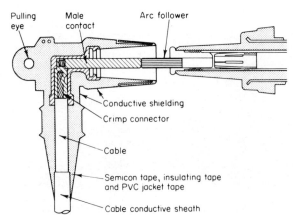

Fig. 34-33   Cross-sectional view of load-break terminator with component parts identified. *(Courtesy R. T. E. Corporation)*

Fig. 34-34   Load-break elbow termination in position to be connected to bushing on a pad-mounted transformer. Lineman holds energized terminator in position with hot-line tool. *(Courtesy I. T. T. Blackburn Co.)*

Fig. 34-35   Load-break elbow terminator installed or closed in on a bushing of pad-mounted transformer by lineman using hot-line tool. *(Courtesy I. T. T. Blackburn Co.)*

Fig. 34-36   Pad-mounted transformer installed on concrete pad located on rear property line.

Fig. 34-37   Plastic pad kept in position with screw anchors being installed by lineman's helper. *(Courtesy A. B. Chance Co.)*

Fig. 34-38   Plastic pad in position with pad-mounted transformer installed. Ground preparation is being completed by lineman and helper. *(Courtesy A. B. Chance Co.)*

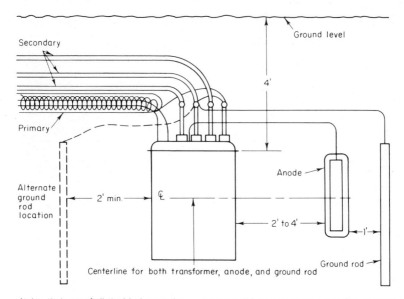

Note: Waterproof all electrical connections   Insulate all bare connectors within 8 ft of transformer

Fig. 34-39 Sketch of suggested direct-buried transformer cathodic protection system. *(Courtesy General Electric Co.)*

The low-voltage leads from the transformer to the customer may go directly as shown schematically in Fig. 34-3, or secondaries may be installed to service pedestals and the services to the customer extended from the service pedestals as shown in Fig. 34-25. If secondaries are installed, they will be in the same trench with the primary cable in most instances, with no definite separation. Figures 34-40 through 34-45 illustrate the installation of a below-grade service enclosure and the secondary and service wire connections completed in the enclosure.

Fig. 34-40 Typical convergence of secondary and service cables in a trench, below grade, before the service pedestal is installed. *(Courtesy I. T. T. Blackburn Co.)*

Fig. 34-41 The service enclosure (not visible) has been placed in the ground. The lineman has slightly "penciled" the cable insulation and is slipping sealing sleeves on the cable. *(Courtesy I. T. T. Blackburn Co.)*

Fig. 34-42 The insulation is being stripped away by the lineman to expose the bare conductor. The stripped length is determined by the size terminal lug to be installed. *(Courtesy I. T. T. Blackburn Co.)*

Fig. 34-43 The conductors must be wire-brushed and covered with inhibitor compound prior to installing the terminal lug. Some terminals have the inhibitor in the barrel of the conductor lug. The lineman is crimping the terminal lug on the conductor. *(Courtesy I. T. T. Blackburn Co.)*

The service wires originating at a below-grade service enclosure or a transformer such as the one illustrated in Fig. 34-46 extend directly to the customer's house or meter pedestal. At the customer's house, the service wires usually enter a conduit and terminate in a meter socket. The diagram (Fig. 34-47) shows a typical method of terminating the service wires at a house.

An above ground metering pedestal for terminating the underground services for two mobile homes is in the foreground of Fig. 34-48.

Street Lights Electric underground residential distribution areas usually have street lights installed on ornamental poles with underground supply conductors. The standard is usually fabricated from aluminum or precast concrete. The luminaire is generally constructed for use with mercury-vapor lamps. The underground supply cables are connected to the terminals of a transformer or the connectors in a service enclosure. Figure 34-49 shows a typical street light installation.

Fig. 34-44 The lineman is applying silicon lubricant to the insulated neck of the bus bar and to the inside of the sealing sleeve to facilitate removal of the sealing sleeve at some future time. *(Courtesy I. T. T. Blackburn Co.)*

Fig. 34-45 The lineman is preparing to bend the completed connections down into the below-grade service enclosure. A stainless steel plate will be locked on top of the pedestal. *(Courtesy I. T. T. Blackburn Co.)*

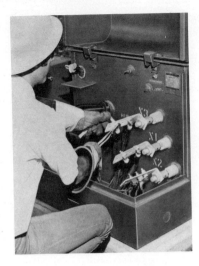

Fig. 34-46 Lineman connecting service wires to the low-voltage terminals of a pad-mounted transformer. *(Courtesy General Electric Co.)*

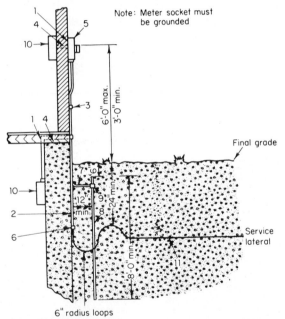

Fig. 34-47 Typical specification diagram for installing underground service wires at customer's house for 120/240-volt three-wire single-phase supply. (1) Service entrance conductors. Minimum recommended capacity 200 amp. (2) Galvanized steel rigid conduit 2½ in. (3) Galvanized steel conduit strap 2½ in. (4) Galvanized steel rigid conduit 2½ in. (5) Outdoor meter socket with 2½-in conduit hubs. Minimum recommended capacity 200 amp. (6) End bushing for 2½-in conduit. (7) Ground wire. (8) Ground rod copperweld ½ in minimum or galvanized water pipe ¾ in minimum. (9) Connector. (10) Fused switch, circuit breaker, or fuse box. Minimum recommended size 200 amp. Two possible locations shown. (11) Underground service conductors. Minimum recommended capacity 200 amp.

Fig. 34-48   Meter pedestal for terminating underground service for two mobile homes.

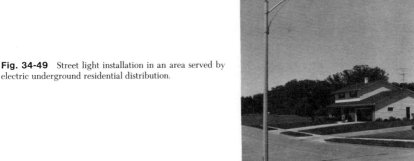

Fig. 34-49   Street light installation in an area served by electric underground residential distribution.

Fault Indicators   When a fault occurs on an underground residential circuit, it is important to locate the source of trouble quickly and restore service with minimum outage time. Fault locators installed at strategic points in the system will aid in identifying the trouble location. The fault locators are usually equipped with a neon light that glows, indicating fault current passed through the cable. The fault locators are waterproof and reset automatically after normal operating current is detected.

Protection   Supply circuits operating above 300 volts to ground or 600 volts between conductors shall be so constructed, operated, and maintained that, when faulted, they shall be promptly deenergized initially or following subsequent protective device operation. Ungrounded supply circuits operating above 300 volts shall be equipped with a ground fault indication system. Communication protective devices shall be adequate for the voltage and currents expected to be impressed on them in the event of contact with the supply conductors. Adequate bonding shall be provided between the effectively grounded supply conductors and the communication cable shield or sheath, preferably at intervals not to exceed 1000 ft. In the vicinity of supply stations where large ground currents may flow, the effect of these currents on communication circuits should be evaluated before communication cables are placed in random separation with supply cables.

Tree Trimming

Line Clearance Objectives   Line clearance is preventive line maintenance to ensure that the utility's service to its customers is not interrupted as the result of tree interference with conductors or circuit equipment by growing trees. Elements include removal of danger trees and overhangs, trimming to clear the conductors, and clearing transmission right-of-way. Natural tree growth and storm-tossed branches can ground or break distribution and transmission lines. The trimming process is intended to anticipate such a possibility by removing this hazard. Lines are checked and cleared on a planned time cycle. The amount of clearance sought should provide hazard-free operations for at least two years. It should be accomplished while maintaining the health and beauty of the trees involved, the goodwill of property owners, and the safety of the trimming crew.

Factors in Trimming Techniques   There are fundamentals essential to safe and competent trimming operations. Primary among these are (1) knowing how to climb and use a rope, (2) knowing how to tie essential knots, (3) a knowledge of tree species—their growth characteristics and wood strength, (4) a knowledge of electrical conductors, (5) a knowledge of alternative trimming methods, (6) knowing how to cut limbs and lower them under full control. Ideally, climbers should be trained to recognize structural problems such as weak crotches and chronic disease symptoms, so that trimming and tree removal decisions result in optimum accomplishment.

Time for Pruning   Trees may be pruned at any time of the year and utility line clearance operations are maintained throughout the year. In the past, work often was concentrated into a particular season. However, today, it is recognized that there are compensating values and factors, so that trimming operations run year round except where there are unique, limiting local considerations.

A Basic Set of Pruning Tools
1. One light power chain saw (16-in cutter bar length)
2. One handsaw and scabbard
3. One 12-ft pole saw
4. One 12-ft pole clip or pruner
5. One ½-in polyester climbing rope (120 ft)
6. Several hand lines (½-in polyester)
7. One ¾-in manila bull rope (150 ft)
8. One wooden 24-ft extension ladder
9. One dielectric hard hat for each crew member
10. Tree wound paint in a pot or spray can
11. Belt snaps for handsaw and paint
12. A set of road signs and flags warning of men working in the trees

Some of the pruning tools appear in Figs. 35-1 and 35-2.

Other items should be considered, especially when working in residential areas where cleanup is vital: a pair of loppers, a sledge and wedges, brush rake, street broom, and scoop shovel.

Climbing Equipment   For safety and proficiency, climbers should always tie in to a crotch before beginning to trim. The elements involved are a body sling or saddle and a climbing rope. Figure 35-3 illustrates the correct position of the body sling. The climbing rope is a ½-in polyester rope 120 ft long, attached to the snap of the sling which is in front at the center and then secured by a taut-line hitch. The method of tying the rope with a taut line hitch is covered in detail in Fig. 35-4. This combination permits the climber to move around a tree freely and with safety. Proper use distributes the climber's weight, so he will tire less easily and be able to move onto branches which would not otherwise support his full weight.

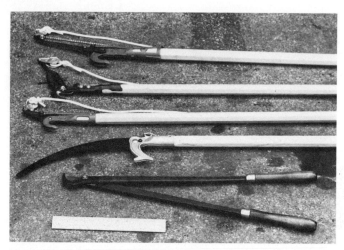

Fig. 35-1   Trimming hand tools. Starting at the top, tools are a center-cut bull clip, a side-cut pole clip, a center-cut pole clip, a pole saw, and a pair of loppers. *(Courtesy Asplundh Tree Expert Co.)*

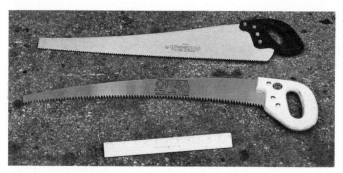

Fig. 35-2   Two types of handsaws. Upper saw cuts on the push, lower saw cuts on the pull. Usage is primarily a matter of personal preference. *(Courtesy Asplundh Tree Expert Co.)*

Permission   No trimming should be started until permission is obtained from the property owner or the state, county, or municipal agency with jurisdiction. Verbal permission is usually satisfactory for routine trimming, but written permission may be advisable for heavy pruning or tree removals. The trimming should then be done under the direction of the person who secured the permission, since he can best interpret any special directions or limitations imposed. To ensure goodwill necessary for future trimming cycles, the work must be professional and the cleanup thorough.

Good and Bad Methods   Before providing the details of specific tree trimming techniques, it will be useful to review good and bad general methods. Pollarding and shearing or rounding over are undesirable because the visual effect is ugly and contrary to normal free form and because these methods represent uneconomical line clearance. Many small cuts take unnecessary time, create an unhealthy tree condition, and stimulate rapid regrowth back into the conductors. Natural trimming is desirable. In this method, branches are flush-cut at a suitable parent limb back toward the center of the tree. This method is also called drop-crotching or lateral trimming. It involves fewer but heavier cuts, usually made with a saw, not a pole pruner. This natural pruning generally permits securing more clearance, so that trimming cycles can be lengthened. In addition, better tree form and value are maintained and regrowth direction can be influenced. Note the following examples in Figs. 35-5 through 35-9.

Fig. 35-3   A typical climber's saddle, illustrating the taut-line hitch and figure-eight safety knot. *(Courtesy Asplundh Tree Expert Co.)*

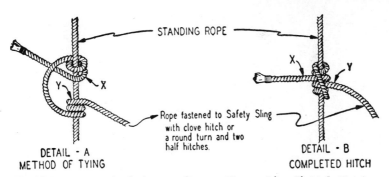

Fig. 35-4   Tying a taut-line hitch to a standing rope. *(Courtesy Edison Electric Institute.)*

Removing Large Branches   The procedure in the removal of a large side limb is shown in Fig. 35-10. The first cut is an undercut 10 to 12 in out from the place where the final flush cut is to be made. This cut should be ¼ to ½ the distance through. The second cut is made about 1 to 6 in farther out on the upper side of the branch. The reason for shifting out a few inches is to prevent the bark from being stripped back beyond the point of the final cut when the limb falls. When the bulk of the weight has been removed, the finish cut is made flush with the parent limb or trunk in two steps. First an undercut to prevent stripping is made and finally the limb is sawed through from above.

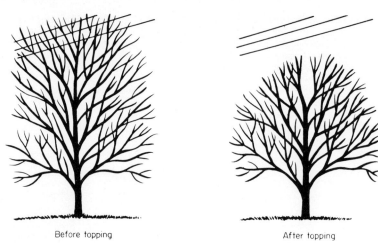

Before topping   After topping

Fig. 35-5   *Topping*. Topping is cutting back large portions of the upper crown of the tree. Topping is often required when a tree is located directly beneath a line. The main leader or leaders are cut back to a suitable lateral. (The lateral should be at least one-third the diameter of the limb being removed.) Most cuts should be made with a saw; the pole pruner is used only to get some of the high lateral branches. For the sake of appearance and the amount of regrowth, it is best not to remove more than one-fourth of the crown when topping. In certain species—sugar maple for example—removal of too much of the crown will result in the death of the tree.

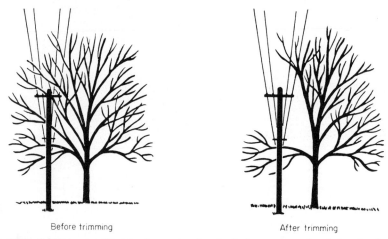

Before trimming   After trimming

Fig. 35-6   *Side Trimming*. Side trimming consists of cutting back or removing the side branches that are threatening the conductors. Side trimming is required where trees are growing adjacent to utility lines. Limbs should be removed at a lateral branch. Unsightly notches in the tree should be avoided if possible. Shortening branches above and below the indented area or balancing the opposite side of the crown will usually improve the appearance of the tree. When trimming, remove all dead branches above the wires, since this dead wood could easily break off and cause an interruption.

Lowering Large Limbs   Large branches that might cause damage to the line conductors or other property in falling should be carefully lowered with tackle. Figure 35-11 illustrates the manner of lowering a large limb which would otherwise fall directly into the line conductors. Before the branch is sawn off, it is supported with two ropes, one at the butt and the other well out toward the end. The ropes are run through crotches in the branches above and secured to the

Before trimming

After trimming

Fig. 35-7 *Under Trimming.* Under trimming involves removing limbs beneath the tree crown to allow wires to pass below the tree. To preserve the symmetry of the tree, lower limbs on the opposite side of the tree should be removed also. All cuts should be flush to avoid leaving unsightly stubs. The natural shape of the tree is retained in this type of trimming and the tree can continue its normal growth. Overhangs are a hazard, however, when a line passes beneath a tree. Overhangs should be removed in accordance with the species of tree, location, and the general policy of the utility that you work for. When trimming, remove all dead branches above the wires, since this dead wood could easily break off and cause an interruption. Many utilities have a removal program set up for trees that overhang important lines.

Before trimming

After trimming

Fig. 35-8 *Through Trimming.* Through trimming is the removal of branches within the crown to allow lines to pass through the tree. It is best suited for secondaries, street-light circuits, and cables, although it is often used on primary circuits where there is no other way of trimming the tree. Cuts should be made at crotches to encourage growth away from the lines.

trunk of the tree. A third rope—known as the guide rope—is fastened to the branch at such a point that it can be used to swing the branch out from over the line conductors. After the sawing is completed, the branch is swung out to clear the line and then gradually lowered by means of the two supporting ropes.

Removing Small Branches   In the case of small branches, it is not necessary to make three separate cuts. One cut with the saw close up against the limb while it is held in place, as shown in Fig. 35-12, is all that is necessary. Small branches can also be cut off with a pole pruner. When trees are being trimmed, all dead branches should be pruned out.

Treating the Wounds   Every pruning cut larger than 1½ in. in diameter should be sprayed or painted with an asphalt-base tree paint to discourage the entry of insects and disease organisms.

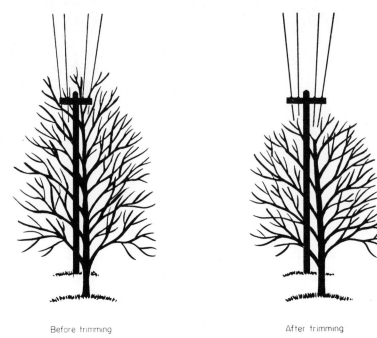

Before trimming   After trimming

Fig. 35-9 *Natural Trimming.* Natural trimming is a method by which branches are cut flush at a suitable parent limb back toward the center of the tree. This method of trimming is sometimes called "drop crotching" or lateral trimming. Large branches should be removed to laterals at least one-third the diameter of the branch being removed. Natural trimming is especially adapted to the topping of large trees, where a great deal of wood must be removed. In natural trimming, almost all cuts are made with a saw and very little pole-pruner work is required. This, when finished, results in a natural looking tree, even if a large amount of wood has been removed. Natural trimming is also directional trimming, since it tends to guide the growth of the tree away from the wires. Stubbing or pole-clip clearance, on the other hand, tends to promote rapid sucker growth right back into the conductors. The big factor to remember is that natural clearance does work and that two or three trimming cycles done in this manner will bring about an ideal situation for both the utility and the tree owner. Most shade trees lend themselves easily to this type of trimming. Elm, Norway maple, red oak, red maple, sugar maple, silver and European linden are our most common street trees, and these species react especially well to natural clearance methods.

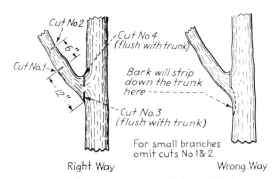

Fig. 35-10 Steps in right and wrong removal of a large side limb. *(Courtesy REA.)*

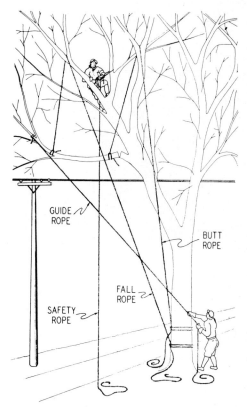

Fig. 35-11 Lowering large or hazardous limb with rope tackle. The two supporting ropes are called butt and fall rope; the third rope is the guide rope. *(Courtesy Edison Electric Institute.)*

Care should be taken to overlap the area around the cut. The material is applied either by brush from a paint pot snapped to the climber's belt (see Fig. 35-13) or by spraying from a can. There are pole applicators for a spray can that can be used to increase both reach and production. The dark-colored tree paint also serves to make the cuts less noticeable. Brushed tree paint is produced in both summer and winter grades to provide for flowability in extreme cold. Paint is generally formulated to spray under all weather conditions. Recent field research suggests that on many tree species the addition of a growth inhibitor—such as 1 percent naphthaleneacetic acid, ethyl ester—to the tree paint will significantly reduce resprouting at the treated cut. Utilization can make second-cycle trimming more economical because fewer cuts are required and there is less brush to manage.

Brush Disposal   All brush and wood trimmed in line clearance operations must be disposed of in such a way as to maintain the goodwill of property owners. In residential trimming, it is

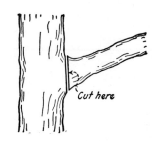

Fig. 35-12   How a small branch can be removed with a single cut.

Fig. 35-13 A typical climber's saddle with saw scabbard and tree paint pot snapped in position.

usually most economical to chip the brush. Note Fig. 35-14. Public relations generally requires that the work area be raked or swept up. Material suitable for fireplace usage should be cut to size and neatly stacked. It should first be offered to the property owner, and if he rejects it, may be left available for the general public. However, some situations may require that the wood be hauled away and dumped. On right-of-way clearance or when trimming in remote areas, it is sometimes possible to pile brush and leave it as game cover. Generally, burning is no longer permitted because of concern about smoke and fly ash.

Power Equipment   Technology responds to operational needs and there are now various pieces of specialized equipment designed to solve special problems or to increase productivity.

Trim-Lift   Wherever a truck can drive, an aerial device can be effectively used to position a man for trimming, thus saving the trouble of climbing. The mechanism involves two connected booms, with a work platform or bucket at one end, mounted on a truck chassis. The upper boom is supported by the lower boom through the use of cables. A hydraulic power system runs off the truck engine and lifts the booms, operates the tools, and lifts the dump bodies to discharge chips and other materials gathered in trimming. Note Figs. 35-15 and 35-16. Mechanical outriggers are

Fig. 35-14 A manual climbing crew in operation. Note the groundman chipping brush behind the climbers. The chute of the chipper directs the chips into the back of the dump van. (*Courtesy Asplundh Tree Expert Co.*)

Fig. 35-15 A trim-lift and chipper in use by crew to trim trees and dispose of brush. *(Courtesy Asplundh Tree Expert Co.)*

used to stabilize the unit during operation. Generally, the booms and the work platform are fully insulated to protect the operator from shock and the truck body from becoming energized if the boom or platform should come in contact with a conductor.

Tree Crane   This type of aerial device is used primarily for taking down trees too tall to reach with a trim-lift. See Fig. 35-17. The equipment utilizes a boom that telescopes and a climber working from a "saddle" and a rope swung from the boom tip.

Power Tools   Chain and circular blade saws are powered by compact gasoline engines and designed in a variety of sizes and weights to adapt to the many conditions encountered, from take-downs to tree trimming to brush cutting. (See Figs. 35-18 and 35-19.) Noise levels generally dictate hearing protection for the operator.

A variety of cutting tools can be hydraulically powered or air-operated by a system connected with the bucket of a trim-lift. These include pole chain saws, circular saws, and limb loppers illustrated in Fig. 35-20. Some situations, such as hospital areas, may require the use of electrically powered tools operated by a portable gasoline generator. Power equipment requires systematic inspection and preventive maintenance to ensure crew safety and efficiency.

Fig. 35-16 A specialized right-of-way trim-lift with the boom mechanism mounted on a diesel-powered, four-wheel-drive vehicle. *(Courtesy Asplundh Tree Expert Co.)*

Fig. 35-17 A tree crane in operation. *(Courtesy Asplundh Tree Expert Co.)*

Fig. 35-18   Several gasoline-powered chain saws, each a size and weight to be used easily by one man. It is unlikely that the model at the top of the picture would be used off the ground. *(Courtesy Asplundh Tree Expert Co.)*

Fig. 35-19   Large portable power-driven saw. *(Courtesy Asplundh Chipper Co.)*

Fig. 35-20 Hydraulic tools of a type used from a trim-lift bucket. *(Courtesy Asplundh Tree Expert Co.)*

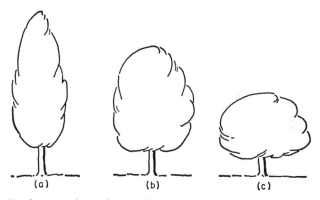

Fig. 35-21 Tree-form types that are basic in planning and maintaining clearance for overhead lines. (*a*) Upright; (*b*) spreading; (*c*) horizontal.

Table 35-1   Characteristics of Various Trees*

Species (common name)	Avg. annual terminal growth,† ft	Approx. mature height, ft	Comparative wood strength values (green),‡ lb per sq in.	Suggestions
Upright growth form:				
Cottonwood (Carolina poplar) §	4.30	85	5,300	Can do any type of trimming. Avoid overbuilding
Elm (American)	2.14	85	7,200	Avoid overbuilding. Can top notch. Best underbuilt
Gum (sweet)	0.66	100	6,800	Slow growing. Can be topped. Best side trimmed or underbuilt
Hickory	0.96	65	11,000	Top side, or under trim. Slow-growing tough wood
Kentucky coffee tree	0.90	80	Unknown	Hard, strong wood. Can be topped. Best underbuilt
Locust (honey)	1.50	80	10,200	Avoid if possible. Should be underbuilt
Maple (silver)	2.14	75	5,800	Vigorous sucker growth when topped too heavily
Oak (pin)..........	2.00	70	8,300	Strong hardwood. Do not top
Poplar (Lombardy) §	3.63	80	Unknown	Avoid or remove. Soft, brashy wood
Sycamore (American)	1.50	75	6,500	Vigorous sucker growth when topped. Best to underbuild
Spreading growth form:				
Basswood	1.50	75	5,000	Soft, weak wood. Overbuild
Beech (American)	1.85	60	8,600	Does not stand severe topping. Side or under trim
Birch (white)	1.80	50	6,400	Do not top. Avoid or trim lightly
Buckeye (horse chestnut)	1.00	60	4,800	Top or side trim. Soft, weak wood
Cherry (black)	1.00	60	8,000	Do not top heavily. Underbuild old trees
Hackberry	1.00	50	6,500	Top or side trim on older trees
Locust (black) §	1.30	60	13,800	Vigorous sucker growth when topped. Side or under trim old trees
Maple (red)	1.50	75	7,700	Top or side trim. Underbuild old trees
Maple (sugar)	1.50	75	9,400	Do not top old trees too heavily
Oak (live)	1.11	60	11,900	Overbuild
Oak (red)	1.50	75	6,900	Top, side or under trim
Oak (white)	0.75	75	8,300	Do not top too heavily. Side or under trim
Sycamore (oriental)	1.80	60	Unknown	Can be topped. Vigorous sucker growth
Walnut (black)	1.00	80	9,500	May die back when topped heavily. Underbuild old trees
Horizontal growth form:				
Box elder §	2.15	50	Unknown	Vigorous sucker growth. Use good pole heights. Top or side trim
Dogwood	0.30	30	8,800	Slow growing, overbuild
Maple (Norway)	1.20	50	Unknown	Overbuild, top trim
Osage orange §	1.40	45	13,700	Thorny, fast growing. Can be topped
Willow (black) §	1.67	50	3,800	Brashy, fast-growing wood. Avoid if possible

NOTE: All species of conifers with exception of white and red cedar and hemlock do not lend themselves well to topping.

*Reprinted with permission from "Line Clearing Manual," Publication K-10 of The Edison Electric Institute.

† Approximate *normal* average rates for Central Western states—subject to variation due to soil, moisture and site conditions. As pruning stimulates growth, these *values* should be approximately tripled to arrive at first year growth rate of topped trees.

‡ U.S. Government standard test for modulus of rupture, static bending of a simple beam (Technical Bulletin No. 479).

§ Remove if possible.

Table 35-2 Tree Trimming Clearances
(in feet)

Clearances	Secondary, 100–600 volts	Primary, 2400–4800 volts	Primary, 7200–13,800 volts
Topping			
Fast growers	6	8	9
Slow growers	4	6	7
Side			
Fast growers	4	6	8
Slow growers	2	4	6
Overhang			
Fast growers	4	8	12°
Slow growers	2	6	12°

SOURCE: Asplundh Tree Expert Co.
°Remove if possible.

Poisonous Plants   When he is trimming trees and removing brush, there is always the possibility that a lineman may encounter plants which may cause skin poisoning. These include the following:

Poison Ivy   This plant sometimes climbs trees and poles. It is also a crawling plant. Poison ivy has broad and glossy leaves that always grow in clusters of three leaflets, two branching from the stem 1 in or less below the center one.

Poison Oak   A low-growing erect plant, poison oak never climbs. The leaves are broad but not always glossy and occur in clusters of three. The dark-green leaves are permanently and very heavily hairy on the undersurface.

Poison Sumac   This shrub or small tree may grow to a height of 20 ft. Its smooth, glossy leaves are in the form of 7 to 13 oblong leaflets (always an odd number). Poison sumac grows in swamps and low areas.

The poisoning from these plants is caused by an oily substance which gets on a person's skin. The poison can be conveyed by smoke, insects, clothing, and direct contact, and it is more likely to occur when the skin is covered with perspiration. No one should ever consider himself immune to these poisons, even after having had an attack.

If a worker suspects that he has been exposed to these plants or their oil, he should wash exposed areas of skin with warm water and ordinary brown laundry soap. Another way to prevent poisoning is to wash the exposed skin in rubbing alcohol and then rinse in clear water and dry the skin. No brush should be used, because it would irritate the skin.

Natural Tree Forms   All trees can be classified as one of three forms:
1. Upright
2. Spreading
3. Horizontal

The tree-form types are shown in Fig. 35-21. As far as possible trees should be trimmed to preserve their natural shape. Table 35-1 lists the trees according to their form and also indicates their annual growth in feet as well as the comparative strength of the wood. Table 35-2 itemizes suggested trimming clearances for various voltages.

Section **36**

Distribution
Transformer
Installation

Transformers constructed to reduce the primary voltage down to the customer's utilization voltage, are commonly referred to as distribution transformers. The distribution transformers may be single phase, normally used for residential customers, or three phase, normally used for commercial or industrial customers. The transformers are constructed for mounting on poles or racks, in vaults, or on pads constructed with concrete, plastic, or fiberglass, and for direct burial in the earth. The construction, theory of operation, and the method of making connections to distribution transformers is described in Section 15 "Distribution Transformers."

Table 36-1   Watts Demand for Residences to Be Used in Determining Size of Transformer Required

Size of house	Watts demand
Small	3,500
Average	6,000
Large	15,000

Table 36-2   Diversity Factor for Selecting Transformer Capacity Required to Serve Several Houses

Number of houses	Diversity factor
1	1.00
2	0.90
4	0.80
6	0.75
8	0.70
12	0.67

Distribution Transformer Size   The transformer capacity required to supply any given load may be estimated as follows.

Residential Loads   Residential loads may be estimated in accordance with Table 36-1, which gives the watt demand per house. As the table shows, a small house is taken as having a demand of 3500 watts, a medium-sized house 6000 watts, and a large house 15,000 watts.

A large home utilizing electricity for heating may have a demand of 20,000 to 30,000 watts. When several houses are served from the same transformer, there will be diversity between the peak use of energy as shown in Table 36-2.

The size of distribution transformer required to serve several houses is determined by totaling the estimated watts for each house obtained from Table 36-1, dividing the total by 1000 to obtain kilowatts, and multiplying by the diversity factor obtained from Table 36-2. If two small, two medium, and two large houses are to be served from one transformer, the size of the transformer would be determined by calculations itemized in Table 36-3.

The nearest size transformer adequate to serve the six houses with the loads shown in Table 36-3 would be 37.5 kVA. A distribution transformer can carry considerable overload without damage. A distribution transformer with a rated capacity of 37.5 kVA should be adequate to serve the six houses for several years even with normal load growth.

Commercial Lighting   In determining the size of transformer necessary for commercial-lighting customers, such as stores of all kinds, it is first necessary to add the total connected load. This requires making a list of all lights, motors, air conditioners, etc., together with the wattage of each. The sum is the total connected load on the transformer. The average demand on the transformer will be 40 to 60 percent of the connected load, depending on individual cases. The demand represents the necessary transformer capacity.

Table 36-3   Sample Calculation for Determining Proper Size of Transformer

	Houses		Watts		Watts
	2	×	3,500	=	7,000
	2	×	6,000	=	12,000
	2	×	15,000	=	30,000
Total	6				49,000
		49,000 watts/1000 =			49 kW
	49 kW × 0.75 diversity factor =				38.8 kW

Table 36-4   Percentages by Which to Multipy Total Connected Motor Load to Obtain Required Transformer Kilovolt-Ampere Capacity
(Percentages for various numbers of motors and magnitude of connected horsepower load)

Total connected load, hp	Percentages for calculating kilovolt-ampere demand from the normal rated capacity of motors					
	Number of motors					
	1	2	3–5	6–10	11–20	20 or more
Less than 10	83	81	75	73		
10 to 50	78	76	70	68	67	65
50 to 100	74	72	65	62	61	60
100 to 300	72	70	61	59	58	57
Over 300	70	69	58	56	55	54

Motor Loads   The transformer capacity required for motor loads is best found from the motor ratings. Add the ratings of all the motors to be supplied from one transformer or bank of transformers. Then from Table 36-4 obtain the proper percentage for the number of motors. Multiply the actual connected load by this percentage. The result is the required transformer capacity. Select the next larger standard rating.

If the transformer bank is to be open-delta connected, multiply the rating as found above by 1.35 before selecting the transformer rating required.

The foregoing load estimates should be used only if there is no local information available for arriving at an approximate transformer rating. It is realized that load demands vary considerably for the various classes of customers throughout the country, and therefore local data should be collected to form a basis for estimating.

Load Checks   When a new subdivison in an urban area is first supplied with electric service, the smallest-size transformer used by the utility is generally installed. This may be a 15- or 25-

kVA size. This is done when the first house is connected to the secondary. When the load has built up to nearly full transformer capacity, load measurements are made on the transformer to see what the actual load is.

The customer's monthly kilowatt hour consumption as determined by meter readings made for billing the customer can be analyzed automatically by a computer to determine the distribution transformer loading. If a transformer is over or under loaded 165 percent or 50 percent respectively, or other values selected, the computer can print out an exception report. The report can be used to initiate action to change out the distribution transformers to the proper size. The distribution transformers must be properly loaded to efficiently utilize the investment and to provide the customer reliable service with good voltage. If a distribution transformer is overloaded, it may be desirable to install an additional transformer and divide the secondary and services to distribute the load properly on each transformer.

The red light on self-protected distribution transformers automatically signals an overload condition. It is fairly general practice to permit overloads of 50 percent in summer and 75 percent in winter in the cooler climates before changing out the transformer or installing an additional transformer and reconnecting the load.

Distribution Transformer Location   The distribution transformer should be installed as near as possible to the center of the load area. The transformer secondary leads and services should be kept as short as possible to minimize the voltage drop and line losses. The transformers are

Fig. 36-1   Pad-mounted distribution transformer installed on rear lot at the corner of the lots. *(Courtesy Iowa-Illinois Gas and Electric Co.)*

usually installed on the rear lot line at the point where the corners of four lots are common, permitting four houses to be served by services direct from the transformer whether the transformer is mounted on a pole, on a pad, or in a vault. If additional houses are to be served from one transformer, the secondaries are extended along the rear lot line.

Commercial loads are usually served by an individual transformer or a three-phase transformer bank. The transformers should be installed as close as possible to the load to minimize the secondary and service voltage drop.

Figure 36-1 illustrates the proper location for installing a distribution transformer on a rear lot line. The transformer pictured is connected to serve six houses and will serve six additional houses when the subdivision is completed. The same plan is followed when the distribution transformers are installed along the street in front of the houses.

Distribution Transformer Grounding   The term "ground" is used in electrical work to refer to the earth as the zero potential. A ground actually consists of an artificial electrical connection to the earth, having a very low resistance to the flow of electric current. To ground a circuit or apparatus, therefore, means to connect it to earth. All distribution transformer tanks and one conductor of all transformer secondary circuits must be grounded to protect life and property. The points to be grounded in the various kinds of secondary circuits are illustrated in Section 15 "Distribution Transformers."

If the distribution transformer has a two-wire secondary, either of the two wires can be grounded. In a three-phase transformer bank with a delta-connected secondary, the center of one of the transformer windings should be grounded. If the secondaries are Y connected, the neutral of the Y is grounded.

The ground connection is made to the secondary circuit at the transformer, and the ground connecting wire is connected onto a rod which has been driven into the ground at least 8 ft. Figure 36-2 illustrates a rod used in making the ground connection. The rod must be driven deep enough so that it is in moist soil all year round.

In addition all service neutrals are grounded at customer's entrance. The usual ground is the customer's water main.

The low-voltage secondary circuit is grounded to guard against an excessive voltage being impressed on that circuit from some external source. A conductor of the primary mains may break and fall upon the secondary conductors below it. The greatest danger, however, exists in the breaking down of the insulation in the distribution transformer, thereby bringing the secondary winding in contact with the high-voltage primary winding. If a person should then touch a bare part of the secondary circuit and should also happen to be in contact with a radiator, bathtub, gas

Fig. 36-2   Typical ground rod and clamp used in making ground connection. *(Courtesy Hubbard & Co.)*

range, or water faucet, he would be likely to receive a very severe shock. Furthermore, any excessive voltage on the low-voltage fittings in lamp sockets, etc., is very apt to cause fire. If the secondary is grounded, however, a breakdown in the transformer insulation or other high-tension cross is very apt to blow the primary transformer fuses, thereby clearing the defective circuit from the source.

Distribution Transformer Mounting   The installation of distribution transformers in vaults, on pads, and buried in the earth is covered in the sections describing underground distribution installations. Small distribution transformers normally installed on overhead circuits are often referred to as pole-type transformers. These transformers may be fastened directly to the pole, hung from crossarms, mounted on racks or platforms suspended from poles, or placed on pads usually concrete, enclosed with a fence. Figure 36-3 illustrates a distribution transformer bolted directly to the pole, Fig. 36-4 illustrates a distribution transformer mounted on a crossarm, Fig. 36-5 illustrates a bank of three transformers supported by means of cluster pole mounts, and Fig. 36-6 illustrates a small power bank mounted on a platform. Platforms are built of any shape or size to suit local conditions. If the structure is to be permanent, the supporting members should be steel. Steel makes a good appearance and requires no maintenance. The supporting members

Fig. 36-3 Distribution transformer mounted directly on pole. Transformer is rated 25 kVA, 7620–120/240 volts.

Fig. 36-4 Illustration showing distribution transformer mounted on crossarm. *(Courtesy Line Material Industries.)*

should be covered with flooring. Very large transformers, however, are set on pads on the ground (Fig. 36-7). A high fence must then be provided to keep persons from coming in contact with the live parts.

Small distribution transformers can be bolted directly to the pole by means of special brackets, Fig. 36-8, or hung from the top crossarm when there is only one crossarm on the pole. As a rule, transformers hung on crossarms should not exceed 25 kVA. If there are two or more crossarms on the pole, the transformer should be hung on the lowest arm, or even lower on separate arms. It is good practice to use double crossarms with each transformer installation.

Fig. 36-5 Three-phase transformer bank supported on pole by means of cluster mount. Eliminates use of platform supported by two poles. Single pole must be set 1 to 3 ft deeper than normal.

Fig. 36-6 Three single-phase transformers connected into a three-phase bank and mounted on a platform supported by two poles. *(Courtesy Iowa-Illinois Gas and Electric Co.)*

Fig. 36-7 Three-phase transformer bank set on ground. Note enclosing fence. *(Courtesy Electrical Engineers Equipment Co.)*

Fig. 36-8 Distribution transformer designed to be supported by direct-to-pole mounting. Transformer is fastened to pole by bolts passing through special brackets and pole. *(Courtesy General Electric Co.)*

Fig. 36-9 Lineman installing distribution transformer on pole. Transformer is lifted in place by boom and winch line of line truck. Transformer is mounted on bracket clamped to the pole. *(Courtesy A. B. Chance Co.)*

Transformers should not be mounted on junction or buck-arm poles, as it makes working on such poles more hazardous for the lineman. The pole-type transformers are normally raised to the mounting position with the boom and winch on the line truck (Fig. 36-9). Installation of distribution transformers on poles located on rear lot lines in residential areas or other locations inaccessible for line trucks requires the use of a block and tackle or a portable winch to raise the transformers (Figs. 36-10 and 36-11).

The distribution transformer should be handled carefully. Bushings and other equipment on the transformer can be damaged easily. The windings of the transformer may be damaged if the transformer is dropped from the truck or severely jolted. The lifting equipment, including slings used on the transformer, should be carefully inspected before the operation is started. The linemen and groundmen should stay in the clear while the transformers are raised into position. Protective equipment should be installed on the pole to make the primary area safe. The linemen must wear rubber gloves, and protectors, and other appropriate personal protective equipment while working near the primary wires.

Fig. 36-10 Special pole-top extension used in hoisting distribution transformers. Note attachment provided for upper support of block and tackle. *(Courtesy James R. Kearney Corp.)*

Fig. 36-11 Raising a transformer with block and tackle. Transformer will be part of a three-phase bank supported on a single pole with cluster-mount brackets. *(Courtesy Joslyn Mfg. and Supply Co.)*

Figure 36-12 illustrates a complete conventional pole-type, single-phase distribution transformer installation. Conventional distribution transformers should always be protected on the primary side with distribution fuses (cutouts) to disconnect the transformer in case of trouble. The cutouts are mounted near the transformer so they will be readily accessible to the lineman.

When all the equipment is mounted in the proper location, the ground and neutral connections should be completed. The secondary connections should be completed last to eliminate the danger of a back feed. The primary connections should be completed by the lineman wearing rubber gloves and rubber sleeves, or using hot-line tools determined by the voltage of the primary circuit and associated working practice. The lineman should not stand on the transformer or other grounded device to work on the energized circuits.

When the transformer has been energized from the primary feeder, the secondary connections can be completed. The lineman should check the secondary with a voltmeter to be sure it is deenergized before completing the connections. If the transformer is to be connected in parallel with an existing transformer, the secondary circuits must be properly phased out using the voltmeter to be sure they can be connected together without causing a short circuit. The voltmeter should always indicate zero voltage between the transformer low-voltage terminal and the secondary wire before the connections are completed. The linemen must be especially careful to be sure the customer's neutral or ground wire is not connected to an energized terminal or wire to prevent a short circuit and damage to the customer's equipment.

A voltmeter or test lamp can be used to check connections if a single-phase transformer is to be connected in parallel with an existing transformer installation. Connect the voltmeter or lamp across a live secondary lead to test for proper operation. After the ground or neutral connection has been completed and the transformer is energized, test for voltage between one of the remaining secondary mains as shown in Fig. 36-13. If this is the proper connection, the voltmeter will read zero or very nearly zero. To avoid confusion, it is best to connect this lead immediately to the secondary main. Now proceed to the last lead and last secondary main. This, of course, should also read zero. After completing this connection, the new transformer is in parallel with other transformers connected to the same secondary mains and will share the load connected thereto.

Figure 36-14 shows the procedure to be followed in connecting a two-wire 120-volt transformer to a three-wire 120/240-volt secondary. All connections should be completed carefully in accordance with the instructions and checked for proper tightness.

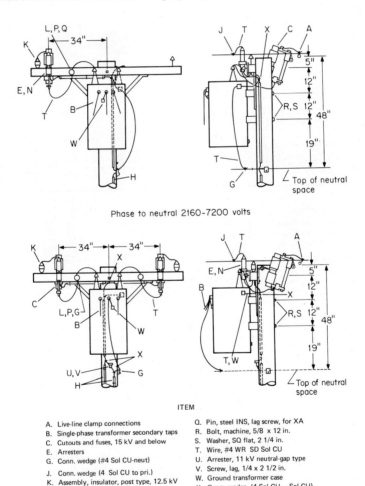

Phase to neutral 2160-7200 volts

ITEM

A. Live-line clamp connections
B. Single-phase transformer secondary taps
C. Cutouts and fuses, 15 kV and below
E. Arresters
G. Conn. wedge (#4 Sol CU-neut)
J. Conn. wedge (4 Sol CU to pri.)
K. Assembly, insulator, post type, 12.5 kV
L. Conductor ties (conventional)
N. Bracket, cutout and arrester, Arm MTG
P. INS, pin, 12 kV top groove

Q. Pin, steel INS, lag screw, for XA
R. Bolt, machine, 5/8 x 12 in.
S. Washer, SQ flat, 2 1/4 in.
T. Wire, #4 WR SD Sol CU
U. Arrester, 11 kV neutral-gap type
V. Screw, lag, 1/4 x 2 1/2 in.
W. Ground transformer case
X. Conn. wedge, (4 Sol CU — Sol CU)

Fig. 36-12 Single-phase, 25 kVA conventional distribution-transformer installation for 13,200-volt or lower voltage-distribution primary-feeder circuit.

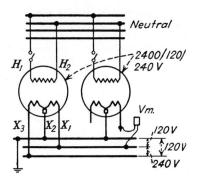

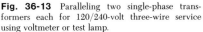

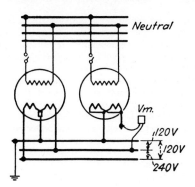

Fig. 36-13 Paralleling two single-phase transformers each for 120/240-volt three-wire service using voltmeter or test lamp.

Fig. 36-14 Paralleling a single-phase transformer two-wire, 120 volts with another connected three-wire, 120/240 volts using voltmeter or test lamp. One of two possible connections.

Three-phase transformer connections must be completed properly to obtain the voltages desired and to prevent an accidental short circuit. The primary connections for a wye-wye, wye-delta, or delta-delta transformer bank can be completed and the transformer energized without danger of a short circuit providing one corner of the delta secondary is left open. A voltmeter must be used on the Y-Y transformer bank to measure the phase-to-phase voltages for proper magnitude and balance. The voltmeter must have a voltage rating equal to twice the secondary line voltage to safely check the delta-delta voltage. The voltmeter or test lamp is connected across the open corner of the secondary delta windings (Fig. 36-15). If the connection of the delta is correct, the lamp will not burn and the voltmeter will read zero or almost zero. But if one of the transformers

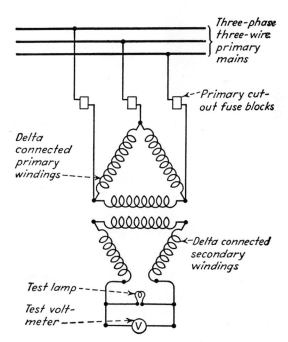

Fig. 36-15 Use of lamp or voltmeter in testing out delta connection on three-phase transformer bank. When lamp tests dark or voltmeter reads zero, the delta connection is correct.

is reversed, the lamp will burn brightly. If the voltmeter is used instead of the lamp, it will indicate a voltage equal to twice the voltage of one of the secondary windings. Under these conditions the delta must not be closed, for a severe short circuit would result. To correct the connection, reverse the leads from one of the secondaries and repeat the test. If this does not correct the connection, reverse the secondary of another transformer and repeat the test. When the lamp fails to burn or the voltmeter reads zero, the delta connection is correct.

On higher voltage secondaries a small step-down potential transformer with a lamp connected to its secondary can be used in place of the lamp or voltmeter mentioned above.

The CSP Distribution Transformer   The completely self-protected (CSP) distribution transformer is a completely self-protected transformer requiring no auxiliary protective equipment such as cutout fuses and lightning arresters. These are incorporated within the transformer and therefore do not have to be supplied or mounted separately. Figure 36-16 shows an exterior view of a CSP transformer, and Fig. 36-17 is a diagrammatic sketch of the transformer and its

Fig. 36-16   Exterior view of a CSP distribution transformer showing primary bushing, lightning arrester, secondary terminals, tank ground connection, and secondary circuit breaker operating handle. *(Courtesy Westinghouse Electric Corp.)*

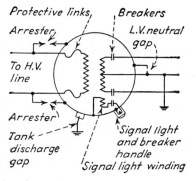

Fig. 36-17   Schematic diagram of connections of a typical CSP distribution transformer. Note primary protective links, secondary circuit breaker, lightning arresters, tank discharge gap, low-voltage neutral gap, signal lamp winding, and breaker handle.

component parts. Such transformers are equipped with brackets for direct to pole mounting, but when provided with hanger irons can also be hung from crossarms.

The following features are incorporated in a CSP transformer:

1. Primary-voltage lightning arresters. These are supported externally on the tank. They may discharge through external discharge gaps.

2. Primary protective links in series with line leads serve to disconnect the transformer in case of internal transformer fault.

3. Overcurrent protection in the secondary leads provided by a low-voltage circuit breaker. This breaker is mounted above the core and coils but operates under oil.

4. Overload signal lamp which becomes lighted when the maximum safe operating load on the transformer is exceeded or when the secondary breaker has tripped automatically.

5. An external operating handle for opening or closing the secondary breaker or for resetting the signal lamp and breaker tripping mechanism.

6. Secondary neutral gap to transformer tank. This is used only if the transformer is not grounded.

7. No-load tap changer. This tap changer provides four 2½ percent voltage taps which give a choice of four values of transformation ratio and compensate for voltage drop on long distribution lines. The operating handle is above the oil level, but the taps and contacts are under oil. The taps should be changed only when the transformer is deenergized.

To install a transformer of this type, it is only necessary to ground the neutral conductor, connect the high-voltage terminals to the primary line, and connect the secondary terminals to the secondary mains. Figures 36-18, 36-19, and 36-20 illustrate typical installation details to obtain single-phase and three-phase-secondary service using CSP single-phase-distribution transformers manufactured for line-to-ground primary connections.

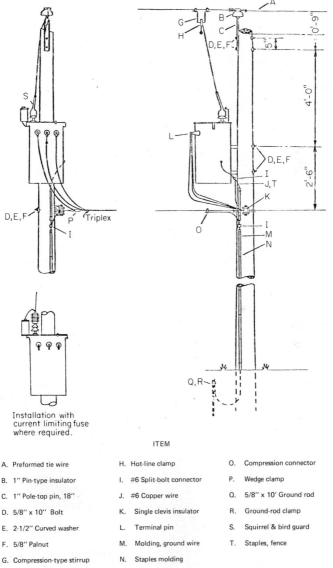

Installation with
current limiting fuse
where required.

ITEM

A. Preformed tie wire	H. Hot-line clamp	O. Compression connector
B. 1″ Pin-type insulator	I. #6 Split-bolt connector	P. Wedge clamp
C. 1″ Pole-top pin, 18″	J. #6 Copper wire	Q. 5/8″ x 10′ Ground rod
D. 5/8″ x 10″ Bolt	K. Single clevis insulator	R. Ground-rod clamp
E. 2-1/2″ Curved washer	L. Terminal pin	S. Squirrel & bird guard
F. 5/8″ Palnut	M. Molding, ground wire	T. Staples, fence
G. Compression-type stirrup	N. Staples molding	

Fig. 36-18 Specifications for installation of a single-phase CSP-distribution transformer connected line to ground on a single-phase tap to a three-phase, four-wire grounded-neutral primary circuit.

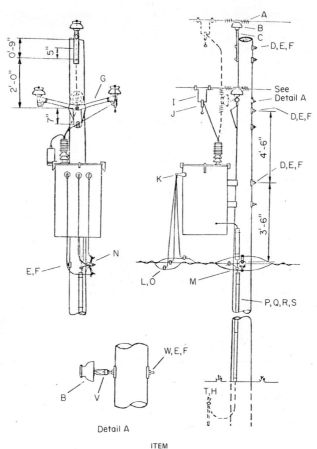

Detail A

ITEM

A. Preformed tie wire
B. 1″ Pin insulator
C. 1″ Pole-top pin, 18″
D. 5/8″ Bolt
E. 2-1/2″ Curved washer
F. 5/8″ Palnut
G. Epoxy-rod bi-unit assembly
H. Ground-rod clamp
I. Stirrup, compression type
J. Hot-line clamp
K. Pin, terminal
L. Compression connector

M. #6 Split-bolt connector
N. Aerial cable clamp
O. Connector cover
P. Staples, fence
Q. #6 Copper wire
R. Molding, ground wire
S. Staples, molding

T. 5/8″ x 10′ Ground rod
V. Transformer lead adapter pin
W. 5/8″ Double-arming bolt
E. 2-1/2″ Curved washer
F. 5/8″ Palnut

Details: Additional material required if transformer is connected to center-phase conductor

Fig. 36-19 Specifications for installation of a single-phase CSP-distribution transformer connected line to ground on a three-phase, four-wire, grounded-neutral primary circuit.

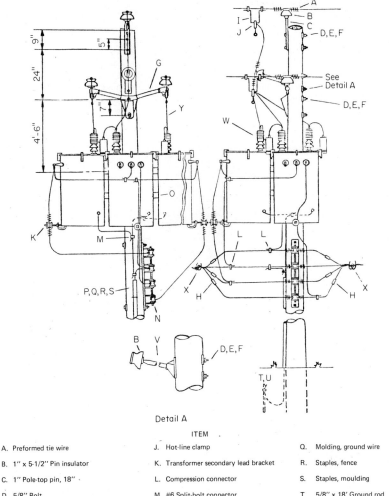

Detail A

ITEM

A. Preformed tie wire	J. Hot-line clamp	Q. Molding, ground wire
B. 1" x 5-1/2" Pin insulator	K. Transformer secondary lead bracket	R. Staples, fence
C. 1" Pole-top pin, 18"	L. Compression connector	S. Staples, moulding
D. 5/8" Bolt	M. #6 Split-bolt connector	T. 5/8" x 18' Ground rod
E. 2-1/2" Curved washer	N. 4 Spool rack	U. Ground-rod clamp
F. 5/8" Palnut	O. Transformer mounting bracket	V. Small Epoxy-glass standoff
G. Epoxy-rod biunit assembly 1-1/2" D rod, 36" between conductors	P. #6 Copper wire	W. Partial-range current-limiting fuse
H. Non-tension sleeve		X. #33 Scotch electrical tape
I. Stirrup, compression type		Y. #4 Standed copper wire

Fig. 36-20 Specifications for installation of single-phase CSP-distribution transformers manufactured for line-to-ground primary connections to obtain three-phase, four-wire grounded-neutral secondary voltages using a wye-wye transformer connection.

Phase Sequence Phase sequence is the sequence or order in which the three voltages of a three-phase system appear. The phase sequence of a three-phase system is often desired in order to:
1. Determine the direction of rotation of polyphase motors.
2. Determine the proper connections for paralleling three-phase transformer banks, generators, and power buses.

3. Determine that the phase sequence is not changed when a three-phase transformer installation is replaced.

4. Determine the proper connections for watthour meters, instruments, and relays.

One type of phase-sequence indicator, like that illustrated in Fig. 36-21, makes use of a small
capacitor connected in Y with two small neon lamps. When used, the terminals of the Y are
connected to the conductors of the three-phase circuit of which the phase sequence is to be determined. A schematic diagram of the connections is shown in Fig. 36-22.

When the phase-sequence indicator is connected across a three-phase line as shown in Fig. 36-
22, an unequal distribution of voltage occurs in the three arms of the Y network. This makes the

Fig. 36-21   Phase-sequence indicator showing two neon lamps and three connecting leads. This particular
model can be used on 120-, 240-, and 480-volt circuits. *(Courtesy General Electric Co.)*

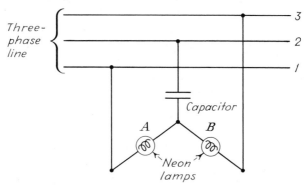

Fig. 36-22   Phase-sequence indicator connected to three-phase line. Indicator consists of a capacitor and two
neon lamps, *A* and *B*, connected in Y.

voltage across one of the neon lamps considerably greater than across the other for a given phase
rotation. In fact, if the voltage across the darker lamp falls below the minimum breakdown or
ignition value, the lamp will cease to glow completely. Consequently, only one of the neon lamps
glows brightly at a time. The one that glows brightest indicates the phase rotation of the line
voltages.

Potential transformers or potential devices must be connected in series with the phase-sequence
instrument to check phase rotation of high-voltage lines.

Phase sequence or rotation is designated as 1-2-3 or 3-2-1. The first order is engraved on the
case under the left-hand (*A*) lamp, and the second order is engraved under the right-hand (*B*)

lamp. The leads which are used to measure the phase sequence of the three-phase line are identified with engraved markings of 1, 2, and 3.

To use the phase-sequence indicator, the voltage-selector switch is first set to the desired voltage. The three leads of the instrument are then connected to the lines to be measured. One lamp should then glow much more brightly than the other. If the left-hand (A) lamp glows brightest, the phase sequence is 1-2-3, and if the right-hand (B) lamp glows brightest, the phase sequence is 3-2-1.

In order to check the sequence indicated, interchange any two of the leads. This changes the order, and therefore the first lamp should become dark and the second lamp should become bright. This also serves as a check on the condition of the lamps. If one of the lamps should fail to glow during this test procedure, it should be replaced with a new neon lamp and the checking procedure repeated.

When it is necessary to change out a three-phase transformer installation, the lineman should always check the phase sequence. When the new transformer installation is completed, the phase sequence should be rechecked before the load is connected. The phase sequence must not be changed. Reversing the phase sequence would reverse the rotation of the customer's three-phase motors and would probably damage the customer's equipment.

Distribution Secondaries   The low-voltage wires from the distribution transformer low-voltage terminals to the terminal point of the customer's service-wire connections at a pole or an

Table 36-5   Approximate Number of Allowable House Spans on One Run of Secondary Main for Any Given Wire Size and Type of Secondary

(The demand per house is assumed at about 3500 watts, and the span length is taken at 125 ft. If larger houses are to be supplied, they can be counted as two or three houses in the calculations. If longer or shorter spans are being used, the number of house spans can be changed in direct proportion. In this way the table can be used for various conditions.)

Wire size, AWG	Maximum allowable number of house spans	
	120-volt, two-wire	120/240-volt, three-wire
No. 4 copper or No. 2 aluminum	2	6
No. 2 copper or No. 1/0 aluminum	3	10
No. 1/0 copperor No. 3/0 aluminum	4	16
No. 2/0 copper or No. 4/0 aluminum	5	20
No. 4/0 copper or 350 mcm aluminum	8	31

underground-system pedestal are called secondaries or secondary mains. In well-designed distribution systems the voltage drop from the transformer secondary terminals to the end of the secondary in any direction should not exceed reasonable values. For secondary lighting mains, this drop is usually limited to 3 percent and for secondary power mains to 5 percent. These percentages apply to the voltage of the circuit. On a 120-volt two-wire circuit a 3 percent drop would equal 3.60 volts, and on a 240-volt circuit a 3 percent drop would equal 7.20 volts. A 5 percent drop would be equal to 12.0 volts on a 240-volt circuit.

The size of wire required to keep within the 3 percent drop in ordinary lighting circuits can be roughly determined from Table 36-5. This table is based on the number of "house spans" on either side of the transformer. By *house spans* is meant the product of the number of houses served from each pole times the number of spans between the pole and the transformer. The services, of course, will have to be connected so as to balance the three-wire secondaries; that is, the same number of houses should be connected between each outside wire and neutral. In addition, the houses connected to both sides should be distributed along the length of the secondary, so as to make the house spans connected to each outside wire also equal if possible.

The estimating chart for size of three-wire secondary mains shown in Fig. 36-23 can be used to find the size of secondary mains required for residential areas.

Make a secondary layout sketch similar to that shown in Fig. 36-24 and show the kilowatt demand at each pole for the residences served from that pole. A demand of 3.5 kW is assumed for each residential customer. The kilowatt demand at each pole is calculated by multiplying the

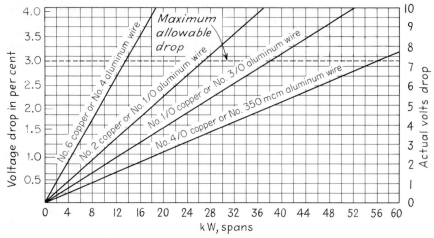

Fig. 36-23 Chart for selecting wire size for 120/240-volt single-phase 60-Hz three-wire secondary mains. Chart is based on the use of conductors spaced 12 in apart. Assumed power factor is 80 percent. Span length is 100 ft. Unbalance factor is 1.2.

demand per customer by the number of services connected to that pole. The amount is shown on the sketch beside the pole. As an example, take the No. 2 pole. This pole serves four customers. At 3.5 kW each, the total demand at that pole is $3.5 \times 4 = 14.0$ kW.

Then the kilowatt spans are calculated for each pole by multiplying the demand at each pole by the number of pole spans between the pole and the transformer. The No. 2 pole in the sketch, for example, has a demand of 14.0 kW and is two spans from the transformer pole. The kilowatt spans therefore are $14.0 \times 2 = 28$ kW spans. These values of kilowatt spans are tabulated as shown in Table 36-6.

Now add the kilowatt spans served to the right and to the left of the transformer pole. The total is 38.5 each, for both left and right, showing that the transformer is correctly located. If these left and right values should not be approximately equal, the transformer should be moved to the next pole in the direction of the larger kilowatt span. The kilowatt spans should be as nearly equal as possible. This is the same as saying that the transformer should be located as nearly as possible in the center of the load.

Now, to determine the size of wire required for the three-wire secondary mains, refer to the chart of Fig. 36-23 and find the value of kilowatt spans along the horizontal scale, which in this case is 38.5. Then move vertically upward on the chart until you reach the line marked "maximum allowable." The No. 1/0 copper or 3/0 aluminum wire is the proper size as it will give a voltage drop below 3 percent, which corresponds to 7.2 volts as shown on the right-hand scale of the chart.

Table 36-6   Kilowatt-Span Calculations for Secondary Layout
(The values shown in the kilowatt-spans column are the products of the kilowatt demands at each pole multiplied by the number of spans from the transformer pole. If any load is connected to the transformer pole, it does not enter the calculation for determining the wire size. However, this load must be considered when selecting the size of the transformer required.)

Left side			Right side		
kW	Spans	kW spans	kW	Spans	kW spans
10.5 ×	1	= 10.5	10.5 ×	1	= 10.5
14.0 ×	2	= 28.0	14.0 ×	2	= 28.0
Total		38.5	Total		38.5

The chart is based on a span length of 100 ft. If the actual average span length varies very much from this, the total kilowatt spans as calculated should be multiplied by the ratio of the actual average span to 100 ft before using the chart. Likewise, if the kilowatt demand per customer is more than 3500 watts, this should also be taken into account by increasing the kilowatts per pole accordingly.

Accurate determination of the size of wire required for the secondary mains of lighting circuits can be made by use of the curves in Figs. 36-25 and 36-26. To use these curves the kilowatt

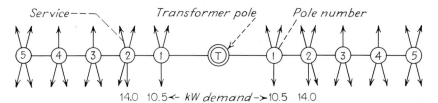

Fig. 36-24 Secondary layout sketch showing transformer pole, services at each pole, and demand at each pole. Demand per customer is assumed to be 3.5 kw, or 3500 watts.

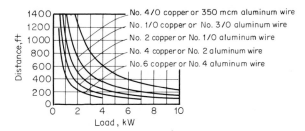

Fig. 36-25 Load curves for 120-volt single-phase lighting. Curves show distance that any load in kilowatts K can be transmitted with 3 percent drop on lines. Voltage at load, 120 volts; spacing of wires, 8 in; power factor of load, 95 percent; 60-Hz current.

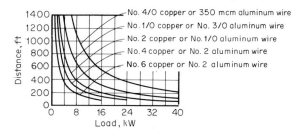

Fig. 36-26 Load curve for 120/240-volt single-phase three-wire lighting. Curve shows the distance in feet D that any load in kilowatts K can be transmitted with 3 percent voltage drop on line, allowing for unbalance of three-wire circuit. Voltage at load, 120/240 volts; spacing of wires, 4 in; power factor of load, 95 percent; 60-Hz current.

demand for each house and the distance in feet between the house and the transformer must be determined. Then the demand in kilowatts and the distance in feet are multiplied to get the kilowatt distance for that house. The kilowatt distance is likewise obtained for each house and the total for all the houses on one side of the transformer. This total may be called KD. Next add the total kilowatt demand on the same side of the transformer and call it K. Now divide KD by K and get D, which is the equivalent distance in feet to which a concentrated load equal to K may be transmitted. With K and D known, refer to Fig. 36-25 for 120-volt two-wire mains or to

Fig. 36-26 for 120/240-volt three-wire mains, and find the intersection of the lines from K and D. If the point of intersection does not fall directly on one of the curves, use the wire size of the curve next above the point. These curves are based on a 3 percent voltage drop.

In calculating the wire size for three-phase power secondaries, the procedure is the same as for lighting secondaries outlined above. In the case of power loads, however, the load can be more easily determined as the motors to be supplied and their ratings are known. The procedure consists in getting the value of each motor load K and the distance D from the transformer. Then obtain the total of the KD products, divide by K, and obtain the equivalent distance D. Now refer to Fig. 36-27 for 240-volt three-phase secondaries, and obtain the intersection of K and D. The curve on which this point of intersection falls is the proper wire size to use for a 5 percent voltage drop.

The procedure for single-phase power secondaries is the same as that already outlined for lighting and three-phase secondaries, except that the curves in Fig. 36-28 must be used. These curves are for 240 volts, single phase, and allow for a 3 percent voltage drop.

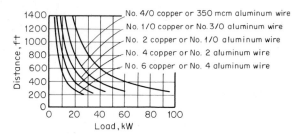

Fig. 36-27 Load curves for 240-volt three-phase power. Curves show distance in feet D that any load in kilowatts K can be transmitted with 5 percent voltage drop on lines. Voltage at load, 240 volts; equivalent spacing, 5.05 in; power factor of load, 80 percent; 60-Hz current.

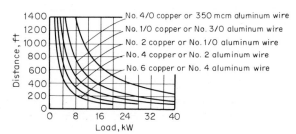

Fig. 36-28 Load curve for 240-volt single-phase power. Curves show distance in feet D that any load in kilowatts K can be transmitted with 3 percent drop on lines. Voltage at load, 240 volts; spacing of wires, 4 in; power factor of load, 94 percent.

Installing Overhead Secondary Mains   Overhead secondary mains are usually installed on vertical racks bolted to the poles (Fig. 36-29) or with triplex or quadruplex cable secured to the poles (Fig. 36-30).

In stringing wires on bracket construction, the conductors are unreeled and passed through the rack, as shown in Fig. 36-31a. When the desired number of pole spans have been laid in place, the conductors are drawn up and dead-ended. As many as 10 spans can be drawn up at one time in this manner. Figure 36-32 shows a lineman pulling up a secondary main by means of block and tackle, and Fig. 36-33 shows the conductors properly dead-ended on the rack. Note the use of the terminal guy to counterbalance the pull of the line conductors.

The conductors are then lifted up to the insulator groove and tied in. The regular Western Union tie is generally used (see Fig. 36-31b and c). If the conductors are to be tied to the outside of the insulator, the Western Union tie is also used (see Fig. 36-31d).

In turning corners and at angles in the line, the position of the line wires on the insulators will be determined by the direction of the strain. They should always be so placed that the conductor is pulled against the insulator and not away from it. Figure 36-34 illustrates the correct positions for corner and angle construction.

Fig. 36-29 Lineman working from bucket mounted on a truck making connection to primary neutral conductor. Secondary mains above the lineman's head are secured to pole with vertical rack. The neutral or grounded conductor of the 120/240-volt three-wire single-phase secondary main is located on the white color center insulator.

Fig. 36-30 Triplex secondary main operating at 120/240-volts single-phase terminates on pole. Triplex services connected to the secondary extend to the customers' houses.

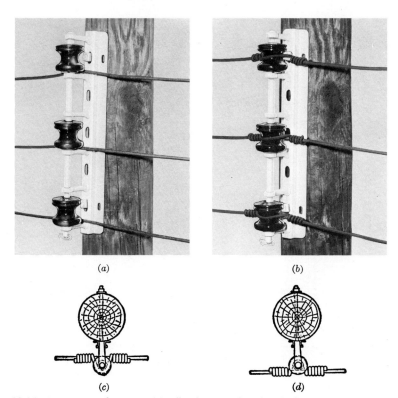

(a) (b)

(c) (d)

Fig. 36-31 Stringing secondary mains. (a) Pulling line wires through rack. (b) Line wires tied to inside of insulator. (c) Top view of line wire tied to inside of insulator. (d) Top view of line wire tied to outside of insulator. *(Courtesy Hubbard & Co.)*

Fig. 36-32   Lineman pulling up a secondary-line wire by means of block and tackle. The pole is a dead-end pole from which a service connection is to be strung. Note terminal guy, which takes up the pull of the second-ary mains. *(Courtesy Hubbard & Co.)*

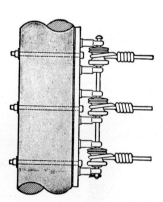

Fig. 36-33 Completed dead end on secondary-rack construction. Two or more through bolts should be used on each rack. *(Courtesy Hubbard & Co.)*

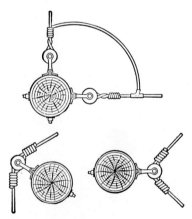

Fig. 36-34 Suggested rack construction on corners and angles. Line conductor is always placed on insu-lator so that the pull is against the insulator as shown. *(Courtesy Hubbard & Co.)*

Distribution Services   The low-voltage wires that connect to the secondary main conductors and extend to the customer's service entrance on his building are called services. The services may be overhead or underground usually depending on the type of distribution system. However, underground services are frequently installed from an overhead secondary main to the customer's building to eliminate the overhead wires across the customer's property. The size of the wire used for the customer's service is determined by the size of the electrical load. Single-phase 120/240-volt three-wire services are installed to all new homes.

Overhead services are usually No. 4 AWG aluminum conductor with a polyethylene insulation assembled in a triplex configuration (Fig. 36-35). Copper conductor No. 6 AWG is used in some instances instead of No. 4 AWG aluminum. Single-conductor wire may be used with supporting racks. The use of single-conductor wires on racks permits the use of weatherproof wire instead of insulated wire for overhead services, reducing wire costs. However, triplexed cables with the close conductor spacing reduces the reactance and improves the voltage for the customer.

Underground services to new homes are usually installed with 2/0 aluminum conductor or larger to provide a minimum capacity of 200 amps. Underground services are installed large enough to prevent the need to replace them when the customer's load increases. Overhead services can be installed appropriately for the customer's load and replaced wtih larger conductors when the load increases minimizing the investment.

Fig. 36-35   Lineman holding ends of three, conductor, No. 4 AWG aluminum, polyethylene-insulated triplex cable. Light-colored conductor will be used for neutral or grounded wire. Similar cable is used for overhead and underground low-voltage services.

Installing Overhead Services   Service brackets are installed on the customer's building, that is, the house, store, or factory, as the case may be, and on the pole to support the service (Figs. 36-36 and 36-37). Service supports should be mounted about 15 ft above the ground. The practice varies slightly with different companies. Figure 36-36 shows the practice of one company which requires 12 ft for residences and 16 ft for business buildings, and Fig. 36-38 illustrates the requirement of another company.

The minimum recommended clearance to ground of the service-drop wire varies with the use made of the ground below the wires. The minimum clearance is 12 ft for voltages from 0 to 750 volts. If an alley or street, the minimum ground clearance is 18 ft, and if they cross a state highway, the clearance must not be less than 22 ft. The span length is limited to 125 ft. Figure 36-39 shows these minimum ground clearances.

The service connection should be made to the nearest distribution-line pole. If this distance should exceed 125 ft, an intermediate support should be provided. This intermediate support should support the wires at least 15 ft above the ground.

After the service wires are unreeled one of the linemen dead-ends the wires onto the service bracket (see Fig. 36-40). As soon as this is done, the other lineman climbs the service pole, pulls up the service wires, and dead-ends them on the distribution pole (see Fig. 36-41). These wires

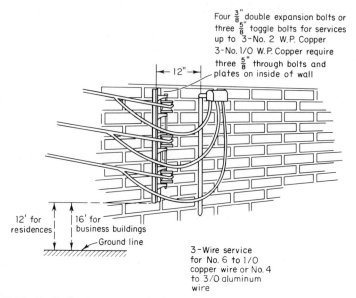

Four $\frac{3}{8}$" double expansion bolts or
three $\frac{5}{8}$" toggle bolts for services
up to 3-No. 2 W.P. Copper
3-No. 1/O W.P. Copper require
three $\frac{5}{8}$" through bolts and
plates on inside of wall

← 12" →

12' for
residences
16' for
business buildings
Ground line

3-Wire service
for No. 6 to 1/O
copper wire or No. 4
to 3/O aluminum
wire

Fig. 36-36 Details of service connection for three-wire services for residences and business buildings.

need be pulled up by hand only, as little strain should be placed on the service supports. If the service-drop wires are pulled otherwise than by hand, too great a strain may be put on the service brackets which may cause them to break or be pulled out of the building structure.

If the service drop consists of a self-supporting service-drop cable (Fig. 36-42), only the messenger wire need be dead-ended. Usually a special feed-through dead-end fixture is employed as shown in Figs. 36-43 and 36-44.

The type of support for the service wires on the distribution pole depends on the type of secondary main construction. If the secondary mains are supported by secondary racks, the same racks may also be used for the support of the service drops, as shown in Fig. 36-45.

The installation of the watthour meter is the next step in the running of the service. The meter is, as a rule, the property of the company and is therefore installed by the company at the time of making the service connection. The details of the installation of the meter socket (Fig. 36-46) will not be discussed here as this work is generally done by an electrician working for the customer.

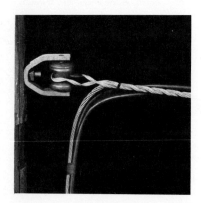

Fig. 36-37 Typical service drop cable terminated on pole near residence with prefabricated grip. *(Courtesy Preformed Line Products Co.)*

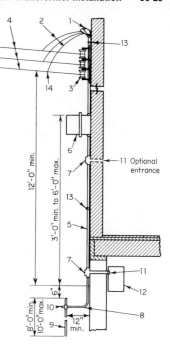

Fig. 36-38 Installation of customer's outside meter and service entrance, three-wire 120/240-volt single-phase, from overhead distribution system. (1) Weatherhead—must extend above service wires. (2) Service wires—size of service wires is determined by application of the National Electric Code. Service wires must extend at least 36 in from weatherhead. The neutral wire must be identified. (3) Service rack. (4) Service wires. (5) Galvanized rigid conduit (not water pipe). (6) Outdoor meter socket. (7) Galvanized conduit fitting with threaded hubs. (8) Galvanized conduit or water pipe of size to carry neutral wire. (9) ½ in or larger copperweld ground rod or ¾ in galvanized water pipe. Must be driven minimum of 8 ft in ground. (10) Solderless connector (copper or bronze) of proper size to connect ground wire to rod or pipe. (11) Galvanized conduit nipple and end bushing. (12) Service entrance fused switch, circuit breaker, or fuse box. (13) Galvanized conduit straps spaced approximately 4 ft apart. *(Courtesy Iowa-Illinois Gas and Electric Company.)*

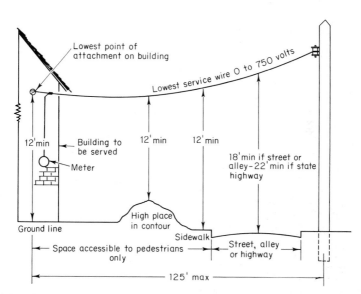

Fig. 36-39 Sketch showing minimum ground clearances of service drop wires having voltages of 0 to 750 volts above open spaces, sidewalk, and street, alley, or highway, as recommended by the National Electrical Safety Code. Note that span length is limited to 125 ft.

Fig. 36-40   Lineman preparing to insert watthour meter in newly connected customer's residence. *(Courtesy Kansas City Light and Power Company.)*

Fig. 36-41   Dead-ending service drop to new house on distribution pole. *(Courtesy American Electric Power System.)*

Fig. 36-42   Self-supporting service drop cable insulated with polyethylene or neoprene. Messenger serves as neutral conductor. *(Courtesy Reynolds Metals Company.)*

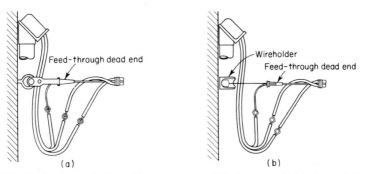

Fig. 36-43   Dead-ending service drop cable on consumer's building by means of (*a*) feed-through dead-end fixture attached to eye bolt, (*b*) feed-through dead-end fixture attached to wire holder.

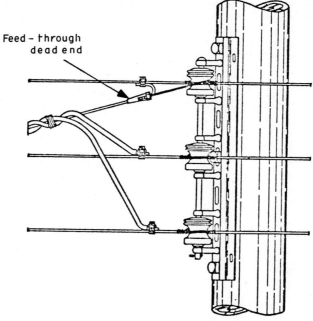

**Feed – through
dead end**

Fig. 36-44   Dead-ending service drop cable on distribution service pole by means of special feed-through dead end attached to secondary rack insulator spool.

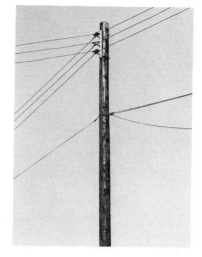

Fig. 36-45   Service drop wires supported on secondary-main bracket. In this instance neutral conductor of three-wire service is middle conductor. *(Courtesy Iowa-Illinois Gas and Electric Company.)*

Testing Service Wires for Polarity and Ground   Before the service wires can be connected onto the secondary mains, they must be tested with a voltmeter or *lamped out* for polarity. This must be done to avoid connecting the customer ground to either of the *hot* wires of the secondaries. To do so would produce a *short circuit* on the secondary mains. The customer ground wire, usually white or gray color or bare, should always be connected to the neutral or grounded wire of the service.

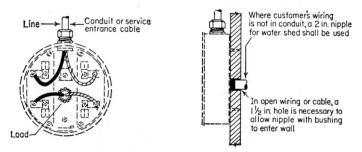

Fig. 36-46   Details of typical meter socket installation. Leave 12 in of wire or cable for socket connections which will be made at the time meter is installed. Entrance and outlet to sockets shall be with conduit or cable and watertight connector terminating in such a manner as to be weather- and insect-proof.

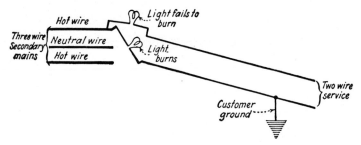

Fig. 36-47   Lamp test for polarity on two-wire service. The service drop wire that makes the lamp burn is the grounded wire which should be connected to the neutral or grounded wire of the secondary mains.

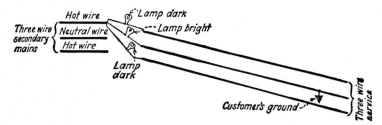

Fig. 36-48   Lamp test for polarity on three-wire service. The service drop wire that makes the lamp burn is the grounded wire which should be connected to the neutral or grounded wire of the secondary mains.

To perform the polarity, or *lamping-out*, test, an incandescent lamp or voltmeter is used. This should have a voltage rating equal to the voltage across the *hot* wires of the secondary. If the service is a two-wire service, the test is made as shown in Fig. 36-47. The service wire which is grounded in the customer's premises should be connected to the neutral of the secondary main. To determine which of the service-drop wires is grounded, the lamp is connected from one of the

Fig. 36-49 Connector used for service connection. *(Courtesy James R. Kearney Corp.)*

Fig. 36-50 Split bolt type of connector. *(Courtesy James R. Kearney Corp.)*

hot or *live* secondary main wires to one of the service wires. If the lamp burns, the service-drop wire to which the lamp is then connected is grounded. If the lamp does not burn, that service-drop wire is not grounded. If the service is a three-wire service, the procedure is the same except that three wires of the service must be *lamped out*. See Fig. 36-48. All one needs to remember is that the service-drop wire that makes the lamp burn is the grounded wire and should be connected to the secondary main neutral.

Connecting Service Drop to Secondary Mains When the service-drop leads are lamped out, they are ready to be permanently connected to the secondary mains.

Connectors are used to obtain a good electrical connection. Taps to secondary mains, service entrance connections, jumper loop connections, etc., are generally made with so-called "connectors."

Two types of connector are in general use, namely:
1. Bolted or clamp type
2. Compression sleeve type

The bolted type (Figs. 36-49 to 36-51) depends on the use of bolts and nuts to clamp or press the overlapped and parallel conductors together. The connector can be a split bolt or a U bolt if only one bolt is employed, or two or three bolts may be used if the connector takes the form of a clamp. The compression-type connector is a form of open sleeve connector (Figs. 36-52 and 36-53) that is installed with a compression tool.

Fig. 36-51 Service drop cable connected to customer's service entrance loop with split-bolt connector. Note bare neutral conductor serving as messenger supporting service drop cable. *(Courtesy Burndy Corporation.)*

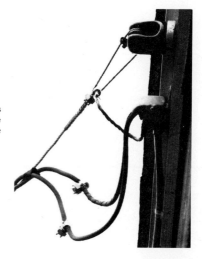

Fig. 36-52 Lineman connecting service drop cable onto service entrance loop with compression sleeve and power compressor. Note compression sleeve on bare neutral conductor and covered sleeve joint on left conductor. Bare neutral conductor also serves as messenger supporting cable. *(Courtesy A. B. Chance Co.)*

Fig. 36-53 Lineman connecting service drop cable to secondary mains by use of crimpit tap sleeve and compression tool. Note bare neutral, which also serves as messenger wire and supports service drop. *(Courtesy Burndy Corporation.)*

Connectors are easily and quickly applied. If the conductor has a weatherproofed covering, this is first removed and the exposed conductor scraped clean and bright. The two ends are then inserted in the bolt or clamp, and the nuts are drawn up tight. If the connector is of the compression type, the proper compression tool is used to crimp the sleeve around the two conductors. Then the connection may be coated with inhibitor grease and taped so that it will conform to the service-drop wires and secondary mains.

Electrical
Drawing Symbols

Diagrams of electrical circuits show the manner in which electrical devices are connected. Since it would be impossible to make a pictorial drawing of each device shown in a diagram, the device is represented by a symbol.

A list of some of the most common electrical symbols used is given below. It will be noted that the symbol is a sort of simplified picture of the device represented.

Electrical device	Symbol	Electrical device	Symbol
Air circuit breaker		Cable termination	
Air circuit breaker drawout type, single pole		Capacitor	
three pole		Coil Electromagnetic actuator	
Air circuit breaker with magnetic-overload device			
Air circuit breaker with thermal-overload device		Contact Make-before-break	
		Normally closed	
Ammeter		Normally open	
Autotransformer			
Battery		Contact (spring-return) Pushbutton operated Circuit closing (make)	
Battery cell		Circuit opening (break)	
		Two circuit	

Electrical device	Symbol	Electrical device	Symbol
Contactor Electrically operated with series blowout coil		Lightning arrester with Gap Valve Ground	
Contactor Electrically operated three pole with series blowout coils and auxiliary contacts— one closed, two open		Magnetic blowout coil	
Current transformer		Network protector	
Current transformer with polarity marking		Motor	
		Oil circuit breaker	O.C.B.
		Oil circuit recloser	O.C.R.
Fuse		Polarity markings Negative Positive	− +
Fused disconnect switch		Potential transformer	
Gap Horn		Potential transformer with polarity markings	
Horn with ground			
Gap protective		Rectifier	
Generator	G	Relay Circuit closing	
Ground connection		Circuit opening	
Inductance Fixed		Resister Fixed	
Adjustable		Tapped	
Lamp Red	R	Variable	
Green	G	Variable tapped	

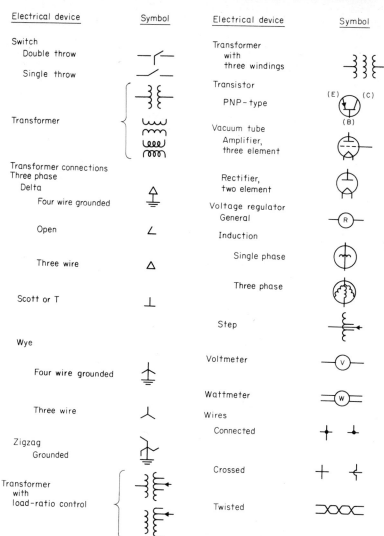

Electrical device	Symbol	Electrical device	Symbol
Switch		Transformer with three windings	
Double throw			
Single throw		Transistor	
Transformer		PNP-type	
		Vacuum tube Amplifier, three element	
Transformer connections Three phase			
Delta		Rectifier, two element	
Four wire grounded		Voltage regulator General	
Open		Induction	
Three wire		Single phase	
Scott or T		Three phase	
		Step	
Wye			
Four wire grounded		Voltmeter	
Three wire		Wattmeter	
Zigzag		Wires	
Grounded		Connected	
Transformer with load-ratio control		Crossed	
		Twisted	

Single-Line Diagrams

Purpose Most electric transmission and distribution circuits are three-phase and require three lines to show them schematically in detail. For many functions it is not necessary to have all the detailed information shown on the schematic diagram. Therefore single-line diagrams are utilized to represent three-phase circuits and equipment. A single-line diagram provides a quick method whereby the lineman or cableman may analyze the complete facilities in schematic form.

Distribution Unit Substation Single-Line Diagram A 13,200-4160Y/2400-volt unit-type distribution substation is illustrated by Fig. 38-1 in single-line form. Analyzing the unit substation single-line diagram, it can be determined that the substation has as its source an overhead 13,200-volt three-phase subtransmission electric circuit designated 13–61–T–1. The overhead line connects to the substation transformer through 300 MCM copper conductor, paper-insulated, lead-covered cables with pothead terminations on each end of the cable circuit. A dotted line representing the cable circuit indicates that it is installed underground. The power transformer has three potential transformers connected to the primary leads through fuses.

The high-voltage winding of the power transformer is connected delta, has a tap changer, and the low-voltage winding is connected wye with the neutral grounded. The transformer is rated 13,200–4160Y/2400 volts, 5000/6250 kVA. The capacity varies with the type of cooling. Oil-air cooling provides 5000 kVA of capacity, and forced-air cooling provides 6250 kVA of capacity. The tap changer operates underload. The current transformers in series with the 4160-volt bus connections to the power transformer provide a means of measuring the current flow in the power transformer secondary connections.

Cubicle No. 1 of the metal-clad switchgear contains a potential transformer that connects to the 4160-volt leads through a fuse. Cubicle No. 2 of the switchgear has three fused potential transformers that connect to the transformer secondary leads and three current transformers rated 1000/5 amps utilized for relaying and metering circuits. The power transformer's secondary leads connect to a 1200-amp air circuit breaker through draw-out contacts. The transformer's secondary circuit breaker provides a means of isolating the 4160-volt bus from the power transformer. Switchgear cubicle No. 3 contains an operating transformer to provide low-voltage power for auxiliary devices and three potential transformers complete with fuses for metering and relaying purposes. Sections 4, 5, 6, and 7 of the metal-clad switchgear each contain a feeder air circuit breaker and current transformers. Each of the feeder circuits exits from the substation through underground cables to supply distribution circuits operating at 4160Y/2400 volts.

Distribution Substation Single-Line Diagrams Figure 38-2 is a single-line diagram of a distribution substation supplied by two 69,000-volt three-phase subtransmission circuits designated 66–58–59–1 and 66–58–A–1. The subtransmission circuits of 336,400 circular mil aluminum conductor connect to the high-voltage structure of the substation overhead. Lightning arresters and potential transformers in the substation structure connect to the overhead line terminals. The potential transformers provide inputs for meters, relays, and auxiliary devices.

The subtransmission circuits are switched in and out of service utilizing the 69,000-volt 1200-amp oil circuit breakers connected in series with group-operated disconnect switches. The group-operated disconnect switches provide a means of isolating the oil circuit breakers. The disconnect switches are operated after the oil circuit breaker contacts have been opened. The substation 69,000-volt bus has a normally closed group-operated disconnect switch, making it possible to manually isolate the bus into two sections. A 69,000-volt group-operated disconnect switch provides a means for connecting a mobile substation to the 69,000-volt bus. The fused potential transformer connected to the bus provides a means of measuring bus voltage.

The two power transformers in the substation (rated 69,000–13,200Y/7620 volts, 20,000 kVA oil-air cooled, 26,700 kVA forced-oil air cooled with tap changing underload equipment) are each

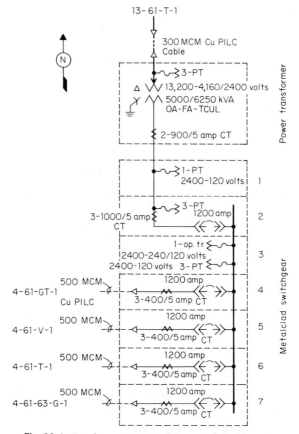

Fig. 38-1   Distribution-unit-type substation single-line diagram.

switched in and out of service by an oil circuit breaker connected to the high-voltage winding and an air circuit breaker in the metal-clad switchgear connected to the low-voltage winding. The 13,200Y/7620-volt leads from the power transformers connect to the circuit breakers in the metal-clad switchgear through 750 MCM copper conductor, paper-insulated lead covered cables installed under ground and terminated with potheads. The current transformers in series with each power transformer air circuit breaker provide a means of measuring the current magnitudes and detecting short circuits with protective relays.

Cubicle No. 1 of the metal-clad switchgear has a group-operated disconnect switch and draw-out power fuses that connect to underground cables terminated with potheads providing a source to the house service transformers rated 13,200–480Y/277 volts. Cubicle No. 7 of the metal-clad

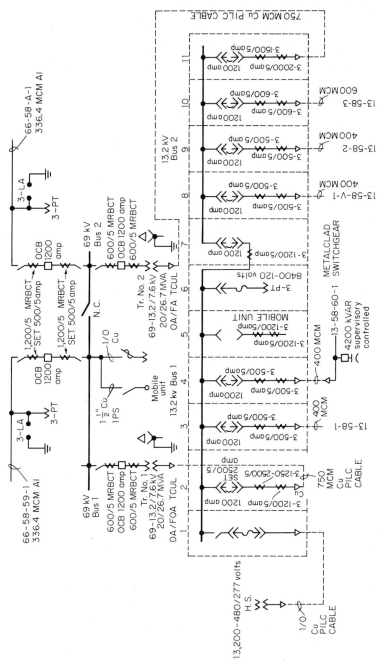

Fig. 38-2 Distribution substation single-line diagram.

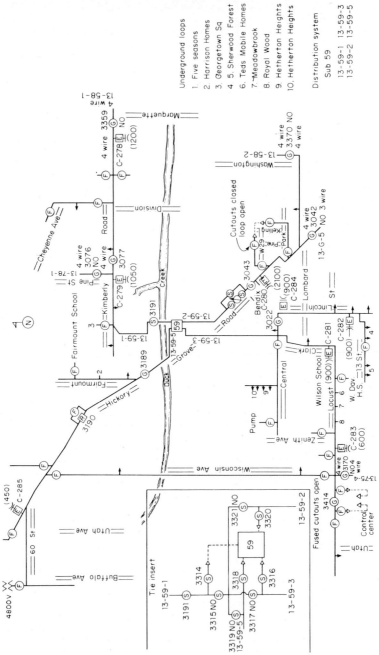

Fig. 38-3 Distribution circuit single-line diagram.

switchgear contains a bus tie air circuit breaker providing a means for sectionalizing the bus for maintenance or to isolate a bus section in case of a bus fault. Switchgear units 3, 4, 8, 9, and 10 each provide a means of switching a distribution circuit in and out of service. Each distribution circuit connects to an air circuit breaker through underground cables. The feeder circuit originating in cubicle No. 4 has a capacitor bank connected to it in the yard of the substation.

Switchgear unit No. 5 provides a means for connecting the 13,200-volt bus to a mobile substation with a power transformer rated 69,000–13,200Y/7620 volts. The mobile substation is utilized to replace one of the power transformers for maintenance work or in the case of a transformer failure.

Distribution Circuit Single-Line Diagram   The single-line diagram of Fig. 38-3 shows the 13,200Y/7620-volt three-phase four-wire distribution circuits originating at substation 59 in single-line form. A square in the center of the figure with numeral 59 represents the substation. Four circuits originating at the substation extend in different directions to supply electrical power to the customers in the surrounding area. The major physical terrain details—such as Duck Creek, Kimberly Road, Hickory Grove Road, Lombard Street, and Wisconsin Avenue—give the schematic diagram physical significance.

The four circuits originating at substation 59 are designated 13–59–1, 13–59–2, 13–59–3, and 13–59–5. All four of the circuits exit from the substation through underground cables. Each of the circuits can be connected to an adjacent circuit at the substation perimeter to provide a means of supplying the customer load in case of a cable failure, as shown on the schematic tie insert at the left edge of the figure. Circuit 13–59–1 exits from the substation by underground cables that terminate in potheads on the riser pole, as shown on the tie insert schematic. The circle in circuit 13–59–1 with an S in the middle represents single-pole disconnect switches operated with hot-line tools.

The number 3314 is installed near the switches on the cable riser pole to identify the switches. The single-pole disconnect switches designated 3314 are used to isolate the underground cables from the overhead circuit in case of a cable failure. The circuit extends from the riser pole through single-pole disconnect switches number 3191. Single-pole disconnect switches 3315 are normally open, as designated by NO on the tie insert schematic. Circuit 13–59–1 can be tied to circuit 13–59–5 by closing the normally open single-pole disconnects numbers 3315 and 3319.

Circuit 13–59–1 extends across Duck Creek to Kimberly Road. The circle with an F followed by the number 3 indicates a fused three-phase tap to the main circuit extending north from Kimberly Road. At Pine Street a capacitor bank number C–279 with the capacity of 1050 kVA is connected to the main circuit through switching equipment. The circuit extending north along Pine Street terminates in a group-operated disconnect switch designated by the circle with the letter G in the middle. The group-operated disconnect switch number 3076 is normally open, designated by NO.

Circuit 13–78–1 terminates on the north side of the group-operated disconnect switch. This switch provides a method of connecting circuit 13–78–1 to circuit 13–59–1. The main feeder portion of circuit 13–59–1 extends east on Kimberly Road through group-operated switch 3077 and terminates at group-operated switch 3359, which is normally open. Group-operated switch 3077 provides a means for sectionalizing circuit 13–59–1 in case of a sustained fault on the circuit that causes the circuit breaker at substation 59 to remain open.

Group-operated switch 3359 provides a method for connecting circuit 13–59–1 to circuit 13–58–1. The distribution circuit single-line diagram provides the linemen and cablemen with a means of understanding switching instructions and locating major pieces of equipment with respect to known landmarks. Circuit 13–59–2 has a voltage regulator installed in it along Hickory Grove Road, southeast of the substation. The voltage regulator is represented by a circle with an arrow through it. The voltage regulator has single-pole disconnect switches in parallel with it for by-passing the regulator and single-pole switches in series with the regulator to isolate it.

Circuit 13–59–5 has a reclosing device in the line along Hickory Grove Road northwest of the substation. The three-phase reclosing device has fused bypass switches in parallel with it. The fused bypass switches are utilized to permit removing the recloser from service for maintenance purposes. The distribution circuit single-line diagram is very valuable for analyzing the circuits to assist the lineman and cableman to complete their work safely and effectively.

Schematic Diagrams

Purpose   The purpose of a schematic diagram is to make it possible to picture the operation of an entire control circuit for a piece of equipment, so that it will be easy to think in terms of the entire circuit instead of in terms of a particular component part of it. Once the control schematic diagram is understood, it is easy to apply the information contained on it to some portion of the associated wiring diagram.

Feeder Circuit Control Schematic   The control schematic diagram for a distribution substation feeder circuit breaker installed in a metal-clad switchgear unit (Fig. 39-1) is traced utilizing the numbers appearing by the component parts of equipment. Each piece of equipment and the purpose it serves is described.

The schematic diagram of Fig. 39-1 is drawn for the condition of all isolating devices open and the circuit breaker in the open position. This is an accepted standard method of drawing schematic control diagrams. The symbol ———<—— represents a circuit breaker draw-out contact in the switchgear. All equipment shown between the draw-out contacts is located on the feeder circuit breaker.

The control source for closing the feeder circuit breaker is 240 volts alternating current. The lead designations for the control power source are X_1 and Y_1. Starting with wire X_1, a circuit extends to item (1), a knife switch which isolates and deenergizes the control circuit. From the knife switch, the circuit is completed to item (2), which is a low-voltage fuse used for isolating the control circuit for a fault in the control wiring, protecting the control power source and preventing a fault in the wiring of one circuit breaker's control wiring from causing all circuit breakers to be inoperative.

From the fuse, the circuit is completed through wire X_2 and $4X_2$ to item (3), 86, which is a contact on the lockout relay operated by the bus fault relays. The purpose of the 86 contact is to open the closing circuit for the feeder breaker to prevent closing the breaker after a bus fault until the lockout relay has been manually reset. From the lockout relay contact, the circuit is complete through wires 4W and W to item (4), CSO, the control-switch-off contact. This contact on the control switch is closed when the control switch handle is in the "off" or straight up-and-down position. The purpose of this contact is to keep the 152Y relay coil deenergized until the control switch is returned from the closed position to the "off" position if the breaker tripped immediately after a close operation.

From the control-switch-off contact, the circuit is completed through wire 2a and the breaker draw-out contact to item (5), 152b, a contact on the circuit breaker auxiliary switch which is closed when the circuit breaker is open. This contact opens this portion of the control circuit so that current will not flow through the resistor continuously while the breaker is closed and completes the circuit to energize the 152Y relay when the breaker is in the open position. From the 152b auxiliary switch contact, the circuit is completed through the breaker draw-out contact, wire 1a, and a resistor to item (6), 152Y, closing cutoff relay coil. The purpose of the 152Y closing

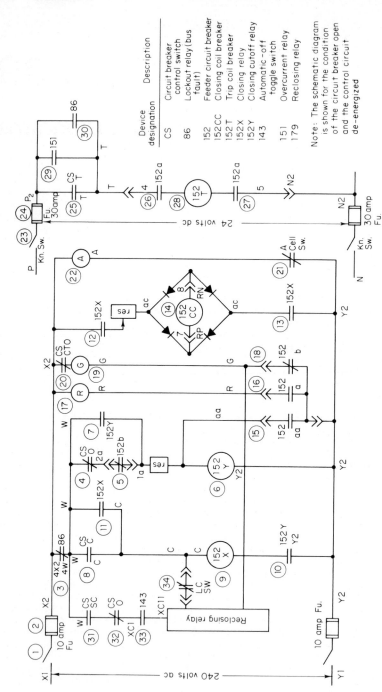

Fig. 39-1 Unit-type substation, feeder circuit, control schematic diagram.

Device designation	Description
CS	Circuit breaker control switch
86	Lockout relay (bus fault)
152	Feeder circuit breaker
152CC	Closing coil breaker
152T	Trip coil breaker
152X	Closing relay
152Y	Closing cutoff relay
143	Automatic-off toggle switch
151	Overcurrent relay
179	Reclosing relay

Note: The schematic diagram is shown for the condition of the circuit breaker open and the control circuit de-energized

cutoff relay is to make the control scheme antipump, or electrically trip free. From the 152Y relay coil, the circuit is completed through wire Y2, a fuse and a knife switch to wire Y1, the other side of the alternating-current control power source.

If the knife switches are closed, a current will flow through the 152Y closing cutoff relay coil, operating the associated relay contacts, items (7) and (10). Item (7), 152Y, closing cutoff relay contact, completes the circuit through wires W and 1a to establish a second path for the current to flow through the 152Y relay coil. The purpose of this 152Y relay contact is to keep the 152Y closing cutoff relay coil energized when the control switch is turned to the closed position or when the breaker starts to close until the breaker is completely closed.

Starting with wire X_1, one side of the alternating-current source, a circuit exists through items (1), (2), (3) and wires 4W and W to item (8), CSC, which is the control-switch-close contact. This contact on the control switch is closed when the control switch handle is turned to the closed position. This contact provides a means of closing the circuit breaker by operating the control switch manually. From the control-switch-close contact, a circuit is completed through wire C to item (9), 152X, the closing relay coil. The purpose of this relay is to energize the circuit-breaker-closing solenoid. From the closing relay coil, the circuit is complete to item (10), 152Y which is a contact on the closing cutoff relay. The purpose of this contact is to make the control scheme anti-pump or electrically trip-free.

From the closing cutoff relay contact, the circuit is completed through wire Y2, the fuse, and the knife switch to wire Y1, the other side of the alternating-current source. If the control-switch-close contact is closed, the closing relay coil 152X will have a current flow through it, since the closing cutoff relay is energized and the 152Y contact, item (10), is in the closed position. When a current flows through the 152X closing relay coil, the relay will operate its contacts, items (11), (12), and (13). Item (11), 152X, is a closing relay contact which short-circuits the control-switch-close contact to keep the 152X closing relay energized once the breaker closing operation is initiated until the breaker has closed.

Starting with wire X_1, one side of the alternating-current source, the circuit is completed through the knife switch, item (1); the fuse, item (2); and wire X_2 to item (12), 152X, which is the closing relay contact. The purpose of this contact is to supply a high current contact capable of energizing the circuit-breaker closing coil. From the 152X contact, a circuit extends through the resistor, the rectifier elements, and the circuit-breaker draw-out contact to item (14), 152cc, the circuit-breaker closing coil. The purpose of this coil is to supply the energy to close the main contacts of the circuit breaker.

From the circuit-breaker closing coil, a circuit is completed to the breaker draw-out contact, the copper oxide rectifier, and to item (13), 152X, a closing relay contact which serves the same purpose as item (12). From the closing relay contact, the circuit is established to wire Y2, the other side of the alternating-current source. With the closing relay energized and its associated contacts, items (12) and (13), closed, a current flows through the circuit-breaker closing coil, causing the circuit breaker to close its main contacts.

Located on the circuit-breaker mechanism is an auxiliary switch which is connected to the linkage which operates the circuit-breaker main contacts. The auxiliary switch has contacts which are referred to as "a" contacts and "b" contacts. The auxiliary switch "a" contacts operate the same as the circuit-breaker main contacts; that is, when the circuit-breaker main contacts are closed, the "a" contacts are closed, and when the circuit-breaker main contacts are open, the "a" contacts are open. The auxiliary switch "b" contacts operate opposite to that of the circuit-breaker main contacts; that is, when the circuit-breaker main contacts are closed, the "b" contacts are open, and when the circuit-breaker main contacts are open, the "b" contacts are closed.

Starting with wire Y1, one side of the alternating-current source, a circuit is established through the knife switch, fuse, wire Y2, and the circuit-breaker draw-out contact to item (15), 152aa, a circuit-breaker auxiliary "a" switch contact adjusted to close late in the closing cycle. The 152aa switch serves the purpose of cutting off the circuit-breaker closing power. The 152aa switch is adjusted to close late in the closing operation to provide adequate power for closing the circuit breaker in case a fault should exist in the distribution line supplied through the breaker. From the 152aa contact, the circuit is completed through the breaker draw-out contact to a point between the resistor and the 152Y closing cut-off relay coil.

With the 152aa auxiliary switch contact closed, the 152Y closing cutoff relay coil is short-circuited and the 152Y relay contacts, items (7) and (10), will be opened. The 152X closing cutoff relay will be deenergized and will open its contacts, items (12) and (13), deenergizing the 152cc circuit-breaker closing coil. The 152X closing relay, as described above, was deenergized even though the control-switch-close contact, CSC, was maintained closed by the operator; therefore the circuit breaker cannot reclose if it trips out on a closing operation due to a fault until the control-switch handle is returned to the "off" position, closing contact CSO, item (4), and thus reenergizing the 152Y closing cutoff relay coil, item (6).

From the Y1 wire of the alternating-current source through the knife switch, the fuse, wire Y2, the breaker draw-out contact, a circuit is established to item (16), 152a, which is an "a" contact on the circuit-breaker auxiliary switch. This contact serves the purpose of energizing the red light when the circuit breaker closes. From the 152a contact, a circuit continues through the breaker draw-out contact, wire R, to item (17), R, which is the red indicating light. The red indicating light serves the purpose of informing the operator that the circuit breaker is closed. From the red light, the circuit is completed through wire X_2, the fuse, and the knife switch to wire X_1 the other side of the alternating-current source.

When the circuit breaker closes and the 152a auxiliary switch contact closes, the red light will be energized, indicating that the circuit breaker is closed. From wire Y1, one side of the alternating-current source, a circuit exists through the knife switch, the fuse, wire Y2, the breaker draw-out contact to item (18), 152b, which is a circuit-breaker auxiliary switch "b" contact. The 152b contact serves the purpose of energizing the green light and the reclosing relay. From the 152b contact, the circuit is complete through a circuit-breaker draw-out contact, wire G, to item (19), G, which is the green indicating light.

The green indicating light serves the purpose of telling the operator that the circuit breaker is open. From the green light, a circuit extends to item (20), CS CTO, which is the control-switch-lamp cut-out contact. This contact on the control switch is closed except when the control-switch handle is turned to the trip position and pulled out, placing the control switch in the lockout position. The purpose of the CS CTO contact is to deenergize the green light so that the green indicating light will not be lit when the circuit breaker is out of service due to a planned outage.

From the CS CTO contact, the circuit is complete to the X side of the alternating-current source. When the circuit breaker is open and the 152b contact is closed, the green light will be energized, indicating that the circuit breaker is open. From the Y side of the alternating-current source, a circuit is established to item (21), cell switch, which is a contact that is closed when the circuit breaker is in the operating position. The purpose of the cell switch is to energize the amber light.

From the cell switch, the circuit is complete through wire A to item (22), A, the amber light. The purpose of the amber light is to indicate that the circuit breaker is in the operating position. From the amber light, the circuit is complete to the X side of the alternating-current source. When the circuit breaker is in the jacked-in or operating position, the cell switch will be closed and the amber light will be lit, indicating the breaker is in the operating position.

From the wire P, the positive side of the 24-volt direct-current control power, a circuit extends to item (23), the knife switch. The knife switch can be used to isolate the circuit-breaker trip circuit. From item (23), the circuit is completed to item (24), a fuse. The fuse serves the purpose of protecting the direct-current control source from a fault in the trip circuit control wiring of the circuit breaker. From the fuse, the circuit is complete through wire P2 to item (25), CST, the control-switch trip contact. The purpose of the control-switch trip contact is to initiate a trip operation when the control-switch handle is turned to the trip position.

From the control-switch trip contact, the circuit is complete through wire T, the breaker draw-out contact, and wire 4, to item (26), 152a, a circuit-breaker auxiliary switch "a" contact. The purpose of the circuit-breaker auxiliary switch "a" contact is to deenergize the circuit-breaker trip coil after the tripping operation has been initiated and the circuit breaker starts to open. From the 152a contact, the circuit is complete to item (28), 152T, which is the circuit-breaker trip coil. The circuit-breaker trip coil serves the purpose of unlatching the circuit-breaker mechanism so that the circuit-breaker main contacts will open.

From the 152T trip coil, the circuit is complete to item (27), 152a contact, which serves the same purpose as item (26). From the 152a contact, the circuit extends through wire 5, the breaker draw-out contact, wire N2, the fuse, the knife switch, and wire N to the negative side of the 24-volt direct-current control source. If the control-switch handle is turned to the trip position when the circuit breaker is closed, the control-switch trip contact, CST, will close and the trip coil will be energized, initiating a tripping operation. Item (29), 151, is a trip contact on an overcurrent protective relay.

The overcurrent relay trip contact is connected to wire P2 and T so that when the contact is closed and the circuit breaker is closed, a tripping operation will be initiated. The purpose of the overcurrent relay trip contact is to trip the circuit breaker when a fault exists on the feeder connected to the circuit breaker. Item (30), 86, is the lock-out relay trip contact which serves the purpose of tripping the feeder circuit breaker when a bus fault occurs in the substation. The lock-out relay trip contact is wired in parallel with the overcurrent relay trip contact.

Starting with the X_1 wire of the alternating-current source, a circuit is completed through the knife switch, the fuse, wire X_2, wire $4X_2$, 86 lock-out relay contact and wire 4W to item (31), CS SC, which is the control switch sliding contact. This contact on the control switch closes when the control-switch handle is turned to the closed position, and it remains closed until the control-switch handle is turned to the trip position. The purpose of the CS SC control-switch contact is to open the reclosing circuit to prevent an automatic reclose when the circuit breaker is tripped by the control switch.

From the CS SC contact, a circuit is established to item (32), CSO, an off contact on the control switch. This contact on the control switch is closed when the control-switch handle is in the "off" or the straight up and down position. The purpose of the CSO contact is to open the reclosing circuit when the circuit breaker is closed by the operation of the circuit breaker control switch, thus preventing the circuit breaker from reclosing automatically on a fault when the circuit breaker is closed manually.

From the CSO contact, the circuit is complete through wire XC1 to item (33), 143, which is the automatic "on-off" toggle switch. The purpose of the 143 toggle switch is to deenergize the automatic reclosing circuit. From the 143 contact, the circuit is complete through wire XC11 to the reclosing relay. Item (34), LCSW, is the latch check switch contact. The latch check switch contact is open except when the circuit breaker is in the full open position. The latch check switch is connected between the 152X closing relay and the automatic reclosing relays to prevent an automatic reclosure after a breaker trip operation until the breaker has completely opened. Wire G connects from the 152b, auxiliary switch contact, item (18), to the reclosing relay to energize the relay and initiate a reclosing operation when the circuit breaker opens due to a fault on the line.

Voltage Regulator or Substation Transformer Tap Changer Control Schematic Distribution substation transformers are supplied by the subtransmission system. The high-voltage windings of most of the transformers have a no-load tap changer and an underload tap changer.

The purpose of the no-load tap changer is to change the number of turns in the primary winding when the transformer is not energized, so that the number of turns of the winding is in accordance with the general range of voltage that exists on the system at the location of the transformer. For example, if the voltage at a particular site varies from 71,000 to 67,000 volts, the no-load tap changer would probably be set on a tap which is rated 69,000 volts. This means that the actual turns utilized are in proportion to 69,000 volts.

The purpose of the tap changing underload equipment is to further change the number of turns in the primary winding to automatically keep the voltage on the secondary of the transformer at the proper value when the primary supply voltage varies to some value other than the value for which the no-load tap changer is adjusted. The underload tap changer consists of a switching arrangement designed to operate when the transformer is energized and carrying load. This switching arrangement is operated through an insulated mechanism by means of a motor.

The schematic control diagram (Fig. 39-2) illustrates the operation of the control equipment necessary to operate the tap changer motor which supplies the energy to operate the tap-changing underload switching equipment. Each item of equipment on the schematic control diagram (Fig. 39-2) is numbered. The circuit is traced, each piece of equipment is described, and the purpose each piece of equipment serves is outlined utilizing the numbers on the diagram. The circuit elements are shown for the condition of the transformer deenergized and all operating devices open. The control schematic for a distribution line voltage regulator would be similar.

Item (1) is a potential transformer connected to the regulated voltage supply. The purpose of the potential transformer is to reduce the voltage level so that it can be safely connected to relays and other control equipment. The output voltage of the transformer will normally be approximately 120 volts. From terminal V7 on the secondary of the potential transformer, the circuit is completed to item (2) AB1, which is an air-circuit breaker. The circuit breaker protects the potential transformer from a short circuit in the control wiring. From item (2), the circuit is completed to item (3), FTS1, which is a test switch. The test switch provides a means of testing the voltage regulating relay. From item (3), the circuit is completed to item (4), NV, which is the coil of the no-voltage relay.

The coil of the no-voltage relay operates its contacts to prevent operation of the tap changer if the regulated voltage supply is removed from the voltage regulating relay coil. From item (4) the circuit is complete to item (5), FTS1, another test switch and item (6), AB1, another pole of the

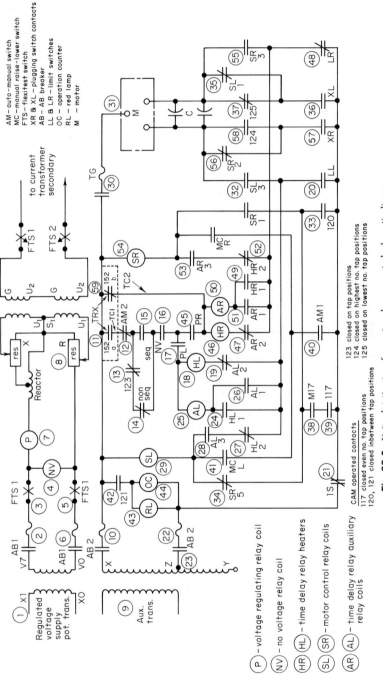

Fig. 39-2 Unit substation transformer tap changer control schematic diagram.

air-circuit breaker. From item (6), the circuit is complete to terminal VO on the other end of the potential transformer secondary. As long as the potential transformer is energized and the air-circuit breaker is closed, the no-voltage relay coil will be energized and the relay contacts will be closed to permit automatic operation of the tap changer.

From item (3), the circuit is also complete to item (7), P, the voltage regulating relay coil. The voltage regulating relay coil actuates the raising and lowering contacts which controls the direction of operation of the tap changer motor. From item (7) the circuit is complete to item (8), the compensator. The compensator elements connect also to the secondary of a current transformer, which has its primary winding connected in series with the load.

The compensator varies the voltage applied to the voltage regulating relay coil, in accordance with the settings of the resistance and reactance elements, and the current flowing in the secondary of the transformer. Therefore it is possible to have the regulated voltage at a higher level during times of heavy load conditions to compensate for line voltage drops without readjusting the voltage regulating relay. From item (8) the circuit is completed to item (5), etc., previously described.

Item (9) is an auxiliary power transformer. This transformer supplies the power to operate the tap changer motor and other equipment. From terminal X on the secondary of the auxiliary power transformer, the circuit is complete to item (10), AB2, an air-circuit breaker which protects the auxiliary power transformer from a fault in the tap changer control equipment. From item (10), the circuit is complete through wire TRX to item (11), 152a, a circuit-breaker auxiliary switch contact located on the transformer secondary breaker. The 152a switches are closed when the circuit breaker is closed and open when the circuit breaker is open.

The purpose of the auxiliary switch 152a contact is to prevent automatic voltage regulation when the transformer secondary breaker is open and the transformer is not connected to a load. From item (11) the circuit is completed through wire TC1 to item (12), AM2, an automatic-manual switch contact. The purpose of this contact is to prevent automatic operation of the tap changer controls when the tap changer is being operated by the manual control switch. From item (12) the circuit is completed to item (13), 123, a contact operated by a cam on the tap changer motor drive. This contact is closed when the tap changer is on a tap position and open when the tap changer is between tap positions.

The purpose of the 123 contact is to deenergize the tap changer control circuit after each tap changer operation is initiated to give nonsequential operation. From item (13) the circuit is complete to item (14), non SEQ, which is the nonsequential contact on the transfer switch. The purpose of this contact is to place the 123 cam switch in the circuit. The circuit is complete from item (14) to item (15) and also from item (12) to item (15), SEQ, the sequential contact on the transfer switch. The purpose of this contact is to bypass the 123 cam switch so that the tap changer will keep running once an operation is initiated until the requirements of the voltage regulating relay are fulfilled.

If the transformer is operated in a network, it is best to have the tap changers set on nonsequential operation to allow first one transformer and then another to operate its tap changer to fulfill the requirements of the voltage regulating relays. From item (15) the circuit is complete to item (16), NV. the contact on the no-voltage relay previously discussed. The purpose of this contact is to prevent automatic operation of the tap changer if the voltage is removed from the coil of the voltage regulating relay, thus preventing operation of the tap changer to raise the voltage excessively if the protective devices operate on the regulated potential supply.

From item (16) the circuit is complete to item (17), PL, the contact on the voltage regulating relay, which closes when the regulated voltage rises above the value for which the voltage regulating relay is adjusted.

The purpose of this contact is to initiate a tap-changing operation to lower the regulated voltage level. From item (17) the circuit is complete to item (18), HL, the heating element of the timing relay for the tap changer lower operation.

The heating element supplies heat to a bimetal strip which closes a contact after time delay. From item (18) the circuit is complete to item (19), AL2, an auxiliary relay contact, which is a component part of the timing relay. The purpose of this contact is to deenergize the heating element and permit the bimetal strip to return to its original position opening contact HL1 and closing contact HL2 again. From item (19) the circuit is complete to item (20), LL, the tap changer lower limit switch. The purpose of this limit switch is to prevent the tap changer motor from operating to change taps in the lower direction after the lowest tap has been reached. From item (20) the circuit is complete to item (21), IS, the interlock switch, which is opened when the crank is inserted in its socket to manually operate the tap changer.

The purpose of the interlock switch is to prevent simultaneous operation of the tap changer electrically and mechanically. From item (21) the circuit is completed to item (22), AB2, another pole of the air circuit breaker. From item (22) the circuit is completed to item (23), Z, the mid-point on the auxiliary power transformer, thus completing a circuit. Therefore if all the conditions described previously are fulfilled for an automatic lower operation, a current will flow through item (18), HL, causing it to close its contact item (24), HL1, the time delay relay heating element contact that closes after time delay. The purpose of the HL1 contact is to provide a circuit to energize the lower time delay relay auxiliary relay coil item (25), AL.

The circuit is then complete from item (17) through items (25) and (24) to item (20), and a current will flow through the auxiliary relay coil, AL, operating its contacts. The purpose of the auxiliary relay is to permit the timing relay to complete its cycle. From item (25) the circuit is complete to item (26), AL1, a contact on the auxiliary relay which closed when the AL coil was energized. The purpose of this contact is to complete a circuit in parallel with the HL1 contact to keep the AL coil energized. When the AL coil was energized, item (19), AL2, contact opened, deenergizing the HL heating element, which causes contact item (24), HL1 to open, and causes contact HL2 item (27) to close after time delay.

The HL2 contact is a time delay relay heating element contact, which is closed when the heating element is deenergized. The purpose of this contact is to help energize the relay which energizes the tap changer motor to operate in the lower direction. Item (28), AL3, is another auxiliary relay contact. The purpose of this contact is to energize the tap changer motor relay with the help of item (27), HL2 contact.

From item (10) the circuit is complete to item (29), SL, the tap changer lower motor control relay coil. The purpose of this coil is to operate the relay contacts to energize the tap changer motor. From item (29) the circuit is complete through item (28), AL3 contact, which closed when the timing cycle was halfway completed; item (27), HL2 contact, which opened when the timing cycle was initiated by the voltage regulating relay and closed when the timing cycle was completed; and item (20), etc., previously discussed.

From item (10) a circuit is established to item (30), TG, the motor thermal guard contact and heater element. The purpose of the thermal guard is to protect the motor if it is overloaded or the limit switches fail, and the motor is stopped by the mechanical stops. From item (30) the circuit is complete to item (31), the tap changer motor, which supplies the power to operate the tap changer. From item (31) the circuit is complete to item (32), SL3 contact on the motor control relay. The purpose of the SL3 contact is to energize the motor so that it will operate the tap changer to a lower tap position. From item (32) the circuit is complete to item (20), etc., as previously described. When the tap changer has moved off position, item (33), 120 cam switch contact, closes.

The purpose of the cam switch is to keep the motor control relay energized until the tap changer is back on position, at which time the 120 cam switch opens again. This prevents a partial tap change due to the voltage regulating relay opening its contacts or due to the 123 cam switch item (13) opening if the control is switched to nonsequential operation. This is accomplished through item (34), SR5, a contact on the raise motor control relay which is closed when the raise relay coil is deenergized. The purpose of this contact is twofold; first, to prevent operation of the SL and SR motor control relays simultaneously through the 120 cam switch and, second, to cause the tap changer to run to the next lower tap if it is stopped in a middle position between taps.

Item (35), SL1, a lower motor control relay contact, is opened when the motor is energized to operate the tap changer in the lower direction and closes as the tap change is completed. The purpose of this contact is to assist in placing power on the tap changer motor in the reverse direction to brake the motor when the operation is completed. Item (36), XL, is the lower direction plugging switch. This contact closes when the motor is operated in the direction to lower taps. Therefore when the lower direction motor control relay coil, SL, is deenergized, power is supplied through contacts XL and SL1, which tends to rotate the motor in the reverse direction, serving the purpose of a brake.

When the motor stops, contact XL, item (36), opens, removing power from the motor. If the motor is energized in the lower direction beyond the lowest tap position, item (20), LL, the lower limit switch contact, opens, deenergizing the motor, and item (37), 125 cam switch, closes to supply braking power. The purpose of the 125 cam switch which closes after the tap changer has reached the lowest position is to apply braking power immediately to prevent operation of the tap changer against the mechanical stop.

Item (38), M17, is a contact on a transfer switch. The purpose of the M17 transfer switch is to energize item (39), 117 cam switch. The 117 cam switch is open only on the odd-numbered tap positions. The purpose of items (39) and (38) is to short-circuit the 120 cam switch on the even-numbered tap positions so that when a tap change is initiated, it will complete two operations instead of one.

Item (40), AM1, is an automatic-manual transfer switch contact. When the transfer switch is operated to manual, contact item (12) is opened and item (40), AM1, is closed. The purpose of this is to permit manual electrical control of the tap changer. From item (29), SL, the lower motor control relay coil, a circuit is completed to item (41), MCL, the manual control switch lower contact. The purpose of this switch is to operate the tap changer electrically by means of the manually operated control switch. From item (41) the circuit is completed through items (40), (21), etc.

From item (10) a circuit is established to item (42), 121, a cam switch which is closed between each tap position. The purpose of this switch is to energize the inter-position light and the operation counter. From item (42) a circuit is complete to item (43), RL, a red light. The purpose of the light is to indicate that the tap changer is between positions. From item (43), the circuit is complete to item (22) and the other side of the control power source. From item (42), a circuit is established to item (44), OC, the operation counter. The purpose of the operation counter is to record each tap changer operation to help determine the need for maintenance work and operational inspections.

The tap changer is operated in the raised direction in similar manner as previously described for the lower direction. A circuit exists from the auxiliary transformer through the circuit breaker, item (10), the circuit-breaker 152 a auxiliary switch contact, item (11), through wire TC1 to item (12), the automatic manual switch contact, through items (13), 123 cam switch, and item (14), the nonsequential transfer switch, or item (15), the sequential switch contact; then through item (16), the no-voltage relay contact, to item (45), PR, the contact on the voltage regulating relay which closes when the regulated voltage decreases below the value for which the voltage regulating relay is adjusted.

The purpose of this contact is to initiate a tap-changing operation to raise the regulated voltage level. From item (45) the circuit is complete to item (46), HR, the heating element of the timing relay for the tap changer raise operation. The heating element supplies heat to a bimetal strip which closes a contact after time delay. From item (46) the circuit is complete to item (47), AR2, an auxiliary relay contact which is a component part of the timing relay. The purpose of this contact is to deenergize the heating element and permit the bimetal strip to return to its original positions, opening contact HR1 and closing contact HR2 again. From item (47) the circuit is complete to item (48), LR, the tap changer raise limit switch.

The purpose of the limit switch is to prevent a tap changer motor from operating to change taps in the raise direction after the highest tap has been reached. From item (48) the circuit is complete to item (21), the interlock switch, item (22), the air-circuit breaker contact, and item (23), the secondary of the operating transformer. Therefore if all the conditions described previously are fulfilled for an automatic raise operation, a current will flow through item (46), HR, causing it to close its contact, item (49), HR1, the time delay relay heating element contact that closes after a predetermined time delay.

The purpose of the HR1 contact is to provide a circuit to energize the raise time delay relay, auxiliary relay coil item (50), AR. The circuit is complete from item (45) through items (50) and (51) to item (48), and a current will flow through the auxiliary relay coil AR, causing it to operate its contacts. The purpose of the auxiliary relay is to permit the timing relay to complete its cycle. From item (50) the circuit is complete to item (51), AR1, a contact on the auxiliary relay which closed when the AR coil was energized. The purpose of this contact is to complete a circuit in parallel with the HR1 contact to keep the AR coil energized. When the AR coil was energized, item (47), AR2 contact, opened, deenergizing the HR heating element, which opened contact, item (49), HR1, and closed contact HR2, item (52), after time delay. The HR2 contact is the time delay relay heating element contact which is closed when the heating element is deenergized.

The purpose of this contact is to help energize the relay which energizes the tap changer motor to operate in the raise direction. Item (53), AR3, is another auxiliary relay contact. The purpose of this contact is to energize the tap changer motor relay with the help of item (52), HR2 contact. From item (10), the circuit is complete to item (54), SR, the tap changer raise motor control relay coil. The purpose of this coil is to operate the relay contacts to energize the tap changer motor. From item (54) the circuit is complete through item (53), AR3 contact, which closed when the timing cycle was halfway completed, item (52), HR2 contact, which opened when the timing cycle was initiated by the voltage regulating relay and closed when the timing cycle was completed and item (48), etc., previously discussed.

As previously described, a circuit is established from item (10) through item (30), TG, the motor thermal guard contact and heater element, item (31), M, the tap changer motor, to item (55), SR3 contact on the motor control relay. The purpose of the SR3 contact is to energize the motor so that it will operate the tap changer to raise the regulated voltage. From item (55) the circuit is complete to item (48), LR, etc., as previously described. When the tap changer motor has moved off position, item (33), 120 cam switch contact, closes to keep the raise motor control relay energized until the tap changer is back on position, at which time the 120 cam switch opens. This prevents a partial tap change due to the voltage regulating relay opening its contact prior to the time the tap changer completes its operation.

Item (56), SR2 raised motor control relay contact, is opened when the motor is energized and closes as the tap change is completed. The purpose of this contact is to assist in placing power on the tap changer motor in the reverse direction to brake the motor when the operation is completed. Item (57), XR, is the raise direction plugging switch. Therefore when the raise motor control relay coil, SR, is deenergized, power is supplied through contacts XR and SR 2, which tends to rotate the motor in the reverse direction, serving the purpose of a brake. When the motor stops, contact XR, item (57), opens, removing the power from the motor.

If the motor is energized in the raise direction beyond the highest tap position, item (48), LR, the raise limit switch contact, opens, deenergizing the motor, and item (58), 124 cam switch contact, closes to supply braking power. The purpose of the 124 cam switch contact which closes after the tap changer has reached the highest position is to apply braking power to stop the rotation of the tap-changer motor to prevent operation of the tap changer against the mechanical stop.

Item (59), 152b, is a breaker auxiliary switch contact located on the power transformer secondary breaker. This contact is closed when the secondary breaker is open. The purpose of the contact is, if the control power is on, to operate the tap changer in the raise direction and thus to assist the secondary breaker network relays to close the breaker automatically after an outage of the circuit serving as a source to the power transformer if the transformer is operated in a network system.

Voltage Regulation

Voltage regulation on any electrical system is defined as the difference in voltage between periods of no load and periods of full load on the system. Characteristics of the system and the amount of load will determine the extent of the voltage regulation. In order to provide good electric service to customers served from the distribution system, it is necessary to maintain voltage variations within a very limited range. The voltage variation to residences has been standardized from a high limit of 126 volts to a low limit of 114 volts for a nominal 120-volt secondary service.

Electrical equipment is designed to operate on a certain voltage and best operation is obtained when that rated voltage is applied. The effect of low voltage on heating type appliances such as heaters, toasters, irons, and ranges is an increase in heating time. Low-voltage conditions cause motors to operate at decreased efficiency. They draw more current from the line, have less starting and running torque, and may overheat to the extent that they might experience thermal failure. Incandescent lights operate less efficiently, give less light, and in the case of fluorescent lights may not operate at all.

High-voltage conditions are also detrimental to the operation of electrical equipment. Heating-type elements, light bulbs, and electron tubes have a shorter life span because of high voltage. Motors and other magnetic equipment will have excess speed or torque. Television sets are particularly susceptible to damage from high voltage.

The best service to customers and the greatest amount of revenue would result from a distribution system which provides rated voltage at every point of the system. Since this is not economically feasible, a compromise is made and the voltage variation is confined to the standard limits established.

The regulation or control of voltage on a distribution system is usually accomplished by a combination of two different methods. One method utilizes transformer tap changing equipment, step voltage regulators, or induction voltage regulators. With this equipment, distribution system line voltage may be held constant while its supply voltage varies over a certain range. This equipment may also raise or lower the line voltage to compensate for load changes on the circuit. Still another function of this equipment is to provide, in the case of a distribution tie line, some control of kilovar flow over the tie line. To accomplish any of these functions with the above voltage regulating equipment, an in-phase voltage or approximately in-phase voltage is either added or subtracted from the normal line voltage. Voltage control is accomplished over an approximate ±10 percent range in easily controlled steps without interruption in service to the customer.

The second method of voltage control normally used on distribution systems makes use of capacitors. The control of voltage with capacitors is accomplished by the ability of a capacitor to supply a leading current to a distribution line to counteract the lagging current which the line must supply to magnetize motors, transformers, and other magnetic equipment. The leading current and the lagging current combine to produce a smaller current more nearly in phase with the voltage. The smaller current produces less voltage drop on the line. Capacitors are usually switched on and off in banks and therefore the control of voltage with capacitors is less flexible than with voltage regulating equipment. However, capacitors have other advantages, probably the most important of which is their ability to reduce the kVA load on lines transformers and generators, making them available for additional load growth.

The use of voltage regulating equipment is made necessary because of a characteristic common to all electrical circuits. That characteristic is called "impedance." Impedance is defined as the opposition to the flow of current in an electrical circuit. When current flows through an electrical circuit, a voltage drop occurs between the input and output terminals of the circuit. This voltage drop is a function of the impedance in the circuit as well as the amount of current flowing in the circuit.

In a circuit having a constant impedance, the voltage drop will be proportional to the current flowing in the circuit. This can be determined by the relation

$$V = IZ$$

where V = voltage drop
       I = current flowing in the circuit
       Z = impedance of the circuit

The voltage drop through a transformer can be visualized by considering a transformer energized on the primary side with 100 percent voltage and no connected load on the secondary terminals. For this particular condition, the voltage on the secondary terminals would be 100

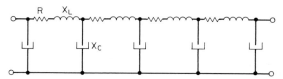

Fig. 40-1   Representation of a transmission or distribution line.

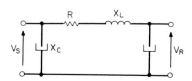

Fig. 40-2   Representation of a transmission line.

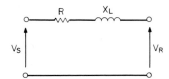

Fig. 40-3   Short line circuit.

percent rated voltage also. If a load were now applied to the secondary terminals of the transformer while the primary voltage remained constant, something other than rated voltage would be obtained on the secondary terminals. This change in voltage would be caused by current flowing through the internal impedance of the transformer.

The voltage drop on a transmission or distribution line is probably a little easier to visualize because no transformation ratio is involved as in the case of a transformer. A transmission or distribution line has an impedance distributed along its entire length. This impedance is made up of resistance and inductive and capacitive reactance. Each foot of conductor of a line has a certain amount of resistance depending on the conductor size and the material used to make up the conductor (copper, aluminum, etc.). Each 1-ft length of a line has a certain amount of inductive reactance because of the spacing between conductors which is required for electrical clearance. Also, each foot length of a line has a certain amount of capacitive reactance due to the conductor size, distance between conductors, and the height of the conductor above ground. The capacitive reactance of a distribution line is small compared to the inductive reactance and is usually neglected. Current flowing through the line impedance causes a voltage drop so that the voltage at the load end of the line is less than the voltage at the source end.

Load current flowing through the impedance of a line produces a voltage drop. To help illustrate the effect of the voltage drop, an electrical representation of a line is shown in Fig. 40-1.

Distributed along the line are resistance, inductance, and capacitors. In order to simplify the representation still further the values of resistance, inductance, and capacitance can be lumped together to form a network as shown in Fig. 40-2.

If the line is a short transmission line or a distribution feeder or tie line, the capacitive effect of the line can usually be neglected. This results in a still further simplification of the representation of a line as shown in Fig. 40-3.

The voltage drop, which would occur on this line because of the load current and impedance, is shown in the following vector diagram, which assumes a lagging power factor load, Fig. 40-4.

From the vector diagram (Fig. 40-4) it is apparent that the sending end voltage is of a greater magnitude than the receiving end voltage and the differences in the two voltages is the IZ or impedance voltage drop. The IZ component is made up of two parts—the drop due to the current flowing through the resistance and the drop due to the current flowing through the line reactance.

As has previously been noted, capacitance exists between conductors of a line and between the conductors and ground. On a long line that is lightly loaded, the capacitive charging current may exceed the load current and result in the line operating with a leading power factor. In such cases, the receiving end voltage will rise, and may exceed the voltage at the sending end of the line.

From the vector diagram (Fig. 40-4) several important points can be seen. The first is that the receiving end voltage of a line can be made more nearly equal to the sending end voltage by reducing the current on the line and thereby reducing the impedance drop. This is often done to

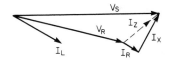

Fig. 40-4   Voltage on a short line.

V_S = Sending end or source end voltage

V_R = Receiving end voltage

I_L = Load current

I_R = Voltage drop due to the resistance in the line

I_X = Voltage drop due to the reactance in the line

a distribution feeder when the load on the line becomes so great that excessive voltage drop is encountered. A portion of the load is then transferred to some other line which is not so heavily loaded.

The second point, which is apparent from the diagram, is that the impedance drop may be reduced by decreasing the impedance of a line. To reduce the resistance, wires of a larger cross sectional area would have to be installed. To reduce the reactance of a line would mean either decreasing the conductor spacing or using a series capacitor, neither of which is too practical.

A third point, which can be seen from the vector diagram, is that the power factor of the load could be improved. This could be done by installing shunt capacitors near the load. This method of voltage improvement is used quite extensively. For the best effect, the capacitors should be switched off or on as load changes. This is done with time and temperature controls. The effect of adding capacitors to a circuit is shown in Fig. 40-5.

There are times and conditions where the voltage regulations on a line, that is the difference between the no load voltage and the full load voltage, becomes so great that other means must be used to maintain suitable voltages. Under these conditions step-type voltage regulators, tap-changing transformers, or induction voltage regulators are employed. In general, it can be stated that voltage regulation of substation buses is accomplished through the use of under-load tap-changing equipment on transformers, while the voltage regulation on distribution circuits is controlled by shunt capacitors or step-voltage regulators. The step-voltage regulator is primarily an autotransformer with tap-changing under-load switching available to change the ratio between the primary and the secondary windings.

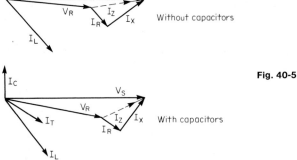

Fig. 40-5   Effect of installing capacitors.

The purpose of under-load tap-changing (LTC) equipment is to change the primary to secondary turns ratio of a transformer winding while the transformer is carrying load. The turns ratio of a transformer is equal to its voltage ratio. That is, a transformer with a turns ratio of 100:1 would also have a voltage ratio of 100:1. This means that by controlling the turns ratio of a transformer one can regulate the voltage on its "low" side.

The heart of LTC equipment is the automatic control which operates a motor driven mechanical tap changer thus maintaining a controlled voltage at the load as the input voltage to the power transformer fluctuates or as the load conditions vary.

As the distribution-line load increases, the voltage at its load end drops due to transformer and line impedances (both reactive and resistive components). This line voltage is monitored at the low voltage side of the power transformer, and when the voltage changes by a prescribed amount, this automatic control device senses the change and causes the under-load tap-changing equipment to either increase or decrease the turns ratio, thus changing the voltage and bringing it back to its desired value.

There are compensating devices in the automatic control unit that correct for the reactive and resistive loss components of the distribution line. This "line drop compensating device" raises the voltage at the transformer as the line current increases, thus maintaining a nearly constant voltage at the regulating point on the line.

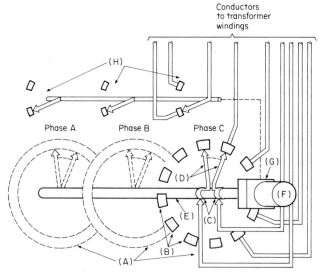

Fig. 40-6   Transformer under-load tap-selector diagram.

The typical under-load tap selector (see Fig. 40-6) consists of three (one per phase) dial-type switches (A). Each switch has stationary contacts (B) which are connected to the taps in the transformer winding. These contacts are mounted on an insulating panel in a circle around two concentric slip rings (C). The two movable contact fingers (D) for each switch are mounted as an assembly on the main rotor shaft (E), each finger making contact with its own slip ring. The rotor shaft which extends through all three phases of the tap selector is turned by the tap-changer motor (F) through reduction gears and switch controls (G). The two fingers move together, stopping either with both on the same contact or with one on each of two adjacent contacts.

Each of the three dial-type tap-selector switches is provided with a separate single-pole double-throw switch (H) used to reverse the polarity of the under-load tap-changing windings.

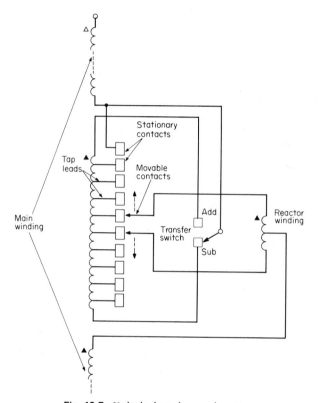

Fig. 40-7   Under-load tap-changer schematic.

Tap leads from the transformer winding are connected to stationary contacts (see Fig. 40-7) on the tap selector. When the motor operates the selector, it will change taps to vary the transformer turns ratio and thereby regulate the voltage.

Two movable contacts are used on the tap selector so that one can be moved to an adjacent tap while the other contact maintains the power circuit. These movable contacts are connected together through a reactor winding to prevent excessive circulating current through the bridged section of the winding. By designing the winding and reactor to share the circulating current, the bridged position can be used as an intermediate operating position, thus doubling the number of tap-change steps.

A two-position switch, integral with the tap selector, is used to double the number of steps again. This switch is used to reverse the polarity taken from the tapped section of the transformer, permitting the tapped section to be used twice—one time adding to and the other time subtracting from the number of turns on the main winding.

When the automatic control unit calls for a change of tap position, the drive motor is energized and, through a series of gears, the tap selector is operated. In the first step in a sequence of operation, the movable fingers are slid toward the next stationary contact. The first finger breaks contact and the current is carried by the second finger (see Fig. 40-8). In the second step, the movable fingers are slid so that the first finger is on the next stationary contact while the second finger is still on the original contact. In this step the currents are shared by both fingers and the center tapped reactor to which the fingers are connected through the slip rings providing for an inter-

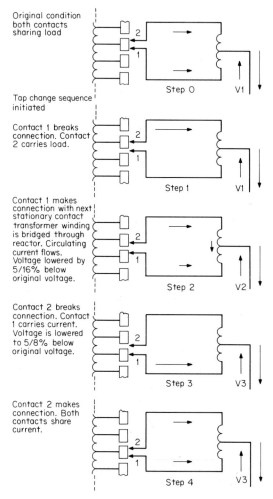

Fig. 40-8   Under-load tap-changer operation sequence.

mediate voltage between the two stationary contacts. In the third step, the second finger is slid off the original contact and the current is carried by the first finger. In the last step in the sequence of operation, the fingers are slid so that both fingers are on the second stationary contact, thus completing the sequence.

Typical automatic controls which govern operation of the load tap changer consist of a line drop compensating unit, a voltage sensing unit, and a time delay unit.

Line Drop Compensator   Sometimes it is desirable to maintain a constant voltage at some other point on the circuit than at the source end of the line. This point, for example, may be the center of the load on the circuit. It would be possible to hold a constant voltage at this point by placing a potential transformer on the line and extending a pair of wires back to the voltage regulating relay for the secondary voltage. This method is not too practical, and a device called a "line drop compensator" is used instead. The line drop compensator consists of a resistance and a reactance element wired in series with a current transformer which is in series with one of the regulator terminals as shown in Fig. 40-9. In this circuit the values of resistance and reactance are so adjusted that they form a miniature line. Current flow through the current transformer primary winding to the load causes a proportional current to flow through the compensator. This current causes a voltage to appear across the resistance and reactance elements which is proportional to the voltage drop appearing on the line at some previously selected point. The voltages produced across the resistance and reactance elements are then introduced into the voltage regulating relay circuit. The resistor voltage is applied so as to subtract from the voltage applied to the regulating relay by the potential transformer. The reactance element voltage may be applied so that it either subtracts from or adds to the potential transformer voltage depending on what condition the element is set for. Normally it would be set to subtract.

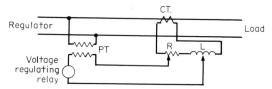

Fig. 40-9   Simplified connections of a line drop compensator.

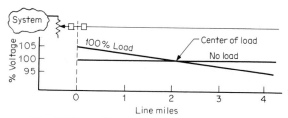

Fig. 40-10   Regulated line using line drop compensation.

With this type of equipment, the voltage regulation of a distribution-system feeder for various load conditions is shown graphically in Fig. 40-10. From figure it can be seen that this method of regulation is a good solution to a regulation problem because it supplies rated voltage or voltage within a very limited range to all customers on the line.

Static Voltage Regulating Relay   The line drop compensator unit, the voltage sensing unit, and the time delay unit, and all auxiliary functions except the final output to the motor can be performed with solid-state devices shown on schematic control diagrams (Figs. 40-11 through 40-14). The operation of the control circuits will be described.

The power transformer secondary or distribution system primary output voltage is stepped down to 120 volts by the control potential supply transformer and is applied to the input of the automatic static-load tap-changer control (Fig. 40-11). At this point, a normal test power supply switch permits the input of a test power supply for calibration and maintenance. Past the circuit breaker, a test rheostat is provided to permit variation of the voltage applied to the sensing circuit of the control without changing the transformer output voltage. The test rheostat is left in the "auto" position for normal operation.

A 5:1 ratio transformer (T1) is used to reduce the voltage applied to the line drop compensator and sensing circuit. This line drop compensator consists of a tapped line drop reactance compensating reactor, a compensating polarity switch, and a line drop resistance compensating rheostat.

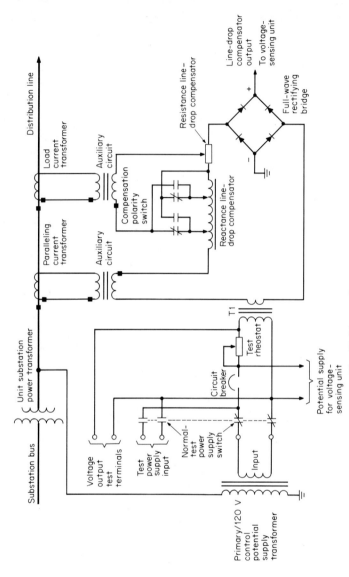

Fig. 40-11 Line drop compensating unit.

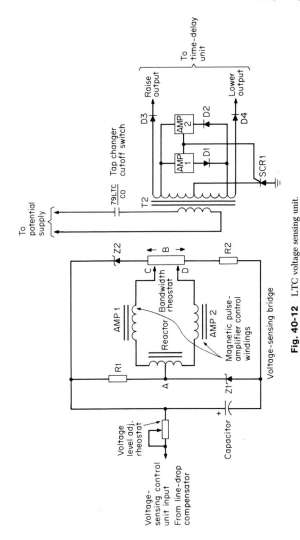

Fig. 40-12 LTC voltage sensing unit.

These devices are fed by the paralleling current transformer, the load current transformer, and transformer T1.

The purpose of the line drop compensating unit is to develop voltages proportional to the reactive (IX) and resistive (IR) drops in the distribution line fed by the transformer. The final settings for the line drop compensator may be made by field adjustment; however, initial settings may be determined from line data by using the following expressions

where d = miles from the unit to load center or regulating point
$\quad\quad r$ = resistance per conductor from unit to load center in ohms per mile
$\quad\quad x$ = inductive reactance per conductor from unit to load center in ohms per mile
$\quad\quad I$ = load current transformer primary current rating
$\quad\quad E$ = control potential supply transformer ratio
$\quad\quad R$ = resistance dial setting
$\quad\quad X$ = reactance dial setting

For a three phase balanced unit then

$$R = \frac{rdI}{E}$$
$$X = \frac{xdI}{E}$$

For normal operation, the compensation polarity switch is left in the "normal" position. Reversed reactance compensation is obtained by placing it in the "reverse" position. This reverses the current supplied by the load current transformer in the reactance element.

"Reversed reactance" compensation is a method used to reduce the circulating current that might flow when two or more transformers are paralleled. Instead of running toward opposite extreme positions, tap changers having reversed reactance compensation tend to move toward whatever position causes the least amount of transformer circulating current to flow. This is accomplished at some sacrifice to the normal line drop compensation but is generally satisfactory when units paralleled are not located in close proximity to each other or where the supply is from different sources.

For the paralleling of two transformers located in close proximity to each other and supplied from the same source, the circulating currents are used to control the line drop compensator through the paralleling current transformer causing the tap changers to operate in such a direction so as to reduce the transformer circulating currents to a minimum.

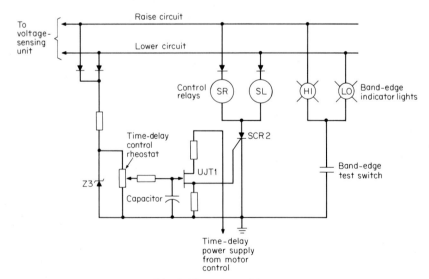

Fig. 40-13   LTC time delay unit.

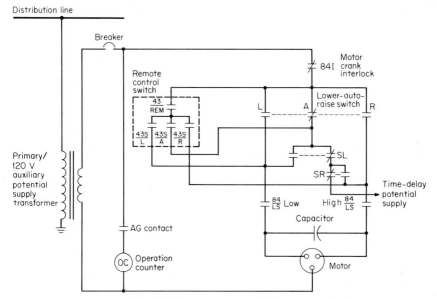

Fig. 40-14   LTC motor control schematic.

For normal compensation applications, the voltage developed across the line drop compensator is subtracted from the output of transformer T1. This forces the output of T1 to be higher for a given voltage into the sensing circuit and thus across the transformer to compensate for the drop created by increased current in the line.

The output of the line drop compensator is rectified and filtered into a dc voltage by the full-wave bridge rectifier. At this point the voltage-level-adjust rheostat provides for transformer voltage level settings between about 90 percent and 110 percent of nominal line voltage.

The adjusted voltage is fed into the voltage sensing unit which consists of a zener diode bridge network and the control windings of two magnetic pulse amplifiers (Fig. 40-12). The basic elements of the bridge are resistors R1 and R2, zener diodes Z1 and Z2, the minimum-band-width rheostat, and the ganged-band-width rheostat.

The zener diodes serve as reference voltage devices, and any voltage variations are reflected as differences in voltages across the resistors. The zener diode bridge is in a balanced state when the potential at the junction of R1 and Z1 (point A) is equal to the center of the active portion of the minimum-band-width rheostat (point B). As the applied voltage increases, the potential at point B rises with respect to that at point A. As the voltage decreases, the reverse occurs.

The output of the voltage sensing bridge is fed through the control windings of a pair of magnetic amplifiers (Amp 1 and Amp 2). At balance, the current in the control winding of Amp 1 is from point C to point A and the current in the control winding of Amp 2 is from point A to point D. Current applied in this direction causes each amplifier to produce negative pulses to the gate of SCR 1. Since positive gate voltage is required to make the silicon controlled rectifier (SCR) conduct, there will be no output from the control. At equilibrium the ac windings of the magnetic amplifier are of low impedance; thus the transformer secondary is short circuited, and the "raise" and "lower" circuits are deenergized.

When the applied input voltage increases, the potential at point B rises with respect to that at point A. When this increase is sufficient to cause the potential at point D to become equal to that at point A, the current in the control winding of Amp 2 will cut off. When the control current is cut off, the ac winding of Amp 2 changes impedance from low to high. This causes the voltage at the gate of SCR 1 to become positive and causes it to conduct on the half-cycle. Since the secondary of transformer T2 is shorted out in only one direction, through diode D1, the voltage builds up across the ac winding of Amp 2 and voltage is applied to the "lower" portion of the relay circuit.

This voltage is applied through diode D4 to the time delay unit (Fig. 40-13)—the time delay circuit input voltage is regulated by zener diode Z3—and the time delay control rheostat varies the amount of time delay. This time delay unit has a range of about 10 to 100 sec. and is usually set at 30 sec.

The time delay circuit consists basically of a unijunction transistor and an RC charging network. When the time delay capacitor has charged sufficiently to cause the unijunction transistor to conduct, the positive pulse created by this conduction is applied to the gate of SCR 2 causing it to conduct. The conduction of SCR 2 energizes relay SL which in turn energizes the tap changer operating motor in the direction which will lower the output voltage (Fig. 40-14).

A similar sequence may be followed through when the voltage applied to the sensing circuit decreases. In this case, the current through the control winding of Amp 1 will cut off when the potential at point C equals that at point A. (Note that the setting of the band-width rheostat determines the amount of deviation from balance required to initiate a controlling action).

Since Amp 1 functions when the top of T2 is positive with respect to the bottom, SCR 1 is caused to conduct on the half-cycle which energizes the "raise" portion of the circuit, this time through diode D3. Under these conditions it will energize relay SR which will in turn operate the drive motor to raise the output voltage.

The tap changer motor is operated from an auxiliary potential supply transformer and has the following operation sequence.

Potential is supplied through the 84I crank-handle interlocking switch. When the crank handle is used to operate the tap changer manually, the 84I contact is open preventing simultaneous operation of the tap changer manually and electrically. Potential is supplied from this point to the "lower-auto-raise" switch and to the "remote control" switch. When the remote switch is activated, the tap changer can be raised, lowered, or set to automatic control from a point outside the tap-changer control compartment on the transformer.

The "lower-auto-raise" switch permits the tap changer to be raised, lowered, or set to automatic control at the tap changer.

In the "auto" position, the SL and SR contacts of the automatic static control device are employed. If the "lower" circuit is energized by the sensing unit and the SL relay picks up, the SL contact to the "lower" terminal of the motor is closed, completing a circuit through the 84/LS lower end limit switch which cuts off motor operation if the tap changer has advanced to its lowest position. The motor is now energized and will continue to run until it lowers the transformer voltage to satisfy the conditions set by the voltage sensing unit.

When SL is energized, the contact supplying potential to the SR contacts are opened preventing simultaneous raise and lower operations. When SL or SR is energized, the power supply to the time delay unit is cut off permitting it to reset.

In a similar manner, when the "raise" circuit is energized by the sensing unit and the SR relay picks up, the SR contact to the "raise" terminal of the motor is closed completing a circuit through the 84/LS upper end limit switch which cuts off motor operation if the tap changer has advanced to its highest position. The motor is again energized and will continue to run until it raises the transformer voltage to satisfy the conditions set by the voltage sensing unit.

When a tap change occurs, the AG contact is closed thus causing the operation counter OC to record that an operation has taken place.

Electrical Formulas and Calculations

General   In this section the most common electrical formulas are given and their use is illustrated with a problem. In each case the formula is first stated using the customary symbols, the same expression is then stated in words, then a problem is given, and finally the substitutions are made for the symbols in the formula from which the answer is calculated. Only those formulas which the lineman is apt to have use for are given. The formulas are divided into three groups: direct-current circuits, alternating-current circuits, and electrical apparatus.

DIRECT-CURRENT CIRCUITS

Ohm's Law   The formula for Ohm's law is the most fundamental of all electrical formulas. It expresses the relationship that exists in an electrical circuit containing resistance only between the current flowing in the resistance, the voltage impressed on the resistance, and the resistance of the circuit, thus

$$I = \frac{E}{R}$$

where I = current, amp
E = voltage, volts
R = resistance, ohms

Expressed in words, the formula states that current equals voltage divided by resistance, or

$$\text{Amperes} = \frac{\text{volts}}{\text{ohms}}$$

Example: How much direct current will flow through a resistance of 10 ohms if the pressure applied is 120 volts direct current?

Solution:
$$I = \frac{E}{R} = \frac{120}{10} = 12 \text{ amp}$$

Ohm's law involves three quantities, namely, current, voltage, and resistance. Therefore, when any two are known, the third one can be found. The procedure for solving for current has already been illustrated. To find the voltage required to circulate a given amount of current through a known resistance, Ohm's law is written as follows:

$$E = IR$$

In words the formula says that the voltage is equal to the current multiplied by the resistance, thus

$$\text{Volts} = \text{amperes} \times \text{ohms}$$

Example: How much direct-current voltage is required to circulate 5 amp direct current through a resistance of 15 ohms?

Solution: $\qquad\qquad E = IR = 5 \times 15 = 75 \text{ volts}$

In like manner if the voltage and current are known and the value of resistance is to be found, the formula is stated as follows:

$$R = \frac{E}{I}$$

Expressed in words the formula says that the resistance is equal to the voltage divided by the current, thus

$$\text{Ohms} = \frac{\text{volts}}{\text{amperes}}$$

Example: What is the resistance of an electrical circuit if 120 volts direct current causes a current of 5 amp to flow through it?

Solution: $\qquad\qquad R = \frac{E}{I} = \frac{120}{5} = 24 \text{ ohms}$

Power Formula   The expression for the power drawn by a direct-current circuit is

$$P = EI$$

where E and I are the symbols for voltage in volts and current in amperes and P is the symbol for power in watts.

Expressed in words the formula says that the power in watts drawn by a direct-current circuit is equal to the product of volts and amperes, thus

$$\text{Power} = \text{volts} \times \text{amperes}$$

Problem: How much power is taken by a 120-volt direct-current circuit when the current flowing is 8 amp direct current?

Solution: $\qquad\qquad P = EI = 120 \times 8 = 960 \text{ watts}$

The power formula given above also contains three quantities or terms, namely, watts, volts, and amperes. Therefore, when any two of the three are known, the third one can be found. The procedure for finding the power when the voltage and current are given has already been illustrated. When the power and voltage are given and the current is to be found, the formula is written thus:

$$I = \frac{P}{E}$$

In words the formula states that the current equals the power divided by the voltage, thus

$$\text{Amperes} = \frac{\text{watts}}{\text{volts}}$$

Example: How much direct current would a 1000-watt load draw when connected to a 120-volt direct-current circuit?

Solution: $\qquad\qquad I = \frac{P}{E} = \frac{1000}{120} = 8.33 \text{ amp}$

In like manner, when the power and current are known, the voltage can be found by writing the formula thus

$$E = \frac{P}{I}$$

Expressed in words the formula says that the voltage is equal to the power divided by the current, thus

$$\text{Volts} = \frac{\text{watts}}{\text{amperes}}$$

Example: What direct-current voltage would be required to deliver 660 watts with 6 amp direct current flowing in the circuit?

Solution:
$$E = \frac{P}{I} = \frac{660}{6} = 110 \text{ volts}$$

Line Loss or Resistance Loss   The formula for computing the power lost in a resistance when current flows through it is

$$P = I^2 R$$

where the symbols have the same meaning as in the foregoing formulas.

Expressed in words the formula says that the power in watts lost in a resistance is equal to the square° of the current in amperes multiplied by the resistance in ohms, thus

$$\text{Watts} = \text{amperes squared} \times \text{ohms}$$

Example: Compute the watts lost in a line having a resistance of 4 ohms when 8 amps direct current is flowing in the line.

Solution:
$$P = I^2 R = 8 \times 8 \times 4 = 256 \text{ watts}$$

ALTERNATING-CURRENT CIRCUITS

Ohm's Law   Ohm's law is the same for resistance circuits when alternating voltage is applied as when direct voltage is applied, namely,

$$I = \frac{E}{R}$$

and
$$E = RI$$

and
$$R = \frac{E}{I}$$

In the above, E is the "effective" value of the alternating voltage and I the "effective" value of the alternating current. (See Direct-Current Circuits for examples.)

Ohm's Law for Other Than Resistance Circuits   When alternating currents flow in circuits, these circuits may exhibit additional characteristics besides resistance. They may exhibit inductive reactance or capacitive reactance or both. The total opposition offered to the flow of current is then called "impedance" and is represented by the symbol Z. Ohm's law then becomes

$$I = \frac{E}{Z}$$

and
$$E = IZ$$

and
$$Z = \frac{E}{I}$$

where I = current, amp
E = voltage, volts
Z = impedance, ohms

Example: Find the impedance of an alternating-current circuit if 120 volts alternating current causes a current of 30 amp alternating current to flow.

Solution:
$$Z = \frac{E}{I} = \frac{120}{30} = 4 \text{ ohms impedance}$$

Impedance   The impedance of a series circuit is given by the expression
$$Z = \sqrt{R^2 + (X_L - X_c)^2}$$

°Square means to multiply by itself.

where Z = impedance, ohms
      R = resistance, ohms
      X_L = inductive reactance, ohms
      X_c = capacitive reactance, ohms

Example: If an alternating-current series circuit contains a resistance of 5 ohms, an inductive reactance of 10 ohms, a capacitive reactance of 6 ohms, what is its impedance in ohms?

$$\begin{aligned} Z &= \sqrt{R^2 + (X_L - X_c)^2} \\ &= \sqrt{5 \times 5 + (10 - 6)^2} \\ &= \sqrt{25 + (4 \times 4)} \\ &= \sqrt{41} = 6.4 \text{ ohms} \end{aligned}$$

Note: The values of inductive and capacitive reactance of a circuit depend upon the frequency of the current, the size, spacing, and length of the conductors making up the circuit, etc. For distribution circuits and transmission lines, values may be found in appropriate tables.

It will be observed in the expression for Z that, when neither inductive reactance nor capacitive reactance is present, Z reduces to the value R, thus

$$Z = \sqrt{R^2 + (0 - 0)^2} = \sqrt{R^2} = R$$

Likewise when only resistance and inductive reactance are present in the circuit, the expression for Z becomes

$$Z = \sqrt{R^2 + (X_L - 0)^2} = \sqrt{R^2 + X_L^2}$$

Example: Compute the impedance of an alternating-current circuit containing 3 ohms of resistance and 4 ohms of inductive reactance connected in series.

Solution:

$$\begin{aligned} Z &= \sqrt{R^2 + X_L^2} = \sqrt{3^2 + 4^2} \\ &= \sqrt{3 \times 3 + 4 \times 4} = \sqrt{9 + 16} \\ &= \sqrt{25} = 5 \text{ ohms impedance} \end{aligned}$$

Line Loss or Resistance Loss   The formula for computing the power lost in a resistance or line is

$$P = I^2 R$$

where the symbols have the same meaning as above. Expressed in words the formula says that the power in watts lost in a resistance is equal to the current in amperes squared multiplied by the resistance in ohms, thus

$$\text{Watts} = \text{amperes squared} \times \text{ohms}$$

Example. Compute the power lost in a line having a resistance of 3 ohms, if 20 amp are flowing in it.

Solution: $P = I^2 R = 20 \times 20 \times 3 = 1200$ watts

Power Formula   The power formula for single-phase alternating-current circuits is

$$P = E \times I \times \text{pf}$$

where P = power, watts
      E = voltage, volts
      I = current, amp
      pf = power factor of circuit

Expressed in words the formula says that the power in watts drawn by a single-phase alternating-current circuit is equal to the product of volts, amperes, and power factor. If the power factor is unity or one, the product is simply that of volts and amperes.

Example: How much power is delivered to a single-phase alternating-current circuit operating at 120 volts if the circuit draws 10 amp at 80 percent power factor?

Solution:

$$\begin{aligned} P &= E \times I \times \text{pf} \\ &= 120 \times 10 \times 0.80 \\ &= 960 \text{ watts} \end{aligned}$$

The power formula given above contains four quantities, namely, P, E, I, and pf. Therefore, when any three are known, the fourth one can be determined. The case where the voltage, current, and power factor are known has already been illustrated. When the voltage, power, and power factor are known, the current can be computed from the expression

$$I = \frac{P}{E \times \text{pf}}$$

Example: How much current is drawn by a 10-kW load from a 220-volt alternating-current circuit if the power factor is 0.80?

Solution:
$$I = \frac{P}{E \times \text{pf}}$$
$$= \frac{10,000}{220 \times 0.80} = 5.7 \text{ amp}$$

In like manner if the power, voltage, and current are known, the power factor can be computed from the following formula:

$$\text{pf} = \frac{P}{E \times I}$$

Example: What is the power factor of a 4-kW load operating at 230 volts if the current drawn is 20 amp?

Solution:
$$\text{pf} = \frac{P}{E \times I} = \frac{4,000}{230 \times 20}$$
$$= 87 \text{ percent power factor}$$

Three Phase   The power formula for a three-phase alternating-current circuit is

$$P = \sqrt{3} \times E \times I \times \text{pf}$$

where the symbols have the same meaning as for single phase. The additional quantity in the three-phase formula is the factor $\sqrt{3}$, expressed "square root of three" the value of which is

$$\sqrt{3} = 1.73$$

Example: How much power in watts is drawn by a three-phase load at 230 volts if the current is 10 amp and the power factor is 0.80 percent?

Solution:
$$P = \sqrt{3} \times E \times I \times \text{pf}$$
$$= \sqrt{3} \times 230 \times 20 \times 0.80$$
$$= 3183 \text{ watts}$$

The three-phase power formula also contains four quantities as in the single-phase formula. Therefore, if any three are known, the fourth one can be determined. The case where the voltage, current, and power factor are known was illustrated above. When the power, voltage, and power factor are known, the current in the three-phase line can be computed from the formula

$$I = \frac{P}{\sqrt{3}E \times \text{pf}}$$

Example: If a 15-kW three-phase load operates at 2300 volts and has a power factor of 1, how much current flows in each line of the three-phase circuit?

Solution:
$$I = \frac{P}{\sqrt{3}E \times \text{pf}}$$
$$= \frac{15,000}{\sqrt{3} \times 2300 \times 1.0} = 3.75 \text{ amp}$$

If the only quantity not known is the power factor, this can be found by using the following formula:

$$\text{pf} = \frac{P}{\sqrt{3} \times E \times I}$$

Example: Find the power factor of a 100-kW three-phase load operating at 2300 volts if the line current is 40 amp.

Solution:
$$\text{pf} = \frac{P}{\sqrt{3} \times E \times I}$$
$$= \frac{100{,}000}{\sqrt{3} \times 2300 \times 40}$$
$$= 62.5 \text{ percent power factor}$$

Volt-Amperes   Often loads are given in volt-amperes instead of watts or in kVA instead of kW. The relationships then become

and
$$\begin{aligned} \text{VA} &= E \times I & \text{for single phase} \\ \text{VA} &= \sqrt{3}E \times I & \text{for three phase} \end{aligned}$$

In using these quantities it is not necessary to know the power factor of the load.

Line Loss or Resistance Loss (Three Phase)   The power lost in a three-phase line is given by the expression

$$P = 3 \times I^2R$$

where I = current in each line wire
 R = resistance of each line wire

The factor 3 is present so that the loss in a line wire will be taken three times to account for the loss in all three wires.

Example: If a three-phase line carries a current of 12 amps and each line wire has a resistance of 3 ohms, how much power is lost in the resistance of the line?

Solution: $P = 3 \times I^2R = 3 \times 12 \times 12 \times 3 = 1296 \text{ watts}$

ELECTRICAL APPARATUS

Motors

Motor Direct Current   The two quantities that are usually desired in an electric motor are its output in horsepower, hp, and its input current rating at full load. The following expression for current is used when the motor is a direct-current motor:

$$I = \frac{\text{hp} \times 746}{E \times \text{eff}}$$

The formula says that the full load current is obtained by multiplying the horsepower by 746 and dividing the result by the product of voltage and percent efficiency. (The number of watts in 1 hp is 746.)

Example: How much current will a 5-hp 230-volt direct-current motor draw at full load if its efficiency at full load is 90 percent?

Solution: $I = \dfrac{\text{hp} \times 746}{E \times \text{eff}} = \dfrac{5 \times 746}{230 \times 0.90} = 18.8 \text{ amp}$

Motor Alternating Current, Single Phase   If the motor is a single-phase alternating-current motor the expression for current is almost the same, except that it must allow for power factor. The formula for full load current is

$$I = \frac{\text{hp} \times 746}{E \times \text{eff} \times \text{pf}}$$

Example: How many amperes will a ¼-hp motor take at full load if the motor is rated 110 volts and has a full load efficiency of 85 percent and a full load power factor of 80 percent?

Solution: $I = \dfrac{\text{hp} \times 746}{E \times \text{eff} \times \text{pf}} = \dfrac{0.25 \times 746}{110 \times 0.85 \times 0.80} = 2.5 \text{ amp}$

Motor Alternating Current, Three Phase The formula for full load current of a three-phase motor is the same as for a single-phase motor except that it has the factor $\sqrt{3}$ in it, thus:

$$I = \frac{hp \times 746}{\sqrt{3} \times E \times eff \times pf}$$

Example: Calculate the full load current rating of a 5-hp three-phase motor operating at 220 volts and having an efficiency of 80 percent and a power factor of 85 percent.

Solution:
$$I = \frac{hp \times 746}{\sqrt{3} \times E \times eff \times pf} = \frac{5 \times 746}{\sqrt{3} \times 220 \times 0.80 \times 0.85}$$
$$= 14.4 \text{ amp}$$

Motor Direct Current To find the horsepower rating of a direct-current motor if its voltage and current rating are known, the same formula is used as for current except the quantities are rearranged, thus:

$$hp = \frac{E \times I \times eff}{746}$$

Example: How many horsepower can a 220-volt direct-current motor deliver if it draws 15 amp and has an efficiency of 90 percent?

Solution:
$$hp = \frac{E \times I \times eff}{746} = \frac{220 \times 15 \times 0.90}{746} = 4.0 \text{ hp}$$

Motor Alternating Current, Single Phase The horsepower formula for a single-phase alternating-current motor is

$$hp = \frac{E \times I \times pf \times eff}{746}$$

Example: What is the horsepower rating of a single-phase motor operating at 480 volts and drawing 25 amp at a power factor of 88 percent if it has a full load efficiency of 90 percent?

Solution:
$$hp = \frac{E \times I \times pf \times eff}{746}$$
$$= \frac{480 \times 25 \times 0.88 \times 0.90}{746}$$
$$= 5.08 \text{ hp}$$

Motor Alternating Current, Three Phase The horsepower formula for a three-phase alternating-current motor is

$$hp = \frac{\sqrt{3} \times E \times I \times pf \times eff}{746}$$

Example: What horsepower does a three-phase 240-volt motor deliver if it draws 10 amp, has a power factor of 80 percent, and has an efficiency of 85 percent?

Solution:
$$hp = \frac{\sqrt{3} \times E \times I \times pf \times eff}{746}$$
$$= \frac{\sqrt{3} \times 240 \times 10 \times 0.80 \times 85}{746} = 3.9 \text{ hp}$$

Alternating-Current Generator

Frequency The frequency of the voltage generated by an alternating-current generator depends on the number of poles in its field and the speed at which it rotates, thus:

$$f = \frac{p \times rpm}{120}$$

where f = frequency (hertz)
  p = number of poles in field
  rpm = number of revolutions the field rotates per minute

Example: Compute the frequency of the voltage generated by an alternator having two poles and rotating at 3600 rpm.

Solution:
$$f = \frac{p \times \text{rpm}}{120} = \frac{2 \times 3600}{120} = 60 \text{ Hz}$$

Speed   To determine the speed at which an alternator should be driven to generate a given frequency, the following expression is used:

$$\text{rpm} = \frac{f \times 120}{p}$$

Example: At what speed must a four-pole alternator be driven to generate 60 Hz?

Solution:
$$\text{rpm} = \frac{f \times 120}{p} = \frac{60 \times 120}{4} = 1800 \text{ rpm}$$

Number of Poles   If the frequency and speed of an alternator are known, the number of poles in its field can be calculated by use of the following formula:

$$p = \frac{f \times 120}{\text{rpm}}$$

Example: How many poles does an alternator have if it generates 60 Hz at 1200 rpm?

Solution:
$$p = \frac{f \times 120}{\text{rpm}} = \frac{60 \times 120}{1200} = 6 \text{ poles}$$

Transformer (Single Phase)

Primary Current   The full load primary current can be readily calculated if the kVA rating of the transformer and the primary voltage are known, thus:

$$I_p = \frac{\text{kVA} \times 1000}{E_p}$$

where E_p = rated primary voltage
I_p = rated primary current

Example: Find the rated full load primary current of a 10-kVA 2300-volt distribution transformer.

Solution:
$$I_p = \frac{\text{kVA} \times 1000}{E_p} = \frac{10 \times 1000}{2300} = 4.3 \text{ amp}$$

Secondary Current   The expression for secondary current is similar to that for primary current, thus:

$$I_s = \frac{\text{kVA} \times 1000}{E_s}$$

where I_s = rated secondary current
E_s = rated secondary voltage

Example: A 3-kVA distribution transformer is rated 2300 volts primary and 110 volts secondary. What is its full load secondary current?

Solution:
$$I_s = \frac{\text{kVA} \times 1000}{E_s}$$
$$= \frac{3 \times 1000}{110} = 27 \text{ amp secondary}$$

SUMMARY OF ELECTRICAL FORMULAS

Current

1. *Direct Current*   To find the current when the voltage and resistance are given:

$$I = \frac{E}{R}$$

2. *Alternating Current*   To find the current when the voltage and resistance are given:

$$I = \frac{E}{R}$$

3. *Alternating Current*   To find the current when the voltage and impedance are given:

$$I = \frac{E}{Z}$$

4. *Direct Current*   To find the current when the voltage and the power are given:

$$I = \frac{P}{E}$$

5. *Alternating Current, Single Phase*   To find the current when the voltage, the power, and the power factor are given:

$$I = \frac{P}{E \times \text{pf}}$$

6. *Alternating Current, Three Phase*   To find the current when the voltage, the power, and the power factor are given:

$$I = \frac{P}{\sqrt{3}E \times \text{pf}}$$

7. *Alternating Current, Single Phase*   To find the current when the volt-amperes and the volts are given:

$$I = \frac{\text{VA}}{E}$$

and
$$I = \frac{\text{kVA} \times 1000}{E}$$

8. *Alternating Current, Three Phase*   To find the current when the volt-amperes and the volts are given:

$$I = \frac{\text{VA}}{\sqrt{3}E}$$

and
$$I = \frac{\text{kVA} \times 1000}{\sqrt{3}E}$$

Voltage
9. *Direct Current*   To find the voltage when the current and resistance are given:

$$E = IR$$

10. *Alternating Current*   To find the voltage when the current and resistance are given:

$$E = IR$$

11. *Alternating Current*   To find the voltage when the current and impedance are given:

$$E = IZ$$

Resistance
12. *Direct Current*   To find the resistance when the voltage and the current are given:

$$R = \frac{E}{I}$$

13. *Alternating Current*   To find the resistance when the voltage and the current are given:

$$R = \frac{E}{I}$$

Power Loss
14. *Direct Current*   To find the power loss when the current and resistance are given:

$$P = I^2 R$$

15. *Alternating Current, Single Phase*   To find the power loss when the current and the resistance are given:

$$P = I^2 R$$

16. *Alternating Current, Three Phase*   To find the power loss when the current and the resistance are given:

$$P = 3 \times I^2 R$$

Impedance

17. *Alternating Current*   To find the impedance when the current and the voltage are given:

$$Z = \frac{E}{I}$$

18. *Alternating Current*   To find the impedance when the resistance and the inductive reactance are given:

$$Z = \sqrt{R^2 + X_L^2}$$

19. *Alternating Current*   To find the impedance when the resistance, inductive reactance, and capacitive reactance are given:

$$Z = \sqrt{R^2 + (X_L - X_c)^2}$$

Power

20. *Direct Current*   To find the power when the voltage and the current are given:

$$P = EI$$

21. *Alternating Current, Single Phase*   To find the power when the voltage, the current, and the power factor are given:

$$P = E \times I \times \text{pf}$$

22. *Alternating Current, Three Phase*   To find the power when the voltage, the current, and the power factor are given:

$$P = \sqrt{3}E \times I \times \text{pf}$$

Volt-Amperes

23. *Alternating Current, Single Phase*   To find the volt-amperes when the voltage and the current are given:

$$VA = E \times I$$

and

$$kVA = \frac{E \times I}{1000}$$

24. *Alternating Current, Three Phase*   To find the volt-amperes when the voltage and the current are given:

$$VA = \sqrt{3}E \times I$$

and

$$kVA = \frac{\sqrt{3}E \times I}{1000}$$

Power Factor

25. *Alternating Current, Single Phase*   To find the power factor when the power, the volts, and the amperes are given:

$$\text{pf} = \frac{P}{E \times I}$$

26. *Alternating Current, Three Phase*   To find the power factor when the power, the volts, and the amperes are given:

$$\text{pf} = \frac{P}{\sqrt{3} \times E \times I}$$

Motor

27. *Direct-Current Motor*   To find the current when the horsepower of the motor, the voltage, and the efficiency are given:

$$I = \frac{hp \times 746}{E \times eff}$$

28. *Alternating-Current Motor, Single Phase*   To find the current when the horsepower of the motor, the voltage, the efficiency, and the power factor are given:

$$I = \frac{hp \times 746}{E \times pf \times eff}$$

29. *Alternating-Current Motor, Three Phase*   To find the current when the horsepower of the motor, the voltage, the power factor, and the efficiency are given:

$$I = \frac{hp \times 746}{\sqrt{3}E \times pf \times eff}$$

30. *Direct-Current Motor*   To find the horsepower of the motor when the voltage, the current, and the efficiency are given:

$$hp = \frac{E \times I \times eff}{746}$$

31. *Alternating-Current Motor, Single Phase*   To find the horsepower of the motor when the voltage, the current, the power factor, and the efficiency are given:

$$hp = \frac{E \times I \times pf \times eff}{746}$$

32. *Alternating-Current Motor, Three Phase*   To find the horsepower of the motor, when the voltage, the current, the power factor, and the efficiency are given:

$$hp = \frac{\sqrt{3}E \times I \times pf \times eff}{746}$$

Generator

33. *Alternating-Current Generator*   To find the frequency of the generator when the number of poles and the rpm are given:

$$f = \frac{p \times rpm}{120}$$

34. *Alternating-Current Generator*   To find the rpm of the generator when the number of poles and the frequency are given:

$$rpm = \frac{f \times 120}{p}$$

35. *Alternating-Current Generator*   To find the number of poles of the generator when the frequency and the rpm are given:

$$p = \frac{f \times 120}{rpm}$$

Transformer

36. *Alternating-Current Transformer, Single Phase*   To find the primary current when the kVA and the primary voltage are given:

$$I_p = \frac{kVA \times 1,000}{E_p}$$

37. *Alternating-Current Transformer, Single Phase*   To find the secondary current when the kVA and the secondary voltage are given:

$$I_s = \frac{kVA \times 1,000}{E_s}$$

Definition of
Electrical Terms*

air switch   A switch in which the interruption of the circuit occurs in air. Same as air circuit breaker (ACB).

ammeter   Instrument used to measure electrical current.

ampere   Practical unit of electrical current. That produced by a pressure of 1 volt across a circuit having a resistance of 1 ohm.

ampere-hour   One ampere flowing for 1 hr.

ampere-turn   Magnetizing force produced by a current of 1 amp flowing through a coil of one turn.

annunciator   audible or visible alarm or signal initiated electrically.

anode   Positive electrode in a cell; the positive pole; attracts negative ions.

apparent load   Current in amperes times emf in volts gives apparent load in volt-amperes. Used for alternating-current circuits because the current flow is not always in phase with the emf; hence, amperes times volts does not give the real energy load.

arcing contacts   Contacts on which the arc is drawn after the main contacts of a switch or circuit breaker have parted.

arcing rings (grading shields)   Metal rings or projections located at each end of an insulator string to equalize the voltage gradient or distribution over the insulator string.

Armor rod (preformed)   A spiral-formed aluminum rod, a group of which is placed around a conductor at the point of suspension to minimize vibration and to protect the conductor from burning in case of a flashover.

astronomic dial   A device which changes the daily time of "on" or "off" operation of street-light time switches to correspond to the daily change of sunset and sunrise.

autotransformer   A transformer in which the primary and secondary have a common winding.

auxiliary relay   Relay which operates in response to current to assist another relay or device in the performance of a function.

baker board   Wooden platform which attaches to the side of a pole in a position horizontal to the ground.

battery   A group of several cells connected together as a unit for furnishing electrical current.

battery earphone set   Instrument for checking continuity in low-voltage circuits.

boatswain's chair   A small wooden seat approximately 2 ft wide supported by four ropes secured to a ring or tied at a common point above the workman's head.

*Taken largely from "Glossary of Terms" prepared by the Department of Water and Power of the City of Los Angeles.

boost   Raise or attempt to raise voltage.

brush   Conducting material, usually made of carbon, bearing against the commutator or slip rings through which the current flows in or out of the rotor of a machine.

buck   Lower or attempt to lower voltage.

bull wheel   A reel device used to hold tension on a transmission conductor during stringing operations.

Burndy Hysplice   A type of compression of splicing equipment.

bus   Conductor or group of conductors in a switch-gear assembly which serves as a common connection for three or more circuits.

capacitor   An electrical device for storing a charge of electricity and returning it to the line. It is used to balance the inductance of a circuit, since its action is opposite in phase to that of inductive apparatus; i.e., it throws the current ahead of the emf in phase. It is made of alternate plates of tinfoil and insulating material, the size of plates and thickness of insulation determining the capacity for holding electric charge. Capacity is measured, practically, in microfarads, or millionths of a farad.

capstan   Winch used for stringing rope or cable.

carrier current   High frequency electrical impulse superimposed on the transmission or telephone lines and used for transmission of intelligence between stations and for relaying.

cascaded circuits   One or more series street-light circuits controlled from another series street-light circuit by a 6.6-amp magnetic switch.

catenary curve   A parabolic-type curve which is formed by a conductor strung between two points.

cathode   The electrode in a cell (voltaic or primary) that attracts the positive ions and repels the negative ions; the negative pole.

charging current of a transmission line   Current that flows into the capacitance of a transmission line when voltage is applied at its terminals.

choke coil   A coil of relatively low ohmic resistance and comparatively high impedance to alternating current.

circuit breaker   A device for interrupting a circuit between separable contacts under normal or abnormal conditions. May be operated manually or automatically for circuit control by overload or other selected conditions.

circuit-breaker mechanism   An assembly of levers, cranks, and other parts which actuate the moving contacts of a circuit breaker.

circular mil   An area equal to that of a circle with a diameter of 0.001 in. It is used for designating the cross-sectional area of wire.

clamp-on volt-ammeter   Portable instrument which can be used to measure current in cables and wires without disconnecting the circuit.

clearance   A statement from a person having proper authority that a particular circuit or piece of electrical apparatus is disconnected from all sources of electrical energy.

climbers   Used for climbing or support on a pole.

closing coil   The electromagnet or solenoid which supplies power for closing a circuit breaker or other device.

collector rings (slip rings)   Metal rings suitably mounted on the rotating member of an electric machine with stationary brushes bearing thereon to conduct current into or out of the rotating member.

commutator   That part of the rotor of a direct-current motor or generator which is used to conduct electrical energy to the rotor windings from the stationary brushes with which the commutator is in contact.

compensator   See *Autotransformer*.

conductance   The reciprocal of electrical resistance.

conductivity   The ease with which a substance transmits electricity.

conductor   A carrier of electric current.

conservator   An auxiliary tank, normally only partly filled with oil or other cooling liquid and connected to the completely filled main tank.

constant-current transformer   A transformer which when supplied from a constant potential source automatically maintains a constant current in its secondary circuit under varying conditions of load impedance.

contactor   An electrically operated device for energizing and deenergizing electrical equipment, usually low voltage.

control circuit   A circuit carrying a low voltage used to operate various pieces of equipment such as circuit breakers, contactors, or valves, usually from a remote location.

control switch   A switch for controlling electrically operated devices.

core   A mass of iron placed inside a coil to decrease the reluctance to magnetic lines of force.

core-type transformer   Core takes the form of a single ring of laminated steel encircled by two or more groups of primary and secondary windings.

corona   A phenomenon which occurs around a conductor when the conductor potential is raised above the dielectric strength of the surrounding air. Corona appears as a bluish brush discharge around the conductor with occasional streamers into the surrounding air.

counter emf   Counter electromotive force, an emf induced in a coil or armature by self-induction. It opposes the applied voltage.

counterpoise   Ground wires which are connected to tower footings to provide an adequate lightning current path to ground.

current transformer   A small transformer used in conjunction with a meter for measuring heavy currents in power leads. The primary is in series with the power lead, and full rated current in the primary gives 5 amp in the secondary. Never open the secondary circuit of a current transformer when the primary is energized.

damper   A device used to inhibit the vibration of conductors on a transmission line.

dashpot   A device using a gas or liquid to absorb energy or retard the movement of moving parts, such as on a circuit breaker.

dead-end ladder   Aluminum ladder used horizontally or vertically in working on dead-end insulators and hardware on transmission towers.

delta   A type of connection for a three-phase electrical machine or for transformer windings.

demand meter   A watthour meter containing an extra dial which measures the maximum demand over a specified period of time, usually 15 min.

dielectric   A nonconducting or insulating material. The dielectric value of a material is measured by comparing it with that of a like thickness of air taken as unity: glass, 3 to 8; porcelain, 4.4; treated paper, 2 to 4; paraffin and various forms of rubber, 2.2 to 2.5; mica, 6; water, 80.

differential relay   A relay which functions by reason of the difference between two quantities of the same nature such as current or voltage.

direct current   An unidirectional electric current.

directional relay   A relay which functions in conformance with the direction of power, voltage, current, phase relation, etc.

disconnecting switch   A form of air switch used for changing connections in a circuit or system or for isolating purposes.

distribution panel   Electric panel with circuit breakers which energize other panels or circuits.

distribution transformer   Transformers rated at 500 kVA and below used for voltage transformation on the distribution system.

double-circuit station   An electrical station which is supplied from two lines. Usually refers to industrial stations.

dry-type transformer   A transformer which operates without oil as a cooling medium and designed with insulation to withstand higher temperatures.

duct sheet   Shows duct arrangements in manholes and vaults.

easement   The right of one person to use the land of another for a specific purpose.

eddy-current loss   Loss in core of transformer due to currents being induced in and flowing around core. Also induced in conductors, where it causes heating.

electrical operation   A switch or circuit breaker power-operated by electricity. *Manual operation* of a switch or circuit breaker is operation by hand without the use of any other source of power.

electrode   The terminal by which a current leaves or enters an electrolytic cell. An electric terminal.

electrolyte   A substance that conducts a current of electricity by the movement of ions.

electromagnet   A magnet made by winding coil of wire around a soft iron core. Current passed through this winding produces magnetic line of force in the coil.

electromotive force   The electrical pressure that moves or tends to move electrons.

electron   The smallest particle of negative electricity.

enclosed bus   A bus having its conductors enclosed in an insulating or metal enclosure.

excitation   Current flow that causes magnetic flux.

exciter   An auxiliary direct-current generator which supplies energy for the field excitation of another electric machine.

exciting current   Current used for energizing the field coils of a motor or generator to create magnetic flux in the pole pieces. It also refers to the no-load current drawn by a transformer.

fault (wire or cable)   A partial or total local failure in the insulation or continuity of a conductor.

feedback   Electrical flow usually from the output of a device back to its input. Used for stabilizing, measurement, amplifying, etc.

feeder   A conductor or group of conductors connecting two generating stations, two substations, a generating station and a substation or feeding point, or a substation and a feeding point.

feeder regulator   See *Induction voltage regulator*.

field   The region around a magnet or electric charge in which the magnet or charge is capable of exerting its influence.

field coil   Coil used to excite a field magnet.

field magnet   A magnet used to produce a magnetic field.

flux   The magnetic lines of force that flow from the north pole of a magnet and return to the south pole.

frequency   The number of cycles per second of an alternating current or hertz.

frequency changer   A machine which converts the power of an alternating-current system from one frequency to another with or without a change in the number of phases or in the voltage.

frequency control   The regulation of the frequency of a generating station within a narrow range.

fuse   A part of a circuit made of a low-melting-point material so that it will melt and break the circuit when a specified current is exceeded. Always the weakest point in the circuit.

galloping phenomenon   A type of vibration which results in large elliptical-shaped loops along the conductors.

galvanometer   An instrument used to indicate and measure small electric currents.

generator   Machine that converts mechanical energy into electrical energy.

ground bus   A bus used to connect a number of grounding conductors to one or more grounding electrodes.

ground current   Any current flowing in or to the earth.

ground detector   Instrument used for indicating the presence of a ground on an ungrounded system.

grounded (earthed)   Connected to earth or to some conducting body which serves in place of the earth.

grounded system   A system of conductors in which at least one conductor or point (usually in the middle wire or neutral point of transformer or generator windings) is intentionally grounded either solidly or through a current-limiting device.

grounding switch   A form of air switch by means of which a circuit or piece of apparatus may be connected to ground.

grounding transformer   A transformer intended primarily for developing a neutral point for grounding purposes.

henry   The inductance of a closed circuit which produces an emf of 1 volt when the electric current varies uniformly at 1 amp per second.

hertz   The unit of frequency equal to one cycle per second of an alternating current.

high pot   To apply high potential to electrical machine or equipment. Normally done during insulation testing.

high side   The higher voltage electrical system of two systems connected by a transformer.

high voltage   Above 600 volts.

horsepower   A unit of power equal to 746 watts.

hot   Energized electrically.

hot-line tools (hot sticks)   A group of special tools used primarily on the transmission and subtransmission systems for replacing circuit components or for other types of work with the circuit energized.

hydrometer   An instrument used to measure the specific gravity of liquids.

hystersis   Loss in core of transformer due to molecular friction.

IL transformer   A two-winding insulating transformer connected with its primary in series with a 6.6-amp street-light circuit and used to insulate an individual lamp from the high-voltage circuit.

impedance   The opposition a circuit offers to the flow of alternating current.

induce   To produce a force in a body by exposing it to an influence such as magnetic force, electric force, or changing current.

inductance   The coefficient of self-induction of a circuit. The effect is to cause current to lag behind emf in phase. It depends on the size and shape of the circuit, cross section and shape of conductor, magnetic properties of the conductor and the surrounding medium, the frequency of the current reversal, and resistance of the conductor. It is, therefore, difficult to compute except for simple wire circuits suspended in air. The unit for measurement is the henry or the millihenry (1/1000 henry).

induction coil   An arrangement of two coils such that a changing current in the first produces a voltage in the second.

induction motor   One in which rotor current is set up by transformer action from alternating currents supplied to the stator windings. The rotor currents are induced by those in the stator, hence the name "induction motor."

induction voltage regulator (two types)   A form of transformer in which the output voltage may be varied by changing the relative positions of the primary and secondary windings. An autotransformer in which the voltage is regulated by automatically selecting turns ratio steps in the windings.

instrument switch   A switch used to disconnect an instrument or to transfer it from one circuit or phase to another.

instrument transformer   Used for measuring and control purposes; provides currents and voltages representative of the primary components but of such magnitude that there is less danger to instruments and personnel.

interlock   An electrical or mechanical arrangement that prevents one operation or sequence of operations from taking place until another prerequisite operation or condition has been satisfied.

interrupter   A device to open an electrical circuit quickly.

Jacob's ladder   A portable rope ladder.

jumper harness   A safety device which is to be worn when patrolmen are working on a string of insulators on a dead-end tower.

kilovolt-ampere   The unit of apparent power in alternating-current circuits as distinguished from kilowatt, which represents the true power.

kilowatt   One thousand watts. The true power being used in an alternating-current circuit.

kilowatthour   A unit of energy equal to 1000 watthours.

kVA   Abbreviation for kilovolt-amperes.

lag   The number of degrees an alternating current lags behind the voltage.

lead   The number of degrees an alternating current leads the voltage.

leads   Conductors to or from a piece of electrical equipment.

lightning arrester   A device which has the property of reducing the voltage of a surge applied to its terminals by passing the surge current to ground. It is capable of interrupting follow current if present and restores itself to its original operating conditions.

lightning rod   A device used to divert a lightning surge to a ground path.

line of force   A line in a field of force that indicates the direction of the force.

line loss   Power used up in overcoming the resistance of the transmission line to flow of electric current. It varies with the resistance of the line and as the square of the current flowing. Also known as I^2R loss.

link   Semifixed means of opening an electrical conductor. Not used in normal switching procedures.

load factor   The ratio of the average load over a given period of time to the peak load occurring in that period.

loop feeder   Consists of a number of tie feeders in series, forming a closed loop. *Note:* There are two routes by which any point on a loop feeder can receive electric energy so that the flow can be in either direction.

low voltage   600 volts and lower.

magnetic flux   The total number of lines of force issuing from a pole.

magnetic switch   Either a solenoid-operated primary oil switch or a magnetic-coil-operated air-break contactor with 6.6-amp operating coils. Generally used for cascading street-light circuits.

magneto   An alternating-current generator in which the field is supplied by a permanent magnet.

magnetomotive force   The force responsible for the magnetic flux in a magnetic circuit. It is proportional to the ampere turns.

maximum demand   The greatest of all the demands of an installation or system which have occurred during a given period of time, usually 15 min.

megavar   One million vars.

megawatt   One million watts.

megger   A high-voltage insulation-resistance-testing device.

megohm   One million ohms.

meter loop   Wiring within the meter fixture. Includes wiring to current transformers in built-up meters.

meter test switches   Switches which provide a convenient method of shorting the secondary winding of the current transformer so that the meter can be tested and the transformer protected while service continuity is maintained.

microfarad   One-millionth of a farad. The unit of capacity.

mil   One-thousandth, usually of an inch or an ampere.

motor generator set   A machine which consists of one or more motors mechanically coupled to one or more generators.

motoring   This usually refers to a generator with the driving energy removed from the prime mover but still connected to system electrically. The generator is then acting as a motor and continues to rotate.

multiple circuits   Street-light circuits which are two-wire 120 constant voltage or three-wire 120/240-volt.

mutual induction   The inducing of an emf in a circuit by a changing current in a nearby circuit.

ohm   The unit of electrical resistance about equivalent to the resistance of 1000 ft of No. 10 copper wire.

ohmmeter   Instrument used to measure electrical resistance.

oil circuit breaker   A high-voltage circuit-disconnecting device with its interrupting contacts submerged in oil.

oil switch   See *Oil circuit breaker*.

on the line   Synchronized and connected electrically with the system. Usually refers to the main generator.

operation   As applied to a switch or circuit breaker, i.e., the method provided for its normal functioning.

out of step   Different speeds of electrical phase sequence, i.e., different frequency.

overhead ground wires   Wires which are used in transmission lines to protect the lines by providing a ground path for lightning.

overload relay   A relay that operates on excessive current.

peak load   The maximum load consumed or produced by a unit or group of units.

permeability   A property of matter that indicates the ease with which it is magnetized.

phase   The relative time of change in values of current or emf. Values which change exactly together are in phase. Difference in phase is expressed in degrees, a complete cycle or double reversal being taken as 360 deg. A 180-deg phase difference is complete opposition in phase.

photoelectric control   Sunlight control of series street-light circuits. Used in connection with a remote-control switch.

pilot exciter   Small direct-current generator used to control the field of the main exciter.

pilot light   A small light usually near the control switch to indicate the condition of the circuit being controlled.

polarity   Refers to the direction of current flow in a closed circuit.

pole   One of the ends of a magnet where most of its magnetism is concentrated.

pole-top station   A distributing station which usually consists of a transformer bank mounted on a platform with or without feeder regulators and with pole-top disconnects to open the circuit on the high-voltage side.

pot   Potential transformer.

pothead   Point where separate or overhead electrical conductors come together and continue as a cable.

potential   Voltage or difference in electric pressure between two parts of a circuit or between a circuit and an outside point.

potential transformer   A small transformer used in connection with high-voltage measurement. The primary is connected to the high-voltage circuit. The low-voltage secondary provides the measuring potential for a voltmeter.

power   The amount of work done in a given interval of time.

power factor   The relation of real to apparent power in an alternating-current circuit. It depends on the difference in phase between current and emf and is equal to the cosine of the angle of phase lag or lead of the current.

power transformer   The name power transformer is a misnomer as most transformers pass "power" in larger or smaller quantities from one circuit to another. The description "power transformer" is usually meant to indicate the larger high-voltage transformers found on the system.

primary winding   The winding of a transformer to which the electrical energy is supplied.

protective relay   A relay the principal function of which is to protect service from interruption or to prevent or limit damage to apparatus.

quick break   A switch or circuit breaker that has a high contact-opening speed which is independent of the operator.

rated kVA of a transformer   The output which can be delivered for the time specified at rated secondary voltage and rated frequency without exceeding the specified temperature limitations.

rating   A designated limit of operating characteristics based on definite conditions.

ratio of a transformer   The turns ratio between the windings of a transformer unless otherwise specified.

reactance grounded   A system connected to ground through an inductive reactance.

rectifier   A device which converts alternating current into unidirectional current by permitting appreciable flow of current in only one direction. Types: (1) mechanical, (2) electromagnetic, (3) electrolytic, (4) mercury vapor or mercury arc.

rectify   To change an alternating current to a unidirectional or direct current.

regulation   Steadiness of maintaining the value of emf or current. Controlled by generator or transformer action or by a special regulator. Good regulation should hold voltage variation within 3 percent of normal.

relay   An electrically operated device for the closing and opening of a circuit.

repulsion motor   An alternating-current motor which operates by repulsion of armature conductors by the stator field.

resistance   The opposition of a conductor to a flow of electric current. Resistance is normally expressed as ohms or megohms.

resistivity   The specific resistance of a substance, the resistance in ohms of a centimeter cube of the material to flow of current between opposite faces.

resonance   That condition in an alternating-current circuit which exists when the capacitive reactance exactly balances the inductive reactance.

rheostat   An adjustable resistor so constructed that its resistance can be changed without opening the circuit in which it may be connected.

right-of-way   The right of one person to pass over the land of another. The strip of land over which an electric power transmission line passes.

rotation meter   A small three-phase meter used to determine the direction of phase rotation on circuits or transformer banks.

rotor   The rotating part of an alternating-current motor. Usually the secondary of an induction motor but may be applied to the revolving field of a synchronous motor.

safety harness   Used to raise or lower an injured person to a safe location.

sag   In transmission lines, the distance that the lowest point of a span is below the straight line between supports.

secondary winding   The winding of a transformer from which the load is supplied.

selector switch   A form of air switch arranged so that a conductor or conductors may be connected to any one of several other conductors.

self-contained meter   Either single or three-phase watthour meter which has sufficient current-carrying capacity to meet the specific demand for which it is designed without the need for a current transformer.

series circuit   Street-light circuit which has a current value maintained at 6.6 amp by a constant-current transformer and with a circuit voltage equal to the sum of the voltages across the individual luminaires plus the line drop.

series connection   Any arrangement of cells, generators, condensers, resistors, conductors, etc., so that each carries the entire current of the circuit.

series wound   Applies to generators or motors having the armature and field windings connected in series.

shell-type transformer   The primary and secondary windings take the form of a common ring which is encircled by two or more rings of magnetic material.

short circuit   An abnormal connection of relative low resistance, whether made accidentally or intentionally, between two points of different potential in a circuit.

shunt wound   Applies to generators or motors having the field winding in shunt across the armature.

single-feed station   An electrical station which is supplied from only one circuit.

SL transformer   A two-winding 1 to 1 ratio, 6.6- to 6.6-amp insulating transformer connected with its primary in series with the street-light circuit and used to insulate a group of lamps connected in a series loop from the high-voltage circuit.

slip   The difference between the synchronous speed and the operating speed of an induction machine. May be expressed as a percentage of synchronous speed.

slip ring   A solid ring with brush bearing on it for conveying alternating current to or from the armature or rotor of an alternating-current machine. Also used for feeding current to the revolving field of a synchronous motor or generator.

solenoid   An electromagnet coil which when electrically energized produces a mechanical force by acting upon a free armature in the coil axis.

solidly grounded (directly grounded)   A circuit or equipment grounded through an adequate ground connection in which no impedance has been inserted.

span   The distance between adjacent supports on a line.

spinning reserve   That reserve-generating capacity connected to the system and ready to take load.

squirrel cage   Winding for a rotor of an induction motor made of solid bars joined to connecting rings at each end.

stability   Maximum stable point in carrying capacity of conductors or the pull-out point of a generator.

standard three-phase rotation-test set   Portable device which provides for both phasing and rotation tests.

statiscope   A small instrument used for indicating high-voltage alternating-current or pulsating direct-current potentials and high-voltage static or a high voltage field adjacent to conductors energized at 1000 volts and above.

stator   The part of a machine which contains the stationary parts of the magnetic circuit with their associated windings.

step-down transformer   A transformer in which the primary is the high-voltage winding and the secondary is the low-voltage winding.

step-up transformer   A transformer in which the primary is the low-voltage winding and the secondary is the high-voltage winding.

street-light schedules   Refers to the time of energizing and deenergizing a street-light circuit for each day in the year.

supervisory control   System under which a distributing or other station can be operated unattended but under the positive control of an operator in an attended station.

synchronism   Matched in electrical speed.

synchronous condenser   A special synchronous motor running without mechanical load, the field excitation of which can be varied so as to modify the power factor of the system or to influence the load voltage.

synchronous motor   An alternating-current motor having the field excited by direct current. It runs at a constant speed determined by the number of poles and frequency of the current supply.

target   A supplementary device used in conjunction with a relay or circuit breaker to indicate that it has functioned.

telemetering   Transmission of intelligence such as meter readings fairly long distances, usually from stations to the dispatcher's office, by direct wire or carrier current.

telephone test set   A portable field telephone set used to determine whether or not communication lines are working. May be used for communication between work locations.

testolite   A small instrument used for indicating alternating- or direct-current potentials from 120 to 600 volts. Generally used to locate grounded leads of a two-wire or three-wire distribution circuit of 600 volts or less.

thermal trip   A heat-sensitive device located in the leads to a circuit breaker or contactor that recognizes the heat caused by excessive current and causes the circuit to be opened.

three phase   A term applied to circuits or machines carrying three voltages 120 deg apart in phase.

timer controller   A device used to control solenoid-operated primary oil switches.

transformer   A stationary device for transferring alternating-current electrical energy from one circuit to another circuit by electromagnetic means.

transformer efficiency   Ratio of useful power output to its total power input.

transmission line   A line used for electric-power transmission.

transposition   The physical rotation of the location of conductors on transmission towers.

traveling chair   A fabricated aluminum two-wheeled trolley with an attached chair.

traveling ladder   Wooden ladder with fiber rollers generally used when work or inspection has to be done on transmission hardware or conductors.

trip   An accessory used for tripping. The act of causing a piece of equipment to remove itself or other equipment from the source of energy.

trip coil   An electromagnet used for opening a circuit breaker.

two phase   A term applied to circuits or machines carrying two voltages 90 deg apart in phase.

utilitarian system   Street-lighting system where the fixtures are generally supported on or by the same type of poles and crossarms that carry the distribution lines.

var   Reactive volt-amperes.

vibration   A periodic motion which repeats itself at certain intervals of time.

volt   The practical unit of electrical pressure. The pressure which will cause a current of 1 amp to flow against a resistance of 1 ohm.

voltage regulator   A regulator which functions to maintain the voltage of a synchronous generator, synchronous condenser, or motor at a predetermined plan.

voltage relay   A relay which functions at a predetermined value of voltage.

voltmeter   Instrument used to measure voltage.

watt   The unit of electric power. To find the watts consumed in a given electrical circuit, such as a lamp, multiply the volts by the amperes by the power factor.

watthour   A unit of electrical energy equal to 1 watt power acting for 1 hr.

wattless component   That part of the current in an alternating-current circuit assumed to be 90 deg out of phase with the emf, hence resulting in no useful work.

wye   Star or Y connection of a three-phase electrical machine or transformer windings.

Section **43**

Rope, Knots, Splices, and Gear

ROPE

Manila Rope   Manila rope is made from manila fiber which is spun into a yarn with a twist from left to right. A number of yarns are drawn through a tube to form a single strand. During this drawing, the twist is from right to left. Three of these strands are "laid" to form a rope with the twist from left to right. Because the successive twists in the yarn, strand, and rope are in opposite directions, one twist offsets the other and the rope holds together. Figure 43-1 shows the parts of the rope and a cross section of the rope. The finished rope is light yellow in color. It is hard, but pliant, and has a smooth waxy surface.

Rope sizes and lengths commonly used in linework are as given in Table 43-1. The last column lists the manner in which the rope ends are made up. The meaning of eye splice, back lash, and whipped end is illustrated in Fig. 43-2.

Various Uses of Rope   *Bull Rope*   Bull ropes are used for raising or lowering heavy pieces of equipment, for temporary guys, for setting poles, for holding out heavy transformers, and for lowering large limbs and trunks of trees.

Hand Lines   Hand lines are used for raising and lowering light material and tools or for holding small transformers away from a pole while the latter is being raised. In addition, the ⅜-in hand line is used as a throw line.

Running Line   The running line is used for pulling in several span lengths of wire at one time.

Safety Line   The safety line is used only for lowering a man to the ground.

Slings   Slings are used for lashing tools or material in place, for attaching blocks and snatch blocks to a pole, for lashing an old pole to a new pole temporarily, and for tying line wires up temporarily.

Care of Rope   The life of rope depends upon the care and handling of it. Care begins from the time a new coil of rope is unwrapped and prepared for using.

To avoid kinks in uncoiling new rope, the coil should be laid flat on the floor with the inside end down. Pull the inside end up through the center of the coil and unwind counterclockwise. A new coil usually has a tag on the end giving directions on removing rope from it. To remove the wrong end or pull it through the coil in the wrong direction results in placing many kinks in the rope and consequent excess strain on the fibers. The kinks will remain in the line after many days of use.

Rope should never be dragged over the ground or over sharp objects. Dragging rope over another rope also will result in damage to both.

When tying rope to an object that has sharp corners, the rope should be padded to prevent cutting the fibers.

Rope that has become wet should not be permitted to freeze. If a rope has become full of mud or sand, it should be flushed with a hose and permitted to dry.

Table 43-1   Sizes of Manila Rope Used for Various Purposes

Name	Size, in	Length, ft	How ends are made up
Bull rope	¾	100	Both ends whipped
	1	200	One end whipped, eye splice in other end
Hand line	⅜	75	Eye splice in both ends
	½	75	Eye splice in both ends
	½	150	Eye splice in both ends
Running line	½	600	Eye splice in both ends
Safety line	½	100	One end backlashed, and painted red, eye splice in other end
Sling	½	20	Both ends whipped
	¾	20	Both ends whipped
	1	20	Both ends whipped
	1	30	Both ends whipped
	1⅛	20	Both ends whipped

Never hang up rope where it will be exposed to heat, as high temperatures will harm the rope. Learn how to do up hand lines and blocks and hang them up in the truck in their proper places.

When storing large rope, it is best to place it on wood gratings above the floor where it is well ventilated.

All rope used in line work should be kept dry. Dry rope is a fairly good insulator and is frequently thrown over energized conductors. A wet rope is not a good insulator and would certainly be dangerous both to personnel and equipment, if so used. Rope should never be used on conductors over 5000 volts even when dry.

Safe Loadings   The safe working loads for various sizes of manila rope are given in Table 43-2. These loads are based on a factor of safety of 5 for new rope. Rope in service for more than 6 months should have the safe working loads reduced to one-half of the values shown.

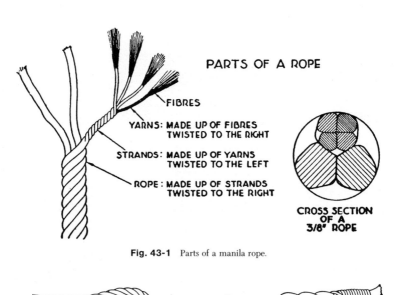

PARTS OF A ROPE

FIBRES

YARNS: MADE UP OF FIBRES TWISTED TO THE RIGHT

STRANDS: MADE UP OF YARNS TWISTED TO THE LEFT

ROPE: MADE UP OF STRANDS TWISTED TO THE RIGHT

CROSS SECTION OF A 3/8" ROPE

Fig. 43-1   Parts of a manila rope.

Eye splice   Backlash   Whipped

Fig. 43-2   Three ways of finishing the ends of a rope.

Table 43-2   Safe Work Loads for Various Sizes of Manila Rope

Rope size, diam, in	Safe work load, lb		Length per lb		Full coil (approx)	
	New	6 months	ft	in	Length, ft	Weight, lb
⅜	270	135	29		1450	50
½	530	265	13	6	1200	90
⅝	880	440	8	7	1200	141
¾	1080	540	6	1	1200	200
⅞	1540	770	4	5	1200	270
1	1800	900	3	8	1200	324
1¼	2700	1350	2	4	1200	502
1½	3700	1850	1	7	1200	720

Column 2 is for new rope, and column 3 is for rope that has been in use for 6 months or more. It should also be remembered that a knot has 50 percent of the strength of the rope and a splice has 80 percent of the strength of the rope.

KNOTS AND KNOT TYING

Knots are used for fastening a rope to an object or for joining two ends of a rope. The knot or hitch used must hold the strain to be applied without damaging the rope or the load. The knot used must also be one that can be tied or loosened easily and quickly.

Terms Used in Knot Tying   All knots and hitches are a combination of the three different kinds of bends: the bight, the loop, and the round turn (see Fig. 43-3).

For convenience in describing the method of making various knots, the following terms will be used: "standing part," "bight," and "running end" (see Fig. 43-3a). The standing part is the principal portion, or longest part of the rope; the bight is a loop formed with the rope so that the two parts lie alongside each other; and the running end of a rope is the free end that is used in forming the knot. Figures 43-4 through 43-25 illustrate various kinds of knots.

SPLICES

Eye Splice   When it is desirable to have a permanent eye at the end of a rope, such as a hand line, it can be formed by splicing the end of the rope into its side, thereby making an eye or side splice.

The steps in making the eye splice are illustrated and described in Fig. 43-26.

Short Splice   The short splice is sometimes called a butt splice because the line is unlaid and the ends are butted together. This splice can be used where it is desirable to splice together two ropes which are not required to pass over a pulley. This splice can be made quickly and is nearly as strong as the rope. As the diameter of the rope is nearly doubled, this type of splice is too bulky to pass through a sheave block.

The steps in making the short splice are illustrated and described in Fig. 43-27.

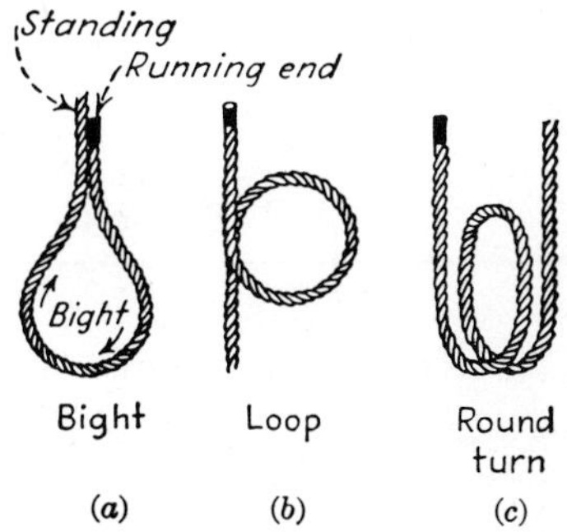

Fig. 43-3   Terms used in knot tying.

Standing   Running end

Bight

Bight   Loop   Round turn

(a)   (b)   (c)

Fig. 43-4 Overhand knot. The overhand knot is the simplest knot made and forms a part of many other knots. This knot is often tied in the end of a rope to prevent the strands from unraveling or as a stop knot to prevent rope from slipping through a block.

Fig. 43-5 Half hitch. A half hitch is used to throw around the end of an object to guide it or keep it erect while hoisting. A half hitch is ordinarily used with another knot or hitch. A half hitch or two thrown around the standing part of a line after tying a clove hitch makes a very secure knot.

Fig. 43-6 Two half hitches. This knot is used in attaching a rope for anchoring or snubbing. It is easily and quickly made and easily untied.

Fig. 43-7 Square knot. The square knot is used to tie two ropes together of approximately the same size. It will not slip and can usually be untied even after a heavy strain has been put on it. Linemen use the square knot to bind light leads, lash poles together on changeovers, on slings to raise transformers, and for attaching blocks to poles and crossarms.

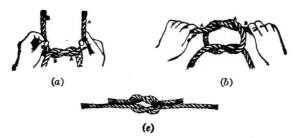

Fig. 43-8 Method of making square knot. (a) Passing left end A over right end B and under. (b) Passing right end B over left end A and under. (c) The completed knot drawn up. *(Courtesy Plymouth Cordage Co.)*

Fig. 43-9 Granny knot. Care must be taken that the standing and running parts of each rope pass through the loop of the other in the same direction, i.e., from above downward, or vice versa; otherwise a granny knot is made, which will not hold.

Fig. 43-10 Thief knot. In tying the square knot the standing part of both ropes must cross, as otherwise a useless knot known as the thief knot is formed.

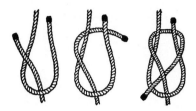

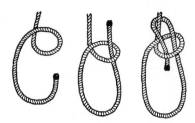

Fig. 43-11 Single sheet bend. The single sheet bend is used in joining ropes, especially those of unequal size. It is more secure than the square knot but is more difficult to untie. It is made by forming a loop in one end of a rope; the end of the other rope is passed up through the loop and underneath the end and standing part, then down through the loop thus formed.

Fig. 43-12 The bowline. The bowline is used to place a loop in the end of a line. It will not slip or pull tight. Linemen use the bowline to attach come-alongs (wire grips) to rope, to attach tail lines to hook ladders, and as a loose knot to throw on conductors to hold them in the clear while working on poles.

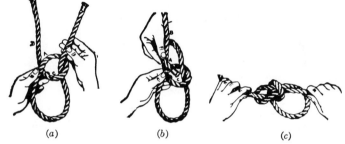

(a)   (b)   (c)

Fig. 43-13 Method of making bowline knot. (a) Threading the bight from below. (b) Leading around standing part and back through bight C. (c) The completed bowline. (*Courtesy Plymouth Cordage Co.*)

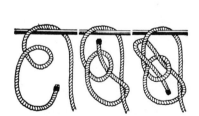

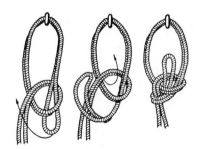

Fig. 43-14 Running bowline. This knot is used when a hand line or bull rope is to be tied around an object at a point that cannot be safely reached, such as the end of a limb.

Fig. 43-15 Double bowline. This knot is used to form a loop in the middle of a rope that will not slip when a strain is put upon it.

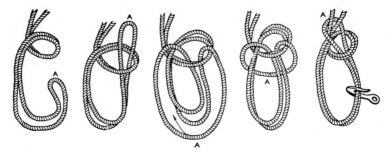

Fig. 43-16 Bowline on a bight. The bowline on a bight is used to place a loop in a line somewhere away from the end of the rope. It can be used to gain mechanical advantage in a rope guy by doubling back through the bowline on a bight much as a set of blocks. The bowline on a bight also makes a good seat for a man when he is suspended on a rope.

To tie this bowline, take the bight of the rope and proceed as with the simple bowline; only instead of tucking the end down through the bight of the knot, carry the bight over the whole and draw up, thus leaving it double in the knot and double in the standing part. The loop under the standing part is single.

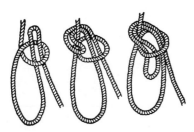

Fig. 43-17 Single intermediate bowline. This knot is used in attaching rope to the hook of a block where the end of the rope is not readily available.

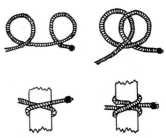

Fig. 43-18 Clove hitch. The clove hitch is used to attach a rope to an object such as a crossarm or pole where a knot that will not slip along the object is desired. Linemen use the clove hitch for side lines, temporary guys, and hoisting steel.

To make this hitch, pass the end of the rope around the spar or timber, then over itself, over and around the spar, and pass the end under itself and between the rope and spar as shown.

Fig. 43-19 Clove hitch used for lifting. (*Courtesy General Electric Co.*)

Fig. 43-20 Timber hitch. The timber hitch is used to attach a rope to a pole when the pole is to be towed by hand along the ground in places where it would be impossible to use a truck or its winch line to spot it. The timber hitch is sometimes used to send crossarms aloft. This hitch forms a secure temporary fastening which may be easily undone. It is similar to the half hitch but is more secure. Instead of the end being passed under the standing part of the rope once, it is wound around the standing part three or four times, as shown in the figure.

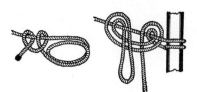

Fig. 43-21 Rope timber hitch and half hitch. The timber hitch will not slip under a steady pull but may slip when slack. To make the timber hitch more secure, a single half hitch may be taken a little farther along on the spar. *(Courtesy General Electric Co.)*

Fig. 43-22 Snubbing hitches. Knots used for attaching a rope for anchoring or snubbing purposes.

Fig. 43-23   Taut rope hitch. This knot is used in attaching one rope to another for snubbing a load.

Fig. 43-24   Rolling bend. The rolling bend is often used for attaching the rope to wires that are too large for the wire-pulling grips. It is also used for skidding poles or timber.

Fig. 43-25 Blackwall hitch. This hitch consists of a loop with the end of the rope passed under the standing part and across the hook. Under load the hauling part jams the end against the hook. This hitch should be used only where the strain is steady and there is no hazard if the hitch slips. It is exceedingly useful when the hitch has to be made quickly or where the hitch must be changed frequently. *(Courtesy General Electric Co.)*

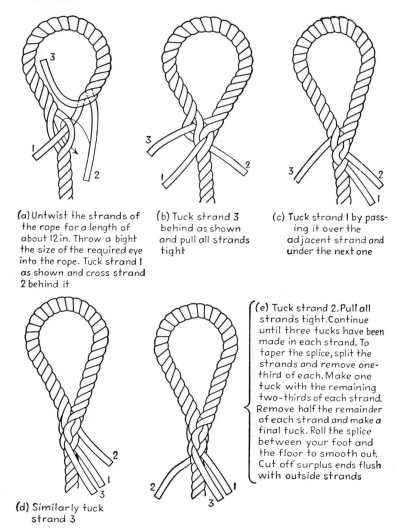

(a) Untwist the strands of the rope for a length of about 12 in. Throw a bight the size of the required eye into the rope. Tuck strand 1 as shown and cross strand 2 behind it

(b) Tuck strand 3 behind as shown and pull all strands tight

(c) Tuck strand 1 by passing it over the adjacent strand and under the next one

(d) Similarly tuck strand 3

(e) Tuck strand 2. Pull all strands tight. Continue until three tucks have been made in each strand. To taper the splice, split the strands and remove one-third of each. Make one tuck with the remaining two-thirds of each strand. Remove half the remainder of each strand and make a final tuck. Roll the splice between your foot and the floor to smooth out. Cut off surplus ends flush with outside strands

Fig. 43-26 Eye splice, fiber rope.

Long Splice   The long splice is sometimes known as a running splice. It is used to splice rope where it is undesirable to increase the diameter of the rope. A good long splice is hard to detect without close examination and will run freely through a sheave or blocks. Using a running splice, a damaged part of a long line can be removed from the line without detracting from its usefulness. The strength will be reduced, however.

The steps in making the long splice are illustrated and described in Fig. 43-28.

Reduction in Strength Due to Knots and Splices   Table 43-3 shows the extent to which the strength of manila rope is reduced when used in connection with some of the common knots and splices.

ROPE GEAR

Slings   The simplest sling consists of a short piece of rope whose ends are spliced together to form an endless piece of rope. Slings are used for lashing tools or material in place, for attaching

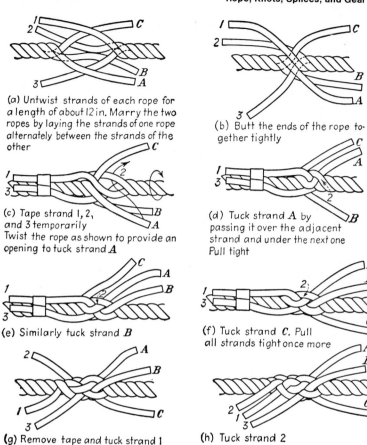

(a) Untwist strands of each rope for a length of about 12 in. Marry the two ropes by laying the strands of one rope alternately between the strands of the other

(b) Butt the ends of the rope together tightly

(c) Tape strand 1, 2, and 3 temporarily
Twist the rope as shown to provide an opening to tuck strand A

(d) Tuck strand A by passing it over the adjacent strand and under the next one
Pull tight

(e) Similarly tuck strand B

(f) Tuck strand C. Pull all strands tight once more

(g) Remove tape and tuck strand 1

(h) Tuck strand 2

(i) Tuck strand 3. Pull all strands tight. Continue until three tucks have been made in each strand. To taper the splice, split the strands and remove one-third of each. Make one tuck with the remaining two-thirds of each strand. Remove half of the remainder of each strand and make a final tuck. Roll the splice between your foot and the floor to smooth out. Cut off surplus ends flush with outside strands

Fig. 43-27 Short splice, fiber rope.

Table 43-3 Percentage Strength of Spliced or Knotted Rope

Type of splice or knot	Percentage strength
Straight rope	100
Eye splice	90
Short splice	80
Timber hitch or half hitch	65
Bowline or clove hitch	60
Square knot or sheet bend	50
Overhand knot (half or square knot)	45

blocks or snatch blocks to a pole, for lashing an old pole to a new pole temporarily, and for tying line wires up temporarily. Slings should be 6 to 10 ft long, depending, of course, on their planned use. Slings can also be made with an eye on one end and a dog knot on the other end.

Slings are also made of three-strand manila rope, spliced for a hook at one end and a hook or ring at the other.

Safe Loads   Before attempting to lift a load with a manila-rope sling, the weight of the load should be carefully estimated and a sling of the proper size selected. The greatest load can be lifted when all legs of the sling are in a vertical position and the least when the legs are nearly horizontal. Table 43-4 shows the safe loads that can be lifted with slings of different diameter rope when the legs are vertical and when they make angles of 60, 45, and 30 deg with the horizontal. When a rope sling has been in service for 6 months or more, even though it shows no signs of wear or damage, the loads placed on it should be limited to one-half of those shown.

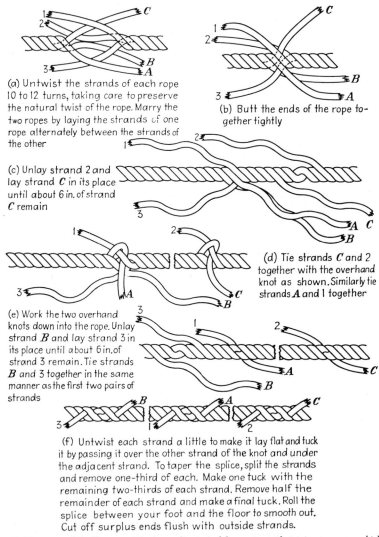

(a) Untwist the strands of each rope 10 to 12 turns, taking care to preserve the natural twist of the rope. Marry the two ropes by laying the strands of one rope alternately between the strands of the other

(b) Butt the ends of the rope together tightly

(c) Unlay strand 2 and lay strand *C* in its place until about 6 in. of strand *C* remain

(d) Tie strands *C* and 2 together with the overhand knot as shown. Similarly tie strands *A* and 1 together

(e) Work the two overhand knots down into the rope. Unlay strand *B* and lay strand 3 in its place until about 6 in. of strand 3 remain. Tie strands *B* and 3 together in the same manner as the first two pairs of strands

(f) Untwist each strand a little to make it lay flat and tuck it by passing it over the other strand of the knot and under the adjacent strand. To taper the splice, split the strands and remove one-third of each. Make one tuck with the remaining two-thirds of each strand. Remove half the remainder of each strand and make a final tuck. Roll the splice between your foot and the floor to smooth out. Cut off surplus ends flush with outside strands.

Fig. 43-28   Long splice, fiber rope. The long splice is used for permanently joining two ropes which must pass through a close fitting pulley.

Figure 43-29 illustrates the manner in which the strain on the rope sling is increased when the angle with the horizontal decreases. Thus at 60 deg the strain in the legs of the sling is only 578 lb for a 1000-lb weight, whereas it is 1930 for a 1000-lb load when the angle is reduced to 15 deg. This is almost twice as great as the weight of the load lifted. At 30 deg the strain is 1000 lb, which is the same as the weight of the load. Sling angles smaller than 30 deg are therefore not recommended.

If the sling is attached around a sharp corner, it should be carefully padded to prevent cutting.

Block and Tackle   Block and tackle are used for applying tension to line conductors when sagging in, for applying tension to guy wires, when hoisting transformers, and for other general-purpose hoisting (see Fig. 43-30).

The use of block and tackle has two advantages: (1) the user can stand on the ground and pull downward while hoisting or lifting a load; (2) the manual force applied need only be a fractional part of the load lifted.

Table 43-4   Safe Load for Slings When Rope Is New

Approximate diameter of rope, in	Safe load for single-leg vertical sling, lb	Safe load for double (two-leg) sling, lb, degree of angle with horizontal		
		60 deg	45 deg	30 deg
½	475	820	670	475
¾	970	1675	1375	970
1	1620	2800	2290	1620

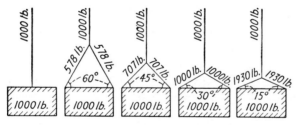

Fig. 43-29   Sample loading of parts of lifting slings. For loads other than 1000 lb, use direct ratios. Sling angles less than 30 deg from horizontal are not recommended.

Fig. 43-30   A typical three-part block and tackle. (*Courtesy A. B. Chance Co.*)

LOAD LIFTED	DIAGRAM OF·RIGGING	
2 times safe load on rope (approx.) **	Single 2 parts Single	block block
3 times safe load on rope (approx.) **	Double 3 parts Single	block block
4 times safe load on rope (approx.) **	Double 4 parts Double	block block
5 times safe load on rope (approx.) **	Triple 5 parts Double	block block
6 times safe load on rope (approx.) **	Triple 6 parts Triple	block block
7 times safe load on rope (approx.) **	Quadruple 7 parts Triple	block block

Fig. 43-31   Lifting capacity of block and tackle. °°Less 20 percent approximately for friction.

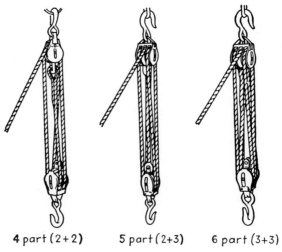

4 part (2+2)   5 part (2+3)   6 part (3+3)

Fig. 43-32   Four-, five-, and six-part blocks and tackle.

Table 43-5   Safe Loads on Block and Tackle Using ½- and ¾-in Manila Rope

Tackle	Safe load, lb	
	4-in block (½-in rope)	6-in block (¾-in rope)
4 part (2 + 2)	800	1650
5 part (2 + 3)	1000	2100
6 part (3 + 3)	1250	2350

Mechanical advantages are 4, 5, and 6.

Mechanical Advantage   To find the pull required to lift a given weight, divide the weight by the number of ropes running from the movable block. The lead line or fall line is not to be counted. However, there is always some friction loss around the sheaves. This can be estimated at 20 percent and added to the load to be lifted. The sketches in Fig. 43-31 show the various combinations of block and tackle employed to give mechanical advantages from 2 to 7. As mentioned above, the ratio of load to pull on the fall line is given by the number of ropes running from the movable block. The load that may be lifted is therefore the mechanical advantage times the safe load on the rope. For block and tackle using ½- and ¾-in rope the safe loads are as given in Table 43-5. Normal uses for 4-in blocks are sagging No. 4 and No. 6 conductors and raising distribution transformers rated up to 37½ kVA single phase. The 6-in blocks are used for heavier work up to the limit of the tackle. Blocks having mechanical advantages of 4, 5, and 6 are illustrated in Fig. 43-32.

Use and Care of Pole Climbing Equipment*

Pole Climbing Equipment Pole climbing equipment consists of a body belt and safety strap and a pair of climbers. Figure 44-1 shows a workman wearing this equipment.

The equipment allows a workman to climb, stand, or change position on a pole when no other suitable means of support is available. It also allows the free use of both hands while in any position on the pole.

Suggested specifications for pole climbing equipment are detailed in Edison Electric Institute Publication No. AP-2-1973. General descriptions of components of the equipment and their use and care follow.

Fig. 44-1 Climbing equipment.

The Body Belt The body belt consists of a cushion section, a belt section with tongue and buckle ends, a tool saddle, and D rings which are attached solidly to the cushion, or on shifting D-ring belts, attached solidly to a D-ring saddle. The body belt usually has provisions made for a holster which will carry one or more tools in addition to the tools which are carried in the tool loops.

Tool loops should be of proper size to prevent the tools from slipping through the loops and falling. There should be no tool loops for 2 in on either side of the center in the back, in accordance with EEI Specifications for Linemen's Climbing Equipment, AP-2-1973.

*Reprinted with the permission of the Edison Electric Institute.

The belt, as a general rule, is marked in "D" sizes. The "D" size is the distance between the heels of the D rings when the belt is laid flat. (See Fig. 44-2.)

The waist size measurement of the belt is found by measuring from the roller on the buckle to the center hole on the tongue end. (See Fig. 44-2.)

Proper measurement for the D-ring size is the distance around behind the body at the point where the belt will be worn between the prominent points of the hip bones, *plus one inch.* (See Fig. 44-3.)

The proper waist size of body belt is determined by measuring the distance around the body at the point where the belt will be worn. The measurement should be made outside of any clothing normally worn while working. (See Fig. 44-4.)

The body belt should be worn snugly but not too tightly. The end of the strap should always be passed through the keeper and kept clear of the D ring when the belt is being worn.

Manufacturers have standardized on a relationship between "D" sizes and waist sizes. In the event your measurements do not coincide with these standard sizes, the belt should be ordered by "D" size, as the waist size is adjustable.

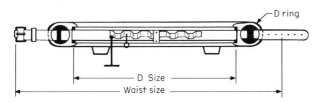

Fig. 44-2   Body Belt.

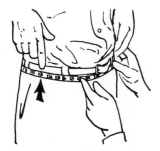

Fig. 44-3   Measuring individual for D size.

Fig. 44-4   Measuring individual for waist size.

The Safety Strap   The safety strap is used for support while working on poles, towers, or platforms. Snap hooks on each end are provided for attachment to the D rings in the body belt.

When climbing poles, under normal conditions, both snaps should be engaged in the *same* D ring for safety. The snap on the double end should have the keeper facing outward and the other snap should face inward. (See Fig. 44-5.) Right-handed men and some left-handed men usually carry the strap on the left as shown.

When in use, one snap hook should be securely engaged in each D ring, never both snaps in the same D ring. The user should look to be sure that snaps are properly engaged. He should never depend on sound or feel for security.

Safety straps are adjustable for length by means of a buckle in the strap to suit the workman and the size of the pole. When in use, the side of the strap to which the buckle is attached should be next to the pole with the buckle tongue outward.

The Climbers   Climbers (Fig. 44-6), are used for ascending, descending, and maintaining the working position on poles when no other means of support is available. The condition, length, and shape of the gaffs of the climbers are of the utmost importance. The gaffs support the workman as he climbs or does his work. Defective gaffs are dangerous and can cause inadvertent electric contacts or falls from poles.

Climbers are made in adjustable or fixed lengths from 14 to 20 in by ½-in steps. Gaffs are furnished in either solid or replaceable type. Proper fit requires a leg iron to reach about ½ in below the inside prominence of the knee joint.

Climbers are secured to the workman's legs by foot and leg straps. These straps should be drawn up to a snug fit, but *not* so tight as to be tiring. High-top shoes with heavy soles and heels should be worn for climbing.

The buckle on the foot strap should lie just outside the shoe lacing. Several types of pads are furnished for the upper end of the leg irons. One type of pad is shown in Fig. 44-7. All leg and

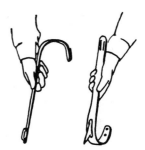

Fig. 44-5   Position of belt and strap when climbing.

Fig. 44-6   Climbers.

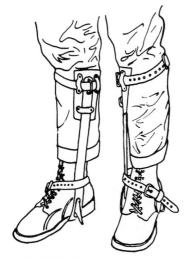

Fig. 44-7   Climbers properly worn.

Fig. 44-8   Adjusting pants leg.

foot strap ends should be snugged down in their keepers after buckling, and the strap ends should point to the rear and outside.

Before the leg straps are fastened, pull up the pants legs so that they bag at the knees and do not bind. Fold the pants leg snugly against the calf, toward the outside, as shown in Fig. 44-8. This prevents the pants leg from tripping the workman while climbing. With climbers properly adjusted the workman will feel comfortable and confident.

Climbing   Arms and hands should not be bare, but properly protected when climbing.

Before ascending a pole, inspect it carefully for unsafe conditions, such as rake, rotted places, nails, tacks, cracks, knots, foreign attachments, pole steps, or ice. Remove rocks and other objects from the ground at the base of the pole. Unauthorized attachments, such as signs, radio aerials, or clothes lines, should be reported to the workman's supervisor or removed, as company instructions may require.

Inspect the pole as you ascend and descend to avoid placing gaffs in cracks, knots, woodpecker holes, etc., which might cause a fall.

When ascending the pole, keep the arms and body relaxed with the hips, shoulders, and knees a comfortable distance away from the pole. Take it easy; favor short steps (step length should be natural for each workman); use the hands and arms for balance only. Climb with the legs, (do not be tempted to "pull up" with the hands or arms); let the legs take the initiative over the hands (the legs "push" the hands). It is necessary that the gaffs be directed toward the center (or heart) of the pole in a natural manner. The size of the pole and the length of leg between hip and knee will determine automatically the amount of gaff separation on the pole. (See Fig. 44-9.)

The effective leg stroke is that angle or stroke which will cause the gaff to cut effectively into the pole wood without side thrust of any sort. The effective stroke results when the knee is thrown comfortably away from the pole (without straining the hip), the gaff is aimed at the target (the imaginary line down the center of the pole), and the leg force and travel is made to parallel the climber shank until proper penetration is accomplished. Kicking or slapping the gaffs against the

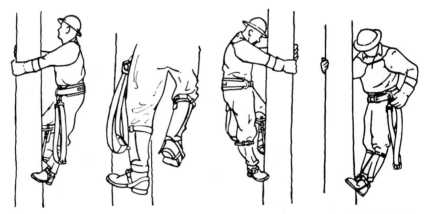

Fig. 44-9 Position while climbing.   **Fig. 44-10** Ascending pole.   **Fig. 44-11** Descending pole.   **Fig. 44-12** Releasing safety strap.

pole should be avoided. The hands and feet should work in coordination; right hand is raised with right foot and left hand with left foot. Weight should be shifted gradually and easily from one foot to the other. (See Fig. 44-10.)

When descending, the hands are lowered first. Each leg is relaxed and straightened before lowering. When the straightened and relaxed leg is "lined up" with the center of the pole and the body weight has been shifted above the gaff, drop the gaff into the pole. In descent, the leg is not stroked; it is merely lowered into position with the body weight behind it. The hands and arms take the initiative over the feet. The hands "push" the feet, which is opposite from ascending the pole when the feet "push" the hands. Keep hips, shoulders, and knees away from the pole. Do not take long steps or try to coast or slide when descending. (See Fig. 44-11.)

When ascending, gaff removal is facilitated by a twisting action of the ankle (outward) and slight prying action of the inside of the footwear against the pole. When descending, the climber gaffs should naturally break out with the outward and lowering movement of the knee. Removal of the climber when the last step to ground is taken is accomplished by a slight twisting and prying action as in ascending.

Wear climbers only when necessary for climbing or working on poles. Always remove them when working on the ground, riding in or driving a vehicle, etc. Gaff guards should be installed when climbers are not being worn.

No one should stand at the base of a pole while a man is ascending, descending, or working on it. There is always a possibility that the workman may fall or drop something. All persons, especially children, should be warned to keep away.

If a second man is to ascend the pole, he should wait until the first man has placed his safety in his working position. When descending a pole, one man should remain in his working position with his safety in use until the other has reached the ground and is out of the way.

Whenever possible, a slippery pole or one partly coated with snow or ice should be ascended with the gaffs in the slippery side and the hand-holds on the less slippery side. Under very slippery conditions or when a strong wind is blowing, the safety strap may be placed around the pole and worked upward or downward in ascending or descending.

Do not hold to pins, crossarm braces, insulators, and other hardware in ascending, descending, or changing position on a pole.

Always ascend and descend on the high side of a leaning, raked, or bent pole.

Use of the Safety Strap   In placing the safety strap around a pole for support, the following steps constitute the best procedure.

Place both climber gaffs firmly in the pole at or near the same level. Keep the knees and hips away from the pole and unsnap the single end of the safety strap with the left hand. (See Fig. 44-12.)

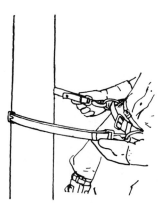

Fig. 44-13 Passing strap to right hand.

Fig. 44-14 Position of srap in use.

Fig.44-15 Moving safety strap position.

Pass the single end of the safety strap to the right hand around the back of the pole and grasp the pole with the left hand. With the right hand, carry the snap hook to the right D ring and engage the snap with the D ring with the keeper facing outward. (See Figs. 44-13 and 44-14.) Both snap keepers should face outward and the strap should lie flat, without twists, against the pole, with the buckle tongue side out when in use. (See Fig. 44-14.) Always visually check and be sure that the snaps are securely and properly engaged in the D rings before trusting your weight on the safety strap. Never depend on the sound of the snap keeper. It will make the same sound if the snap is engaged in an insecure attachment, such as a plier handle or wire hook.

Keep plier pockets and other objects well clear of D rings to avoid accidentally hooking the safety belt snap into them. There should be a minimum of 4 in of clearance between D rings and tool pockets. There should be no wire hooks used on body belts.

Always use the belt and safety strap when working aloft on a pole or structure. It is good insurance against falling and provides a safe place to carry tools normally needed. *Never* place the safety strap around the top of a pole above the top crossarm or in any other place where it can accidentally slip off. If it is necessary to place the safety strap high on a new bare pole, place a long through-bolt in the top gain hole to keep the safety strap from slipping off the pole.

To move up or down on a pole with the safety strap in use, hold the pole with one hand, as you release tension on the strap, and move the strap up or down with the other hand. (See Fig. 44-15.)

Before a person is permitted to climb, he should be taught the proper method of handling a safety strap. This can be accomplished by standing on the ground at the base of a pole without wearing climbers and going through the operations shown in Figs. 44-12, 44-13, and 44-14.

Practice at the base of a pole until the proper climbing position for the feet, arms, and body is acquired. Place the feet on the ground at about a 90-deg angle with each other and with the side of the arch of each foot against the sides of the pole. Extend both arms out forward in a horizontal plane from the shoulders and hold on to the back of the pole with the hands. Thrust the hips well back from the pole with the legs and arms straight until the body is in a position as shown in Fig. 44-16.

Practice climbing on a medium-size, smooth pole which has not been badly cut by previous climbing. Start using climbers by first confining climbing to a section between the ground and 3 ft above the ground. This permits one to learn to climb without fear of falling. Before starting to climb, inspect the pole from the ground for cracks, knots, holes, tacks, and nails, as the presence of any of these may deflect a gaff and obtain an insecure foothold. (See Fig. 44-17.)

Practice placing the safety strap around the pole and fastening it to the D ring while on the pole. Stand on the pole with both legs straight and both gaffs firmly in the pole at or near the same level and at a point approximately 2 ft above the ground level for the first trials. This is the position for placing the strap when in the working position at any location on the pole. Then proceed to place it around the pole as described. [See Figs. 44-12 (or 44-18), 44-13 and 44-14.]

Fig. 44-16 Practice position.

Fig. 44-17 Practice climbing.

Fig. 44-18 Ready to place strap.

Inspection and Care   The user of climbing equipment should carefully inspect it before each use.

Inspect leather or fabric parts for cuts, cracks, tears, enlarged buckle tongue holes, narrowing down due to stretch, or for hard and dry leather. Inspect stitching for broken, ragged, or rotted threads. Inspect metal parts for breaks, cracks, loose attachments, or wear that might affect strength. Care should be taken to determine that the keepers of the snaps are firmly and securely seated in the recess at the end of the hooks and that there is a reasonable resistance to depressing the keeper. Any distortion of the hook may change the tension of the spring so that the keeper is not in firm contact with, or securely seated in, the nose of the hook. If the keeper is not securely seated, it is possible that the D ring may become disengaged by slipping between the nose of the hook and the keeper and be the cause of a serious fall. Check the gaffs and determine whether the cutting edges are properly sharpened and shaped.

Defective equipment should be repaired or replaced and gaffs properly shaped as soon as possible. If defects affect safety of the equipment, it should be removed from service until repaired or replaced.

Cleaning and Dressing Leather   Leather parts of climbing equipment should be cleaned approximately every 3 months and dressed approximately every 6 months and on return from vacation periods to keep them pliable and in good condition. Cleaning and dressing may be done more often if necessary due to excessive moisture or perspiration. Paint stains should be removed before the paint dries. Never, however, use gasoline or similar solvent on leather, as this will dry out and severely damage the leather.

To Clean Leather   Wipe off surface dirt with a damp sponge.

Using a neutral soap, such as castile, and a clean moist sponge, work up a good lather to remove imbedded dirt and perspiration. Wipe off with a clean cloth.

Work up a good lather with saddle soap and a clean sponge and rub it well into all parts of the leather. Wipe off with a clean cloth.

To Dress Leather   Never dress or oil leather before cleaning as just outlined.

While the leather is still damp after cleaning apply a good leather dressing or neatsfoot oil, working it into the leather with the hands. Do not use an excess amount of oil, as this will saturate and weaken the leather.

Allow the leather parts to dry in a cool, shady place for 24 hours. Never dry leather near a source of heat, as heat destroys leather.

Remove any excess dressing by rubbing vigorously with a clean soft cloth. Never use any mineral oil or grease for dressing leather.

Cleaning Nylon Fabric Safety Straps   Since there are differences in the methods of fabricating nylon safety straps, it is recommended that each manufacturer be consulted on the proper cleaning procedure to be used in cleaning his type nylon safety straps.

Storage   Store leather equipment away from sources of heat, such as stoves, radiators, steam pipes, or open fires. Keep leather goods away from sharp-edged objects. Before storing the climb-

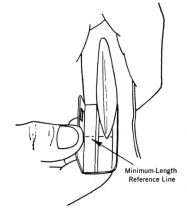

Fig. 44-19   Checking for minimum length.

Minimum-Length
Reference Line

ers it is good practice to install approved climber gaff guards to protect the gaff points and edges and to protect other objects and persons from accidental injury or damage.

Inspection and Maintenance of Climber Gaffs   Gaffs of climbers should be inspected and checked frequently for length, width, thickness, profile of the point, and sharpness of the cutting edges. If a cutout or any other difficulty is experienced in getting the gaffs to hold in a pole, investigate immediately to determine the cause. The trouble may be that the gaffs have been accidentally damaged since the last inspection.

The gaff gauge can be used for checking climber gaffs and as an aid in shaping the gaff properly. One type of gaff gauge is shown in the following sketches (Figs. 44-19 through 44-29).

Be sure to check the instructions issued by the manufacturers of the gauge you are using since there may be differences in the exact procedures to be followed.

The reference line across the gauge indicates the minimum length of the gaff, measured on the underside from the heel of the gaff to the point. (See Fig. 44-19.)

Company requirements as to minimum length of gaffs should be observed.

The "TH" slots are used to check the thickness of the gaff 1 in and ½ in from the point. The face of the gauge should lie flat against the back or ridge of the gaff. If within acceptable limits, the point of the gaff will lie on or between one pair of the reference lines on the gauge, as indicated in Figs. 44-20 and 44-21.

The "W" slots are used to check the width of the gaff 1 in and ½ in from the point, as shown in Figs. 44-22 and 44-23. The face of the gauge should be flat against the back or ridge of the gaff, as in the previous check. If within acceptable limits, the point of the gaff will lie on or between one pair of the reference lines on the gauge.

If necessary to shape the gaff for proper thickness, it should be filed on the flat underside which lies between the gaff and the stirrup. Care should be used not to cut or notch the leg iron or stirrup with the file. Also there should be no file marks left on the gaff. (See Fig. 44-24.) A 10-in mill bastard flat file or an 8-in smooth knife file will do a nice job, if kept clean by frequent use of a file card brush. File marks on a gaff, particularly if they are crosswise, may weaken the steel and result in a broken gaff. Final filing, therefore, should be carefully done with light strokes, lengthwise of the gaffs.

If necessary to shape the gaff to proper gauge width, the filing should be done on the two outside rounded surfaces, as shown in Fig. 44-25. Care should be taken not to file the ridge or back edge of the gaff, for it is important to keep it absolutely straight, as shown in Fig. 44-26. The back of the gaff gauge can be used as a straight edge to check this.

If the back edge is rounded off, as shown in Fig. 44-27, the gaff will have a tendency to cut out of the pole.

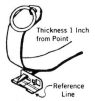
Thickness 1 Inch from Point
Reference Line

Fig. 44-20 Checking thickness of gaff.

Thickness ½ Inch from Point
Reference Line

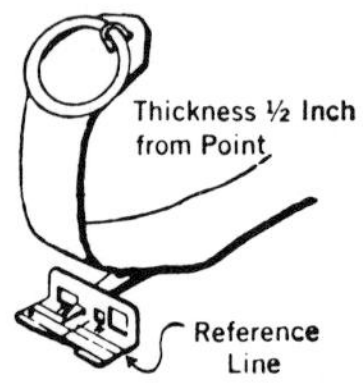

Fig. 44-21 Checking thickness of gaff.

Width 1 Inch from Point
Reference Line

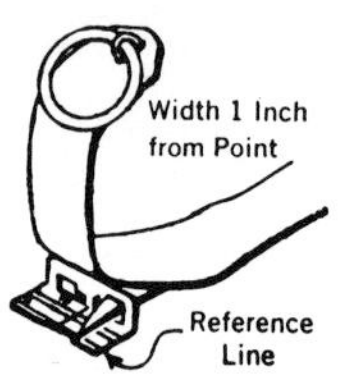

Fig. 44-22 Checking width of gaff.

Width ½ Inch from Point
Reference Line

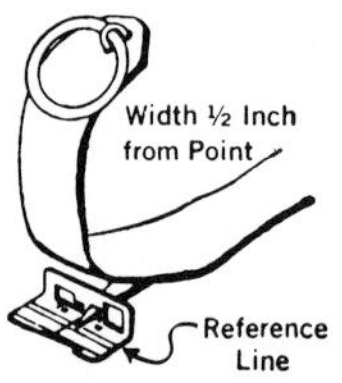

Fig. 44-23 Checking width of gaff.

File Lengthwise, Not Crosswise

Fig. 44-24 Shaping the gaff.

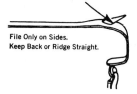
File Only on Sides.
Keep Back or Ridge Straight.

Fig. 44-25 Shaping the gaff.

As the filing progresses, the work should be checked with the gaff gauge until the gaff is within the indicated gauge limits.

The point of a gaff should function as a chisel (see Fig. 44-26) cutting its way into a pole, instead of as a spike or needle (see Fig. 44-28) to be driven into the pole. Spike-pointed gaffs will cut out easier than chisel points, if the knee is brought too close to the pole for any reason. Spike points also reduce the cross-sectional area near the tip and may result in gaff breakage.

The chisel point penetrates the pole easier and deeper with less effort and, if properly shaped, with the proper gaff angle, will not cut out.

To obtain a chisel point, the flat undersurface should be rounded off toward the point with a slight curve as shown in Fig. 44-26. This filing should be done with light even strokes.

If the gauge has a profile recess, use it to check the work, as in Fig. 44-29. If not, use Fig. 44-29 as a guide to the proper profile.

After filing and shaping is completed and the gaff checks with all the proper gauge dimensions, the flat underside of the gaff and the underside of the point should be honed, using a small carborundum pocket stone. Honing removes any burrs on the edges of the gaff and sharpens the edges for easier penetration into the pole. It is this penetration of the wood which supports the workman as he ascends, descends, or does his work on the pole. The edges of the flat underside should be knife sharp and *not* rounded.

Testing Climber Gaffs for Proper Shape and Sharpness *The Plane Test* The plane test may be used to determine that the climber gaff is properly sharpened to cut into the pole like a chisel.

Place the climber on a soft pine or cedar board as shown in Fig. 44-30. Holding it upright, but with *no* pressure on the stirrup, push the climber forward along the board. If the gaff is properly shaped and sharpened, and if the gaff angle with the wood is sufficient, the gaff point will dig into the wood and begin to hold within a distance of approximately one in.

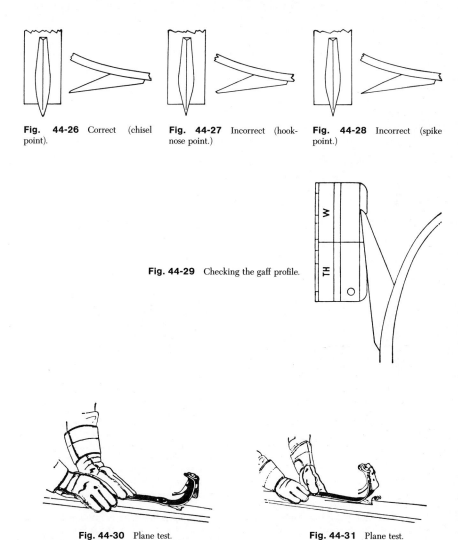

Fig. 44-26 Correct (chisel point).

Fig. 44-27 Incorrect (hook-nose point.)

Fig. 44-28 Incorrect (spike point.)

Fig. 44-29 Checking the gaff profile.

Fig. 44-30 Plane test.

Fig. 44-31 Plane test.

If the climber gaff slides along the wood without digging in, or merely leaves a mark or groove in the wood as shown in Fig. 44-31, the gaff is either not properly sharpened and shaped or the gaff angle is too small. The gaff angle is built into the climber by the manufacturer and should be between 9 and 16 deg with the leg iron placed parallel to the surface of the wood, according to EEI Specifications for Linemen's Climbing Equipment, AP-2-1973.

The Cutout Test   The cutout test may be made to determine whether the gaff will cut out of the wood when the angle between the leg iron and the surface of the wood is reduced beyond the critical point.

When the gaff is not properly sharpened or has a hook-nosed point or a convex ridge, it will cut out.

To make the cutout test, jab the climber gaff into the soft pine or cedar board at an angle of about 30 deg and with sufficient force to penetrate the wood to a depth of ¼ in or more, as shown in Fig. 44-32. Hold the gaff in place in the wood and bring the leg iron down against the wood surface, applying forward pressure on the stirrup with one hand while holding the leg iron with the other hand. If the gaff is in proper condition, it will tend to cut in deeper rather than cut out.

If the gaff is not properly shaped and sharpened it will cut out within a horizontal distance of 3 in or less, leaving a typical gaff cut-out mark or groove in the wood.

Fig. 44-32   Cutout test.

Fig. 44-33   Pole cutout test.

Fig. 44-34   Pole cutout test.

The Pole Cutout Test   In addition to the above mentioned tests, the pole cutout test may be used to check the climber gaffs in the field.

To make the test, place the climber on the leg, holding the sleeve with the hand, palm facing the pole, as shown in Fig. 44-33. With the leg at about a 30-deg angle to the pole and the foot about 12 in off the ground, lightly jab the gaff into the pole to a distance of approximately ¼ in. Keeping enough pressure on the stirrup to keep the gaff in the pole but not so much as to cause the gaff to penetrate any deeper, push the climber and the hand toward the pole by moving the knee until the strap loop of the sleeve is against the pole, as shown in Fig. 44-34. Making certain that the strap loop is held against the pole with pressure from the leg, gradually exert full pressure of the foot straight down on the stirrup without raising the other foot off the ground, so as to maintain balance if the gaff does not hold.

The point of the gaff should cut into the wood and hold (dig itself in) within a distance of not more than 2 in, measured from the point of gaff entry into the pole to the bottom of the cut on the pole surface.

CAUTION: To prevent hand injury from wood splinters, good work gloves should be worn when testing climbers.

Causes of Climber Cutouts   The causes of climber cutouts can be grouped under the following five general headings:

1. Condition of Poles
2. Conditions on Poles
3. Clothing
4. Climbing Practices
5. Climbers

A more detailed listing of "causes" is given in the following discussion of each of the five general types. Also, typical accidents are cited and preventive measures discussed.

1. Condition of Poles

   a. Specific Causes

      (1) Knots
      (2) Knotholes
      (3) Broken poles
      (4) Weather cracks
      (5) Wet creosoted poles
      (6) Excessive gaff cuts
      (7) Splinters
      (8) Poles coated with ice
      (9) Rotted outer surface
     (10) Poles hardened by preservative treatments
     (11) Crooked or leaning poles
     (12) Bird holes
     (13) Improperly guyed poles
     (14) Pole splintered by lightning
     (15) Old gains and drilled holes
     (16) Flat surfaces on poles
     (17) Small and limber poles
     (18) Cemented bird holes
     (19) Checks

   b. Examples of Accidents

      (1) As a lineman was descending a distribution pole, a climber gaff cut out and he fell to the ground, breaking one leg. The pole surface was badly checked and covered with deep gaff cuts.

      (2) After a lineman had unsnapped his safety strap to descend a pole, a piece of the shell of the pole in which one of his gaffs was placed broke off. He fell to the ground, breaking a leg, spraining a wrist, and bruising his chest. The pole was a native cedar and was weathered and checked.

      (3) A lineman's gaff struck a knot in a chestnut pole as he was climbing and he cut out and fell 12 ft. He injured his right leg so severely that he may be permanently crippled.

   These are typical instances in which the condition of poles contributes to accidents.

   c. Prevention

   Prevention of accidents due to condition of poles can be aided in several ways. For example, poles can be purchased to meet better specifications which will help to eliminate some of the defects. A good inspection schedule will help to eliminate other causes. Poles that are to be climbed frequently can perhaps be equipped with steps. Workmen themselves should become alert to these hazards, and be continually on the lookout for the conditions on the poles which they climb which might contribute to cutouts and injuries.

2. Conditions on Poles

   a. Specific Causes

      (1) Conduit or cable on pole
      (2) Strain plates
      (3) Nails and tacks
      (4) Telephone wires and brackets
      (5) Ground wires

    (6) Ground wire molding
    (7) Signs and posters
    (8) Metal pole numbers
    (9) Dating nails
    (10) Street light controls
    (11) House knob screws
    (12) Insect nests
    (13) Metering equipment on pole
    (14) Guy hooks, pole plates, and lag screws no longer in use
    (15) Tree limbs against pole
    (16) Inadequate climbing space
    (17) Telephone company attachments
    (18) Excessive number of service drops
    (19) Telephone terminal boxes
    (20) Vines on poles
    (21) Fences attached to poles
    (22) Clothes lines and radio aerials attached to poles
    (23) Loose and bent pole steps

  b. Examples of Accidents

    (1) A lineman climbed a pole to a height of 14 ft, where one gaff cut out and he fell. His right heel struck the curb and was fractured. He also suffered a fracture of the right leg. The pole was excessively gaff marked and no steps were provided. Climbing space was restricted by a street lighting riser cable conduit running up the pole and a government mailbox mounted on the lower part of the pole.

    (2) A new cedar pole had been set and supply attachments transferred to it. The old chestnut pole was lashed against the new pole pending transfer of the telephone attachments. As the lineman descended the new pole, one of his climber gaffs struck a strain plate and he cut out and fell 12 ft to the ground. He landed on his right side and rolled over on his back. His back was severely injured by the impact of a hammer he carried in his belt.

  c. Prevention

Preventing cutouts and falls from these causes requires the attention of both management and workmen. Many of these causes can be removed, or at least minimized, by planning installations and setting construction standards with due regard for the men who must climb the poles, sometimes under very adverse conditions. When a pole is called a "clean pole" by men who climb, it is inferred that hardware, fixtures, and conductors are installed with an eye to adequate and safe climbing space. However, when unsafe conditions are found, they should be properly reported and extra precautions taken if the pole must be climbed.

3. Clothing

These cases illustrate how falls can occur because of unsafe conditions of the workman's clothing.

  a. Specific Causes

    (1) Greasy clothing or gloves
    (2) Worn or loose heels on shoes
    (3) Shoes with nailed soles or heels
    (4) Low-cut shoes
    (5) Loose or ragged clothes
    (6) Wrong size gloves
    (7) Gloves with holes or torn places
    (8) Soft-soled shoes
    (9) Shoes with poor arch or ankle supports
    (10) Buckles on overalls
    (11) Loose soles on shoes
    (12) High-heel boots

  b. Examples of Accidents

    (1) A lineman climbing a pole cut out and fell from about an 18-ft height. The pole was in good condition, but his gloves were greasy. He stated that he lost his hold while he was climbing.

    (2) A lineman was attempting to belt off at his working position. As he reached around the pole to grasp his safety, his left gaff cut out and he fell off the pole. Examination revealed that the whole heel had come off the left shoe.

  c. Prevention

Elimination of cutouts and other accidents due to unsafe condition of clothing is largely a problem of the workman himself. Nevertheless, management and supervisors can be alerted to recognize such hazards and insist that their workers wear suitable and safe clothing. Many individuals will appreciate having their attention called to any unsafe condition of their work clothing. In some cases, conversely, it may be necessary for supervisors to take stronger action.

4. Climbing Practices

  a. Specific Causes

    (1) Climbing or descending too fast

    (2) Using low side of pole

    (3) Fatigue

    (4) Gaffs not stuck in pole hard enough

    (5) Climbing too close to pole

    (6) Climbing through or past unprotected conductors

    (7) Not watching placement of gaffs

    (8) Safety belt too loose

    (9) Aiming gaffs at pole at improper angle

    (10) Using too long steps in climbing or descending

    (11) Not inspecting pole before climbing

    (12) Inattention while ascending or descending

    (13) Improper balancing of body weight on gaffs

    (14) Belting off to pole at wrong position

    (15) Physically unfit for climbing

    (16) Horseplay

    (17) Climbing too close to ropes and handlines

    (18) Catching material thrown from ground (while working on pole)

    (19) Placing climber gaffs on ground wires

    (20) Failure to get a good hand hold

    (21) Holding to braces, etc.

    (22) Climbing in a mechanical manner

    (23) Sliding climber gaffs when climbing (sneaking)

    (24) Showing off (fancy danning)

  b. Examples of Accidents

    (1) A lineman was climbing a 40-ft cedar pole which was stepped nearly all the way up. He said he was climbing as was customary on a stepped pole, that is, he used the steps for hand holds and placed his gaffs between the pole steps. He guessed he cut out at a time when he was reaching for a new hand hold with his right hand and fell 15 ft to the ground. In trying to break his fall he extended both hands. He fractured the radius bone in both arms near the wrist and suffered minor abrasions on the face. The pole, climbers, gaffs, and straps were in safe condition. This lineman was known as a "fast climber."

    (2) A lineman climbed a transformer pole which was raked, or leaning. He climbed the "low" side of the pole. On completing his work, he started to climb down the "low" side of the pole. About 15 ft from the ground he cut out and fell, receiving a broken right leg, right foot, hip, back, and left foot, and other injuries to his left arm, wrist, hand, and knee.

    (3) A lineman had been working for 2 weeks getting transformer nameplate data. He kept a record of the number of poles he climbed each day. The accident occurred late in the afternoon of a very hot day. As he was descending a pole, his right gaff cut out and he fell 18 ft, fracturing his right leg and permanetly disabling his back. He stated that he had climbed more poles that day than ever accomplished by an employee in a single day and that he was just about "all in" before he cut out.

  c. Prevention

Experienced veteran climbers, as well as new apprentices, need to be reminded from time to time about the pitfalls involved in these unsafe climbing practices.

Foremen should watch the climbing practices of their men and aid them in correcting any unsafe practices that are observed before accidents occur.

Elimination of unsafe climbing practices is based on sound thorough training of workmen in the safe methods of conducting themselves while working on poles.

5. Climbers

The importance of the climbers themselves in the prevention of cutout accidents cannot be emphasized too strongly.

  a. Specific Causes
    (1) "Duck bill" gaffs
    (2) Loose gaffs
    (3) Gaffs too short
    (4) Dull gaffs
    (5) Gaffs improperly sharpened
    (6) Straps too tight
    (7) Straps too loose
    (8) Broken straps
    (9) Bent or broken gaffs
    (10) Wrong size climbers
    (11) Foreign objects under gaffs
    (12) Needle-point gaffs
    (13) Faulty leg irons

  b. Examples of Accidents
    (1) A lineman started to climb a pole to attach a guy. About 15 ft above the ground he cut out and fell, breaking his left hip and left arm. The pole appeared to be in reasonably good condition. The man's left gaff had been sharpened in a duck-bill shape and wood splinters were found jammed under the gaff.
    (2) A lineman rearranging dead ends on a pole unfastened his safety belt to swing around to the other side of the pole. When he had almost reached his new working position his right climber gaff apparently cut out and he fell. He struck the ground on his neck, shoulders, and back, suffering a broken neck and left shoulder blade, fractured ribs, punctured lungs, internal hemorrhages, and many other bruises. He died a few minutes after the accident. The gaff of one climber was found to be loose; otherwise his equipment was in good condition.
    (3) As a lineman was preparing to descend to a lower position before removing the line hose from the conductors, he unfastened his safety strap. He then cut out and fell 30 ft to the ground, landing on his hands and arms. Both wrists, both arms, right leg, and lower leg were broken and he suffered multiple cuts and bruises. He will be disfigured for life and partially disabled. Inspection of his climbers after the accident revealed that the gaffs were very short.

  c. Prevention

  Prevention of accidents due to climber defects can be accomplished with the proper effort.
  First   Rigid and frequent inspections of climbers, leg straps, leg irons, etc., with special attention being given to the gaffs, the shape and sharpness of the gaff points, the proper angle of the gaff with the leg iron, etc.
  Second   An acceptable gaff gauge should be standardized. It should be simple and easy to use. Minimum lengths of gaffs should be strictly enforced. The smallest deviation below the minimum length requirement should not be permitted. Gaff shape and size should be similarly dealt with.
  Third   There should be a standard procedure for sharpening gaffs. There is universal agreement that gaffs should *never* be sharpened on ordinary dry grinding wheels, as this destroys the proper heat treatment of the metal. A proper type of file only should be used. A professional job of precision sharpening should be the goal. Gaffs should always be honed after sharpening to remove file marks which may initiate cracks in the gaff steel and to improve the cutting edges of the point for better penetration of poles and ease in climbing.
  Fourth   Workmen who use climbing equipment must be thoroughly trained in the fundamentals for proper care and use of this equipment.
  Fifth   Workmen who use climbing equipment should accept responsibility for inspecting, sharpening, shaping, and, if necessary, rejecting climbing equipment that is unsafe without waiting for supervisors or safety department personnel to make inspections.

Section **45**

Protective Equipment

Linemen and cablemen must use protective equipment properly to safely complete work on energized equipment. Safety rules and work procedures developed from years of experience and corrective measures adopted from the investigation of serious electrical accidents all require the use of proper protective equipment to cover energized conductors and equipment as well as grounded surfaces in the work area. Proper procedures, the right tools, and correctly selected and applied protective equipment are all necessary to complete energized electrical distribution work safely and efficiently.

Each work project and especially those projects requiring linemen and cablemen to handle energized electrical conductors should be planned in detail. The plan should be developed to ensure the safety of the workmen and the reliability of electric service. Hazards to the linemen, the cablemen, and the public should be identified as a part of the planning process. As work procedures are developed, they should include safeguards for all hazards identified. Prior to starting the work, a tailboard conference should be held to communicate the plan and the work procedures to all the crew members. A tailboard conference is pictured in Fig. 45-1. When the work is initiated, the plan must be followed without shortcuts to ensure safe and reliable completion of the assignment.

Plan the work and work the plan for safety, efficiency, and reliability.

The safest way of completing electric distribution work is to deenergize, isolate, test, and ground the facilities to be worked on. Figure 45-2 illustrates the proper application of grounding equipment to an electric distribution circuit. The lineman must correctly identify the circuit to be grounded and obtain from the proper authority assurances that the circuit has been deenergized and isolated.

Hold-off tags must be applied to the switches isolating the circuit and these must be identified with the name of the person in charge of the work. A high-voltage tester mounted on hot-line tools can be used to verify that the circuit is deenergized. The grounding conductors are first connected to a low-resistance ground, usually the common neutral, and then to the deenergized conductors by the lineman with a hot-line tool. Once the work is completed and all personnel are clear of the circuit, the grounds are removed in the reverse sequence.

The public's dependence on continuous electric service to maintain their health, welfare, and safety often makes it impossible to deenergize circuits to complete modifications and perform maintenance. Proper application of protective equipment is necessary for the lineman to safeguard the work area—that is, to prevent electric shock or flash burn injuries. Correct application of the protective equipment described is essential for injury prevention and the maintenance of reliable electric service. The linemen pictured in Fig. 45-3 are properly protected for the work in progress. The linemen are wearing hard hats designed for electrical work, safety glasses, body belts with safety straps attached to the work platform, rubber gloves with protectors, and rubber sleeves. The work is being performed while the men are standing on an insulated platform. Low

Fig. 45-1 Supervisor is reviewing work plan with linemen prior to starting the job. The tailboard conference is being held around the hot-line tool trailer. The selection of the proper protective equipment and the correct tools is an important part of the work plan. *(Courtesy A. B. Chance Co.)*

primary voltages may be worked directly from the pole by linemen. The higher distribution voltages must be worked from an insulated platform attached to the pole or an insulated aerial device. The work area is guarded by conductor covers, insulator covers, crossarm covers, rubber blankets, and dead-end covers. The energized conductors have been elevated and supported clear of the crossarm with hot-line tools.

Rubber Gloves   The most important article of protection for a lineman or a cableman is a good pair of rubber gloves with the proper dielectric strength for the voltage of the circuit to be worked on. Leather protector gloves must always be worn over the rubber gloves to prevent physical damage to the rubber while work is being performed. When the rubber gloves are not in use they should be stored in a canvas bag to protect them from mechanical damage or deterioration from ozone generated by sun rays. Rubber gloves should always be given an air test by the lineman or cableman each day before the work is started or if the workman encounters an object that may have damaged the rubber gloves. The lineman pictured in Fig. 45-4 is properly applying an air test to his rubber gloves before putting them on.

Fig. 45-2 Lineman completing installation of grounding conductors to electric distribution circuit with hot-line tool. The common neutral conductor is used as a ground source. *(Courtesy A. B. Chance Co.)*

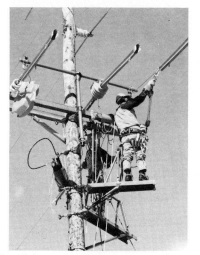

Fig. 45-3 Lineman performing energized circuit work with proper protective equipment. *(Courtesy A. B. Chance Co.)*

Fig. 45-4 Lineman giving his rubber gloves an air test before starting the work to be sure they do not have a hole in them. *(Courtesy W. H. Salisbury & Co.)*

Five classes of rubber gloves are manufactured as itemized in Table 45-1.

The American National Standards Institute standard ANSI/ASTM D 120-77 "Rubber Insulating Gloves" covers lineman's rubber glove specifications.

The proof-test voltage of the rubber gloves should not be construed to mean the safe voltage on which the glove can be used.

The maximum voltage on which gloves can safely be used depends on many factors including the care exercised in their use; the care followed in handling, storing, and inspecting the gloves in the field; the routine established for periodic laboratory inspection and test; the quality and thickness of the rubber; the design of the gloves; and such other factors as age, usage, or weather conditions.

Inasmuch as gloves are used for personal protection and a serious personal injury may result if they fail while in use, an adequate factor of safety should be provided between the maximum voltage on which they are permitted to be used and the voltage at which they are tested.

It is common practice for the user of rubber protective equipment to prepare complete instructions and regulations to govern in detail the correct and safe use of such equipment. These should include provision for proper-fitting leather protector gloves to protect rubber gloves against mechanical injury and to reduce ozone cutting. The lineman in Fig. 45-5 is wearing rubber gloves with leather protectors and rubber insulating sleeves.

Table 45-1 Classes of Rubber Gloves Manufactured, Proof-Test Voltages and Maximum Use Voltages as Specified by ANSI/ASTM D 120-77*

Class of glove	AC proof-test voltage rms	DC proof-test voltage avg.	Maximum use voltage ac rms
0	5,000	20,000	1,000
1	10,000	40,000	7,500
2	20,000	50,000	17,000
3	30,000	60,000	26,500
4	40,000	70,000	36,000

*Reprinted by permission of the American Society for Testing and Materials, 1916 Race Street, Philadelphia, PA 19103. Copyright.

Fig. 45-5 Lineman working on pole with proper personal protective equipment—rubber gloves with leather protectors, rubber sleeves, plastic hard hat, safety glasses, and a body belt with safety strap around the pole. Canvas bag for storage of rubber gloves is snapped to the lineman's body belt. *(Courtesy A. B. Chance Co.)*

Fig. 45-6 Rubber gloves placed in washing machine and washed with warm water and a mild detergent. *(Courtesy Iowa-Illinois Gas and Electric Co.)*

Rubber insulating gloves should be thoroughly cleaned, inspected, and tested by competent personnel regularly (Fig. 45-6). A procedure for accomplishing the work is as follows.

1. Rubber labels showing the glove size, the glove number, and the test date are cemented to each rubber glove.

2. The rubber gloves are washed in a washing machine, using a detergent to remove all dirt, prior to testing.

3. The rubber gloves are inflated utilizing glove inflator equipment, checked for air leaks, and inspected (Fig. 45-7). The inspection includes checking the surface of the rubber gloves for defects such as cuts, scratches, blisters, or embedded foreign material. The cuff area of the inside of the gloves is visually inspected after the glove is removed from the inflator. All rubber gloves having surface defects are rejected or repaired.

4. After the inflated inspection, the rubber gloves are stored for a period of 24 hours to permit the rubber to relax, thus minimizing the possibility of corona damage while the electrical tests are completed (Fig. 45-8).

5. The rubber gloves are supported in the testing tank by means of plastic clothespins on the supporting rack. A ground electrode is placed inside each glove (Fig. 45-9).

6. The rubber gloves to be tested and the testing tank are filled with water. The height of the water in the glove and in the testing tank depends upon the voltage test to be applied (Fig. 45-10).

Class 0 gloves are tested at 5000 volts ac or 20,000 volts dc. The flashover clearances at the top edge of the cuff of the gloves during the test must be 1½ in for both the ac and the dc test.

Class 1 gloves are tested at 10,000 volts ac or 40,000 volts dc. The flashover clearances at the top edge of the cuff of the gloves during the test must be 1½ in for the ac test and 2 in for the dc test.

Class 2 gloves are tested at 20,000 volts ac or 50,000 volts dc. The flashover clearances at the top edge of the cuff of the gloves during the test must be 2½ in for the ac test and 3 in for the dc test.

Class 3 gloves are tested at 30,000 volts ac or 60,000 volts dc. The flashover clearances at the top edge of the cuff of the gloves during the test must be 3½ in for the ac test and 4 in for the dc test.

Fig. 45-7   Rubber gloves inflated for visual inspection. *(Courtesy Iowa-Illinois Gas and Electric Co.)*

Fig. 45-8   Air-drying and storage rack for rubber gloves. Note rubber label cemented on cuff of glove. *(Courtesy Iowa-Illinois Gas and Electric Co.)*

Fig. 45-9   The rubber glove testing electrodes consist of a galvanized steel testing tank mounted on insulators and ground connections which are supported in the rubber gloves in the tank. The ground electrodes are connected to ground through relay contacts which can be used to switch a milliammeter in series with the ground lead. The testing equipment is designed to test six rubber gloves at a time.

Fig. 45-10   Rubber gloves in test tank, filled with water to proper level. *(Courtesy Iowa-Illinois Gas and Electric Co.)*

Fig. 45-11 Sphere gap and series resistor connected to ground is in parallel with high-voltage testing transformer to protect the gloves under test from overvoltages if gap is adjusted correctly. *(Courtesy Iowa-Illinois Gas and Electric Co.)*

Fig. 45-12 The testing equipment is interlocked with gate to prevent energizing the testing transformer until all personnel have left the area and the gate has been closed. *(Courtesy Iowa-Illinois Gas and Electric Co.)*

Class 4 gloves are tested at 40,000 volts ac or 70,000 volts dc. The flashover clearances at the top edge of the cuff of the gloves during the test must be 5 in for the ac test and 6 in for the dc test.

7. Sphere gap setting for the test voltage to be applied (Fig. 45-11).

Test voltage ac, volts	Sphere gap setting, millimeters
5,000	2.5
10,000	5.0
20,000	10.0
30,000	15.0
40,000	22.0

Sphere gap settings will change with variations of temperature and humidity.

8. Connect the high-voltage lead of the testing transformer to the metal tank portion of the rubber glove testing equipment.

9. Connect the ground electrode in each rubber glove to its associated ground lead.

10. Evacuate all personnel from the enclosed testing area (Fig. 45-12).

11. Close the power supply isolating switch to the testing equipment.

12. By means of the remote controls, gradually raise the voltage output of the testing transformer from zero to the proper level at a rate of approximately 1000 volts per second. This voltage should be applied to the rubber gloves for a period of 3 minutes (Fig. 45-13).

13. After the high voltage has been applied to the rubber gloves for 3 minutes, switch the milliammeter to read the leakage current in each ground electrode utilized for each rubber glove. Table 45-2 lists the maximum leakage current for five classes of rubber gloves.

14. The output voltage of the test transformer is gradually reduced to zero at a rate of approximately 1000 volts per second.

15. All rubber gloves which failed during the test or had a current leakage greater than the maximum permissible value are rejected. In some instances the gloves can be repaired and used if they successfully pass the test.

16. All rubber gloves which pass the high-voltage test are dried on a drying rack, powdered, and a tag indicating the leakage current recorded for each glove is placed in each glove. The gloves are then returned to the linemen or cablemen for use. In-service care of insulating gloves is specified in ANSI/ASTM F 496-77 standard.

Table 45-2   Maximum Leakage Current for Five Classes of Rubber Gloves*

Class of glove	Test voltage ac	Maximum leakage current, milliamperes			
		10½ in	14 in	16 in	18 in
Class 0	5,000	8	12	14	16
Class 1	10,000	—	14	16	18
Class 2	20,000	—	16	18	20
Class 3	30,000	—	18	20	22
Class 4	40,000	—	—	22	24

*Reprinted by permission of the American Society for Testing and Materials, 1916 Race Street, Philadelphia, PA 19103. Copyright.

Rubber Sleeves   Rubber sleeves should be worn with rubber gloves to protect the arms and shoulders of the lineman, while he is working on high-voltage distribution circuits, from electrical contacts on the arms or shoulders. Figures 45-3 to 45-5 illustrate the proper use of rubber sleeves by linemen. Rubber insulating sleeves must be treated with care and inspected regularly by the linemen in a manner similar to that described for rubber insulating gloves.

The rubber insulating sleeves should be thoroughly cleaned, inspected, and tested by competent personnel regularly. The inspecting and testing procedures are similar to those described for rubber insulating gloves (Fig. 45-14).

In-service care of insulating sleeves is specified in ANSI/ASTM F 496-77 standard.

Rubber Insulating Line Hose   Primary distribution conductors can be covered with rubber insulating line hose to protect the lineman from an accidental electrical contact. The lineman shown in Fig. 45-15 is installing a line hose on a 4160Y/2400-volt three-phase primary distribution circuit conductor. The line hoses are manufactured in various lengths, with inside diameter measurements that vary from 1 to 1½ in and with different test voltage withstand values. The lineman should be sure that the voltage rating of the line hose provides ample safety factor for the voltage applied to the conductors to be covered.

All line hose should be cleaned and inspected regularly. A hand crank wringer, as illustrated in Fig. 45-16, can be used to spread the line hose to clean and inspect it for cuts or corona damage.

The rubber insulating line hose should be tested for electric withstand in accordance with the specifications at scheduled intervals (Fig. 45-17).

In-service care of insulating line hose and covers is specified in ANSI/ASTM F 478-76 standard.

Rubber Insulating Insulator Hoods   Pin-type or post-type distribution primary insulators can be covered by hoods (Fig. 45-18).

Fig. 45-13   Control panel for test operation. *(Courtesy Iowa-Illinois Gas and Electric Co.)*

Fig.   45-14   Rubber   insulating   sleeve   being inspected while inflated after cleaning by washing. *(Courtesy Iowa-Illinois Gas and Electric Co.)*

Fig. 45-15 Rubber insulating line hose used by lineman to cover primary conductors. *(Courtesy Iowa-Illinois Gas and Electric Co.)*

Fig. 45-16 Rubber insulating line hose is being inspected and cleaned. *(Courtesy Iowa-Illinois Gas and Electric Co.)*

The insulator hood properly installed will overlap the line hose, providing the lineman with complete shielding from the energized conductors (Fig. 45-19).

Insulator hoods, like all other rubber insulating protective equipment, must be treated with care, kept clean, and inspected at regular intervals. Canvas bags of the proper size attached to a hand line should be used to raise and lower the protective equipment when it is to be installed and removed. The lineman pictured in Fig. 45-20 has partially completed the installation of the rubber protective equipment and the canvas bags are being lowered to the ground to permit the groundman to provide additional line hose and insulator hoods.

Conductor Covers   A conductor cover, fabricated from high-dielectric polyethylene, clips on and covers conductors up to 2 in. in diameter. A positive air gap is maintained by a swinging latch that can be loosened only by a one-quarter turn with a clamp stick (Fig. 45-21).

Insulator Covers   Insulator covers are fabricated from high-dielectric polyethylene and are designed to be used in conjunction with two conductor covers. The insulator cover fits over the insulator and locks with a conductor cover on each end. A polypropylene rope swings under the crossarm and hooks with a clamp stick, thus preventing the insulator cover from being removed upward by bumping or wind gusts (Fig. 45-22).

Fig. 45-17 Rubber insulating line hose mounted on high-voltage electrode with ground electrode clamped on it for withstand test. *(Courtesy Iowa-Illinois Gas and Electric Co.)*

Fig. 45-18 Rubber insulating insulator hood provides protection for lineman. *(Courtesy Iowa-Illinois Gas and Electric Co.)*

Fig. 45-19 Rubber insulating line hose and insulator hoods properly installed to cover primary conductors and insulators. *(Courtesy Iowa-Illinois Gas and Electric Co.)*

Fig. 45-20 Canvas bags used to raise rubber insulating line hose and insulator hoods to lineman on pole completing repairs to circuit after ice storm. *(Courtesy L. E. Myers Co.)*

Fig. 45-21 Conductor cover fabricated from high-dielectric polyethylene rated 26,000 volts. *(Courtesy A. B. Chance Co.)*

Fig. 45-22 Insulator cover fabricated from high-dielectric polyethylene rated 26,000 volts. *(Courtesy A. B. Chance Co.)*

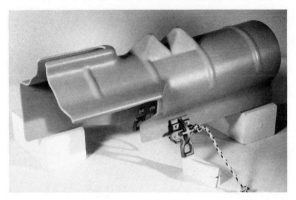

Fig. 45-23 Crossarm cover fabricated from high-dielectric strength polyethylene rated 26,000 volts. *(Courtesy A. B. Chance Co.)*

Crossarm Covers High-dielectric-strength polyethylene crossarm covers are used to prevent tie wires from contacting the crossarm when conductors adjacent to insulators are being tied or untied. It is designed for single- or double-arm construction, with slots provided for the double arming bolts. Flanges above the slots shield the ends of the double arming bolts (Fig. 45-23).

Pole Covers Polyethylene-constructed pole covers are designed to insulate the pole in the area adjacent to high-voltage conductors. The pole covers are available in various lengths. Positive-hold polypropylene rope handles are knotted through holes in the overlap area of the cover (Figs. 45-24 and 45-25).

Rubber Insulating Blankets Odd-shaped conductors and equipment can usually be covered best with a rubber insulating blanket. The blankets, like other protective equipment, must receive careful treatment. Figure 45-26 shows a rubber blanket on a metal ground electrode covered with felt and a high-voltage test lead connected to a copper disk lying on the wet felt in preparation for testing the dielectric strength of the blanket.

Fig. 45-24 Pole covers provide protection from ground to linemen working on insulated platform. *(Courtesy A. B. Chance Co.)*

Fig. 45-25 Pole, crossarm, and conductors not being worked on have been covered by the linemen. *(Courtesy A. B. Chance Co.)*

Fig. 45-26 Rubber insulating blanket prepared for dielectric test after it has been cleaned and visually inspected. *(Courtesy Iowa-Illinois Gas and Electric Co.)*

Fig. 45-27 Rubber insulating blankets in use to cover bushing, lightning arrester, and top of distribution transformer. The blankets are held in place with large wooden clamps. Lineman is installing rubber line hose on primary conductor connecting to transformer. *(Courtesy W. H. Salisbury & Co.)*

The rubber insulating blankets are stored in canvas rolls or metal canisters to protect them when they are not in use. The blankets can be held in place by ropes or large wooden clamps (Fig. 45-27).

In-service care of insulating blankets is specified in ANSI/ASTM F 479-76 standard.

Safety Hat   Hard hats, or safety hats, are worn by linemen, cablemen, and groundmen to protect the workman against the impact from falling or moving objects and against accidental electrical contact by the head with energized equipment, as well as to protect the workman from sun rays, cold, rain, sleet, and snow. The first combined impact-resisting and electrical insulating hat was introduced in 1952. The hat was designed "to roll with the punch" by distributing the force of a blow over the entire head. This is accomplished by a suspension band which holds the hat about an inch away from the head and lets the hat work like a shock absorber (Fig. 45-28).

Fig. 45-28 Cutaway view of electrical worker's protective helmet showing suspension band which assures proper crown clearance.

The hat is made of fiberglass, or plastic material, and has an insulating value of approximately 20,000 volts. New helmets are manufactured to withstand a test of 30,000 volts without failure. The actual voltage the hat will sustain while being worn depends upon the cleanliness of the hat, weather conditions, the type of electrode contacted, and other variables. The wearing of safety hats by linemen and cablemen has greatly reduced electrical contacts.

Physical injuries to the head have been practically eliminated as a result of workmen on the ground wearing protective helmets. The Occupational Safety and Health Act of 1970, and most companies' safety rules require linemen, cablemen, and groundmen to wear safety hats while

Fig. 45-29 The linemen illustrated are working on a 13,200Y/7620-volt energized distribution circuit utilizing proper protective equipment and tools. The linemen are wearing rubber gloves, rubber sleeves, insulated hard hats, safety glasses, and body belts with safety strap secured. They are working on an insulated platform. Hot-line tools are in use to maintain proper clearances. The protective cover-up equipment installed includes conductor covers, crossarm covers, pole covers, dead-end insulator covers, and rubber blankets. *(Courtesy A. B. Chance Co.)*

performing physical work. Specifications for safety hats are found in ANSI standard Z89.2-1971 "Safety Requirements for Industrial Protective Helmets for Electrical Workers, Class B."

Summary Protective equipment provides the lineman a margin of safety for error without injury while working on energized distribution conductors. Application of the protective equipment should be planned like all other aspects of the work. When the plan has been established, it should be followed without shortcut. The safe way of completing the work is the most efficient and reliable method. The skilled lineman and cableman is an expert in the application of protective equipment. Figure 45-29 pictures two skilled linemen utilizing protective equipment to good advantage while working on energized distribution circuits.

Safety Rules

OCCUPATIONAL SAFETY AND HEALTH ACT OF 1970*

Subpart V—Power Transmission and Distribution

§ 1926.950 General Requirements

(a) *Application.* The occupational safety and health standards contained in this Subpart V shall apply to the construction of electric transmission and distribution lines and equipment.

(1) As used in this Subpart V the term "construction" includes the erection of new electric transmission and distribution lines and equipment, and the alteration, conversion, and improvement of existing electric transmission and distribution lines and equipment.

(2) Existing electric transmission and distribution lines and electrical equipment need not be modified to conform to the requirements of applicable standards in this Subpart V, until such work as described in subparagraph (1) of this paragraph is to be performed on such lines or equipment.

(3) The standards set forth in this Subpart V provide minimum requirements for safety and health. Employers may require adherence to additional standards which are not in conflict with the standards contained in this Subpart V.

(b) *Initial inspections, tests, or determinations.* (1) Existing conditions shall be determined before starting work, by an inspection or a test. Such conditions shall include, but not be limited to, energized lines and equipment, conditions of poles, and the location of circuits and equipment, including power and communication lines, CATV and fire alarm circuits.

(2) Electric equipment and lines shall be considered energized until determined to be deenergized by tests or other appropriate methods or means.

(3) Operating voltage of equipment and lines shall be determined before working on or near energized parts.

(c) *Clearances.* The provisions of subparagraph (1) or (2) of this paragraph shall be observed.

(1) No employee shall be permitted to approach or take any conductive object, without an approved insulating handle, closer to exposed energized parts than shown in Table V-1, unless:

(i) The employee is insulated or guarded from the energized part (gloves or gloves with sleeves rated for the voltage involved shall be considered insulation of the employee from the energized part), or

(ii) The energized part is insulated or guarded from him and any other conductive object at a different potential, or

(iii) The employee is isolated, insulated, or guarded from any other conductive object(s), as during live-line bare-hand work.

Federal Register, vol. 37, no. 227, November 23, 1972.

TABLE V-1

ALTERNATING CURRENT—MINIMUM DISTANCES

Voltage range (phase to phase) kilovolt	Minimum working and clear hot stick distance
2.1 to 15	2 ft. 0 in.
15.1 to 35	2 ft. 4 in.
35.1 to 46	2 ft. 6 in.
46.1 to 72.5	3 ft. 0 in.
72.6 to 121	3 ft. 4 in.
138 to 145	3 ft. 6 in.
161 to 169	3 ft. 8 in.
230 to 242	5 ft. 0 in.
345 to 362	[1] 7 ft. 0 in.
500 to 552	[1] 11 ft. 0 in.
700 to 765	[1] 15 ft. 0 in.

[1]NOTE: For 345–362 kV., 500–552 kV., and 700–765 kV., the minimum working distance and the minimum clear hot stick distance may be reduced provided that such distances are not less than the shortest distance between the energized part and a grounded surface.

(2) (i) The minimum working distance and minimum clear hot stick distances stated in Table V-1 shall not be violated. The minimum clear hot stick distance is that for the use of live-line tools held by linemen when performing live-line work.

(ii) Conductor support tools, such as link sticks, strain carriers, and insulator cradles, may be used: *Provided*, That the clear insulation is at least as long as the insulator string or the minimum distance specified in Table V-1 for the operating voltage.

(d) *Deenergizing lines and equipment.* (1) When deenergizing lines and equipment operated in excess of 600 volts, and the means of disconnecting from electric energy is not visibly open or visibly locked out, the provisions of subdivisions (i) through (vii) of this subparagraph shall be complied with:

(i) The particular section of line or equipment to be deenergized shall be clearly identified, and it shall be isolated from all sources of voltage.

(ii) Notification and assurance from the designated employee shall be obtained that:

(a) All switches and disconnectors through which electric energy may be supplied to the particular section of line or equipment to be worked have been deenergized;

(b) All switches and disconnectors are plainly tagged indicating that men are at work;

(c) And that where design of such switches and disconnectors permits, they have been rendered inoperable.

(iii) After all designated switches and disconnectors have been opened, rendered inoperable, and tagged, visual inspection or tests shall be conducted to insure that equipment or lines have been deenergized.

(iv) Protective grounds shall be applied on the disconnected lines or equipment to be worked on.

(v) Guards or barriers shall be erected as necessary to adjacent energized lines.

(vi) When more than one independent crew requires the same line or equipment to be deenergized, a prominent tag for each such independent crew shall be placed on the line or equipment by the designated employee in charge.

(vii) Upon completion of work on deenergized lines or equipment, each designated employee in charge shall determine that all employees in his crew are clear, that protective grounds installed by his crew have been removed, and he shall report to the designated authority that all tags protecting his crew may be removed.

(2) When a crew working on a line or equipment can clearly see that the means of disconnecting from electric energy are visibly open or visibly locked-out, the provisions of subdivisions (i), and (ii) of this subparagraph shall apply:

(i) Guards or barriers shall be erected as necessary to adjacent energized lines.

(ii) Upon completion of work on deenergized lines or equipment, each designated employee in charge shall determine that all employees in his crew are clear, that protective grounds installed by his crew have been removed, and he shall report to the designated authority that all tags protecting his crew may be removed.

(e) *Emergency procedures and first aid.* (1) The employer shall provide training or require that his employees are knowledgeable and proficient in:

(i) Procedures involving emergency situations, and

(ii) First-aid fundamentals including resuscitation.

(2) In lieu of subparagraph (1) of this paragraph the employer may comply with the provisions of § 1926.50(c) regarding first-aid requirements.

(f) *Night work.* When working at night, spotlights or portable lights for emergency lighting shall be provided as needed to perform the work safely.

(g) *Work near and over water.* When crews are engaged in work over or near water and when danger of drowning exists, suitable protection shall be provided as stated in § 1926.104, or § 1926.105, or § 1926.106.

(h) *Sanitation facilities.* The requirements of § 1926.51 of Subpart D of this part shall be complied with for sanitation facilities.

(i) *Hydraulic fluids.* All hydraulic fluids used for the insulated sections of derrick trucks, aerial lifts, and hydraulic tools which are used on or around energized lines and equipment shall be of the insulating type. The requirements for fire resistant fluids of § 1926.302(d) (1) do not apply to hydraulic tools covered by this paragraph.

§ 1926.951   Tools and Protective Equipment

(a) *Protective equipment.* (1) (i) Rubber protective equipment shall be in accordance with the provisions of the American National Standards Institute (ANSI), ANSI J6 series, as follows:

Item	*Standard*
Rubber insulating gloves	J6.6–1971.
Rubber matting for use around electric apparatus	J6.7–1935 (R1971).
Rubber insulating blankets	J6.4–1971.
Rubber insulating hoods	J6.2–1950 (R1971).
Rubber insulating line hose	J6.1–1950 (R1971).
Rubber insulating sleeves	J6.5–1971

(ii) Rubber protective equipment shall be visually inspected prior to use.

(iii) In addition, an "air" test shall be performed for rubber gloves prior to use.

(iv) Protective equipment of material other than rubber shall provide equal or better electrical and mechanical protection.

(2) Protective hats shall be in accordance with the provisions of ANSI Z89.2-1971 Industrial Protective Helmets for Electrical Workers, Class B, and shall be worn at the jobsite by employees who are exposed to the hazards of falling objects, electric shock, or burns.

(b) *Personal climbing equipment.* (1) Body belts with straps or lanyards shall be worn to protect employees working at elevated locations on poles, towers, or other structures except where such use creates a greater hazard to the safety of the employees, in which case other safeguards shall be employed.

(2) Body belts and safety straps shall meet the requirements of § 1926.959. In addition to being used as an employee safeguarding item, body belts with approved tool loops may be used for the purpose of holding tools. Body belts shall be free from additional metal hooks and tool loops other than those permitted in § 1926.959.

(3) Body belts and straps shall be inspected before use each day to determine that they are in safe working condition.

(4) (i) Life lines and lanyards shall comply with the provisions of § 1926.104.

(ii) Safety lines are not intended to be subjected to shock loading and are used for emergency rescue such as lowering a man to the ground. Such safety lines shall be a minimum of one-half-inch diameter and three or four strand first-grade manila or its equivalent in strength (2,650 lb.) and durability.

(5) Defective ropes shall be replaced.

(c) *Ladders.* (1) Portable metal or conductive ladders shall not be used near energized lines or equipment except as may be necessary in specialized work such as in high voltage substations where nonconductive ladders might present a greater hazard than conductive ladders. Conductive or metal ladders shall be prominently marked as conductive and all necessary precautions shall be taken when used in specialized work.

(2) Hook or other type ladders used in structures shall be positively secured to prevent the ladder from being accidentally displaced.

(d) *Live-line tools.* (1) Only live-line tool poles having a manufacturer's certification to withstand the following minimum tests shall be used:

(i) 100,000 volts per foot of length for 5 minutes when the tool is made of fiberglass; or

(ii) 75,000 volts per foot of length for 3 minutes when the tool is made of wood; or

(iii) Other tests equivalent to subdivision (i) or (ii) of this subparagraph as appropriate.

(2) All live-line tools shall be visually inspected before use each day. Tools to be used shall be wiped clean and if any hazardous defects are indicated such tools shall be removed from service.

(e) *Measuring tapes or measuring ropes.* Measuring tapes or measuring ropes which are metal or contain conductive strands shall not be used when working on or near energized parts.

(f) *Hand tools.* (1) Switches for all powered hand tools shall comply with § 1926.300(d).

(2) All portable electric hand tools shall:

(i) Be equipped with three-wire cord having the ground wire permanently connected to the tool frame and means for grounding the other end; or

(ii) Be of the double insulated type and permanently labeled as "Double Insulated"; or

(iii) Be connected to the power supply by means of an isolating transformer, or other isolated power supply.

(3) All hydraulic tools which are used on or around energized lines or equipment shall use nonconducting hoses having adequate strength for the normal operating pressures. It should be noted that the provisions of § 1926.302(d) (2) shall also apply.

(4) All pneumatic tools which are used on or around energized lines or equipment shall:

(i) Have nonconducting hoses having adequate strength for the normal operating pressures, and

(ii) Have an accumulator on the compressor to collect moisture.

§ 1926.952   Mechanical Equipment

(a) *General.* (1) Visual inspections shall be made of the equipment to determine that it is in good condition each day the equipment is to be used.

(2) Tests shall be made at the beginning of each shift during which the equipment is to be used to determine that the brakes and operating systems are in proper working condition.

(3) No employer shall use any motor vehicle equipment having an obstructed view to the rear unless:

(i) The vehicle has a reverse signal alarm audible above the surrounding noise level or:

(ii) The vehicle is backed up only when an observer signals that it is safe to do so.

(b) *Aerial lifts.* (1) The provisions of § 1926.556, Subpart N of this part, shall apply to the utilization of aerial lifts.

(2) When working near energized lines or equipment, aerial lift trucks shall be grounded or barricaded and considered as energized equipment, or the aerial lift truck shall be insulated for the work being performed.

(3) Equipment or material shall not be passed between a pole or structure and an aerial lift while an employee working from the basket is within reaching distance of energized conductors or equipment that are not covered with insulating protective equipment.

(c) *Derrick trucks, cranes and other lifting equipment.* (1) All derrick trucks, cranes and other lifting equipment shall comply with Subpart N and O of this part except:

(i) As stated in § 1926.550(a) (15) (i) and (ii) relating to clearance (for clearances in this subpart see Table V-1) and

(ii) Derrick trucks (electric line trucks) shall not be required to comply with § 1926.550(a) (7) (vi), (a) (17), (b) (2), and (e).

(2) With the exception of equipment certified for work on the proper voltage, mechanical equipment shall not be operated closer to any energized line or equipment than the clearances set forth in § 1926.950(c) unless:

(i) An insulated barrier is installed between the energized part and the mechanical equipment, or

(ii) The mechanical equipment is grounded, or

(iii) The mechanical equipment is insulated, or

(iv) The mechanical equipment is considered as energized.

§ 1926.953   Material Handling

(a) *Unloading.* Prior to unloading steel, poles, cross arms and similar material, the load shall be thoroughly examined to ascertain if the load has shifted, binders or stakes have broken or the load is otherwise hazardous to employees.

(b) *Pole hauling.* (1) During pole hauling operations, all loads shall be secured to prevent displacement and a red flag shall be displayed at the trailing end of the longest pole.

(2) Precautions shall be exercised to prevent blocking of roadways or endangering other traffic.

(3) When hauling poles during the hours of darkness, illuminated warning devices shall be attached to the trailing end of the longest pole.

(c) *Storage.* (1) No materials or equipment shall be stored under energized bus, energized lines, or near energized equipment, if it is practical to store them elsewhere.

(2) When materials or equipment are stored under energized lines or near energized equipment, applicable clearances shall be maintained as stated in Table V-1; and extraordinary caution shall be exercised when moving materials near such energized equipment.

(d) *Tag line.* Where hazards to employees exist tag lines or other suitable devices shall be used to control loads being handled by hoisting equipment.

(e) *Oil filled equipment.* During construction or repair of oil filled equipment the oil may be stored in temporary containers other than those required in § 1926.152, such as pillow tanks.

(f) *Framing.* During framing operations, employees shall not work under a pole or a structure suspended by a crane, A-frame or similar equipment unless the pole or structure is adequately supported.

(g) *Attaching the load.* The hoist rope shall not be wrapped around the load. This provision shall not apply to electric construction crews when setting or removing poles.

§ 1926.954   Grounding for Protection of Employees

(a) *General.* All conductors and equipment shall be treated as energized until tested or otherwise determined to be deenergized or until grounded.

(b) *New construction.* New lines or equipment may be considered deenergized and worked as such where:

(1) The lines or equipment are grounded, or

(2) The hazard of induced voltages is not present, and adequate clearances or other means are implemented to prevent contact with energized lines or equipment and the new lines or equipment.

(c) *Communication conductors.* Bare wire communication conductors on power poles or structures shall be treated as energized lines unless protected by insulating materials.

(d) *Voltage testing.* Deenergized conductors and equipment which are to be grounded shall be tested for voltage. Results of this voltage test shall determine the subsequent procedures as required in § 1926.950(d).

(e) *Attaching grounds.* (1) When attaching grounds, the ground end shall be attached first, and the other end shall be attached and removed by means of insulated tools or other suitable devices.

(2) When removing grounds, the grounding device shall first be removed from the line or equipment using insulating tools or other suitable devices.

(f) Grounds shall be placed between work location and all sources of energy and as close as practicable to the work location, or grounds shall be placed at the work location. If work is to be performed at more than one location in a line section, the line section must be grounded and short circuited at one location in the line section and the conductor to be worked on shall be grounded at each work location. The minimum distance shown in Table V-1 shall be maintained from ungrounded conductors at the work location. Where the making of a ground is impracticable, or the conditions resulting therefrom would be more hazardous than working on the lines or equipment without grounding, the grounds may be omitted and the line or equipment worked as energized.

(g) *Testing without grounds.* Grounds may be temporarily removed only when necessary for test purposes and extreme caution shall be exercised during the test procedures.

(h) *Grounding electrode.* When grounding electrodes are utilized, such electrodes shall have a resistance to ground low enough to remove the danger of harm to personnel or permit prompt operation of protective devices.

(i) *Grounding to tower.* Grounding to tower shall be made with a tower clamp capable of conducting the anticipated fault current.

(j) *Ground lead.* A ground lead, to be attached to either a tower ground or driven ground, shall be capable of conducting the anticipated fault current and shall have a minimum conductance of No. 2 AWG copper.

§ 1926.955 Overhead Lines

(a) *Overhead lines.* (1) When working on or with overhead lines the provisions of subparagraphs (2) through (8) of this paragraph shall be complied with in addition to other applicable provisions of this subpart.

(2) Prior to climbing poles, ladders, scaffolds, or other elevated structures, an inspection shall be made to determine that the structures are capable of sustaining the additional or unbalanced stresses to which they will be subjected.

(3) Where poles or structures may be unsafe for climbing, they shall not be climbed until made safe by guying, bracing, or other adequate means.

(4) Before installing or removing wire or cable, strains to which poles and structures will be subjected shall be considered and necessary action taken to prevent failure of supporting structures.

(5) (i) When setting, moving, or removing poles using cranes, derricks, gin poles, A-frames, or other mechanized equipment near energized lines or equipment, precautions shall be taken to avoid contact with energized lines or equipment, except in bare-hand live-line work, or where barriers or protective devices are used.

(ii) Equipment and machinery operating adjacent to energized lines or equipment shall comply with § 1926.952(c) (2).

(6) (i) Unless using suitable protective equipment for the voltage involved, employees standing on the ground shall avoid contacting equipment or machinery working adjacent to energized lines or equipment.

(ii) Lifting equipment shall be bonded to an effective ground or it shall be considered energized and barricaded when utilized near energized equipment or lines.

(7) Pole holes shall not be left unattended or unguarded in areas where employees are currently working.

(8) Tag lines shall be of a nonconductive type when used near energized lines.

(b) *Metal tower construction.* (1) When working in unstable material the excavation for pad- or pile-type footings in excess of 5 feet deep shall be either sloped to the angle of repose as required in § 1926.652 or shored if entry is required. Ladders shall be provided for access to pad- or pile-type footing excavations in excess of 4 feet.

(2) When working in unstable material provision shall be made for cleaning out auger-type footings without requiring an employee to enter the footing unless shoring is used to protect the employee.

(3) (i) A designated employee shall be used in directing mobile equipment adjacent to footing excavations.

(ii) No one shall be permitted to remain in the footing while equipment is being spotted for placement.

(iii) Where necessary to assure the stability of mobile equipment the location of use for such equipment shall be graded and leveled.

(4) (i) Tower assembly shall be carried out with a minimum exposure of employees to falling objects when working at two or more levels on a tower.

(ii) Guy lines shall be used as necessary to maintain sections or parts of sections in position and to reduce the possibility of tipping.

(iii) Members and sections being assembled shall be adequately supported.

(5) When assembling and erecting towers the provisions of subdivisions (i), (ii), and (iii) of this subparagraph shall be complied with:

(i) The construction of transmission towers and the erecting of poles, hoisting machinery, site preparation machinery, and other types of construction machinery shall conform to the applicable requirements of this part.

(ii) No one shall be permitted under a tower which is in the process of erection or assembly, except as may be required to guide and secure the section being set.

(iii) When erecting towers using hoisting equipment adjacent to energized transmission lines, the lines shall be deenergized when practical. If the lines are not deenergized, extraordinary caution shall be exercised to maintain the minimum clearance distances required by § 1926.950(c), including Table V-1.

(6) (i) Erection cranes shall be set on firm level foundations and when the cranes are so equipped outriggers shall be used.

(ii) Tag lines shall be utilized to maintain control of tower sections being raised and positioned, except where the use of such lines would create a greater hazard.

(iii) The loadline shall not be detached from a tower section until the section is adequately secured.

(iv) Except during emergency restoration procedures erection shall be discontinued in the event of high wind or other adverse weather conditions which would make the work hazardous.

(v) Equipment and rigging shall be regularly inspected and maintained in safe operating condition.

(7) Adequate traffic control shall be maintained when crossing highways and railways with equipment as required by the provisions of § 1926.200(g) (1) and (2).

(8) A designated employee shall be utilized to determine that required clearance is maintained in moving equipment under or near energized lines.

(c) *Stringing or removing deenergized conductors.* (1) When stringing or removing deenergized conductors, the provisions of subparagraph (2) through (12) of this paragraph shall be complied with.

(2) Prior to stringing operations a briefing shall be held setting forth the plan of operation and specifying the type of equipment to be used, grounding devices and procedures to be followed, crossover methods to be employed, and the clearance authorization required.

(3) Where there is a possibility of the conductor accidentally contacting an energized circuit or receiving a dangerous induced voltage buildup, to further protect the employee from the hazards of the conductor, the conductor being installed or removed shall be grounded or provisions made to insulate or isolate the employee.

(4) (i) If the existing line is deenergized, proper clearance authorization shall be secured and the line grounded on both sides of the crossover or, the line being strung or removed shall be considered and worked as energized.

(ii) When crossing over energized conductors in excess of 600 volts, rope nets or guard structures shall be installed unless provision is made to isolate or insulate the workman or the energized conductor. Where practical the automatic reclosing feature of the circuit interrupting device shall be made inoperative. In addition, the line being strung shall be grounded on either side of the crossover or considered and worked as energized.

(5) Conductors being strung in or removed shall be kept under positive control by the use of adequate tension reels, guard structures, tielines, or other means to prevent accidental contact with energized circuits.

(6) Guard structure members shall be sound and of adequate dimension and strength, and adequately supported.

(7) (i) Catch-off anchors, rigging, and hoists shall be of ample capacity to prevent loss of the lines.

(ii) The manufacturer's load rating shall not be exceeded for stringing lines, pulling lines, sock connections, and all load-bearing hardware and accessories.

(iii) Pulling lines and accessories shall be inspected regularly and replaced or repaired when damaged or when dependability is doubtful. The provisions of § 1926.251(c) (4) (ii) (concerning splices) shall not apply.

(8) Conductor grips shall not be used on wire rope unless designed for this application.

(9) While the conductor or pulling line is being pulled (in motion) employees shall not be permitted directly under overhead operations, nor shall any employee be permitted on the crossarm.

(10) A transmission clipping crew shall have a minimum of two structures clipped in between the crew and the conductor being sagged. When working on bare conductors, clipping and tying crews shall work between grounds at all times. The grounds shall remain intact until the conductors are clipped in, except on dead end structures.

(11) (i) Except during emergency restoration procedures, work from structures shall be discontinued when adverse weather (such as high wind or ice on structures) makes the work hazardous.

(ii) Stringing and clipping operations shall be discontinued during the progress of an electrical storm in the immediate vicinity.

(12) (i) Reel handling equipment, including pulling and braking machines, shall have ample

capacity, operate smoothly, and be leveled and aligned in accordance with the manufacturer's operating instructions.

(ii) Reliable communications between the reel tender and pulling rig operator shall be provided.

(iii) Each pull shall be snubbed or dead ended at both ends before subsequent pulls.

(d) *Stringing adjacent to energized lines.* (1) Prior to stringing parallel to an existing energized transmission line a competent determination shall be made to ascertain whether dangerous induced voltage buildups will occur, particularly during switching and ground fault conditions. When there is a possibility that such dangerous induced voltage may exist, the employer shall comply with the provisions of subparagraphs (2) through (9) of this paragraph in addition to the provisions of paragraph (c) of this § 1926.955, unless the line is worked as energized.

(2) When stringing adjacent to energized lines the tension stringing method or other methods which preclude unintentional contact between the lines being pulled and any employee shall be used.

(3) All pulling and tensioning equipment shall be isolated, insulated, or effectively grounded.

(4) A ground shall be installed between the tensioning reel setup and the first structure in order to ground each bare conductor, subconductor, and overhead ground conductor during stringing operations.

(5) During stringing operations, each bare conductor, subconductor, and overhead ground conductor shall be grounded at the first tower adjacent to both the tensioning and pulling setup and in increments so that no point is more than 2 miles from a ground.

(i) The grounds shall be left in place until conductor installation is completed.

(ii) Such grounds shall be removed as the last phase of aerial cleanup.

(iii) Except for moving type grounds, the grounds shall be placed and removed with a hot stick.

(6) Conductors, subconductors, and overhead ground conductors shall be grounded at all dead-end or catch-off points.

(7) A ground shall be located at each side and within 10 feet of working areas where conductors, subconductors, or overhead ground conductors are being spliced at ground level. The two ends to be spliced shall be bonded to each other. It is recommended that splicing be carried out on either an insulated platform or on a conductive metallic grounding mat bonded to both grounds. When a grounding mat is used, it is recommended that the grounding mat be roped off and an insulated walkway provided for access to the mat.

(8) (i) All conductors, subconductors, and overhead ground conductors shall be bonded to the tower at any isolated tower where it may be necessary to complete work on the transmission line.

(ii) Work on dead-end towers shall require grounding on all deenergized lines.

(iii) Grounds may be removed as soon as the work is completed: *Provided,* That the line is not left open circuited at the isolated tower at which work is being completed.

(9) When performing work from the structures, clipping crews and all others working on conductors, subconductors, or overhead ground conductors shall be protected by individual grounds installed at every work location.

(e) *Live-line bare-hand work.* In addition to any other applicable standards contained elsewhere in this subpart all live-line bare-hand work shall be performed in accordance with the following requirements.

(1) Employees shall be instructed and trained in the live-line bare-hand technique and the safety requirements pertinent thereto before being permitted to use the technique on energized circuits.

(2) Before using the live-line bare-hand technique on energized high-voltage conductors or parts, a check shall be made of:

(i) The voltage rating of the circuit on which the work is to be performed;

(ii) The clearance to ground of lines and other energized parts on which work is to be performed; and

(iii) The voltage limitations of the aerial-lift equipment intended to be used.

(3) Only equipment designed, tested, and intended for live-line bare-hand work shall be used.

(4) All work shall be personally supervised by a person trained and qualified to perform live-line bare-hand work.

(5) The automatic reclosing feature of circuit interrupting devices shall be made inoperative where practical before working on any energized line or equipment.

(6) Work shall not be performed during the progress of an electrical storm in the immediate vicinity.

(7) A conductive bucket liner or other suitable conductive device shall be provided for bonding the insulated aerial device to the energized line or equipment.

(i) The employee shall be connected to the bucket liner by use of conductive shoes, leg clips, or other suitable means.

(ii) Where necessary, adequate electrostatic shielding for the voltage being worked or conductive clothing shall be provided.

(8) Only tools and equipment intended for live-line bare-hand work shall be used, and such tools and equipment shall be kept clean and dry.

(9) Before the boom is elevated, the outriggers on the aerial truck shall be extended and adjusted to stabilize the truck and the body of the truck shall be bonded to an effective ground, or barricaded and considered as energized equipment.

(10) Before moving the aerial lift into the work position, all controls (ground level and bucket) shall be checked and tested to determine that they are in proper working condition.

(11) Arm current tests shall be made before starting work each day, each time during the day when higher voltage is going to be worked and when changed conditions indicate a need for additional tests. Aerial buckets used for bare-hand live-line work shall be subjected to an arm current test. This test shall consist of placing the bucket in contact with an energized source equal to the voltage to be worked upon for a minimum time of three (3) minutes. The leakage current shall not exceed 1 microampere per kilovolt of nominal line-to-line voltage. Work operations shall be suspended immediately upon any indication of a malfunction in the equipment.

(12) All aerial lifts to be used for live-line bare-hand work shall have dual controls (lower and upper) as required by subdivisions (i) and (ii) of this subparagraph.

(i) The upper controls shall be within easy reach of the employee in the basket. If a two basket type lift is used access to the controls shall be within easy reach from either basket.

(ii) The lower set of controls shall be located near base of the boom that will permit over-ride operation of equipment at any time.

(13) Ground level lift control shall not be operated unless permission has been obtained from the employee in lift, except in case of emergency.

(14) Before the employee contacts the energized part to be worked on, the conductive bucket liner shall be bonded to the energized conductor by means of a positive connection which shall remain attached to the energized conductor until the work on the energized circuit is completed.

(15) The minimum clearance distances for live-line bare-hand work shall be as specified in Table V-2. These minimum clearance distances shall be maintained from all grounded objects

TABLE V-2

MINIMUM CLEARANCE DISTANCES FOR LIVE-LINE
BARE-HAND WORK (ALTERNATING CURRENT)

Voltage range (phase-to-phase) kilovolts	Distance in feet and inches for maximum voltage	
	Phase to ground	Phase to phase
2.1–15	2'0"	2'0"
15.1–35	2'4"	2'4"
35.1–46	2'6"	2'6"
46.1–72.5	3'0"	3'0"
72.6–121	3'4"	4'6"
138–145	3'6"	5'0"
161–169	3'8"	5'6"
230–242	5'0"	8'4"
345–362	[1] 7'0"	[1] 13'4"
500–552	[1] 11'0"	[1] 20'0"
700–765	[1] 15'0"	[1] 31'0"

[1]NOTE: For 345–362 kV., 500–552 kV., and 700–765 kV., the minimum clearance distance may be reduced provided the distances are not made less than the shortest distance between the energized part and a grounded surface.

and from lines and equipment at a different potential than that to which the insulated aerial device is bonded unless such grounded objects or other lines and equipment are covered by insulated guards. These distances shall be maintained when approaching, leaving, and when bonded to the energized circuit.

(16) When approaching, leaving, or bonding to an energized circuit the minimum distances in Table V-2 shall be maintained between all parts of the insulated boom assembly and any grounded parts (including the lower arm or portions of the truck).

(17) When positioning the bucket alongside an energized bushing or insulator string, the minimum line-to-ground clearances of Table V-2 must be maintained between all parts of the bucket and the grounded end of the bushing or insulator string.

(18) (i) The use of handlines between buckets, booms, and the ground is prohibited.

(ii) No conductive materials over 36 inches long shall be placed in the bucket, except for appropriate length jumpers, armor rods, and tools.

(iii) Nonconductive-type handlines may be used from line to ground when not supported from the bucket.

(19) The bucket and upper insulated boom shall not be overstressed by attempting to lift or support weights in excess of the manufacturer's rating.

(20) (i) A minimum clearance table (as shown in Table V-2) shall be printed on a plate of durable nonconductive material, and mounted in the buckets or its vicinity so as to be visible to the operator of the boom.

(ii) It is recommended that insulated measuring sticks be used to verify clearance distances.

§ 1926.956   Underground Lines

(a) *Guarding and ventilating street opening used for access to underground lines or equipment.* (1) Appropriate warning signs shall be promptly placed when covers of manholes, handholes, or vaults are removed. What is an appropriate warning sign is dependent upon the nature and location of the hazards involved.

(2) Before an employee enters a street opening, such as a manhole or an unvented vault, it shall be promptly protected with a barrier, temporary cover, or other suitable guard.

(3) When work is to be performed in a manhole or unvented vault:

(i) No entry shall be permitted unless forced ventilation is provided or the atmosphere is found to be safe by testing for oxygen deficiency and the presence of explosive gases or fumes;

(ii) Where unsafe conditions are detected, by testing or other means, the work area shall be ventilated and otherwise made safe before entry;

(iii) Provisions shall be made for an adequate continuous supply of air.

(b) *Work in manholes.* (1) While work is being performed in manholes, an employee shall be available in the immediate vicinity to render emergency assistance as may be required. This shall not preclude the employee in the immediate vicinity from occasionally entering a manhole to provide assistance, other than emergency. This requirement does not preclude a qualified employee, working alone, from entering for brief periods of time, a manhole where energized cables or equipment are in service, for the purpose of inspection, housekeeping, taking readings, or similar work if such work can be performed safely.

(2) When open flames must be used or smoking is permitted in manholes, extra precautions shall be taken to provide adequate ventilation.

(3) Before using open flames in a manhole or excavation in an area where combustible gases or liquids may be present, such as near a gasoline service station, the atmosphere of the manhole or excavation shall be tested and found safe or cleared of the combustible gases or liquids.

(c) *Trenching and excavating.* (1) During excavation or trenching in order to prevent the exposure of employees to the hazards created by damage to dangerous underground facilities efforts shall be made to determine the location of such facilities and work conducted in a manner designed to avoid damage.

(2) Trenching and excavation operations shall comply with §§ 1926.651 and 1926.652.

(3) When underground facilities are exposed (electric, gas, water, telephone, etc.) they shall be protected as necessary to avoid damage.

(4) Where multiple cables exist in an excavation, cables other than the one being worked on shall be protected as necessary.

(5) When multiple cables exist in an excavation, the cable to be worked on shall be identified by electrical means unless its identity is obvious by reason of distinctive appearance.

(6) Before cutting into a cable or opening a splice, the cable shall be identified and verified to be the proper cable.

(7) When working on buried cable or on cable in manholes, metallic sheath continuity shall be maintained by bonding across the opening or by equivalent means.

§ 1926.957 Construction in Energized Substations

(a) *Work near energized equipment facilities.* (1) When construction work is performed in an energized substation, authorization shall be obtained from the designated, authorized person before work is started.

(2) When work is to be done in an energized substation, the following shall be determined:

(i) What facilities are energized, and

(ii) What protective equipment and precautions are necessary for the safety of personnel.

(3) Extraordinary caution shall be exercised in the handling of busbars, tower steel, materials, and equipment in the vicinity of energized facilities. The requirements set forth in § 1926.950(c), shall be complied with.

(b) *Deenergized equipment or lines.* When it is necessary to deenergize equipment or lines for protection of employees, the requirements of § 1926.950(d) shall be complied with.

(c) *Barricades and barriers.* (1) Barricades or barriers shall be installed to prevent accidental contact with energized lines or equipment.

(2) Where appropriate, signs indicating the hazard shall be posted near the barricade or barrier. These signs shall comply with § 1926.200.

(d) *Control panels.* (1) Work on or adjacent to energized control panels shall be performed by designated employees.

(2) Precaution shall be taken to prevent accidental operation of relays or other protective devices due to jarring, vibration, or improper wiring.

(e) *Mechanized equipment.* (1) Use of vehicles, gin poles, cranes, and other equipment in restricted or hazardous areas shall at all times be controlled by designated employees.

(2) All mobile cranes and derricks shall be effectively grounded when being moved or operated in close proximity to energized lines or equipment, or the equipment shall be considered energized.

(3) Fenders shall not be required for lowboys used for transporting large electrical equipment, transformers, or breakers.

(f) *Storage.* The storage requirements of § 1926.953(c) shall be complied with.

(g) *Substation fences.* (1) When a substation fence must be expanded or removed for construction purposes, a temporary fence affording similar protection, when the site is unattended, shall be provided. Adequate interconnection with ground shall be maintained between temporary fence and permanent fence.

(2) All gates to all unattended substations shall be locked, except when work is in progress.

(h) *Footing excavation.* (1) Excavation for auger, pad and piling type footings for structures and towers shall require the same precautions as for metal tower construction (see § 1926.955(b) (1)).

(2) No employee shall be permitted to enter an unsupported auger-type excavation in unstable material for any purpose. Necessary clean-out in such cases shall be accomplished without entry.

§ 1926.958 External Load Helicopters

In all operations performed using a rotorcraft for moving or placing external loads, the provisions of § 1926.551 of Subpart N of this part shall be complied with.

§ 1926.959 Lineman's Body Belts, Safety Straps, and Lanyards

(a) *General requirements.* The requirements of paragraphs (a) and (b) of this section shall be complied with for all lineman's body belts, safety straps, and lanyards acquired for use after the effective date of this subpart.

(1) Hardware for lineman's body belts, safety straps, and lanyards shall be drop forged or pressed steel and have a corrosive resistive finish tested to American Society for Testing and Materials B117-64 (50-hour test). Surfaces shall be smooth and free of sharp edges.

(2) All buckles shall withstand a 2,000-pound tensile test with a maximum permanent deformation no greater than one sixty-fourth inch.

(3) D rings shall withstand a 5,000-pound tensile test without failure. Failure of a D ring shall be considered cracking or breaking.

(4) Snaphooks shall withstand a 5,000-pound tensile test without failure. Failure of a snaphook shall be distortion sufficient to release the keeper.

(b) *Specific requirements.* (1) (i) All fabric used for safety straps shall withstand an A.C. dielectric test of not less than 25,000 volts per foot "dry" for 3 minutes, without visible deterioration.

(ii) All fabric and leather used shall be tested for leakage current and shall not exceed 1 milliampere when a potential of 3,000 volts is applied to the electrodes positioned 12 inches apart.

(iii) Direct current tests may be permitted in lieu of alternating current tests.

(2) The cushion part of the body belt shall:

(i) Contain no exposed rivets on the inside;

(ii) Be at least three (3) inches in width;

(iii) Be at least five thirty-seconds (%₂) inch thick, if made of leather; and

(iv) Have pocket tabs that extend at least 1½ inches down and three (3) inches back of the inside of circle of each D ring for riveting on plier or tool pockets. On shifting D belts, this measurement for pocket tabs shall be taken when the D ring section is centered.

(3) A maximum of four (4) tool loops shall be so situated on the body belt that four (4) inches of the body belt in the center of the back, measuring from D ring to D ring, shall be free of tool loops, and any other attachments.

(4) Suitable copper, steel, or equivalent liners shall be used around bar of D rings to prevent wear between these members and the leather or fabric enclosing them.

(5) All stitching shall be of a minimum 42-pound weight nylon or equivalent thread and shall be lock stitched. Stitching parallel to an edge shall not be less than three-sixteenths (³⁄₁₆) inch from edge of narrowest member caught by the thread. The use of cross stitching on leather is prohibited.

(6) The keeper of snaphooks shall have a spring tension that will not allow the keeper to begin to open with a weight of 2½ pounds or less, but the keeper of snaphooks shall begin to open with a weight of four (4) pounds, when the weight is supported on the keeper against the end of the nose.

(7) Testing of lineman's safety straps, body belts and lanyards shall be in accordance with the following procedure:

(i) Attach one end of the safety strap or lanyard to a rigid support, the other end shall be attached to a 250-pound canvas bag of sand:

(ii) Allow the 250-pound canvas bag of sand to free fall 4 feet for (safety strap test) and 6 feet for (lanyard test); in each case stopping the fall of the 250-pound bag:

(iii) Failure of the strap or lanyard shall be indicated by any breakage, or slippage sufficient to permit the bag to fall free of the strap or lanyard. The entire "body belt assembly" shall be tested using one D ring. A safety strap or lanyard shall be used that is capable of passing the "impact loading test" and attached as required in subdivision (i) of this subparagraph. The body belt shall be secured to the 250-pound bag of sand at a point to simulate the waist of a man and allowed to drop as stated in subdivision (ii) of this subparagraph. Failure of the body belt shall be indicated by any breakage, or slippage sufficient to permit the bag to fall free of the body belt.

§ 1926.960. Definitions for Subpart V

(a) *Alive or live (energized).* The term means electrically connected to a source of potential difference, or electrically charged so as to have a potential significantly different from that of the earth in the vicinity. The term "live" is sometimes used in place of the term "current-carrying," where the intent is clear, to avoid repetition of the longer term.

(b) *Automatic circuit recloser.* The term means a self-controlled device for automatically interrupting and reclosing an alternating current circuit with a predetermined sequence of opening and reclosing followed by resetting, hold closed, or lockout operation.

(c) *Barrier.* The term means a physical obstruction which is intended to prevent contact with energized lines or equipment.

(d) *Barricade.* The term means a physical obstruction such as tapes, screens, or cones intended to warn and limit access to a hazardous area.

(e) *Bond.* The term means an electrical connection from one conductive element to another for the purpose of minimizing potential differences or providing suitable conductivity for fault current or for mitigation of leakage current and electrolytic action.

(f) *Bushing.* The term means an insulating structure including a through conductor, or providing a passageway for such a conductor, with provision for mounting on a barrier, conducting or otherwise, for the purpose of insulating the conductor from the barrier and conducting current from one side of the barrier to the other.

(g) *Cable*. The term means a conductor with insulation, or a stranded conductor with or without insulation and other coverings (single-conductor cable) or a combination of conductors insulated from one another (multiple-conductor cable).

(h) *Cable sheath*. The term means a protective covering applied to cables.

NOTE: A cable sheath may consist of multiple layers of which one or more is conductive.

(i) *Circuit*. The term means a conductor or system of conductors through which an electric current is intended to flow.

(j) *Communication lines*. The term means the conductors and their supporting or containing structures which are used for public or private signal or communication service, and which operate at potentials not exceeding 400 volts to ground or 750 volts between any two points of the circuit, and the transmitted power of which does not exceed 150 watts. When operating at less than 150 volts no limit is placed on the capacity of the system.

NOTE: Telephone, telegraph, railroad signal, data, clock, fire, police-alarm, community television antenna, and other systems conforming with the above are included. Lines used for signaling purposes, but not included under the above definition, are considered as supply lines of the same voltage and are to be so run.

(k) *Conductor*. The term means a material, usually in the form of a wire, cable, or bus bar suitable for carrying an electric current.

(l) *Conductor shielding*. The term means an envelope which encloses the conductor of a cable and provides an equipotential surface in contact with the cable insulation.

(m) *Current-carrying part*. The term means a conducting part intended to be connected in an electric circuit to a source of voltage. Non-current-carrying parts are those not intended to be so connected.

(n) *Dead (deenergized)*. The term means free from any electrical connection to a source of potential difference and from electrical charges: Not having a potential difference from that of earth.

NOTE: The term is used only with reference to current-carrying parts which are sometimes alive (energized).

(o) *Designated employee*. The term means a qualified person delegated to perform specific duties under the conditions existing.

(p) *Effectively grounded*. The term means intentionally connected to earth through a ground connection or connections of sufficiently low impedance and having sufficient current-carrying capacity to prevent the buildup of voltages which may result in undue hazard to connected equipment or to persons.

(q) *Electric line trucks*. The term means a truck used to transport men, tools, and material, and to serve as a traveling workshop for electric power line construction and maintenance work. It is sometimes equipped with a boom and auxiliary equipment for setting poles, digging holes, and elevating material or men.

(r) *Enclosed*. The term means surrounded by a case, cage, or fence, which will protect the contained equipment and prevent accidental contact of a person with live parts.

(s) *Equipment*. This is a general term which includes fittings, devices, appliances, fixtures, apparatus, and the like, used as part of, or in connection with, an electrical power transmission and distribution system, or communication systems.

(t) *Exposed*. The term means not isolated or guarded.

(u) *Electric supply lines*. The term means those conductors used to transmit electric energy and their necessary supporting or containing structures. Signal lines of more than 400 volts to ground are always supply lines within the meaning of the rules, and those of less than 400 volts to ground may be considered as supply lines, if so run and operated throughout.

(v) *Guarded*. The term means protected by personnel, covered, fenced, or enclosed by means of suitable casings, barrier rails, screens mats, platforms, or other suitable devices in accordance with standard barricading techniques designed to prevent dangerous approach or contact by persons or objects.

NOTE: Wires, which are insulated but not otherwise protected, are not considered as guarded.

(w) *Ground. (Reference)*. The term means that conductive body, usually earth, to which an electric potential is referenced.

(x) *Ground (as a noun)*. The term means a conductive connection whether intentional or accidental, by which an electric circuit or equipment is connected to reference ground.

(y) *Ground (as a verb)*. The term means the connecting or establishment of a connection, whether by intention or accident of an electric circuit or equipment to reference ground.

(z) *Grounding electrode (ground electrode).* The term grounding electrode means a conductor embedded in the earth, used for maintaining ground potential on conductors connected to it, and for dissipating into the earth current conducted to it.

(aa) *Grounding electrode resistance.* The term means the resistance of the grounding electrode to earth.

(bb) *Grounding electrode conductor (grounding conductor).* The term means a conductor used to connect equipment or the grounded circuit of a wiring system to a grounding electrode.

(cc) *Grounded conductor.* The term means a system or circuit conductor which is intentionally grounded.

(dd) *Grounded system.* The term means a system of conductors in which at least one conductor or point (usually the middle wire, or neutral point of transformer or generator windings) is intentionally grounded, either solidly or through a current-limiting device (not a current-interrupting device).

(ee) *Hotline tools and ropes.* The term means those tools and ropes which are especially designed for work on energized high voltage lines and equipment. Insulated aerial equipment especially designed for work on energized high voltage lines and equipment shall be considered hot line.

(ff) *Insulated.* The term means separated from other conducting surfaces by a dielectric substance (including air space) offering a high resistance to the passage of current.

NOTE: When any object is said to be insulated, it is understood to be insulated in suitable manner for the conditions to which it is subjected. Otherwise, it is within the purpose of this subpart, uninsulated. Insulating covering of conductors is one means of making the conductor insulated.

(gg) *Insulation (as applied to cable).* The term means that which is relied upon to insulate the conductor from other conductors or conducting parts or from ground.

(hh) *Insulation shielding.* The term means an envelope which encloses the insulation of a cable and provides an equipotential surface in contact with cable insulation.

(ii) *Isolated.* The term means an object that is not readily accessible to persons unless special means of access are used.

(jj) *Manhole.* The term means a subsurface enclosure which personnel may enter and which is used for the purpose of installing, operating, and maintaining equipment and/or cable.

(kk) *Pulling tension.* The term means the longitudinal force exerted on a cable during installation.

(ll) *Qualified person.* The term means a person who by reason of experience or training is familiar with the operation to be performed and the hazards involved.

(mm) *Switch.* The term means a device for opening and closing or changing the connection of a circuit. In these rules, a switch is understood to be manually operable, unless otherwise stated.

(nn) *Tag.* The term means a system or method of identifying circuits, systems or equipment for the purpose of alerting persons that the circuit, system or equipment is being worked on.

(oo) *Unstable material.* The term means earth material, other than running, that because of its nature or the influence of related conditions, cannot be depended upon to remain in place without extra support, such as would be furnished by a system of shoring.

(pp) *Vault.* The term means an enclosure above or below ground which personnel may enter and is used for the purpose of installing, operating, and/or maintaining equipment and/or cable.

(qq) *Voltage.* The term means the effective (rms) potential difference between any two conductors or between a conductor and ground. Voltages are expressed in nominal values. The nominal voltage of a system or circuit is the value assigned to a system or circuit of a given voltage class for the purpose of convenient designation. The operating voltage of the system may vary above or below this value.

(rr) *Voltage of an effectively grounded circuit.* The term means the voltage between any conductor and ground unless otherwise indicated.

(ss) *Voltage of a circuit not effectively grounded.* The term means the voltage between any two conductors. If one circuit is directly connected to and supplied from another circuit of higher voltage (as in the case of an autotransformer), both are considered as of the higher voltage, unless the circuit of lower voltage is effectively grounded, in which case its voltage is not determined by the circuit of higher voltage. Direct connection implies electric connection as distinguished from connection merely through electromagnetic or electrostatic induction.

(Secs. 6(b) and 8(g), 84 Stat. 1593, 1600; 29 U.S.C. 655, 657; sec. 107, 83 Stat. 96; 40 U.S.C. 333)

[FR Doc. 72-20118 Filed 11-22-72; 8:45 am]

§ 1926.556   Aerial Lifts

(a) *General Requirements.* (1) Unless otherwise provided in this section aerial lifts acquired for use on or after the effective date of this section shall be designed and constructed in conformance with the applicable requirements of the American National Standard for "Vehicle Mounted Elevating and Rotating Work Platforms," (ANSI) ANSI A92.2-1969, including appendix. Aerial lifts acquired before the effective date of this section, which do not meet the requirements of ANSI A92.2-1969, may not be used after January 1, 1976, unless they shall have been modified so as to conform with the applicable design and construction requirements of ANSI A92.2-1969. Aerial lifts include the following types of vehicle-mounted aerial devices used to elevate personnel to job-sites above ground: (i) Extensible boom platforms, (ii) aerial ladders, (iii) articulating boom platforms, (iv) vertical towers, and (v) a combination of any of the above. Aerial equipment may be made of metal, wood, fiberglass reinforced plastic (FRP), or other material, may be powered or manually operated; and are deemed to be aerial lifts whether or not they are capable of rotating about a substantially vertical axis.

(2) Aerial lifts may be "field modified" for uses other than those intended by the manufacturer provided the modification has been certified in writing by the manufacturer or by any other equivalent entity, such as a nationally recognized testing laboratory, to be in conformity with all applicable provisions of ANSI A92.2-1969 and this section and to be at least as safe as the equipment was before modification.

(b) *Specific requirements*—(1) *Ladder trucks and tower trucks.* Aerial ladders shall be secured in the lower traveling position by the locking device on top of the truck cab, and the manually operated device at the base of the ladder before the truck is moved for highway travel.

(2) *Extensible and articulating boom platforms.* (i) Lift controls shall be tested each day prior to use to determine that such controls are in safe working condition.

(ii) Only authorized persons shall operate an aerial lift.

(iii) Belting off to an adjacent pole, structure, or equipment while working from an aerial lift shall not be permitted.

(iv) Employees shall always stand firmly on the floor of the basket, and shall not sit or climb on the edge of the basket or use planks, ladders, or other devices for a work position.

(v) A body belt shall be worn and a lanyard attached to the boom or basket when working from an aerial lift.

(vi) Boom and basket load limits specified by the manufacturer shall not be exceeded.

(vii) The brakes shall be set and when outriggers are used, they shall be positioned on pads or a solid surface. Wheel chocks shall be installed before using an aerial lift on an incline, provided they can be safely installed.

(viii) An aerial lift truck shall not be moved when the boom is elevated in a working position with men in the basket, except for equipment which is specifically designed for this type of operation in accordance with the provisions of subparagraphs (1) and (2) of paragraph (a) of this section.

(ix) Articulating boom and extensible boom platforms, primarily designed as personnel carriers, shall have both platform (upper) and lower controls. Upper controls shall be in or beside the platform within easy reach of the operator. Lower controls shall provide for overriding the upper controls. Controls shall be plainly marked as to their function. Lower level controls shall not be operated unless permission has been obtained from the employee in the lift, except in case of emergency.

(x) Climbers shall not be worn while performing work from an aerial lift.

(xi) The insulated portion of an aerial lift shall not be altered in any manner that might reduce its insulating value.

(xii) Before moving an aerial lift for travel, the boom(s) shall be inspected to see that it is properly cradled and outriggers are in stowed position except as provided in subdivision (viii) of this subparagraph.

(3) *Electrical tests.* All electrical tests shall conform to the requirements of ANSI A92.2-1969 section 5. However equivalent d.c. voltage tests may be used in lieu of the a.c. voltage specified in A92.2-1969; d.c. voltage tests which are approved by the equipment manufacturer or equivalent entity shall be considered an equivalent test for the purpose of this subparagraph (3).

(4) *Bursting safety factor.* The provisions of the American National Standards Institute standard ANSI A92.2-1969, section 4.9 Bursting Safety Factor shall apply to all critical hydraulic and pneumatic components. Critical components are those in which a failure would result in a free fall or free rotation of the boom. All noncritical components shall have a bursting safety factor of at least 2 to 1.

Move
right

Left arm extended horizontally; right
arm sweeps upward to position over head.

Hold
hover

The signal "Hold" is executed by placing
arms over head with clenched fists.

Move
left

Right arm extended horizontally; left
arm sweeps upward to position over head.

Takeoff

Right hand behind back; left hand
pointing up.

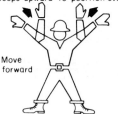

Move
forward

Combination of arm and hand movement
in a collecting motion pulling toward body.

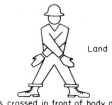

Land

Arms crossed in front of body and
pointing downward.

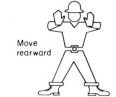

Move
rearward

Hands above arm, palms out using a
noticeable shoving motion.

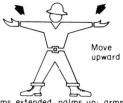

Move
upward

Arms extended, palms up; arms
sweeping up.

Release
sling
load

Left arm held down away from body.
Right arm cuts across left arm in a
slashing movement from above.

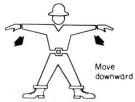

Move
downward

Arms extended, palms down; arms
sweeping down.

Fig. 46-1 Helicopter hand signal.

(5) *Welding standards.* All welding shall conform to the following standards as applicable:

(i) Standard Qualification Procedure, AWS B3.0-41.

(ii) Recommended Pratices for Automotive Welding Design, AWS D8.4-61.

(iii) Standard Qualification of Welding Procedures and Welders for Piping and Tubing, AWS D10.9-69.

(iv) Specifications for Welding Highway and Railway Bridges, AWS D2.0-69.

§ 1926.551   Helicopters

(a) *Helicopter regulations.* Helicopter cranes shall be expected to comply with any applicable regulations of the Federal Aviation Administration.

(b) *Briefing.* Prior to each day's operations a briefing shall be conducted. This briefing shall set forth the plan of operation for the pilot and ground personnel.

(c) *Slings and tag lines.* Load shall be properly slung. Tag lines shall be of a length that will not permit their being drawn up into rotors. Pressed sleeve, swedged eyes, or equivalent means shall be used for all freely suspended loads to prevent hand splices from spinning open or cable clamps from loosening.

(d) *Cargo hooks.* All electrically operated cargo hooks shall have the electrical activating device so designed and installed as to prevent inadvertent operation. In addition, these cargo hooks shall be equipped with an emergency mechanical control for releasing the load. The hooks shall be tested prior to each day's operation to determine that the release functions properly, both electrically and mechanically.

(e) *Personal protective equipment.* (1) Personal protective equipment for employees receiving the load shall consist of complete eye protection and hard hats secured by chinstraps.

(2) Loose-fitting clothing likely to flap in the downwash and thus be snagged on hoist line, shall not be worn.

(f) *Loose gear and objects.* Every practical precaution shall be taken to provide for the protection of the employees from flying objects in the rotor downwash. All loose gear within 100 feet of the place of lifting the load, depositing the load, and all other areas susceptible to rotor downwash shall be secured or removed.

(g) *Housekeeping.* Good housekeeping shall be maintained in all helicopter loading and unloading areas.

(h) *Operator responsibility.* The helicopter operator shall be responsible for size, weight, and manner in which loads are connected to the helicopter. If, for any reason, the helicopter operator believes the lift cannot be made safely, the lift shall not be made.

(i) *Hooking and unhooking loads.* When employees are required to perform work under hovering craft a safe means of access shall be provided for employees to reach the hoist line hook and engage or disengage cargo slings. Employees shall not perform work under hovering craft except when necessary to hook craft except when necessary to hook or unhook loads.

(j) *Static charge.* Static charge on the suspended load shall be dissipated with a grounding device before ground personnel touch the suspended load, or protective rubber gloves shall be worn by all ground personnel touching the suspended load.

(k) *Weight limitation.* The weight of an external load shall not exceed the manufacturer's rating.

(l) *Ground lines.* Hoist wires or other gear, except for pulling lines or conductors that are allowed to "pay out" from a container or roll off a reel shall not be attached to any fixed ground structure, or allowed to foul on any fixed structure.

(m) *Visibility.* When visibility is reduced by dust or other conditions, ground personnel shall exercise special caution to keep clear of main and stabilizing rotors. Precautions shall also be taken by the employer to eliminate as far as practical reduced visibility.

(n) *Signal systems.* Signal systems between aircrew and ground personnel shall be understood and checked in advance of hoisting the load. This applies to either radio or hand signal systems. Hand signals shall be as shown in Fig. 46-1.

(o) *Approach distance.* No unauthorized person shall be allowed to approach within 50 feet of the helicopter when the rotor blades are turning.

(p) *Approaching helicopter.* Whenever approaching or leaving a helicopter with blades rotating, all employees shall remain in full view of the pilot and keep in a crouched position. Employees shall avoid the area from the cockpit or cabin rearward unless authorized by the helicopter operator to work there.

(q) *Personnel.* Sufficient ground personnel shall be provided when required for safe helicopter loading and unloading operations.

(r) *Communications.* There shall be constant reliable communication between the pilot and a designated employee of the ground crew who acts as a signalman during the period of loading and unloading. This signalman shall be distinctly recognizable from other ground personnel.

(s) *Fires.* Open fires shall not be permitted in an area that could result in such fires being spread by the rotor downwash.

Resuscitation

Effects of Electric Shock on the Human Body   The effect of electric shock on a human being is rather unpredictable and may manifest itself in a number of ways:

1. *Asphyxia.* Electric shock may cause a cessation of respiration (asphyxia). Current passing through the body may temporarily paralyze (or destroy) either the nerves or the area of the brain which controls respiration.

2. *Burns—Contact and Flash.* Contact burns are a common result of electric current passing through the body. The burns are generally found at the points where the current entered and left the body and vary in severity, the same as thermal burns. The seriousness of these burns may not be immediately evident because their appearance may not indicate the depth to which they have penetrated.

In some accidents there is a flash or electric arc, the rays and heat from which may damage the eyes or result in thermal burns to exposed parts of the body.

3. *Fibrillation.* Electric shock may disturb the natural rhythm of the heartbeat. When this happens, the muscles of the heart are thrown into a twitching or trembling state and the actions of the individual muscle fibers are no longer coordinated. The pulse disappears, and circulation ceases. This condition is known as "ventricular fibrillation" and is serious.

4. *Muscle Spasm.* A series of erratic movements of a limb or limbs may occur owing to alternating contractions and relaxations of the muscles. This muscle-spasm action on the muscles of respiration may be a factor in the stoppage of breathing.

RESCUE

Because a person may receive electrical shock in many different locations—on the ground, in buildings, on poles, or on steel structures—it is neither possible nor desirable to lay down definite methods of rescue. However, there are certain facts which should be remembered.

Freeing Victim   Because of the muscle spasm at the time of shock, most victims are thrown clear of contact. However, in some instances (usually low-voltage) the victim is still touching live equipment. In either situation, the rescuer must be extremely careful not to get himself in contact with the live equipment or to touch the victim while he is still in contact. He should "free" the victim as soon as possible so that artificial respiration can be applied without hazard. This may involve opening switches or cutting wires so that equipment within reach is deenergized or using rubber gloves or other approved insulation to move the victim out of danger.

If the victim is to be lowered from a pole, tower, or other structure, a hand line of adequate strength, tied or looped around him and placed over a crossarm or tower member, is a simple and satisfactory method. Procedures for lowering a lineman from a pole or substation structure developed by personnel of the Commonwealth Edison Company are illustrated in Sec. 49, "Pole-Top and Bucket Truck Rescue."

Artificial Respiration Methods   Early methods of artificial respiration involved applying heat by building a fire on the victim's stomach, beating the victim with a whip, inverting the victim by suspending him by his feet or rolling the victim over a barrel. Modern methods of artificial respiration include the Schafer prone method, the chest-pressure–arm-lift method, the back-pressure–arm-lift method, and the mouth-to-mouth method. The mouth-to-mouth method is currently advocated by the Edison Electric Institute and the American Red Cross. The mouth-to-mouth method accepted by most utility companies is described in this section. Artificial respiration should be used when breathing stops as a result of electric shock, exposure to gases, drowning, or physical injury. If the heart has stopped functioning as a result of cardiac arrest or ventricular fibrillation, heart-lung resuscitation should be used with mouth-to-mouth resuscitation. Heart-lung resuscitation methods are illustrated in Sec. 48.

Applying Artificial Respiration   To be successful, artificial respiration must be applied *within the shortest time possible* to a victim who is not breathing. The graph below shows the possibility of successful revival for each minute of delay.

Normally, the type of resuscitation in which the victim is placed on the ground or floor is the best because the victim can be given additional treatment (stoppage of bleeding, wrapping in blanket to keep warm, etc.) and the "rescuer" can be easily relieved without an interruption of the breathing rhythm or cycle. However, "pole-top resuscitation" or any adaptation of it should be started if there will be any appreciable delay in getting the victim into the prone position.

In choosing the type of resuscitation to be used, the rescuer must consider the obvious injuries suffered by the victim. Broken ribs, for example, might make inadvisable the use of certain types of resuscitation; burns on the arms or on the face might exclude other types. However, no time should be lost in searching for injuries; artificial respiration should be started at once.

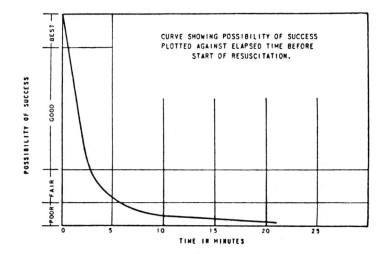

Necessity for Speed in Starting Artificial Respiration   *Artificial respiration must be started at once!* Time is the most important single factor in successful artificial respiration. The human brain can exist for only a few minutes, probably not more than 4 or 5, without oxygenated blood; therefore, every second counts. The greater the delay, the less chance of successful recovery. There must be no delay to loosen clothing, warm the victim, get him down from the pole, or move him to a more comfortable position. However, an immediate check of the victim's mouth should be made by a quick pass of the fingers through the mouth to pull the tongue forward and remove false teeth, tobacco, chewing gum, etc. *After* resuscitation is started, the victim's belt, collar, and other clothing may be loosened, providing this does not interfere with the resuscitation process.

AIR DELIVERY COMPARISONS

NORMAL BREATHING
525 C.C. OF AIR **PER BREATH**

MOUTH-TO-MOUTH
(HORIZONTAL POSITION)
1050 C.C. OF AIR **PER BREATH**

MOUTH-TO-MOUTH
(VERTICAL POSITION)
1300 C.C. OF AIR **PER BREATH**

THE MOUTH-TO-MOUTH (M-M) OR (MOUTH-TO-NOSE) (M-N) METHOD OF ARTIFICIAL RESPIRATION

PREFERRED METHOD

ENDORSED BY:

American Heart Association – American Medical Association
American Red Cross – U.S. Public Health Service
Industrial Medical Association – Edison Electric Institute

In An **EMERGENCY**

- ✓ OBSERVE HAZARDS
- ✓ PROTECT YOURSELF
- ✓ THINK

Then

IF NECESSARY . . .
REMOVE THE INJURED
FROM HAZARDOUS
AREA

IF THE INJURED IS NOT BREATHING

- ✓ OPEN AIRWAY
- ✓ RESTORE BREATHING

**ACT QUICKLY
SECONDS COUNT**

A

TO OPEN AIRWAY

CLEAR MOUTH

TILT HEAD BACK

OBSTRUCTED

OPENED

B

TO RESTORE BREATHING

1 KEEP INJURED'S HEAD TILTED

2 PINCH NOSTRILS CLOSED OR CLOSE MOUTH

3 TAKE A DEEP BREATH

4 PLACE YOUR MOUTH OVER HIS MOUTH (OR NOSE)

5 BLOW FORCEFULLY

6 REMOVE YOUR MOUTH AND ALLOW HIM TO EXHALE

Repeat CYCLE 12 TIMES PER MINUTE

1 KEEP INJURED'S HEAD TILTED

MOUTH-TO-MOUTH

MOUTH-TO-NOSE

ONE HAND ON FOREHEAD

OTHER HAND BEHIND NECK

OTHER HAND HOLDS MOUTH CLOSED

Note

A BABY'S NECK IS VERY PLIABLE...
DON'T EXAGGERATE HEAD TILT

2 PINCH NOSTRILS CLOSED OR CLOSE MOUTH

MOUTH-TO-MOUTH

MOUTH-TO-NOSE

USE THUMB & FOREFINGER OF THE HAND YOU HAVE ON HIS FOREHEAD

MAKE SURE LIPS ARE SEALED

THIS IS THE BASIC DIFFERENCE BETWEEN
MOUTH-TO-MOUTH & MOUTH-TO-NOSE

3
TAKE A DEEP BREATH
✓ ENOUGH AIR FOR YOURSELF AND THE INJURED

4
PLACE MOUTH COMPLETELY OVER HIS MOUTH (OR NOSE)
✓ OPEN YOUR MOUTH WIDELY
✓ MAKE AIRTIGHT SEAL

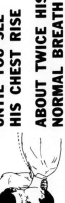

Note ON A BABY PLACE MOUTH OVER BOTH MOUTH & NOSE

5
BLOW FORCEFULLY

UNTIL YOU SEE HIS CHEST RISE

ABOUT TWICE HIS NORMAL BREATH

Note BABIES REQUIRE ONLY SMALL PUFFS OF AIR

Caution YOUR FIRST BLOWING EFFORT MAY REVEAL AN OBSTRUCTION—IF SO...

ADULT
ROLL HIM ON HIS SIDE AND SLAP HIM ON THE BACK

BABY
PICK HIM UP AND LAY HIM OVER YOUR ARM—SLAP HIM ON THE BACK

**REPEAT CYCLE
12 TIMES
PER MINUTE**

RATE FOR BABIES 20 PUFFS PER MINUTE

Check EVERY BREATH BY...

1 SEEING CHEST RISE & FALL

**2 FEELING RESISTANCE OF
LUNGS AS THEY EXPAND**

**3 HEARING AIR ESCAPE
DURING EXHALATION**

6

**REMOVE YOUR MOUTH AND
ALLOW HIM TO EXHALE**

✓ WATCH HIS CHEST FALL

✓ LISTEN FOR AIR ESCAPING
FROM HIS LUNGS

Note

IN MOUTH-TO-NOSE YOU
MAY HAVE TO OPEN HIS
MOUTH TO ALLOW AIR TO ESCAPE

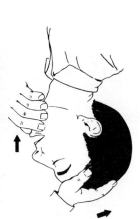

Remember

1 KEEP HEAD TILTED

2 PINCH NOSTRILS (OR CLOSE MOUTH)

3 TAKE A DEEP BREATH

4 SEAL MOUTH COMPLETELY

5 BLOW FORCEFULLY

6 REMOVE YOUR MOUTH

Repeat 12 TIMES A MINUTE

Continue

UNTIL...HE IS BREATHING NORMALLY OR...

HE REACHES A HOSPITAL OR...

A DOCTOR TAKES OVER

DON'T GIVE UP

Heart-Lung Resuscitation

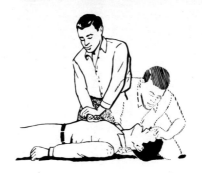

HEART-LUNG RESUSCITATION HAS BEEN

ENDORSED BY:

American Heart Association – American Medical Association

American Red Cross – U.S. Public Health Service

Industrial Medical Association – Edison Electric Institute

HEART-LUNG RESUSCITATION

IS

ARTIFICIAL RESPIRATION
(MOUTH-TO-MOUTH OR MOUTH-TO-NOSE)

PLUS

ARTIFICIAL CIRCULATION
(EXTERNAL CARDIAC COMPRESSION)

IF THE INJURED IS NOT BREATHING —

Open *AIRWAY*

IF THIS DOES NOT HELP –

Restore **BREATHING**

IF HIS HEART HAS STOPPED

Restore **CIRCULATION**

**ACT QUICKLY
SECONDS COUNT**

WE SOMETIMES FACE
EMERGENCIES SUCH AS –

- ELECTRIC SHOCK
- HEART ATTACK
- DROWNING
- SUFFOCATION
- PHYSICAL SHOCK

WHERE DEATH MAY RESULT FROM

CARDIAC ARREST
AND / OR
LACK OF BREATHING

Note

THE SYMPTOMS OF CARDIAC
ARREST AND VENTRICULAR
FIBRILLATION ARE THE SAME

Pulse

CHECK CAROTID ARTERY

■ LOCATED IN NECK ON EITHER SIDE OF WINDPIPE

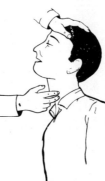

■ USE INDEX AND MIDDLE FINGERS OF ONE HAND

■ PULSE SHOULD BE "FELT" NOT "COMPRESSED"

NO PULSE MEANS NO CIRCULATION

TO RESTORE CIRCULATION USE

- EXTERNAL
- CARDIAC
- COMPRESSION

SIGNS THAT CIRCULATION HAS STOPPED

(HEART STANDSTILL OR VENTRICULAR FIBRILLATION)

ARE –

1. NO CAROTID PULSE
2. WIDELY DILATED PUPILS
3. ASHEN-GRAY SKIN COLOR

BASIC STEPS IN
EXTERNAL CARDIAC COMPRESSION

1. LAY INJURED ON HIS BACK ON A FIRM SURFACE

2. KNEEL AT HIS SIDE

3. PLACE THE HEEL OF ONE HAND ON LOWER
   HALF OF HIS STERNUM (BREASTBONE)

4. PLACE YOUR OTHER HAND ON TOP OF THE FIRST

5. EXERT DOWNWARD PRESSURE

6. RELEASE PRESSURE

Repeat COMPRESSIONS
60 TIMES A MINUTE

Pupils

CHECK BY LIFTING EYELID
TO SEE IF PUPIL CONTRACTS
WHEN EXPOSED TO LIGHT

NORMAL PUPIL   DILATED PUPIL

PUPILS THAT REMAIN WIDELY DILATED
INDICATE LACK OF CIRCULATION

Color

ASHEN-GRAY SKIN COLOR INDICATES
LACK OF OXYGENATED BLOOD

1. LAY INJURED ON HIS BACK ON A FIRM SURFACE

- FOR EFFECTIVE ARTIFICIAL CIRCULATION OF BLOOD – THE HEART MUST BE COMPRESSED BETWEEN STERNUM (BREASTBONE) AND SPINE

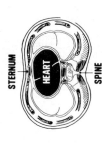

STERNUM

HEART

SPINE

2. KNEEL AT HIS SIDE

- IN LINE WITH LOWER HALF OF HIS CHEST

3. PLACE THE HEEL OF ONE HAND ON THE LOWER HALF OF HIS STERNUM

- FEEL FOR LOWER END OF STERNUM

- PLACE ONLY THE HEEL OF YOUR HAND ON HIS STERNUM

- KEEP YOUR FINGERS ELEVATED DO NOT TOUCH HIS RIBS

INCORRECT HAND POSITION CAN CAUSE INTERNAL INJURIES

Note | ONLY TWO FINGERS ARE REQUIRED TO COMPRESS THE BREASTBONE OF A BABY

4 PLACE YOUR OTHER HAND ON TOP OF THE FIRST

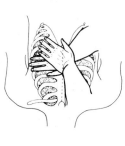

- THIS WILL HELP MAINTAIN HAND POSITION

- IT WILL ALSO BE EASIER TO EXERT PROPER PRESSURE

Note

- THE HEEL OF ONLY ONE HAND WILL PROVIDE ENOUGH PRESSURE TO COMPRESS THE BREASTBONE OF CHILDREN

5 EXERT DOWNWARD PRESSURE

- ROCK FORWARD UNTIL YOUR SHOULDERS ARE ALMOST DIRECTLY ABOVE INJURED'S CHEST

- KEEP YOUR ARMS STRAIGHT

- APPLY ENOUGH PRESSURE (80 TO 100 POUNDS) TO DEPRESS THE STERNUM 1½ TO 2 INCHES

- USE THE WEIGHT OF YOUR UPPER BODY

COMPRESSION FORCES BLOOD OUT OF THE HEART

In An EMERGENCY

✔ OBSERVE HAZARDS

✔ PROTECT YOURSELF

✔ THINK BEFORE YOU ACT

Then

IF NECESSARY

REMOVE THE INJURED

FROM HAZARDOUS AREA

6. RELEASE PRESSURE

- ROCK BACK
- MAINTAIN HAND POSITION - DO NOT LIFT HANDS OFF STERNUM

RELEASE OF PRESSURE ALLOWS HEART TO REFILL WITH BLOOD

Repeat COMPRESSIONS 60 TIMES PER MINUTE (ONCE A SECOND)

Note THE RATE FOR BABIES AND CHILDREN IS 80 TO 100 COMPRESSIONS PER MINUTE

IF THE INJURED IS UNCONSCIOUS AND NOT BREATHING

A PROVIDE AN OPEN AIRWAY

IF THIS DOES NOT RESTORE BREATHING

B GIVE 5 OR 6 QUICK BREATHS BY MOUTH-TO-MOUTH

IF HE RESPONDS, CONTINUE MOUTH-TO-MOUTH RESPIRATION UNTIL HE IS BREATHING WITHOUT HELP

Then

- WATCH HIM CLOSELY - HIS BREATHING MAY STOP AGAIN
- GIVE ANY ADDITIONAL FIRST AID NEEDED
- CALL FOR HELP

IF HE DOES NOT SEEM TO RESPOND TO THE FIRST 5 OR 6 QUICK BREATHS

✓ CHECK FOR PULSE

✓ CHECK PUPIL FOR DILATION

✓ NOTE SKIN COLOR

IF A PULSE IS FELT, PUPIL CONTRACTS AND COLOR IMPROVES -

CONTINUE MOUTH-TO-MOUTH UNTIL HE IS BREATHING WITHOUT HELP

Then

- WATCH HIM CLOSELY - HIS BREATHING MAY STOP AGAIN
- GIVE ANY ADDITIONAL FIRST AID NEEDED
- CALL FOR HELP

IF NO PULSE IS PRESENT, PUPIL DOES NOT CONTRACT AND SKIN COLOR IS NOT NORMAL, HIS HEART HAS STOPPED

C RESTORE CIRCULATION
USE HEART-LUNG RESUSCITATION

MOUTH-TO-MOUTH RESPIRATION

+

EXTERNAL CARDIAC COMPRESSION

=

HEART-LUNG RESUSCITATION

ONE RESCUER | AFTER EVERY 15 COMPRESSIONS GIVE 2 QUICK BREATHS RATIO 15 TO 2

TWO RESCUERS | AFTER EVERY FIFTH COMPRESSION INTERPOSE ONE BREATH RATIO 5 TO 1

CONTINUE
HEART-LUNG RESUSCITATION
UNTIL HE RECOVERS OR HE
REACHES A HOSPITAL OR A
DOCTOR TAKES OVER

Important

- CONTINUE H-L-R ON WAY TO HOSPITAL
- DO NOT ALLOW ANY PRESSURE-CYCLING MECHANICAL RESUSCITATOR TO BE USED WITH CARDIAC COMPRESSION
- DO NOT ALLOW ANYONE (POLICE, FIREMEN OR AMBULANCE CREW) TO TAKE OVER - UNLESS THEY ARE EXPERT IN H-L-R

REMEMBER YOUR **A,B,C**'s
HEART-LUNG RESUSCITATION INVOLVES

AIRWAY OPENED

BREATHING RESTORED

CIRCULATION RESTORED

DOCTOR

**PRACTICE AND COMMON SENSE ARE ESSENTIAL
TO SUCCESSFUL HEART-LUNG RESUSCITATION**

Section **49**

Pole-Top and Bucket Truck Rescue

**POLE
TOP
RESCUE**

**TIME
IS
CRITICAL**

YOU MAY HAVE TO HELP A MAN ON A POLE REACH THE GROUND SAFELY WHEN HE—

- BECOMES ILL
- IS INJURED
- LOSES CONSCIOUSNESS

YOU MUST KNOW—

- WHEN HE NEEDS HELP
- WHEN AND WHY TIME IS CRITICAL
- THE APPROVED METHOD OF LOWERING

BASIC STEPS IN

POLE TOP RESCUE

Evaluate THE SITUATION

Provide FOR YOUR PROTECTION

Climb TO RESCUE POSITION

Determine INJURED'S CONDITION

Then

IF NECESSARY—

- GIVE FIRST AID
- LOWER INJURED
- GIVE FOLLOW-UP CARE
- CALL FOR HELP

Evaluate THE SITUATION

CALL TO MAN ON POLE

IF HE DOES NOT ANSWER

OR APPEARS STUNNED

OR DAZED

• PREPARE TO GO TO HIS AID

**TIME
IS
EXTREMELY
IMPORTANT**

Provide

**FOR YOUR PROTECTION
YOUR SAFETY IS VITAL
TO THE RESCUE**

—PERSONAL TOOLS AND RUBBER GLOVES
(RUBBER SLEEVES, IF REQUIRED)

Check

✓ **EXTRA RUBBER GOODS?**

✓ **LIVE LINE TOOLS?**

✓ **PHYSICAL CONDITION OF POLE?**

  ✓ DAMAGED CONDUCTORS, EQUIPMENT?

  ✓ FIRE ON POLE?

  ✓ BROKEN POLE?

✓ **HAND LINE ON POLE AND IN GOOD CONDITION?**

Determine

THE INJURED'S CONDITION

HE MAY BE . . . .

• CONSCIOUS

• UNCONSCIOUS
  BUT BREATHING

• UNCONSCIOUS
  NOT BREATHING

• UNCONSCIOUS
  NOT BREATHING AND
  HEART STOPPED

Climb TO RESCUE POSITION

• CLIMB CAREFULLY

POSITION YOURSELF—

• TO INSURE YOUR SAFETY

• TO CLEAR THE INJURED FROM HAZARD

• TO DETERMINE THE INJURED'S CONDITION

• TO RENDER AID AS REQUIRED

• TO START MOUTH-TO-MOUTH, IF REQUIRED

• TO LOWER INJURED, IF NECESSARY

THE BEST
POSITION
WILL
USUALLY BE—
SLIGHTLY
ABOVE THE
INJURED

IF THE INJURED IS
CONSCIOUS

- TIME MAY NO LONGER BE CRITICAL
- GIVE NECESSARY FIRST AID ON POLE
- REASSURE THE INJURED
- HELP HIM DESCEND POLE
- GIVE FIRST AID ON GROUND
- CALL FOR HELP, IF NECESSARY

IF THE INJURED IS
UNCONSCIOUS
BUT BREATHING

- WATCH HIM CLOSELY IN CASE BREATHING STOPS
- LOWER HIM TO GROUND
- GIVE FIRST AID ON GROUND
- CALL FOR HELP

IF THE INJURED IS
UNCONSCIOUS AND NOT BREATHING

- PROVIDE AN OPEN AIRWAY
- GIVE HIM 5 OR 6 QUICK BREATHS
- IF HE RESPONDS . . .

CONTINUE MOUTH-TO-MOUTH UNTIL HE IS BREATHING WITHOUT HELP

IF HE DOES NOT RESPOND TO THE FIRST 5 OR 6 QUICK BREATHS

✓ CHECK SKIN COLOR

✓ CHECK FOR PUPIL DILATION

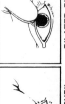

NORMAL PUPIL   DILATED PUPIL

IF PUPIL CONTRACTS AND COLOR IS GOOD

CONTINUE MOUTH-TO-MOUTH UNTIL HE IS BREATHING WITHOUT HELP

Then
- HELP HIM DESCEND POLE
- WATCH HIM CLOSELY—HIS BREATHING MAY STOP AGAIN
- GIVE ANY ADDITIONAL FIRST AID IF NEEDED
- CALL FOR HELP

THE
METHOD
OF
LOWERING AN
INJURED MAN
IS

* SAFE
* SIMPLE
* AVAILABLE

IF PUPIL DOES NOT CONTRACT
AND SKIN COLOR IS BAD

HEART HAS STOPPED

- PREPARE TO LOWER HIM
  IMMEDIATELY

- GIVE 5 OR 6 MORE QUICK
  BREATHS JUST BEFORE LOWERING

- LOWER HIM TO THE GROUND

- START HEART-LUNG RESUSCITATION

- CALL FOR HELP

RESCUER

Position HAND-LINE

OVER ARM OR OTHER PART OF STRUCTURE

VICTIM / FALL LINE

- **POSITION LINE FOR CLEAR PATH TO GROUND**

Note **USUALLY BEST 2 OR 3 FEET FROM POLE**

EQUIPMENT NEEDED

- **1/2 INCH HAND LINE**

PROCEDURE . . .

- *Position* **HAND LINE**
- *Tie* **INJURED**
- *Remove* **SLACK IN HAND LINE**
- *Take* **FIRM GRIP ON FALL LINE**
- *Cut* **INJURED'S SAFETY STRAP**

  Lower **INJURED**

SHORT END OF LINE IS WRAPPED AROUND FALL LINE TWICE. (TWO WRAPS AROUND FALL LINE.)

RESCUER TIES HAND LINE AROUND
VICTIM'S CHEST USING THREE HALF
HITCHES.

Tie INJURED —

PASS LINE AROUND INJURED'S CHEST

TIE THREE HALF-HITCHES

- **KNOT IN FRONT**
- **NEAR ONE ARM PIT**
- **HIGH ON CHEST**
- **SNUG KNOT**

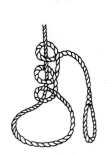

Remove SLACK IN HAND-LINE

- **ONE RESCUER—REMOVES SLACK ON POLE**

- **TWO RESCUERS—MAN ON GROUND REMOVES SLACK**

IMPORTANT
GIVE 5 OR 6 QUICK BREATHS
. . . IF NECESSARY . . .

THEN

Take FIRM GRIP ON FALL-LINE

ONE RESCUER—HOLDS FALL-LINE WITH ONE HAND

TWO RESCUERS—MAN ON GROUND HOLDS FALL-LINE

Cut INJURED'S SAFETY STRAP

CUT STRAP ON SIDE
OPPOSITE DESIRED SWING

Caution
DO NOT CUT YOUR OWN SAFETY
STRAP OR THE HAND-LINE

RIGHT	WRONG

Lower INJURED

ONE RESCUER

- GUIDE LOAD LINE WITH ONE HAND

- CONTROL RATE OF DESCENT WITH THE OTHER HAND

TWO RESCUERS

- MAN ON THE POLE GUIDES THE LOAD LINE

- MAN ON THE GROUND CONTROLS RATE OF DESCENT

IF A CONSCIOUS MAN IS BEING ASSISTED IN CLIMBING DOWN . . . .
THE ONLY DIFFERENCE IS
THAT ENOUGH SLACK IS FED INTO THE LINE TO PERMIT HIM CLIMBING FREEDOM

Remember

THE APPROVED METHOD OF LOWERING AN INJURED MAN IS . . .

- *Position* HAND LINE
- *Tie* INJURED
- *Remove* SLACK IN HAND LINE
- *Take* FIRM GRIP ON FALL LINE
- *Cut* INJURED'S SAFETY STRAP
- *Lower* INJURED

ONE MAN RESCUE

TWO OR MORE MAN RESCUE

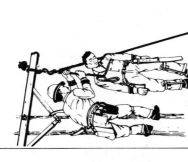

RESCUES DIFFER ONLY IN CONTROL OF THE FALL-LINE

49-15

IF POLE DOES NOT HAVE CROSSARM, RESCUER PLACES HAND-LINE OVER FIBERGLASS BRACKET INSULATOR SUPPORT, OR OTHER SUBSTANTIAL PIECE OF EQUIPMENT SUCH AS A SECONDARY RACK, NEUTRAL BRACKET, OR GUY WIRE ATTACHMENT, STRONG ENOUGH TO SUPPORT THE WEIGHT OF THE INJURED. SHORT END OF LINE IS WRAPPED AROUND FALL LINE TWICE AND TIED AROUND VICTIM'S CHEST USING THREE HALF HITCHES. INJURED'S SAFETY STRAP IS CUT AND RESCUER LOWERS INJURED TO THE GROUND.

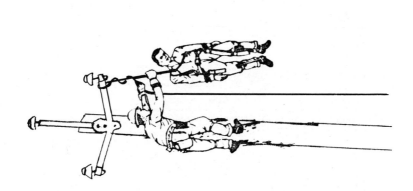

LINE MUST BE REMOVED AND VICTIM EASED ON TO GROUND

LAY VICTIM ON HIS BACK AND OBSERVE IF VICTIM IS CONSCIOUS. IF VICTIM IS CONSCIOUS, TIME MAY NO LONGER BE CRITICAL. GIVE NECESSARY FIRST AID. CALL FOR HELP.

IF THE INJURED IS UNCONSCIOUS AND NOT BREATHING, PROVIDE AN OPEN AIRWAY. GIVE HIM FIVE OR SIX QUICK BREATHS. IF HE RESPONDS, CONTINUE MOUTH-TO-MOUTH RESUSCITATION UNTIL HE IS BREATHING WITHOUT HELP. IF NO PULSE IS PRESENT, PUPIL DOES NOT CONTRACT, AND SKIN COLOR IS NOT NORMAL, HIS HEART HAS STOPPED. RESTORE CIRCULATION. USE HEART-LUNG RESUSCITATION.

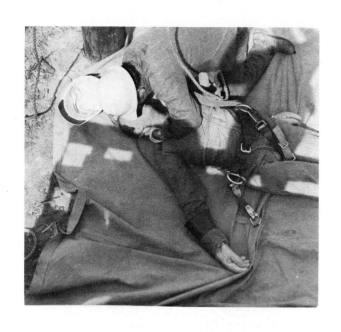

BUCKET TRUCK
RESCUE
TIME
IS
CRITICAL

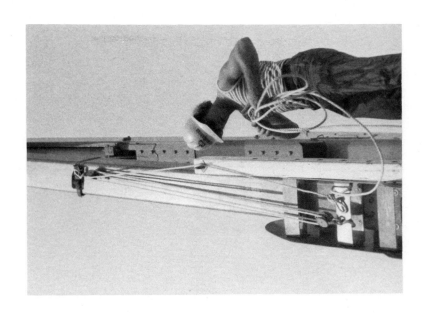

EQUIP PORTION OF INSULATED BOOM OF TRUCK WITH ROPE BLOCKS DESIGNED FOR HOT-LINE WORK.

STRAP IS PLACED AROUND INSULATED BOOM APPROXIMATELY TEN FEET FROM BUCKET TO SUPPORT ROPE BLOCKS. BLOCKS ARE HELD TAUT ON BOOM FROM STRAP TO TOP OF BOOM.

RESCUER ON THE GROUND EVALUATES THE CONDITIONS WHEN AN EMERGENCY ARISES. THE BUCKET IS LOWERED USING THE LOWER CONTROLS. OBSTACLES IN THE PATH OF THE BUCKET MUST BE AVOIDED.

HOOK ON ROPE BLOCKS IS ENGAGED IN A RING ON THE LINEMAN'S SAFETY STRAP. SAFETY STRAP IS RELEASED FROM BOOM OF TRUCK.

ROPE BLOCKS ARE DRAWN TAUT BY RESCUER ON THE GROUND. UNCONSCIOUS VICTIM IS RAISED OUT OF BUCKET WITH ROPE BLOCKS.

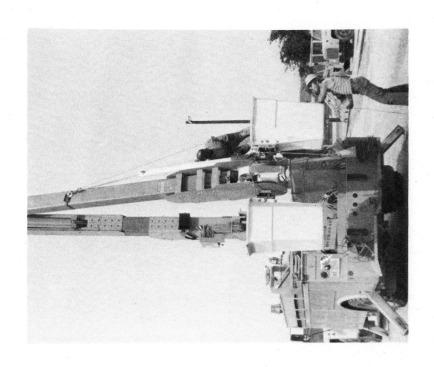

RESCUER EASES VICTIM ON TO THE GROUND. CARE SHOULD BE TAKEN TO PROTECT THE INJURED VICTIM FROM FURTHER INJURY.

RELEASE ROPE
BLOCKS
FROM
VICTIM.

LAY VICTIM ON HIS BACK AND OBSERVE IF VICTIM IS CONSCIOUS. IF VICTIM IS CONSCIOUS, TIME MAY NO LONGER BE CRITICAL. GIVE NECESSARY FIRST AID. CALL FOR HELP.

IF THE INJURED IS UNCONSCIOUS AND NOT BREATHING, PROVIDE AN OPEN AIRWAY. GIVE HIM FIVE OR SIX QUICK BREATHS. IF HE RESPONDS, CONTINUE MOUTH-TO-MOUTH RESUSCITATION UNTIL HE IS BREATHING WITHOUT HELP.

CONTINUE
HEART-LUNG RESUSCITATION
UNTIL HE RECOVERS OR HE
REACHES A HOSPITAL OR A
DOCTOR TAKES OVER

Important

- CONTINUE H-L-R ON WAY TO HOSPITAL

- DO NOT ALLOW ANY PRESSURE-CYCLING MECHANICAL RESUSCITATOR TO BE USED WITH CARDIAC COMPRESSION

- DO NOT ALLOW ANYONE (POLICE, FIREMEN OR AMBULANCE CREW) TO TAKE OVER - UNLESS THEY ARE EXPERT IN H-L-R

IF NO PULSE IS PRESENT, PUPIL DOES NOT CONTRACT AND SKIN COLOR IS NOT NORMAL, HIS HEART HAS STOPPED

RESTORE **CIRCULATION**
USE HEART-LUNG RESUSCITATION

MOUTH-TO-MOUTH RESPIRATION
+
EXTERNAL CARDIAC COMPRESSION
=
HEART-LUNG RESUSCITATION

ONE RESCUER | AFTER EVERY 15 COMPRESSIONS GIVE 2 QUICK BREATHS RATIO 15 TO 2

TWO RESCUERS | AFTER EVERY FIFTH COMPRESSION INTERPOSE ONE BREATH RATIO 5 TO 1

Self-Testing
Questions and Exercises

The section number and page number for the answer to the question or a description of the exercise requested is indicated for each item.

1. Describe the electron theory   1-1
2. Define an electric current.   1-2
3. Define a conductor.   1-2
4. Define an insulator.   1-2
5. What are three major uses of electricity?   1-2
6. What is the unit of electric current?   1-8
7. What instrument is used to measure electric current?   1-8
8. What is the unit of electric pressure?   1-9
9. Draw a schematic diagram to illustrate the proper methods of connecting instruments to a single-phase motor to measure electric current and the electric pressure supplied to the motor.   1-10
10. What is the unit of electric power?   1-10
11. Draw a schematic diagram illustrating the method of connecting a wattmeter in a single-phase circuit to measure the power consumed.   1-11
12. What is the unit of electric energy?   1-11
13. What instrument is used to measure electric energy?   1-11
14. What is the formula expressing Ohm's law?   1-14
15. Define an alternating current.   1-16
16. What is the unit of an alternating-current frequency?   1-16
17. Define power factor of an alternating-current circuit.   1-22
18. Describe the theory of operation of an electric transformer.   1-29
19. Describe a typical electric system itemizing the major parts of the system.   2-2
20. What is a transmission line?   2-12
21. How do the voltage levels of transmission, subtransmission, and distribution lines compare?   2-12, 2-13, 2-21
22. When are direct-current transmission lines used?   2-12
23. List the functions that can be accomplished in an electric substation.   2-13
24. List the component parts of the electric distribution system.   2-16
25. Describe a typical distribution feeder circuit.   2-17
26. What is a primary circuit?   2-21
27. What is the purpose of the distribution transformer installed on an electric distribution circuit?   2-21
28. Describe an electric service.   2-25

29. What is the purpose of an electric substation? 3-1
30. What types of equipment can be found in a typical substation? 3-4
31. What are power transformers used for in an electric substation? 3-5
32. What function is accomplished with circuit breakers installed in electric substations? 3-6
33. Why are disconnect switches installed in series with circuit breakers? 3-10
34. What function is accomplished with the protective relays installed in a substation? 3-11
35. What normally limits the capacity of a transmission line? 4-6
36. What are the advantages of using dc transmission for underground circuits rather than using ac transmission? 4-14
37. What is the relationship between the phase to neutral voltage and the phase to phase voltage of a three-phase four-wire system? 5-4
38. How would the secondary windings of the substation transformer be connected to provide a source for a three-phase four-wire distribution circuit? 5-4
39. Which conductor of a three-phase four-wire distribution system is grounded? 5-7
40. Draw a schematic connection diagram for connecting single-phase, pole-type distribution transformers to a three-phase four-wire primary circuit to obtain three-phase four-wire secondary voltages. 5-7
41. What typical types of information can be obtained from the "Plan and Profile Sheets" furnished with transmission line specifications? 6-8
42. How are wood poles classified? 7-1
43. What types of lines are commonly constructed with wood-pole structures? 7-5
44. What are the advantages of using crossarms on a subtransmission structure? 7-10
45. What types of steel structures are used for transmission lines? 8-5
46. What principles should be followed to locate poles? 9-5
47. Describe a safe method of unloading poles from a railroad flat car. 10-1 through 10-4
48. What is the minimum recommended depth for setting a pole in soil? 11-3
49. What is the rule for determining the setting depth for a pole to be installed in soil? 11-3
50. Where should guys be installed? 12-1
51. When is it necessary to insulate a guy? 12-11
52. Describe an insulator. 13-1
53. Where are double crossarms used? 13-5
54. What type of circuits permit the installation of post-type insulators? 13-7
55. How many porcelain suspension insulator units are required in a string on a circuit operating at 13,200 volts ac? 13-14
56. Where are strain insulators used? 13-14
57. What is the difference between a strain insulator and a suspension insulator? 13-20
58. What factors must be taken into account when selecting line conductors? 14-8
59. What is the purpose of a distribution transformer? 15-1
60. What equipment is used with a self-protected-type distribution transformer? 15-2
61. Draw a schematic diagram of a single-phase distribution transformer connected to a primary circuit and a load through a switch and describe its operation without load and with the switch closed and load connected. 15-3
62. How is the capacity of a distribution transformer determined? 15-6
63. Draw a schematic diagram of a distribution transformer secondary winding connected for 120-volt two-wire operation and 120/240-volt three-wire operation. 15-8
64. How can the polarity of a transformer be determined? 15-12
65. Why are some distribution transformers manufactured with taps on the primary windings? 15-13
66. Describe the theory of operation of an elementary lightning arrester. 16-3
67. Describe the operation of a valve-type lightning arrester. 16-4
68. How can the distribution system be protected from the failure of a distribution-class lightning arrester? 16-9
69. What is the purpose of installing a fuse in a circuit? 17-1
70. Describe the operation of an expulsion fuse cutout. 17-2
71. How do you know when the fuse in an indicating enclosed cutout is blown? 17-4
72. Refer to the time-current curve (Fig. 17-16) and determine the minimum clearing time for a 10K fuse link if 100 amp of current flowed through the fuse link. 17-11
73. Describe the operation of a liquid fuse. 17-13
74. Describe the operation of a current limiting fuse. 17-14

75. What is likely to happen if a disconnect switch is opened while current is flowing through the switch?   18-2

76. Describe the operation of an oil-circuit reclosure.   18-6

77. What is the purpose of using a voltage regulator in a distribution circuit?   19-1

78. Describe the operation of a distribution step voltage regulator.   19-4

79. Why must the voltage regulator be in the neutral position to operate a voltage regulator by-pass switch with the circuit energized?   19-10

80. What is the order of operations for transmission line tower erection?   20-1

81. Why must the surface of high-voltage transmission line conductors be protected from surface scratches while they are being installed?   21-1

82. Why is it necessary to effectively ground conductors while they are being installed?   21-2

83. Describe the procedures to be followed to reconductor distribution lines.   21-8

84. When should conductors be sagged?   22-1

85. Describe how a lineman can use a transit to complete sagging operations.   22-7

86. Why are armor rods installed on some line conductors at the point of support?   22-24

87. Why are spacers installed on bundled conductor transmission lines?   22-25

88. What types of splices can be used on line conductors under tension?   23-1

89. When can bolted connectors be used to join line conductors?   23-16

90. What types of work can be completed to maintain lines while energized with hot-line tools?   24-1

91. Describe the hot-line tool called a Wire Tong.   24-2

92. Describe the method used to phase-out two circuits while they are energized before connecting them together.   24-27

93. Describe the procedure for testing insulators in a string with a high-voltage tester to determine if one of the insulators is defective.   24-29

94. What ways can be used to perform live-line maintenance using an insulated aerial platform?   25-1

95. How deep should a ground rod be driven?   26-3

96. What is the maximum value of resistance allowed by the National Electrical Safety Code for a single ground rod installation?   26-3

97. Protective grounds can guard the lineman from what types of hazards?   27-1

98. How can a high-voltage circuit be tested to be sure it is deenergized?   27-1

99. What is the proper sequence for installing protective grounds?   27-2

100. What are the requirements necessary for a lineman to install an adequate protective ground?   27-3

101. Where should protective grounds be installed to establish the "man-shorted out" concept that provides the best protection for the lineman?   27-7

102. Describe the best procedures for establishing a temporary protective ground on a cable in a dead-front-type pad-mounted distribution transformer.   27-12

103. Define the following street lighting terms: lamp, luminaire, lumen, footcandle, ballast, mast arm.   28-1

104. What types of lamps are in common use for street lighting?   28-9

105. What types of street light lamps require a ballast?   28-10

106. How is the light emitted from a street light controlled?   28-13

107. Name two common street light distribution patterns.   28-13

108. Name the parts of an underground system.   29-1

109. At what minimum voltage should cables have an effective shield?   29-7

110. What is the advantage of sector cable construction?   29-11

111. What is the purpose of having a shielding over the insulation of a high-voltage cable?   29-13

112. Why are potheads installed on cables?   29-20

113. What purpose do vaults serve in an underground system?   31-10

114. What precautions must be taken before cables are installed in a duct?   32-1

115. Describe precautions that must be taken to remove the insulation shielding from a cable to prepare the cable for splicing.   33-5

116. Describe the types of underground residential distribution normally installed.   34-4

117. What are the voltages normally used for underground residential distribution circuits?   34-6

118. Describe the triplexed type cable assemblies used for underground residential distribution secondaries and services.   34-7

119. Identify the component parts of transformers used for underground electric distribution systems.   34-7

120. What is the minimum depth allowable for direct-buried cables operating in the voltage range of 601 to 22,000 volts?   34-17

121. What requirements are necessary to install electric supply cables and communication cables at the same depth with no deliberate separation?   34-17

122. What is the maximum distance permitted between the points where effectively grounded electric cables are bonded to communication cable shields if they are installed with random separation?   34-25

123. What are the objectives of line clearance tree trimming?   35-1

124. What are the fundamentals essential for safe and competent tree trimming operations? 35-1

125. Where should distribution transformers be installed?   36-3

126. Describe the proper location of grounds for various types of distribution transformer secondary wiring.   36-3

127. Describe the procedure to be followed if two single-phase distribution transformers are to be connected in parallel.   36-8

128. When is it desirable to determine the phase sequence of an ac circuit?   36-13

129. What is the minimum height allowable for attaching a low-voltage service to a building?   36-23

130. Describe a method for testing the service wires to avoid connecting the customer ground wire to either of the hot wires of a secondary.   36-26

131. Draw an example of the symbol of a single-pole drawout-type air circuit breaker.   37-1

132. Draw an example of the symbol for a cable termination.   37-1

133. Draw an example of the symbol for a fused disconnect switch.   37-2

134. Draw an example of the symbol for a lightning arrester with gap, valve, and ground.   37-2

135. Draw an example of a symbol for a transformer.   37-3

136. Draw the symbol for a three-phase wye-connected four-wire grounded transformer connection.   37-3

137. Describe the components and significant features shown on the distribution circuit single-line diagram in Fig. 38-3.   38-3

138. Describe the operation and identify the component parts shown on the unit substation transformer-tap-changer control schematic diagram in Fig. 39-2.   39-6

139. Define voltage regulation on an electrical system.   40-1

140. What is the high limit and the low limit of the voltages of an electric service to a residence?   40-1

141. What is the effect of low voltage on residential electric equipment?   40-1

142. What is the effect of high voltage on residential electric equipment?   40-1

143. How can the voltage on an electric distribution system be controlled?   40-1

144. What is the purpose of a line drop compensator on a voltage regulator?   40-7

145. What is the formula for Ohm's law?   41-1

146. What is the formula for calculating the impedance of an alternating-current series circuit?   41-3

147. What is the formula for calculating power for a single-phase alternating-current circuit?   41-4

148. What is the formula for calculating power for a three-phase alternating-current circuit?   41-5

149. What is the horsepower rating of a single-phase motor operating at 480 volts and drawing 25 amp at a power factor of 88 percent if it has a full load efficiency of 90 percent?   41-7

150. What is the definition of an ampere?   42-1

151. What is the name of a transformer which has a common winding for the primary and secondary?   42-1

152. What is a capstand?   42-2

153. Define galloping conductor phenomenon.   42-4

154. What is a grounded system?   42-4

155. What is the definition for the term "high pot"?   42-4

156. What is the name for a motor in which rotor current is set up by transformer action from alternating currents supplied to the stator windings? The rotor currents are induced by those in the stator.   42-5

157. Define the term "power factor."   42-6

158. What is a voltage relay?   42-9

159. Using a piece of rope, tie examples of the following knots:

two half-hitches   43-4

a square knot   43-4

a bowline   43-5

a clove hitch   43-6

160. Describe the proper inspection procedures that should be executed prior to climbing a pole.   44-3

161. Describe the proper methods of inspecting and caring for climbing equipment.   44-7

162. Describe procedures for maintaining climber gaffs.   44-8

163. What is the safest way of completing electric distribution work?   45-1

164. Itemize the protective equipment that should be used by the lineman or cableman when working on energized electric circuits.   45-1

165. How often should rubber gloves be given an air test by the lineman or cableman?   45-2

166. An inspection by the lineman or cableman of existing conditions before starting work should include what items?   46-1

167. Describe the procedures to be followed when deenergizing lines and equipment operated in excess of 600 volts and the means of disconnecting from the electrical energy source is not visibly open.   46-2

168. Describe the inspections required for mechanical equipment.   46-4

169. Describe the proper procedures to install protective grounds to an electric circuit.   46-5

170. What tests are required before entering a manhole without forced ventilation?   46-10

171. How soon must resuscitation be started after a person stops breathing to prevent brain damage?   47-2

172. Describe the proper procedure of performing mouth-to-mouth resuscitation.   47-3

173. Describe the procedures necessary to perform heart-lung resuscitation.   48-2

174. Describe the method one lineman can use to rescue an accident victim from the top of a pole.   49-8

Index

About the Authors

Edwin B. Kurtz, E. E., P. E., Ph.D. (deceased), was Emeritus Professor of Electrical Engineering and Head of Department, University of Iowa. Before this he was a member of the Educational Department, The Wisconsin Electric Power Company. He also was a Fellow, IEEE, Fellow, AAAS, and Member, ASEE.

Thomas M. Shoemaker, P. E., B.S.E.E., is Manager of the Distribution Department, Iowa Illinois Gas and Electric Company. He is also a member, Transmission and Distribution Committee, Edison Electric Institute, and a Senior Member, IEEE. During his service as a captain in the Signal Corps, U.S. Army, he was responsible for related functions described in the Handbook. He has also taught electrical subjects at the Moline Community College, Moline, Illinois.